Nanoporous Materials

Advanced Techniques for
Characterization, Modeling, and Processing

Nanoporous Materials

Advanced Techniques for

Characterization, Modeling, and Processing

Edited by Nick Kanellopoulos

CRC Press
Taylor & Francis Group
Boca Raton London New York

CRC Press is an imprint of the
Taylor & Francis Group, an **informa** business

CRC Press
Taylor & Francis Group
6000 Broken Sound Parkway NW, Suite 300
Boca Raton, FL 33487-2742

First issued in paperback 2017

© 2011 by Taylor and Francis Group, LLC
CRC Press is an imprint of Taylor & Francis Group, an Informa business

No claim to original U.S. Government works

ISBN-13: 978-1-4398-1104-7 (hbk)
ISBN-13: 978-1-138-07655-6 (pbk)

Visit the Taylor & Francis Web site at
http://www.taylorandfrancis.com

and the CRC Press Web site at
http://www.crcpress.com

Contents

Part III   Fundamentals, Recent Advances and Improvements, Membrane, Catalytic and Novel Processes Involving Nanoporous Materials

Part IV   Case Studies of Applications of Advanced Techniques in Involving Nanoporous Materials

Preface

The advancement of materials science, nanotechnology, and biosciences depends on the effective use of advanced characterization and modeling techniques. During the last 30 years, there has been a tremendous improvement in the field of porous materials, with the development of increasing numbers of novel materials. During the same period, there have also been large numbers of significant breakthroughs in the development of advanced characterization and simulation techniques and their combinations. Since all the recent developments are largely scattered in a number of journals and conference proceedings, I believe that the concise information provided in this book covering diverse subjects will be a very useful reference for all scientists involved in the field of porous materials. This book aims to provide academic and industrial researchers of different disciplines and backgrounds with a concise yet comprehensive presentation of the state-of-the-art, recent developments, and expected improvements of advanced characterization and simulation techniques and their applications to optimize processes involving sorbents, membranes, and catalysts.

Nanoporous materials play an important role in chemical processing as, in many cases, they can successfully replace traditional, pollution-prone, and energy-consuming separation processes. These materials are widely used as sorbents, catalysts, catalyst supports, and membranes, and form the basis of innovative technologies, including high-temperature molecular sieve membrane separations and low-temperature reverse sorption membrane separations (hydrogen production, carbon dioxide capture and conversion, alkane/alkene separation, methane conversion, hydrogen storage, FCC catalysis, etc.). This is mainly due to their unique structural or surface physicochemical properties, which can, to an extent, be tailored to meet specific process-related requirements. Any equilibrium or dynamic process taking place within the nanopores of a solid is strongly influenced by the topology and the geometrical disorder of the pore matrix. The complete characterization of nanoporous materials still remains a difficult and frequently controversial problem, even if the equilibrium and transport mechanisms themselves are quite simple and well defined. This is mainly due to the great difficulty in accurately representing the complex morphology of the pore matrix. To this end, the application of combined techniques aided by advanced model analysis is of major importance as it is the most powerful method currently followed. On the other hand, no matter how thorough and complete the characterization, it is quite pointless if it is not related to the process under consideration, since one of the most important parameters in any application is the material's ability to retain its properties over a certain period of time. The "changes" induced on materials during their utilization in specific applications are highly relevant and crucial for the economic viability of many applications (e.g., catalysis and separation processes). In this context, it is necessary to develop skills in establishing advanced combinations of "in situ" and "ex situ" techniques in order to expand our understanding of confinement phenomena in nanopores, to monitor and control the evolution of the properties of nanostructured materials, and to evaluate and optimize the performance of nanoporous sorbents, membranes, and catalysts involved in several important industrial processes.

The book is organized as follows: Part I presents the basic principles and major applications of the most important characterization techniques, ranging from diffraction and spectroscopy to calorimetry, permeability, and other techniques. Part II presents computer simulation techniques, an indispensable complement to the combination of the aforementioned analytical techniques. Part III covers the fundamentals and the recent advances in sorption, membrane, and catalyst processes, while Part IV presents two characteristic "case" studies of emerging areas of application of porous solids in the fields of gas-to-liquid conversion and hydrogen storage.

This book is based on the experience gained from the workshops organized by the network of excellence INSIDE-PORES and is mainly the result of the workshop on NAnoPorous Materials for ENvironmental and ENergy Applications (NAPEN 2008), which was organized in Crete by three cooperating European networks of excellence, namely, the Networks of Excellence IDECAT on catalysis, the NANOMEMBRO on membranes, and the INSIDE-PORES. I would like to thank Professors Gabriele Centi and Gilbert Rios, the coordinators of the cooperating networks of Excellence IDECAT and NANOMEMBRO. I would also thank my colleagues and students from Demokritos, A. Sapalides, G. Romanos, A. Labropoulos, S. Papageorgiou, V. Favvas, N. Kakizis, G. Pilatos, and E. Chatzidaki, for helping with the organization of NAPEN 2008. One of the major characteristics of this book is the impressive list of internationally well-known contributors. I would like to thank each one of them for their invaluable contributions. Special thanks are due to Jill Jurgensen and Allison Shatkin and their colleagues from Taylor & Francis Group for their help and patience. Last but not least, I would like to thank the European Commission for funding the three networks of excellence. I would also like to thank Dr. Soren Bowadt, who was in charge of these networks, for his valuable assistance and patience.

Nick Kanellopoulos

Editor

Nick Kanellopoulos received his PhD from the Department of Chemical Engineering, University of Rochester, Rochester, New York, in 1975, and his diploma in chemical engineering from the National Technical University of Athens, Athens, Greece in 1970. He joined the Mass Transport Laboratory, Institute of Physical Chemistry, National Centre for Scientific Research Demokritos, Attiki, Greece, in 1976, and, since 1992, he has been the head of the "Membranes for Environmental Separations" Laboratory, NCSR Demokritos. His research interests include pore structure characterization, nanoporous membrane, and carbon nanotube systems, and the evaluation of their performance using a combination of in situ and ex situ techniques. Dr. Kanellopoulos is the author and coauthor of more than 140 papers; he is also the editor of *Recent Advances in Gas Separation by Microporous Membranes* (Elsevier Science) and the coeditor of *Nanoporous Materials for Energy and Environment* (Stanford Chong). He has received approximately 12 million euros in funding from over 50 European and national programs and has participated in three high technology companies in the field of nanoporous materials. He participated in the National Representation Committee of Greece for the FP6-NMP and FP7-NMP European programs in nanotechnology from 2001 to 2009. He is the coordinator of the European Network of Excellence in nanotechnology inside-pores.gr, a member of the Committee of the Kurchatov-Demokritos Project Center for Nanotechnology and Advanced Engineering, and the coordinator of the committee for the preparation and submission of the proposal for a Greek national nanotechnology program. He is also a Fulbright scholar and president of the Greek Fulbright Scholars Association.

Contributors

P.M. Adler
Structure et Fonctionnement
  des Systèmes Hydriques
  Continentaux
Université Pierre et Marie Curie
Paris, France

M.C. Campo
Faculty of Engineering
University of Porto
Porto, Portugal

J. Caro
Institute of Physical Chemistry and
  Electrochemistry
Leibniz University of Hannover
Hannover, Germany

Gabriele Centi
Dipartimento di Chimica
  Industriale ed Ingegneria dei
  Materiali

and

Consorzio Interuniversitario
  Nazionale per la Scienza e
  Tecnologia dei Materialis
Laboratorio di Catalisi per una
  Produzione Sostenibile e
  L'energia
Università di Messina
Messina, Italy

G.C. Charalambopoulou
Institute of Nuclear Technology
  and Radiation Protection
National Center for Scientific
  Research "Demokritos"
Athens, Greece

Christian Chmelik
Faculty of Physics and Geosciences
University of Leipzig
Leipzig, Germany

P. Cool
Laboratory of Adsorption and
  Catalysis
Department of Chemistry
University of Antwerpen
Wilrijk, Belgium

Avelino Corma
Instituto Universitario de Tecnología
  Química
Universidad Politécnica de
  Valencia
Valencia, Spain

A.F.P. Ferreira
Faculty of Engineering
University of Porto
Porto, Portugal

Hermenegildo García
Instituto Universitario de Tecnología
  Química
Universidad Politécnica de
  Valencia
Valencia, Spain

Katsumi Kaneko
Department of Chemistry
Graduate School of Science
Chiba University
Chiba, Japan

N.K. Kanellopoulos
Institute of Physical Chemistry
National Centre for Scientific
    Research "Demokritos"
Athens, Greece

F. Kapteijn
Catalysis Engineering
Delft University of Technology
Delft, the Netherlands

G.N. Karanikolos
Institute of Physical Chemistry
National Centre for Scientific
    Research "Demokritos"
Athens, Greece

Jörg Kärger
Faculty of Physics and Geosciences
University of Leipzig
Leipzig, Germany

F.K. Katsaros
Institute of Physical Chemistry
National Centre for Scientific
    Research "Demokritos"
Athens, Greece

Gérald Lelong
Institute de Minéralogie et Physique
    des Milieux Condensés
Université Paris 6

and

Centre National de la Recherche
    Scientifique
Université Paris 7

and

Institut de Physique du Globe de
    Paris
Institut de Recherche pour le
    Développement
Paris, France

Philip L. Llewellyn
Laboratoire Chimie Provence
CNRS-Université de Provence
Marseille, France

A.M. Mendes
Faculty of Engineering
University of Porto
Porto, Portugal

V. Meynen
Laboratory of Adsorption and
    Catalysis
Department of Chemistry
University of Antwerpen
Wilrijk, Belgium

K.G. Papadokostaki
Institute of Physical Chemistry
National Center for Scientific
    Research "Demokritos"
Athens, Greece

Siglinda Perathoner
Dipartimento di Chimica
    Industriale ed Ingegneria dei
    Materiali

and

Consorzio Interuniversitario
    Nazionale per la Scienza e
    Tecnologia dei Materialis
Laboratorio di Catalisi per una
    Produzione Sostenibile e
    L'energia
Università di Messina
Messina, Italy

J.H. Petropoulos
Institute of Physical Chemistry
National Center for Scientific
    Research "Demokritos"
Athens, Greece

David L. Price
Conditions Extrêmes et Matériaux:
    Haute Température et Irradiation
Centre National de la Recherche
    Scientifique
Université d'Orléans
Orléans, France

Ana Primo
Instituto Universitario de Tecnología
    Química
Universidad Politécnica de
    Valencia
Valencia, Spain

F. Rodríguez-Reinoso
Laboratorio de Materiales
    Avanzados
Departamento de Química
    Inorgánica
Instituto Universitario de Materiales
    de Alicante
Universidad de Alicante
Alicante, Spain

G.E. Romanos
Institute of Physical Chemistry
National Centre for Scientific
    Research "Demokritos"
Athens, Greece

Douglas M. Ruthven
Department of Chemical
    Engineering
University of Maine
Orono, Maine

Marie-Louise Saboungi
Centre de Recherche sur la Matière
    Divisée
Centre National de la Recherche
    Scientifique
Université d'Orléans
Orléans, France

A. Sepúlveda-Escribano
Laboratorio de Materiales
    Avanzados
Departamento de Química
    Inorgánica
Instituto Universitario de Materiales
    de Alicante
Universidad de Alicante
Alicante, Spain

J. Silvestre-Albero
Laboratorio de Materiales
    Avanzados
Departamento de Química
    Inorgánica
Instituto Universitario de Materiales
    de Alicante
Universidad de Alicante
Alicante, Spain

K.L. Stefanopoulos
Institute of Physical Chemistry
National Centre for Scientific
    Research "Demokritos"
Athens, Greece

Th.A. Steriotis
Institute of Physical Chemistry
National Center for Scientific
    Research "Demokritos"
Athens, Greece

A.K. Stubos
Institute of Nuclear Technology
    and Radiation Protection
National Center for Scientific
    Research "Demokritos"
Athens, Greece

J.-F. Thovert
Institut PPRIME SP2MI
Futuroscope, France

Anthony A.G. Tomlinson
Consiglio Nazionale delle Ricerche
Istituto per lo Studio dei Materiali
    Nanostrutturati
Rome, Italy

Shigenori Utsumi
Department of Mechanical Systems
    Engineering
Tokyo University of Science
Chino, Japan

E.F. Vansant
Laboratory of Adsorption and
    Catalysis
Department of Chemistry
University of Antwerpen
Wilrijk, Belgium

Part I

Basic Principles, Recent Advances, and Expected Developments of Advanced Characterization Techniques

1

Scattering Techniques

Gérald Lelong, David L. Price, and Marie-Louise Saboungi

CONTENTS

1.1 Introduction

Neutrons produced by reactors and spallation sources, and x-rays produced by synchrotron sources, have wavelengths in the range of 0.1–1 nm, making the scattering experiments a powerful and versatile probe of nanoporous materials. In fact it is hard to find a comprehensive paper on some aspect of confinement on the nm scale that does not include references to scattering measurements. A case in point is the excellent topical review of Alcoutlabi and McKenna[1] dealing with the effects of size and confinement on the melting temperature T_m (always depressed) and the glass transition temperature T_g (may increase, decrease, remain the same, or even disappear). Among other scattering results they refer to the work of Morineau et al.,[2] who measured the density of confined liquid toluene through changes in the Bragg peak intensity in neutron diffraction measurements resulting from the change in contrast with the confined liquid. Whereas little or no changes were observed for confinement in mesoporous silicates with pore sizes of 3.5 nm and above, a decrease in density and an *increase* of 30 K in T_g were observed upon confinement in 2.4 nm pores. Alcoutlabi and McKenna also refer to the inelastic neutron scattering (INS) studies of Zorn et al.[3] who observed a *decrease* in T_g on the confinement of salol in microporous silica glass together with a broadening of the relaxation spectra. These effects were discussed in terms of a cooperativity length scale that, since it cannot become larger than the confining dimensions, leads to an acceleration of the molecular dynamics compared with the bulk.

In this chapter, we summarize the techniques of diffraction—essentially, the study of correlations in atomic arrangements on the scale of 0.1–1 nm, sometimes referred to as wide-angle neutron or x-ray scattering (WANS, WAXS), small-angle scattering (SAS) that measures density fluctuations on the length scale of 5–500 nm, and inelastic scattering that probes dynamical phenomena on the timescale of 0.1 ps to 0.1 ms. We include a brief description of x-ray absorption spectroscopy, often used in conjunction with scattering experiments: for example, extended x-ray absorption fine-structure spectroscopy (EXAFS) provides an element-specific structural probe through the scattering of the emitted photoelectron by neighboring atoms. Figure 1.1 compares the length and timescales probed by different scattering techniques with optical, dielectric, and nuclear magnetic resonance (NMR) spectroscopies.

To flesh out these bare bones, we provide some examples of the application of these techniques from our own work, including selenium absorbed in zeolites, mesoporous silica nanopores, hydrogen adsorbed on carbon nanohorns, and glucose solutions confined in aqueous silica gels.

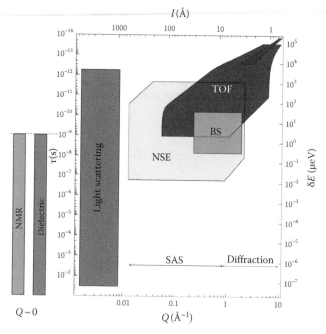

FIGURE 1.1

Schematic representation of accessible length and time scales using light or neutron scattering techniques (time-of-flight (ToF), backscattering (BS), neutron spin echo (NSE), and SAS spectrometers) and spectroscopic methods (nuclear magnetic resonance [NMR] and dielectric spectroscopy).

1.2 Diffraction

Diffraction is generally taken to mean the measurement of atomic or magnetic structure by scattering experiments. In principle, any particle can be used, but for most investigations of atomic structure neutrons, x-rays, and electrons are most common while neutrons and, in certain cases, x-rays can be also used to provide information about magnetic structure.

The neutron is a subatomic particle with, as its name implies, zero charge, mass $m_n = 1.0087$ atomic mass units, spin $I = \frac{1}{2}$ and magnetic moment $\mu_n = -1.9132$ nuclear magnetons. These properties combine to make the neutron a highly effective probe of condensed matter. The zero charge means that its interactions with a sample of a condensed material are confined to the short-ranged nuclear and normally weak magnetic interactions, so that the neutron can usually penetrate into the bulk of the sample.

Thermal neutrons for condensed matter research are usually obtained by slowing down energetic neutrons, produced by a nuclear reaction in either a fission reactor or an accelerator-driven spallation source, by means of inelastic collisions in a moderating material consisting of light atoms. Most of the slow neutrons thus produced will have kinetic energies on the order of k_BT where T is the moderator temperature. Considering the wave nature of the neutron, its wavelength is given by

$$\frac{\hbar^2}{2m_n\lambda^2} = k_BT. \tag{1.1}$$

The neutron mass is such that for $T = 300\,\mathrm{K}$, $\lambda \sim 2$ Å, a distance comparable to the mean atomic separation in a solid or liquid. Such neutrons are therefore ideally suited to studies of the *atomic structure* of condensed matter, discussed below. Furthermore, the kinetic energy of such a neutron is on the order of 25 meV, a typical energy for excitations in solids and liquids. Thus, both wavelength and energy are ideally suited to studies of the *atomic dynamics* of condensed matter in inelastic scattering experiments, discussed in Section 1.5.

The magnetic moment of the neutron makes it a unique probe of *magnetic structure and excitations*: neutrons are scattered from the magnetic moments associated with unpaired electron spins in magnetic materials. Again, the wavelength and energy of a thermal neutron are such that both the magnetic structure and the dynamics of the spin system can be studied in the neutron scattering experiment.

The x-ray is a photon with an energy conventionally taken in the range of keV. It has zero charge, zero magnetic moment, and spin $I = 1$. An x-ray of energy $E = h\nu = hc/\lambda = 12.398\,\mathrm{keV}$ has wavelength $\lambda = 1$ Å, making it also, as is well known, a powerful probe of the structure of condensed matter. The electromagnetic field associated with a moving x-ray makes it, under appropriate circumstances, another probe of magnetic structure. To probe excitations in condensed matter, which typically have energies in the meV range, an energy resolution on the order of 10^{-7} is required in both incident and scattered beams, a formidable challenge that has recently been met in third-generation synchrotron sources.

1.2.1 Diffraction Formalism

We consider a simple scattering experiment shown schematically in Figure 1.2. We suppose that a beam of particles (neutrons, x-rays, or electrons) characterized by a wave vector $\vec{k}_i$ falls on the sample. The magnitude of $\vec{k}_i$ is $2\pi/\lambda$ and its direction corresponds to that of the beam. Usually the sample size is chosen such that most of the beam is transmitted: typically it is ~mm with neutrons, and with x-rays it varies from ~mm for light atoms

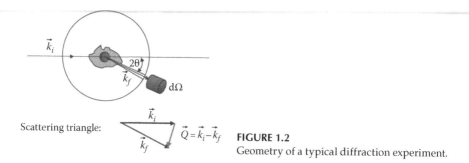

Scattering triangle:

FIGURE 1.2
Geometry of a typical diffraction experiment.

to ~μm for heavy atoms. Some particles are, however, scattered and can be measured with a detector placed, for example, in a direction $\vec{k}_f$. If the incident beam is characterized by a flux Φ (particles crossing unit area per unit time), the sample has N identical atoms in the beam, and the detector subtends solid angle ΔΩ and has efficiency η, we may expect the count rate in the detector to be proportional (if ΔΩ is small enough) to all these quantities. In this case, the constant of proportionality is called the differential cross section and is derived as[4]

$$\frac{d\sigma}{d\Omega} = \frac{C}{\eta\Phi N(\Delta\Omega)}. \tag{1.2}$$

The structural information obtained in a diffraction experiment is normally described by the variation of the intensity of the scattering with the scattering vector $\vec{Q}$:

$$\vec{Q} = \vec{k}_i - \vec{k}_f \tag{1.3}$$

illustrated by the triangle in Figure 1.2. In the case of experiments on samples that are directionally isotropic—polycrystalline solids, glasses, and liquids—the scattering depends only on the magnitude of the scattering vector, the scalar quantity $Q = \|\vec{Q}\|$.

It is usual to fix the directions of $\vec{k}_i$ and $\vec{k}_f$ by means of appropriate collimators, detector placement, etc. and to fix the magnitude of one of these, generally k_i, or sometimes a combination of k_i and k_f, for example one that corresponds to the total time-of-flight from sample to detector in the case of neutron scattering. The total intensity of the scattered particles measured in the detector is normally recorded, irrespective of any energy transfer that may take place, and $\vec{Q}$ is evaluated from Equation 1.3 under the assumption that the scattering is elastic, i.e., there is no energy exchange between the particle and the sample and so $|\vec{k}_i| = |\vec{k}_f|$. In the neutron case, significant inelastic scattering is always present and this can affect the structural interpretation. However, the experiments are usually designed to minimize the errors that result from these approximations, which can usually be taken care of by straightforward corrections. For elastic scattering, Q depends only on k_i and the scattering angle 2θ, corresponding to the Bragg relation $Q = 4\pi \sin \theta / \lambda$.

The nuclear interaction between a slow neutron and an atom can be expressed in a simple form. In the simplest case where the atoms in the sample are both noninteracting and identical, the differential cross section is just a constant:

$$\frac{d\sigma}{d\Omega} = b^2, \tag{1.4}$$

where the scattering length b is normally a constant, depending on the atomic number Z and the atomic weight A of the nucleus, and its spin state relative to that of the neutron. Its magnitude depends on the details of the interaction between the neutron and the components of the nucleus. For this reason both sign and magnitude of b change in an irregular fashion with Z and A.

In the x-ray case, the photon interacts with the electrons in the atom, and since these are distributed in space, the scattering factor is proportional to the total number of electrons and a form factor that represents the Fourier transform of their radial distribution. The differential cross section is then a function of the scattering vector Q:

$$\frac{d\sigma}{d\Omega} = f^2(Q). \tag{1.5}$$

In contrast with the neutron case, the x-ray scattering, which increases monotonically with Z, is independent of isotope and decreases with Q. The scattering length for neutrons and scattering factor for x-rays for two values of θ, and hence of Q, are shown in Figure 1.3.

For convenience, comparison between experiment and theory is usually done in terms of a dimensionless quantity, the *structure factor* $S(Q)$. In the case of neutron diffraction, this is related to the differential cross section by the relation:

$$\frac{d\sigma}{d\Omega} = \left| \sum_{a=1}^{n} c_a \overline{b}_a \right|^2 (S(Q) - 1) + \sum_{a=1}^{n} c_a \overline{b_a^2}, \tag{1.6}$$

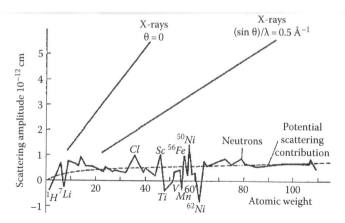

FIGURE 1.3
Scattering lengths for neutrons and scattering factor for x-rays for two values of scattering angle 2θ, as a function of atomic weight A.

where

c_a is the atomic concentration

$\bar{b}_a$ is the average (over isotopes and spin states) of the neutron–nucleus scattering length

$\overline{b_a^2}$ is the mean square scattering length of element a present in the sample.

This can be rewritten as follows:

$$\frac{d\sigma}{d\Omega} = \left| \sum_{a=1}^{n} c_a \bar{b}_a \right|^2 S(Q) + \left[\sum_{a=1}^{n} c_a \left| \bar{b}_a \right|^2 - \left| \sum_{a=1}^{n} c_a \bar{b}_a \right|^2 \right] + \left[\sum_{a=1}^{n} c_a \left(\overline{b_a^2} - \left| \bar{b}_a \right|^2 \right) \right], \quad (1.7)$$

where the leading term contains the structural information that is being sought here. The second term arises from random distributions of different elements (often referred to as *Laue diffuse scattering*) and the third term from random distributions of isotopes and spin states over the atoms belonging to a given element (generally called *incoherent scattering*). It is convenient to define the coherent and incoherent cross sections of element a:

$$\sigma_a^{coh} = 4\pi \left(\bar{b}_a \right)^2 \quad \text{and} \quad \sigma_a^{inc} = 4\pi \left[\overline{b_a^2} - \left(\bar{b}_a \right)^2 \right]. \quad (1.8)$$

It can be seen that the coherent cross sections enter into the first two terms of Equation 1.7 and the incoherent into the third. In the x-ray case, every atom of a given element scatters identically so the incoherent term does

TABLE 1.1

Neutron Scattering Lengths b in Femtometers
(1 fm = 10^{-15} m) and the Respective Coherent
σ_{coh} and Incoherent σ_{inc} Scattering Cross
Sections in Barns (1 barn = 10^{-24} cm^2) for Some
Elements of the Periodic Table

Element	b (fm)	σ^{coh} (Barn)	σ^{inc} (Barn)
^{1}H	−3.742	1.758	80.276
^{2}H	6.674	5.593	2.053
^{6}Li	2.000	0.515	0.465
^{7}Li	−2.222	0.619	0.783
^{12}C	6.653	5.559	0
^{14}N	9.372	11.035	0.501
^{16}O	5.805	4.232	0
^{19}F	5.654	4.017	· 0.001
^{23}Na	3.632	1.662	1.623
^{27}Al	3.449	1.495	0.008
^{28}Si	4.106	2.120	0
^{31}P	5.131	3.307	0.005
^{32}S	2.804	0.988	0
^{70}Ge	10.010	12.63	0

Source: Price, D.L. and Sköld, K., Introduction to neutron scattering, in *Neutron Scattering, Methods
of Experimental Physics*, eds. D.L. Price and
K. Sköld, Vol. 23 (Part A), p. 1, Academic
Press, New York, 1986.

not appear. Table 1.1 gives the values of the neutron scattering lengths and
the coherent and incoherent cross sections for a selection of elements and
isotopes.

By definition, $S(Q)$ tends to unity at large Q, a property that is often used
to normalize the intensities measured in a diffraction experiment. Its low-Q
limit is related to the macroscopic compressibility χ_T:

$$S(0) = \frac{\rho_0}{V}\left(\frac{\partial V}{\partial P}\right)_T k_B T = \rho_0 \chi_T k_B T, \qquad (1.9)$$

where ρ_0 is the number of atoms per unit volume. In between $S(Q)$ exhibits a complex behavior that reflects the detailed atomic structure. In a
crystalline sample—either single crystal or polycrystalline—there are
sharp peaks called *Bragg peaks* that arise from diffraction from parallel
crystallographic planes at Q values corresponding to $2\pi n/d$, where n is an
integer and d the plane spacing. There is also a continuous component,
called *diffuse scattering*, arising from static and/or dynamic disorder. In

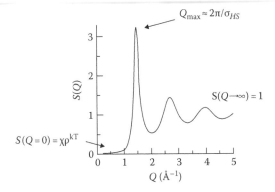

FIGURE 1.4

Typical form for $S(Q)$ in a simple classical liquid or glass. (From Price, D.L., Experimental techniques, in *High-Temperature Levitated Materials*, Cambridge University Press, Cambridge, U.K., p. 45, 2010. With permission.)

fully disordered materials like liquids and glasses, the entire scattering is diffuse, with a generally oscillatory pattern that reflects the short- and intermediate-range order in the sample. A well-defined distance of closest approach between atoms that can be characterized by an equivalent hard-sphere diameter σ_{HS} will be reflected in oscillations in $S(Q)$ with a period $2\pi/\sigma_{HS}$. A typical form for $S(Q)$ in a simple classical liquid is shown in Figure 1.4.

In the case of x-ray diffraction, b_a in the above equations and those that follow is replaced by $f_a(Q)$, the atomic form factor for species a. This results from the fact that the electrons in the atom from which the x-rays are scattered have a spatial distribution, while the nucleus from which the neutrons are scattered can be treated, for the present purposes, as a point object. Since the form factors are generally well tabulated this is not a major problem, but it can complicate the interpretation of the scattering from multicomponent systems.

A *pair correlation function* $g(r)$ that contains the structural information about the sample in real space is then calculated from $S(Q)$ *via* the Fourier transform

$$g(r) = 1 + \frac{1}{2\pi^2\rho_0} \int_0^{Q_{max}} Q(S(Q)-1)\frac{\sin Qr}{r}M(Q)\,dQ, \qquad (1.10)$$

where $M(Q)$ is a modification function that is often used to force the integrand to go smoothly to zero at Q_{max} and reduce the ripples that result from the finite limit of the integration.

For systems with more than one type of atom—different elements, and sometimes different isotopes of the same element—$S(Q)$ is a weighted sum of partial structure factors $S_{ab}(Q)$. Unfortunately, there are a number of alternative definitions of these in the literature: the $S(Q)$ appearing in Equations 1.6 through 1.9 is called the Faber–Ziman definition after its originators.[5] Another definition of partial structure factors for binary systems was proposed by Bhatia and Thornton,[6] where $S_{NN}(Q)$ describes the fluctuations in total particle density, $S_{CC}(Q)$ those in the relative concentrations, and $S_{NC}(Q)$ the cross-correlation of the two. For a two-component system Equation 1.9 applies to $S_{NN}(Q)$. The various definitions are linear combinations of each other and are given in the textbooks, for example, that of March and Tosi.[7]

For a multicomponent system, $g(r)$ is correspondingly a weighted sum of partial pair correlation functions $g_{ab}(r)$, which in the neutron case is given by

$$g(r) = \sum_{a,b} W_{ab} g_{ab}(r) = \sum_{a,b} \frac{c_a c_b b_a b_b}{\left| \sum c_a b_a \right|^2} g_{ab}(r), \tag{1.11}$$

where
 a and *b* are the atom types
 W_{ab} are weighting factors

The partial pair distribution function $g_{ab}(r)$ can be considered as the relative probability of finding a *b* atom at a distance *r* from an *a* atom at the origin. In a one-component system, the indices *a*, *b* disappear and only a single $S(Q)$ and a single $g(r)$ exist. In a system with *n* components, a full structural analysis requires $n(n+1)/2$ different measurements with different coefficients in Equation 1.11: in favorable cases, this may be accomplished with the use of isotope substitution in the case of neutron diffraction,[8] by anomalous x-ray scattering (AXS) near an absorption edge, where the form factor has an additional component that varies rapidly with x-ray energy,[9,10] or by a combination of the neutron and x-ray scattering.[11] With a single measurement, only the average structure factor $S(Q)$ can be determined; nevertheless, this may still contain useful information. For example, if a particular peak *n* in $g(r)$ can be associated uniquely with a coordination shell for a pair of atom types *a*,*b*, the coordination number of *b* atoms about an average *a* atom for that shell is given by

$$C_a^n(b) = \frac{c_b}{W_{ab}} \int_n rT(r)dr, \tag{1.12}$$

where $T(r) = 4\pi\rho_0 rg(r)$ and the integral is taken over the peak *n*, while the centroid of $T(r)$ over the same peak gives the average coordination distance r_{ab}^n.

In a magnetic system, neutrons can also scatter from the magnetic moments associated with the unpaired electrons. In simple cases where the unpaired electrons can be associated with a particular atom, the magnetic scattering can be described by making the substitution

$$b_a \rightarrow g_n r_0 f_{ma}(Q) \frac{1}{2} g_a \mathbf{S}_a \qquad (1.13)$$

in the formalism given above, where $g_n = -1.9132$ is the g-factor for the neutron, $r_0 = 2.8179$ fm is the classical radius of the electron, and $f_{ma}(Q)$, g_a, and $\mathbf{S}_a$ are the magnetic form factor, g-factor, and spin operator of the unpaired electrons on the atoms of element a. It is clear that if there are correlations between the orientations of the magnetic moments in the system with their positions, i.e., some kind of magnetic ordering, there will be a structure-dependent term in the magnetic scattering analogous to the first term in the expression for the nuclear scattering, Equation 1.6. If, on the other hand, the orientations of the magnetic moments are completely random, as in a paramagnetic system, the magnetic scattering is independent of the structure and can be described by a term

$$\frac{2}{3} |g_n r_0 f_{ma}(Q)|^2 \langle \mu_a^2 \rangle \qquad (1.14)$$

analogous to the second term in Equation 1.6, $\vec{\mu}_a = 1/2 g_a \vec{S}_a$ being the magnetic moment of the ath atom.

It is clear from the Fourier transform in Equation 1.10 that long-range structural information will tend to dominate the scattering at low Q and short-range at high Q. Thus, the need to get accurate information about nearest-neighbor correlations, such as bond distances and coordination numbers, has driven the development of diffractometers with a large Q range, exploiting epithermal neutrons from pulsed spallation sources and high-energy x-rays from third-generation synchrotron sources.

1.2.2 Differences between Neutron and X-Ray Scattering

Experimentally, a significant difference is that x-rays with energies available in a laboratory source or in a typical synchrotron beam are more highly absorbing than neutrons, which must be taken into account when designing sample containers and environmental equipment. With the x-ray energies on the order of 100 keV available from third-generation synchrotron sources, the absorption is much less significant. Other differences include the following:

1. Since the form factors depend on Q and fall off as Q increases, measurements at high Q values (e.g., beyond 5 Å^{-1}) become more difficult. Also, the Q dependence has to be taken into account when calculating structure factors as in Equation 1.6. This is a significant problem when the sample has more than one atom.

2. Since the form factors are not significantly isotope dependent, the scattering is always coherent in the sense used in neutron scattering.

3. Form factors are not generally energy dependent but have a strong energy variation near an absorption edge, as well as an imaginary component. This behavior is called *anomalous scattering* and can be exploited to distinguish scattering from a specific element, somewhat like isotope substitution in the neutron case.

4. Since x-ray energies of 10 keV and higher must generally be used to get an adequate Q range for studies of atomic structure and dynamics, it is difficult to get energy resolution in inelastic x-ray scattering (IXS) comparable to that obtainable in INS. At present the limit is about 1 meV. A compensating advantage is that the velocity of x-rays is orders of magnitude higher than any sound velocity in condensed matter, so that the kinematic restrictions that make it hard to make dynamical measurements at low Q and high E with INS do not apply.[4]

5. X-rays do not have a magnetic moment and so the interaction with magnetic moments in condensed matter is much weaker. On the other hand, polarization of an x-ray beam can be exploited for magnetic studies, as in magnetic circular dichroism, for example.

These examples show that neutrons and x-rays have many complementary features, and it is often important to use both techniques, as well as others described below, to investigate a complex material or phenomenon.

1.2.3 Selected Example

An example of a diffraction measurement on a material confined in a porous host is the study of highly loaded Se in a Cu^{2+} ion-exchanged Y zeolite[12] by AXS,[10] which was complemented by diffuse reflectance and Raman spectroscopy measurements. The diffraction measurements were made at two energies, 20 and 300 eV below the K absorption edge of Se at 12,658 eV. Near an absorption edge of an element a, the scattering factors

$$f_a(Q,E) = f_{a0}(Q) + f_a'(Q,E) + i f_a''(Q,E) \tag{1.15}$$

are complex with anomalous terms that vary strongly with energy. Accordingly, the weighting factors W_{aj} involving the element a in the x-ray analogue of Equation 1.11 can be altered by tuning the x-ray energy near the absorption edge. From measurements at two energies below that edge, the difference structure factor $S_a(Q)$ associated with the element can be derived from the relation

$$I(Q,E_1) - I(Q,E_2) = 2c_a \Delta f_a' \langle f(Q) \rangle [S_a(Q) - 1] + 2c_a \Delta f_a' f_a(Q). \tag{1.16}$$

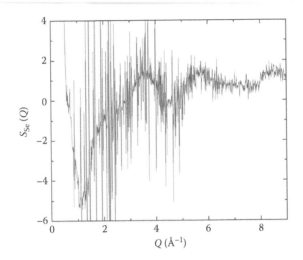

FIGURE 1.5
Difference structure factor $S_{Se}(Q)$ of Cu–Y zeolite loaded with 12.5 Se per supercage derived from two diffraction experiments at 20 and 300 eV below the K absorption edge of selenium. (With permission from Goldbach, A., Saboungi, M.-L., Iton, L.E., and Price, D.L., Approach to band-gap alignment in confined semiconductors, *J. Chem. Phys.*, 115, 11254, 2001.)

The difference pair correlation function $g_a(r)$ can be obtained by Fourier transformation of $S_a(Q)$, through a relation analogous to Equation 1.10.

Figure 1.5 shows the difference structure factor $S_{Se}(Q)$ determined from the diffraction experiments at the two energies, corrected for resonant Raman scattering, dead-time effects, Compton scattering, multiple scattering, and absorption in the sample. It can be seen to consist of a smoothly varying diffuse component with sharp positive and negative spikes superimposed on it. The diffuse component arises from the encapsulated Se, estimated from weight balance to amount to about 12.5 Se atoms per zeolite supercage. The sharp spikes result from a slightly imperfect cancellation of the large Bragg scattering from the zeolite host in the difference of the two measurements. The fact that the Se scattering is diffuse instead of following the Bragg peaks of the zeolite host shows that the Se atoms have a disordered structure, out of registry with the crystalline lattice of the zeolite host.

Figure 1.6 shows the corresponding pair correlation function in real space. For technical reasons, the function $4\pi\rho_0 g_a(r)$ was used in the analysis rather than $g_a(r)$ itself. The spikes that appeared in $S_{Se}(Q)$ due to the imperfect cancellation of the zeolite Bragg peaks do not lead to any observable peaks in $T_{Se}(r)$ because of their very low weight. The oscillations below 2 Å are due to truncation effects caused by the limited Q range of the $S_{Se}(Q)$ measurement. The distinguishable peaks in the region above 2 Å were fitted with Gaussian functions. The first peak centered at $R_{SeSe}(1) = 2.39 \pm 0.02$ Å reflects the intramolecular Se correlation. This distance is significantly longer than the

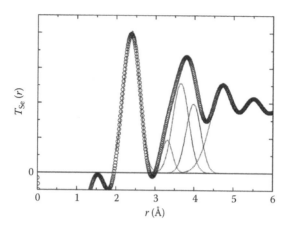

FIGURE 1.6

Difference pair correlation function $T_{Se}(r)$ calculated from the difference structure factor displayed in Figure 1.5. (With permission from Goldbach, A., Saboungi, M.-L., Iton, L.E., and Price, D.L., Approach to band-gap alignment in confined semiconductors, *J. Chem. Phys.*, 115, 11254, 2001.)

corresponding nearest-neighbor distances of Se encapsulates in Nd-Y, La-Y, and Ca-Y zeolites which have values ranging from 2.32 to 2.34 Å derived by a procedure identical to that used for the Cu-Y zeolite,[13,14] pointing to a weakening of the intrachain bonding in comparison to other zeolites. Except for trigonal Se, it also exceeds the values found in bulk Se forms, 2.336 Å in monoclinic Se and 2.356 Å in amorphous Se. The second peak extends between 3.0 and 4.5 Å and contains three types of correlations: secondary Se–Se correlations, Se–O encapsulate framework interactions, and Se–Cu^{2+} pairs. Since the three Gaussian functions fitted to this peak are not completely resolved, the corresponding distances were assigned with significant error bars: $R_{SeSe}(2) = 3.65 \pm 0.10$ Å, $R_{SeO} = 3.95 \pm 0.05$ Å, and, for the small first component, $R_{SeCu} = 3.30 \pm 0.05$ Å. This last component was not observed in the other zeolite studies just mentioned. While it was not possible to obtain absolute coordination numbers from these data, the areas of the peaks at $R_{SeSe}(1)$ and $R_{SeSe}(2)$ were similar, as expected for isolated rings or extended chains, suggesting that intermolecular Se–Se interactions play a minor role in this material as in the other zeolites.

The results presented here, together with the complementary Raman scattering measurements, indicated significant interactions between the incorporated Se and the Cu–Y matrix that modify the semiconductor's electronic structure. The absence of Raman bands characteristic of Se_8 rings suggested the formation of long Se chains inside the voids of the zeolite. The similarity of the values of the peak areas around $R_{SeSe}(1)$ and $R_{SeSe}(2)$ in the AXS measurement showed that these chains are mostly isolated. At $R_{SeSe}(1) = 2.39$ Å the first Se–Se distance is extraordinarily large for a covalent Se bond which points to a weakening of the intrachain bonding in comparison to

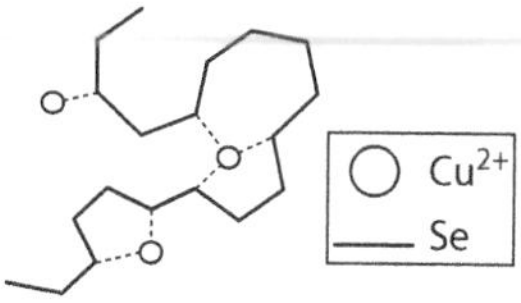

FIGURE 1.7
Schematic illustration of bonding situations between the encapsulated Se and the metal cations in Cu–Y zeolite.

the structure of Se chains in the other zeolites. This conclusion is corroborated by the large red shift of the encapsulate Raman band, while the width of this feature implies strong irregularities within the Se chains. Altogether, these features point to a new type of interaction between the encapsulated Se and the Cu^{2+} ion. This interaction was identified with the short-range correlation at 3.30 Å, which did not appear in the pair distribution functions obtained for other Se zeolite encapsulates. The authors concluded that Cu^{2+} ions could be coordinated to one, two, or even more Se atoms of chain fragments of various length and that these distinct bonding situations could randomly alternate along the chain, as shown schematically in Figure 1.7.

This study demonstrated the possibility of cation-directed band-gap alignments in zeolite-encapsulated semiconductors and established a convenient method for adjusting the electronic levels of clusters and molecules hosted in molecular sieves, which may be expedient for potential technical devices such as lasers or sensors.

1.3 Small-Angle Scattering

SAS is a nondestructive technique and a very effective probe to study geometry and texture of inhomogeneities in the mesoscopic and macroscopic range, i.e., between 5 and 500 nm* according to the IUPAC definition.[15-18] Because of the size range explored, SAS is a perfect complement of scanning and transmission electron microscopy (SEM, TEM) as well as diffraction (Figure 1.8). Small-angle neutron scattering (SANS) and its x-ray analog (SAXS) can be extremely useful in biology, polymer science, materials science, and chemistry. In this case the weaker scattering power gives the neutron measurements an advantage, since the samples are usually of manageable size—1–2 mm thick. Another advantage is that, as we will see later, contrast matching is much easier with neutrons, especially in systems containing light elements and in particular hydrogen atoms.

1.3.1 SAS Spectrometer

A SANS spectrometer is composed first of a monochromator capable of selecting wavelengths λ_0 in the range 5–20 Å, followed by several diaphragms (collimators) used to produce a parallel beam (Figure 1.9). The scattered neutrons

---

* The corresponding scattering vector range is approximately $10^{-3} < Q < 0.1$ Å^{-1}.

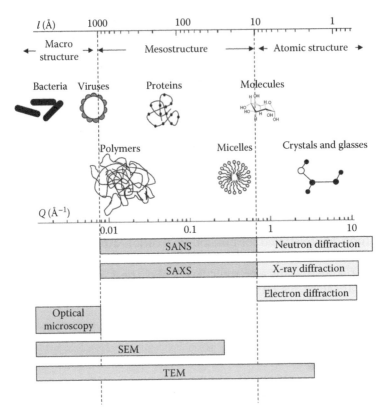

FIGURE 1.8
Examples of structures whose size range is between 1 and 10^4 Å (0.1 nm–1 μm). The comple-
mentarities of SANS and SAXS with x-ray, neutron, and electron diffraction and electron
microscopies (SEM and TEM) are shown in the lower part of the figure.

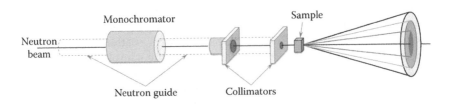

FIGURE 1.9
Schematic representation of a small-angle scattering spectrometer.

resulting from the interaction of neutrons with the sample are collected on
a 2D receptor plate, which counts the number of neutrons C as a function of
the scattering angle $\Delta\Omega$ and, if desired, the azimuthal angle. The detector is
placed far enough from the sample, typically 1–15 m, in order to collect data
at small angles ($\theta < 4°$).

1.3.2 Small Angle Scattering vs. Diffraction

For classical Bragg diffraction, an intense peak is observed when Bragg's law is fulfilled, i.e., when $n\lambda_0 = 2d \sin\theta$, where d is the distance between crystallographic planes, λ_0 the wavelength of the incident beam, and 2θ the scattering angle. In the case of periodic inhomogeneities with period d, the scattering peaks occur at angles where the scattering vector[19] $Q = 4\pi \sin\theta/\lambda_0 = 2\pi n/d$. In order to study objects of a size scale in the range 50–5000 Å, it is important to work at low Q values requiring low scattering angles 2θ (typically $<4°$) and long-wavelength or "cold" neutrons (typically $\lambda > 4\,Å$). For crystalline samples the use of long wavelength has the additional advantage of avoiding multiple processes involving Bragg reflections.

1.3.3 What Does Q Probe?

By definition, Q and d are inversely proportional. Thus, probing a particular Q range is equivalent to taking pictures of the sample at different levels of magnification (Figure 1.10).

For high Q values, the window of observation is so tiny that no contrast variation is observable. It corresponds roughly to the molecular scale, and, in some particular cases, such as crystals or organized materials, sharp diffraction peaks can be observed in this domain. At lower Q, a difference in contrast is observable at interfaces: this is the *Porod domain*, generally exhibiting a Q^{-4} dependence of the intensity, corresponding to a sharp discontinuity between two media. At still lower Q, the total intensity is representative of the shape of the scattering entities and can show an overall dependence ranging from Q^{-1} to Q^{-4}, depending on the shape. At very low Q values, the window is now large enough to probe the structural arrangement of the scattering entities (Figure 1.10). A careful study of the intensity as a function of Q allows us to determine two important quantities: the *form factor* $P(Q)$ that depends on the scattering entities, and the *structure factor* $S(Q)$ that

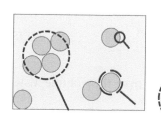

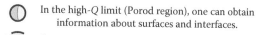
In the high-Q limit (Porod region), one can obtain information about surfaces and interfaces.

For intermediate Q range, information on the shape and size of objects can be deduced by measuring the form factor $P(Q)$.

In the low-Q domain, the arrangement of the objects can be obtained by deducing the structure factor $S(Q)$. In the low-Q limit (Guinier region), the average extent of a group of objects can be determined.

FIGURE 1.10
Schematic representation of spherical objects embedded in a uniform matrix probed at three different values of the scattering vector Q.

depends on the arrangement. We will now give explicit forms for these two functions.

1.3.4 Small-Angle Scattering Formalism

During a SANS experiment, the sample is exposed to a neutron beam and the scattered neutrons are counted as a function of the scattering angle and can be expressed as

$$C = \Phi \Delta\Omega A \eta T d \frac{d\Sigma}{d\Omega}(Q), \tag{1.17}$$

where

    C is the count rate
    Φ is the incident neutron flux
    $\Delta\Omega$ is the considered solid angle
    A is the area of the sample exposed to the beam
    η is the efficiency of the detector
    T is the transmission of the sample
    d is the thickness of the sample
    $\frac{d\Sigma}{d\Omega}(\vec{Q}) = \frac{N}{V}\frac{d\sigma}{d\Omega}(\vec{Q}) = \frac{1}{V}\left|\sum_{i=1}^{N} b_i e^{i\vec{Q}\vec{r}}\right|^2$ is the *macroscopic differential scattering*

    cross section summed over all N atoms in a volume V of the sample.
    V is the volume of the sample
    N is the number of atoms (scatterers)

The aim of the experiment is to determine the differential scattering cross section per atom, which contains all the information on shapes, sizes, and interactions between the scattering entities. In SAS experiments, it is more natural to think in terms of material properties rather than atomic properties, so it is legitimate to define a *scattering length density* ρ (SLD)[20–24]:

$$\rho = \frac{\sum_{i=1}^{N} b_i}{V}. \tag{1.18}$$

By replacing the sum over N atoms with integral over the volume V, we obtain the Rayleigh–Gans equation[25]

$$\frac{d\Sigma}{d\Omega}(\vec{Q}) = \frac{1}{V}\left|\int_V \rho(\vec{r})e^{i\vec{Q}\vec{r}}\,d\vec{r}\right|^2. \tag{1.19}$$

In a medium containing relatively heavy particles in comparison with the neutron mass, the elastic and quasi-elastic scattering, for which $|k_i| = |k_f| = 2\pi/\lambda$, are predominant.[26–29] In this static approximation, one can write

$$\frac{d\Sigma}{d\Omega}(\vec{Q}) = \frac{1}{V}\left\langle \left| \int_V \rho(\vec{r})e^{i\vec{Q}\vec{r}}d\vec{r} \right|^2 \right\rangle,$$

(1.20)

where the brackets $\langle \cdots \rangle$ signify an average over all the atoms.

In general terms, the scattering length b_i depends on both the isotopic state of the nucleus and its spin state. It is then convenient, as in Section 1.2.1, to split the differential scattering cross section per unit volume into two terms containing the coherent and incoherent scattering, respectively:

$$\frac{d\Sigma}{d\Omega}(\vec{Q}) = \left(\frac{d\Sigma(\vec{Q})}{d\Omega}\right)^{\text{coh}} + \left(\frac{d\Sigma(\vec{Q})}{d\Omega}\right)^{\text{inc}},$$

(1.21)

where

$$\left(\frac{d\Sigma(\vec{Q})}{d\Omega}\right)^{\text{coh}} = \frac{1}{V}\left\langle \left| \int \sum_{i=1}^{N} \overline{b}_i e^{i\vec{Q}\vec{r}_i}d\vec{r} \right|^2 \right\rangle \quad \text{and} \quad \left(\frac{d\Sigma(\vec{Q})}{d\Omega}\right)^{\text{inc}} = \frac{1}{V}\sum_{i=1}^{N}\left(\overline{b_i^2} - \overline{b}_i^2\right).$$

The coherent scattering, which is directly related to the spatial distribution of atoms in the sample, gives information on the structure and size of inhomogeneities, while the incoherent scattering gives a flat background independent of Q.

1.3.5 SAS from a Two-Phase System

We consider a two-phase system composed of particles of uniform SLD ρ_1 embedded in a matrix of uniform SLD ρ_2 (Figure 1.11). Each phase is supposed to be incompressible in such a way that the total volume of the sample V is the sum of the respective sub-volumes V_1 and V_2 of the two considered phases 1 and 2. Then, the Rayleigh–Gans equation becomes

$$\frac{d\Sigma(\vec{Q})}{d\Omega} = \frac{1}{V}\left| \int_{V_1} \rho_1 e^{i\vec{Q}\vec{r}}d\vec{r}_1 + \int_{V_2} \rho_2 e^{i\vec{Q}\vec{r}}d\vec{r}_2 \right|^2,$$

(1.22)

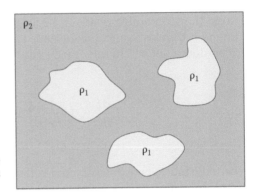

FIGURE 1.11
Example of a sample composed of two incompressible phases of scattering length density ρ_1 and ρ_2.

where $V = V_1 + V_2$. Equation 1.22 can then be rearranged as

$$\frac{d\Sigma(\vec{Q})}{d\Omega} = \frac{1}{V}\left|\rho_1\int_{V_1} e^{i\vec{Q}\vec{r}}d\vec{r}_1 + \rho_2\left(\int_V e^{i\vec{Q}\vec{r}}d\vec{r} - \int_{V_1} e^{i\vec{Q}\vec{r}}d\vec{r}_1\right)\right|^2. \tag{1.23}$$

At nonzero Q values, this reduces to

$$\frac{d\Sigma(\vec{Q})}{d\Omega} = \frac{1}{V}(\rho_1 - \rho_2)^2\left|\int_{V_1} e^{i\vec{Q}\vec{r}}d\vec{r}_1\right|^2, \tag{1.24}$$

$(\rho_1 - \rho_2)^2$ is the so-called *contrast factor* which is an intrinsic property of the material (density, composition). The integral term is indicative of the spatial arrangement of phase 1.

Equation 1.24 shows that $d\Sigma/d\Omega$, and hence $I(\vec{Q})$, is proportional to $(\rho_1 - \rho_2)^2$. Consequently, the higher the contrast factor is, the more intense will be the coherent scattering signal. Moreover, since the contrast factor is the square of the difference in SLD between the particles and the matrix, two homologous structures with reversed SLD will show identical coherent scattering profiles according to Babinet's Principle (Figure 1.12). This property is related to the fact that $d\Sigma/d\Omega$ is proportional to the square of an amplitude, leading to an unavoidable loss of phase information. In the limiting case where the two phases have identical SLD, i.e., when the contrast factor is equal to zero, no coherent scattering is observable since the sample appears homogeneous toward the neutron beam. These properties turn out to be very efficient in the determination of partial structures in multiphase systems.

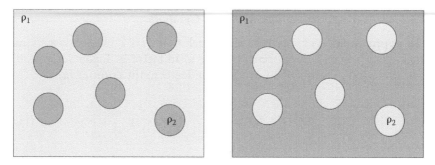

FIGURE 1.12
Babinet's principle: two identical structures with reversed scattering length densities will give the same coherent scattering.

In a *dilute medium*, i.e., when one phase consists of N_p identical particles sufficiently far enough to neglect interparticle interaction between them, Equation 1.24 can be considerably simplified:

$$\frac{d\Sigma(\vec{Q})}{d\Omega} = \frac{N_P}{V}(\rho_1 - \rho_2)^2 \left| \int_{V_P} e^{i\vec{Q}\vec{r}} d\vec{r} \right|^2 = n_P(\rho_1 - \rho_2)^2 V_p^2 P(\vec{Q}), \qquad (1.25)$$

where

$$P(\vec{Q}) = \frac{1}{V_p^2}\left|F(\vec{Q})\right|^2 = \frac{1}{V_p^2}\left| \int_{V_P} e^{i\vec{Q}\vec{r}} d\vec{r} \right|^2. \qquad (1.26)$$

Here $n_p = N_p/V$ and V_p represent the number density of the particles and the volume of each particle, respectively. The scattering amplitude $F(\vec{Q})$ contains all the information about the shape of the particles. $P(\vec{Q})$ is commonly called the *form factor* and represents the interference of neutrons scattered from different parts of the same object. In the limiting case where $Q \to 0$, Equation 1.25 reduces to

$$\frac{d\Sigma}{d\Omega} \xrightarrow{Q \to 0} n_p V_p^2 (\rho_1 - \rho_2)^2, \qquad (1.27)$$

so that the intensity as $Q \to 0$ can be used to measure the volume of the scattering particle V_p or its molecular weight $M_w = \rho V_p N_A$ if its chemical composition and density are known.

1.3.5.1 Guinier Approximation: Radius of Gyration

In the approximation of small Q values such that $Ql \ll 1$, where l is a characteristic particle size, a development of the form factor in Taylor series and a reinterpretation in terms of an exponential leads to the *Guinier law*:

$$P(\vec{Q}) = F^2(Q) = e^{-\frac{(QR_g)^2}{3}} \quad (QR_g < 1), \tag{1.28}$$

where R_g is the *Guinier radius* corresponding to the root-mean-square extension of the particle from its center of scattering density. For a homogeneous particle this is equivalent to the *radius of gyration* given by

$$R_g^2 = \frac{\int_{V_P} r^2 d\vec{r}}{V_P}. \tag{1.29}$$

Knowing that the relative scattering intensity $I(Q) \propto P(Q)$, Equation 1.28 can be formulated in a different way:

$$I(Q) = I(0)e^{-\frac{(QR_g)^2}{3}} \tag{1.30}$$

$$\ln(I(Q)) = \ln(I(0)) - \frac{(QR_g)^2}{3}. \tag{1.31}$$

Using Equation 1.31, it is possible to evaluate the radius of gyration R_g of the scattering entity from the experimental data by plotting $\ln(I(Q))$ vs. Q^2. The Guinier law is a powerful equation to characterize soft matter and more particularly polymers, but it has to be used with caution since the approximation is valid only in dilute systems composed of isotropic particles under the condition $QR_g \ll 1$.

Some examples of radius of gyration for some usual homogeneous particles are given below[30,31]:

Sphere of radius R	$R_g^2 = \dfrac{3R^2}{5}$
Spherical shell with radii $R_1 > R_2$	$R_g^2 = \dfrac{3}{5}\dfrac{R_1^5 - R_2^5}{R_1^3 - R_2^3}$
Ellipse with semiaxes a and b	$R_g^2 = \dfrac{a^2 + b^2}{4}$
Ellipsoid with semiaxes a, b, c	$R_g^2 = \dfrac{a^2 + b^2 + c^2}{5}$

Prism with edges A, B, C	$R_g^2 = \dfrac{A^2 + B^2 + C^2}{12}$
Cylinder with radius R and length l	$R_g^2 = \dfrac{R^2}{2} + \dfrac{l^2}{12}$
Gaussian coil of root-mean-square end-to-end distance L	$R_g^2 = \dfrac{L^2}{6}$

1.3.5.2 Porod Scattering

Where as the Guinier approximation considers the low-Q limit of SAS, the high-Q limit can be described using an approximation due to Porod. At high values of Q, i.e., when $Q \gg 1/l$ where l is the size of the scattering object, the scattering intensity is dominated by the signal coming from the boundaries between the phases of the system. The interface appears as a discontinuity of the SLD leading to the following expressions:

$$I(Q) \propto Q^{-4}, \tag{1.32}$$

and, more specifically,

$$\frac{4\pi^2}{\tilde{Q}} \lim_{Q \to \infty} \left(\frac{d\Sigma}{d\Omega}(Q)Q^4 \right) = \frac{S}{V}, \tag{1.33}$$

where $\tilde{Q}$ is the *scattering invariant*, i.e., the integral of the macroscopic cross section over $\vec{Q}$, and S/V is the *specific surface* corresponding to the total area of interface per unit volume of the sample. The determination of the total specific surface by SAS, whatever the size and shape of the particle, turns out to be useful in the characterization of porous materials and can be an excellent complement to adsorption measurements such as BET since neutron or x-ray beam probes the overall interface surface whereas adsorption is limited to surfaces accessible to the molecules of the working gas.

1.3.5.3 Porod Invariant

In 1952, Porod demonstrated that the total SAS from a sample, which depends on both the contrast and volume fraction, is constant and independent of the dispersion of scattering entities. The *scattering invariant* or *Porod invariant* is representative of the total amount scattered, but not of the detailed structure. By integrating the scattering intensity with respect to $\vec{Q}$, we can demonstrate that in the case of an incompressible two-phase system

$$\tilde{Q} = (2\pi)^3 \varphi_1 (1 - \varphi_1)(\rho_1 - \rho_2)^2$$

$$\tilde{Q} \approx (2\pi)^3 \phi_1 (\rho_1 - \rho_2)^2 \quad \text{if } \phi_1 \ll 1, \tag{1.34}$$

where $\phi_1 = n_p V_p$ is the volume fraction of the particles.

For uncorrelated particles, Equations 1.27 and 1.34 give access to the volume fraction of particles and consequently to the volume of each particle:

$$V_P = \frac{(2\pi)^3}{\tilde{Q}} \frac{d\Sigma}{d\Omega}(0). \tag{1.35}$$

We note that in Equations 1.33 and 1.35 the measured intensity appears in both numerator and denominator, so considerable information can be obtained from measurement of the relative intensity $I(Q)$, without the need for absolute normalization, as long as the Q range is sufficient to obtain accurate values for the integrated intensity and the low- and high-Q limits (not always an easy condition to obtain in practice).

1.3.6 Form Factors

Experimentally, form factors and scattering amplitudes can only be measured directly in the dilute regime where particles are considered as independent and noninteracting. The precise form of $F(\vec{Q})$ depends on the shape and orientation of the scatterer. Isotropic spherical particles represent the simplest case since considerations about particle orientation do not arise. In the case of an ideal spherical particle of radius R and of uniform SLD ρ_1, the form factor can be simply expressed as follows:

$$P(Q) = \left(\frac{3(\sin QR - QR \cos QR)}{(QR)^3} \right)^2. \tag{1.36}$$

If the particle morphology is not spherical, the orientation relative to the scattering vector $\vec{Q}$ must be considered in the evaluation of $P(\vec{Q})$. In the general case, i.e., for particles of arbitrary shape and with no preferential orientation, the form factor has to be averaged over all the orientations taken by the different scatterers:

$$\overline{P(Q)} = \overline{F_i^2(Q)} = \int_0^1 |F_i(Q,\mu)|^2 \, d\mu, \tag{1.37}$$

where $\mu = \cos \alpha$ and α is the angle between an axis of the scatterer and $\vec{Q}$. Form factors of common shapes can be found in the literature.[30] Nonetheless, we give the form factors for two usual geometries:

- For long cylinders of radius R with a length L

$$\overline{P(Q)} = \left[\frac{2}{(QL)} \left(\int_0^{QL} \frac{\sin u}{u} du \right) - \frac{\sin^2 \frac{QL}{2}}{\left(\frac{QL}{2} \right)^2} \right], \tag{1.38}$$

- For thin disks of radius R

$$\overline{P(Q)} = \frac{2}{(QR)^2} \left[1 - \frac{1}{2QR} J_1(2QR) \right], \tag{1.39}$$

where $J_1(x)$ is the first-order Bessel function.

At intermediate values of Q, $P(Q)$ varies as $1/Q$ in the case of a cylindrical shape and as $1/Q^2$ for thin disks (Figure 1.13). This tendency can be expressed as a more general rule: there exists a range of Q for which the measured scattering intensity $I(Q)$ will vary as $1/Q$ for one-dimensional objects (narrow

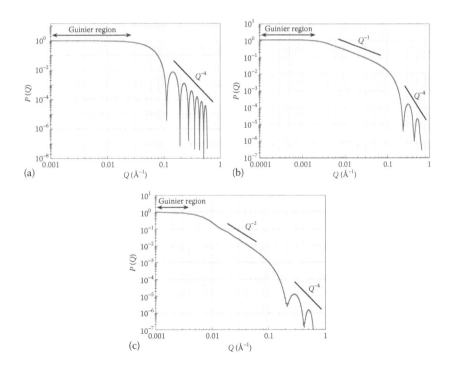

FIGURE 1.13
Form factors of three usual shapes: (a) spheres ($R=40\,\text{Å}$), (b) cylinders ($R=60\,\text{Å}$, $L=34\,\text{Å}$), and (c) thin disks ($R=300\,\text{Å}$, thickness$=30\,\text{Å}$). (From Kline, S.R., *J. Appl. Cryst.*, 39, 895, 2006.)

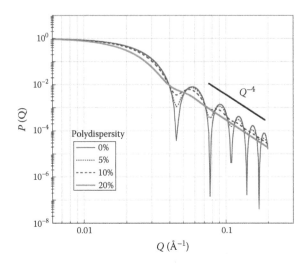

FIGURE 1.14
Evolution of the form factor of solid hard spheres ($R = 100 \,\text{Å}$) as a function of the polydispersity of the radius.

cylinders, wires, …) and as $1/Q^2$ for 2D objects (thin disks, platelets, …). Such dependencies can be graphically detected by plotting log $I(Q)$ vs. log (Q).

The above form factors represent the ideal sample for which the shape and size of particles are perfectly controlled. It is noteworthy that, as good as the synthesis methods are, this perfect scenario is far from the reality, and there are generally distributions in both size and shape. In this case, the form factor has to be averaged over the size distribution leading to a smearing of the principal features. An example of the effect of polydispersity on the global shape of the form factor of solid hard spheres is shown in Figure 1.14.

1.3.7 Multicomponent Systems: Neutron Contrast Matching

So far, we have assumed that our sample is constituted of a single type of particle embedded in a uniform matrix, typically a solvent. In many cases, like in biological systems or porous materials, the sample is composed of different types of subunits with different shapes and/or sizes. If two particle subunits of uniform SLD ρ_1 and ρ_2 give a scattering signal in the studied Q range, the measured differential cross section can be written as follows:

$$\frac{d\Sigma}{d\Omega}(\tilde{Q}) = \frac{1}{V}\left\langle \left| \int_V (\rho(r) - \rho_0)e^{i\vec{Q}\vec{r}}\,d\vec{r} \right|^2 \right\rangle, \tag{1.40}$$

where ρ_0 is the average SLD over the entire sample.

Knowing that $V = V_1 + V_2$, where V_1 and V_2 are the volume of the particles 1 and 2, respectively, the expression for the macroscopic cross section, which is proportional to the scattering intensity, can be rewritten as

$$\frac{d\Sigma}{d\Omega}(\vec{Q}) = \frac{1}{V}\left\langle \left| (\rho_1 - \rho_0) \int_{V_1} e^{i\vec{Q}\vec{r}_1} d\vec{r}_1 + (\rho_2 - \rho_0) \int_{V_2} e^{i\vec{Q}\vec{r}_2} d\vec{r}_2 \right|^2 \right\rangle$$

$$= \left(n_{P_1}(\rho_1 - \rho_0)^2 V_{P_1}^2 \left\langle \left| F_1(Q) \right|^2 \right\rangle + n_{P_2}(\rho_2 - \rho_0)^2 V_{P_2}^2 \left\langle \left| F_2(Q) \right|^2 \right\rangle \right), \quad (1.41)$$

where $F_a(Q)$, $n_{Pa}(Q)$, and $V_{Pa}(Q)$ are the scattering amplitude, number density, and volume of particle type a, respectively.

It appears from Equation 1.41 that the total scattering intensity depends on both differences in SLD between the particles and the matrix, $(\rho_1 - \rho_0)$ and $(\rho_2 - \rho_0)$. Consequently, it is possible to cancel one of the two terms by choosing an appropriate SLD ρ_0 for the matrix (solvent). Then, two measurements are enough to obtain the parameters of interest, i.e., $F_1(Q)$ and $F_2(Q)$ (Figure 1.15). In practice, SANS is more suitable than SAXS to adjust the SLDs because neutrons are very sensitive to the nature of isotopes (cf. Section 1.2). A similar adjustment can be achieved with anomalous scattering or isomorphous substitution in favorable cases. However, the latter may modify the structure of the sample. SANS is also particularly appropriate for light elements such as hydrogen for which the selective deuteration method can lead to spectacular results since the coherent scattering length of hydrogen has a different sign from that of deuterium ($b_H = -0.374 \times 10^{-12}$ cm vs. $b_D = 0.667 \times 10^{-12}$ cm). The H:D substitution technique has turned out to be very powerful since a simple

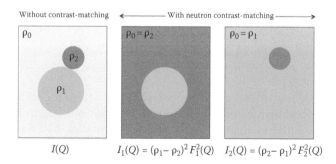

FIGURE 1.15
Schematic representation of the contrast-matching technique. In the present case, with only three measurements and by varying properly the scattering length density of the matrix ρ_0, it is possible to measure separately the two particle subunits. (With permission from Lelong, G., Bhattacharyya, S., Kline, S.R., Cacciaguerra, T., Gonzalez, M.A., and Saboungi, M.-L., Effect of surfactant concentration on the morphology and texture of MCM-41 materials, *J. Phys. Chem. C*, 112, 10674–10680, 2008.)

mixture of D_2O with H_2O is enough to match the contrast of almost every biological system and porous material.

1.3.8 Interacting Scatterers: Structure Factor

Up to now, we have considered the simple case of a gas of randomly dispersed particles, i.e., with no interaction between the scattering entities and no particular tridimensional arrangement. This approximation is valid for dilute systems but does not hold in the case of condensed matter or other systems with interacting particles, for which $d\Sigma/d\Omega$ depends not only on the intraparticle structure but also on the interparticle interactions. In the case of centrosymmetric identical particles, the spatial arrangement leads to the appearance of a new function $S(Q)$ in the expression of the coherent differential scattering cross section:

$$\frac{d\Sigma(Q)}{d\Omega} = n_P(\rho_1 - \rho_2)^2 V_P^2 P(Q) S(Q). \tag{1.42}$$

$P(Q)$ is the *form factor* that we introduced earlier and $S(Q)$ is the *structure factor*, which represents the interference of neutrons scattered from different particles:

$$S(\vec{Q}) = \frac{1}{N_p} \left\langle \sum_{i,j=1}^{N_p} e^{i\vec{Q}(\vec{R}_i - \vec{R}_j)} \right\rangle. \tag{1.43}$$

Thus, for centrosymmetric objects, the scattering arising from one single scattering entity can be dissociated from the signal caused by the interactions between particles. For noninteracting scatterers, as in a dilute medium, one can demonstrate that $S(Q) = 1$ over the whole Q range. In the particular case of an isotropic solution, the interparticle structure factor $S(Q)$ can be expressed as follows[33]:

$$S(Q) = 1 + 4\pi n_p \int_0^\infty [g(r) - 1] \frac{\sin Qr}{Qr} r^2 dr, \tag{1.44}$$

where $g(r)$ is the pair correlation function for the scattering entities and represents the probability of finding an atom at the position r assuming another at the origin. Equation 1.44 is the SAS analogy of Equation 1.10. Subsequently, the product $n_p g(r)$ represents a local density of particles, and will be dependent on both their concentration and their spatial arrangement (Figure 1.16).[33]

In the case of two types of particles with correlations between all of them, $S(\vec{Q})$ will have three partial components:

$$S_{ab}(\vec{Q}) = \frac{1}{(N_{pa} N_{pa})^{1/2}} \left\langle \sum_{i=1}^{N_{pa}} \sum_{j=1}^{N_{pb}} F_i(\vec{Q}) F_j(\vec{Q}) e^{i\vec{Q}.(\vec{R}_i - \vec{R}_j)} \right\rangle. \tag{1.45}$$

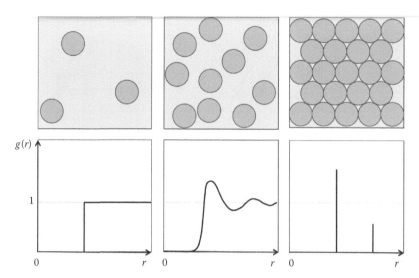

FIGURE 1.16
Schematic representation of the pair correlation function $g(r)$ as a function of the concentration in particles and their 3D organization (dilute, intermediate, and concentrated solutions).

and Equation 1.41 must be replaced by

$$\frac{d\Sigma}{d\Omega}(\vec{Q}) = \left(n_{P_1}(\rho_1 - \rho_0)^2 V_P S_{11}(\vec{Q}) + n_{P_2}(\rho_2 - \rho_0)^2 V_{P_2}^2 S_{22}(\vec{Q}) \right)$$

$$+ 2\left(n_{P_1} n_{P_2} \right)^{1/2} (\rho_1 - \rho_0)(\rho_2 - \rho_0) V_{P_1} V_{P_2} S_{22}(\vec{Q}), \quad (1.46)$$

so that three independent measurements will be needed to determine $S_{ab}(\vec{Q})$.

An example of a structure factor can be given in the case of fractal objects. This represent a particular class of scattering systems that have the property of self-similarity, in which a similar geometric pattern is repeated at every length scale. Two distinct classes of fractal behavior have been defined: the *mass fractal* and the *surface fractal*.[34] A mass fractal is a structure composed of subunits forming branching and crosslinking to create a tridimensional network, while a surface fractal is an object displaying fractal characteristics on its surface but not necessarily in its core (Figure 1.17).

In the case of mass fractals, the particle number density n_p is dependent on the position r and can be expressed as a function of the parameter D_m, the *mass fractal dimension*, generally with values in the range 2–3:

$$n_P(r) = \left(\frac{r}{r_0} \right)^{D_m}. \quad (1.47)$$

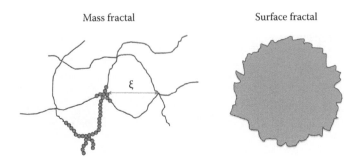

Mass fractal   Surface fractal

FIGURE 1.17
Schematic representation of the two types of fractals: (Left) The mass fractal morphology is composed of subunits of size r_0 arranged in linear chains. The entanglement of the polymeric chains leads to the appearance of a correlation distance ξ. (Right) Example of surface fractal, which is characterized by a surface fractal dimension D_S.

Then, for relatively small Q values, the fractal structure factor $S(Q)$ is given by[35,36]

$$S(Q) = 1 + \frac{\sin\left[(D_m - 1)\tan^{-1}(Q\xi)\right]}{(Qr_0)^{D_m}} \frac{D_m\Gamma(D_m - 1)}{\left[1 + \dfrac{1}{(Q\xi)^2}\right]^{\frac{D_m-1}{2}}}, \qquad (1.48)$$

where
   D_m is the fractal dimension
   r_0 is the radius of the primary particles
   ξ is the correlation length, i.e., a characteristic size of the mass fractal

Clearly, $S(Q)$ shows a main dependence on Q^{-D_a}. This power law decay is a clear signature of fractal objects. Using the Guinier approximation, it is possible to estimate the radius of gyration of a mass fractal knowing its fractal dimension D_m and its correlation length ξ:

$$R_g^2 = \frac{D_m(D_m + 1)\xi^2}{2}. \qquad (1.49)$$

For a surface fractal, the scattering intensity has the form $I(Q) \propto Q^{D_S-6}$, where D_S is the *surface fractal dimension* ranging usually from 2 for a smooth surface to 3–4 in the case of rough surfaces. When $D_S = 2$, the scattering intensity follows a Q^{-4} behavior, in agreement with Porod's law. Fractal dimensions larger than 4 may be found in some cases where there are no sharp interfaces but rather a gradual transition between phases.

1.3.9 SANS vs. SAXS

Small-angle neutron and x-ray scattering are comparable and often complementary techniques. Neutrons and x-rays interact in a different way
with matter, leading to a difference in the sensitivity of both techniques:
neutron scattering measures fluctuations of nuclear density, and x-rays
inhomogeneities in electron density. With the use of isotopic substitution,
SANS can selectively probe different species in a multicomponent system
(cf. Section 1.3.7). In favorable cases, anomalous scattering, exploiting rapid
variation of the x-ray form factor near an absorption edge, can be used with
SAXS to distinguish different elements. Synchrotron radiation sources
deliver a flux highly superior to neutron sources, favoring a fast acquisi-
tion of spectra that allows an in situ monitoring of chemical reactions or fast
structural changes as a function of time. It should also be mentioned that the
interaction of neutrons with unpaired spins can be used to probe magnetic
inhomogeneities, for example, flux line lattices in type-II superconductors.

1.3.10 Selected Examples

Excellent reviews have been published in the last decade showing the main
progresses of the SAS technique in a variety of domains such as magne-
tism,[37] biology,[38,39] polymer,[40–42] or material sciences.[43] We have selected here
some examples showing the capabilities of small-angle neutron and/or x-ray
scattering.

1.3.10.1 Porous Materials

SAS is a natural technique characterizing mesoporous materials since their
characteristic length scales are generally located in the Q window probed
by SAS spectrometers. Furthermore, the richness of the accessible structural data and its complementarity with electron microscopies or adsorption
methods give it a full-fledged legitimacy.

A good example is the family of mesoporous materials called MCM-41,
which has been extensively studied since their discovery in the early 1990s,[44]
and whose particle shape can be easily tailored. A recent work of Lelong et al.[45]
demonstrates the structural evolution of mesoporous silica nanospheres as
a function of the initial surfactant concentration (Figure 1.18). The spherical
silica particles were composed of cylindrical pores arranged in a honeycomb
structure. In view of the different particle shapes, length scales, and the two-
dimensional organization of the pores, a global fit of the scattering intensity
$I(Q)$ over a large Q range is very difficult to achieve. Nevertheless, morphological and structural information were deduced from the scattering profile
by using the form factors of both an ideal sphere $P_{sphere}(Q)$ and an ideal cylinder $P_{cyl}(Q)$, and an appropriate structure factor $S(Q)$. The SANS investigation
reported a decrease in the diameter of the silica particles when the surfactant

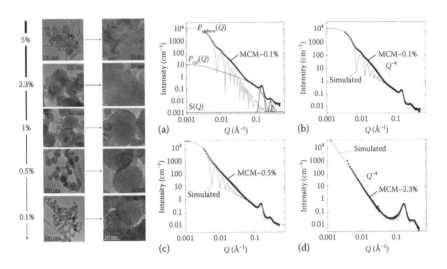

FIGURE 1.18

(Left) TEM images of mesoporous silica spheres as a function of surfactant concentration. (Right) (a) SANS intensity from the MCM–0.1% sample together with the form factors of a cylinder $P_{cyl}(Q)$ and a sphere $P_{sphere}(Q)$. Measured and simulated SANS intensities for the (b) MCM–0.1%, (c) MCM–0.5%, and (d) MCM–2.3%. (With permission from Lelong, G., Bhattacharyya, S., Kline, S.R., Cacciaguerra, T., Gonzalez, M.A., and Saboungi, M.-L., Effect of surfactant concentration on the morphology and texture of MCM-41 materials, *J. Phys. Chem. C*, 112, 10674–10680, 2008.)

concentration was decreased accompanied by a global improvement of both the internal structure and the size dispersion. These data were corroborated by TEM pictures shown in Figure 1.18.

The determination of the total specific area is one of the major assets of SAS since it probes both the opened and closed porosities, which conventional adsorption methods do not do (cf. Section 1.3.5.2). For example, Né and Zemb[46] in a paper of 2003 presented a method to determine experimentally the specific area and the compaction by SAXS, if both the Porod limit and the invariant can be extracted from the scattering profile. The method was applied to a mesostructured material composed of ZrO_2 and a cationic surfactant (CTAB), showing a higher total interface surface and a lower compaction than in the case of crystalline ZrO_2. By applying a careful methodology in the data treatment, they ascribed the origin of these significant differences to the presence of microporosities in the walls separating the mesopores. This discovery explained why this mesoporous material can collapse during removal of the template molecule at high temperature.

In the last few years, some very interesting works have been carried out on porous systems such as the study of the entanglement of labeled single-wall carbon nanotubes[47] to determine the global shape of aggregates, or the combination of SANS with mesoscopic simulation techniques in order to generate 2D or 3D images of the porous structure. The structural model obtained can then be used to understand the macroscopic physicochemical properties.[48]

The development of both high flux neutron and x-ray sources and new environmental chambers[49] opened the way to *in situ* SAS measurements capable of monitoring adsorption or capillary condensations for a multitude of gases.[50] Since the first *in situ* SAS measurements in the middle of the 1990s, many papers were devoted to the study of water or N_2 condensation in mesoporous materials such as Vycor, MCM–41, SBA–15, xerogels. The deformation of the pore walls and/or the pore lattice can then be followed during the intrusion of a fluid inside the mesoporosity.[51,52] Zickler et al. have demonstrated that the sorption of an organic fluid, perfluoropentane (C_5F_2), in SBA-15 is accompanied by a modification of the pore lattice as shown by the shift of the different Bragg peaks as the relative pressure is increased. A reversible deformation of the pore lattice caused by capillary stresses has also been highlighted.

In the same vein, one of the most spectacular evolutions has been the ability to carry out *in situ* time-resolved scattering experiments, opening the way to a vast field of investigation.[53] These new methods turn out to be fruitful in terms of information on transitory states. Fast chemical reactions can then be followed by SANS or SAXS using the *stopped-flow* method in which a small volume of reactants are mixed in a very short period of time. A remarkable example was given by Zholobenko et al.[54,55] who experimentally demonstrated the different stages of the *cooperative self-assembly mechanism* in the case of SBA-15 material. At the very beginning of the reaction, the template molecules (Pluronic P123) assemble into spherical micelles. In a second step, the formation of the organic–inorganic micelles is accompanied by a modification of their overall shape going from spherical to cylindrical. The aggregation of the cylindrical micelles in a 2D hexagonal structure characteristic of the mesoporous material constitutes the last step of the reaction. The void between the cylinders is then progressively replaced by silicate species to finally lead to a complete filling with further condensation (Figure 1.19).

1.3.10.2 Soft Matter

As mentioned earlier in the chapter, neutron scattering has the big advantage of being very sensitive to isotopic substitution. This unique property is particularly well suited to the study of soft matter, since polymers, micelles, gels, colloids, etc. are mainly composed of light elements. The contrast matching is then extensively used to disentangle complex systems.[56]

A good example has been given by Doe et al.,[57] who have fully characterized the different structures of a ternary system composed of a block copolymer (P84 – $(EO)_{19}(PO)_{43}(EO)_{19}$), water, and *p*-xylene. Depending of the water/oil ratio, the sample can show a lamellar or a reverse hexagonal structure. A careful study of the evolution of the relative intensities of the Bragg peaks as a function of the contrast made it possible to determine the structuring in layers of the lamellar phase, and notably the presence of a water-rich layer in between the polar domains. The succession of polar and apolar subdomains has been described as follows: water rich, homogeneous PEO/water mixture,

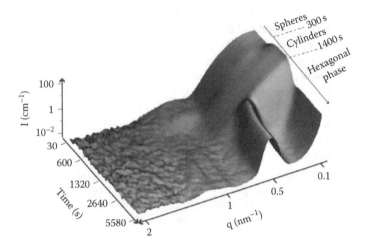

FIGURE 1.19
Time-resolved in situ SANS data for SBA-15 synthesis. (From *Adv. Coll. Interf. Sci.*, 142, Zholobenko, V.L., Khodakov, A.Y., Impéror-Clerc, M., Durand, D., and Grillo, I., Initial stages of SBA-15 synthesis: An overview, 67–74, Copyright 2008, with permission from Elsevier.)

solvent depleted, homogeneous PPO/oil mixture, solvent depleted, homogeneous PEO/water mixture, etc.

The second example deals with sugar solutions confined in silica gel.[58,59] The almost equal and opposite SLD of H_2O ($\rho_{H_2O} = -5 \times 10^{-7}$ Å^{-2}) and D_2O ($\rho_{D_2O} = 6.36 \times 10^{-6}$ Å^{-2}) makes possible the *masking* of the silica network or that of the sugar solution by using different H_2O/D_2O ratios. After determining the SLD of each constituent, two different gels were prepared: one to observe the silica network (*gel–silica*), and one to check the homogeneity of the sugar solution (*gel–sugar*); a sugar solution of same concentration as test sample was also prepared. The intensity profile of the *gel–silica* sample showed a Q^{-2} behavior at intermediate Q range indicative of a mass fractal object and a Q^{-4} behavior in the Porod region, typical of a smooth surface. The test solution and the *gel–sugar* sample showed the same flat intensity profile, indicative of a homogeneous repartition of the sugar molecules on the probed SANS length scale, i.e., between 3 and 100 nm. We should also mention the emergence of new studies on micelles based on rheology measurements monitored by SAS.[60,61]

1.3.10.3 New Developments

For completeness, we should mention the emergence of new *in situ* measurements coupling the SAS technique with one or two other methods of characterization. The most natural one is to use both small-angle and wide-angle x-ray or neutron scattering (WAXS or WANS) in order to increase the probed Q range, especially when objects are in a Q domain located in between SAS and diffraction. More complicated combinations can also be carried out such as SAXS/EXAFS, SAXS/WAXS/FTIR, SAXS/WAXS/Raman, SAXS/WAXS/ Light Scattering, or also SAXS/WAXS/DSC,[62–64] making possible the interpretation of complex phenomena.

1.4 X-Ray Absorption Spectroscopy

X-ray absorption spectroscopy is a general term referring to experiments in which the absorption of an x-ray beam by a sample is measured as a function of incoming energy. Sometimes it is more convenient to measure the fluorescence produced following the absorption rather than the attenuation of the beam. The absorption increases when the energy is raised through and above an absorption edge of one of the elements in the sample. In materials science, the important energy regions are near the edge (x-ray absorption near-edge spectroscopy, XANES) and extending for some range above the edge (EXAFS). Since the absorption edge is associated with the transition of an electron in the sample from a core level to a free state, the detailed energy dependence of the XANES spectrum gives information about the electronic structure of the valence and conduction electrons. EXAFS, on the other hand, is essentially a diffraction phenomenon in which the photoelectron is scattered back from the neighboring atoms: the back-scattered wave interferes with that of the primary photoelectron to produce a change in the absorption probability. The higher the x-ray energy E above the energy E_A of the absorption edge, the larger the wave vector k of the photoelectron and hence the scattering vector $2k$ characterizing the diffraction process:

$$E - E_0 = \frac{\hbar^2 k^2}{2m_e}. \tag{1.50}$$

The spectrum of absorption vs. $2k$ then shows oscillations similar to that of $S(Q)$ in diffraction experiments. This has the advantage that the structural information in the spectrum is element specific, i.e., it relates only to the environment of the absorbing atoms. A disadvantage is that in liquids and glasses it is difficult to get structural information beyond the nearest neighbors of the absorbing atoms. With the powerful x-ray synchrotron sources now available, it is often preferable to use the AXS technique referred to above. However, in that case there is a correlation between the energy of the absorption edge and the maximum magnitude of the scattering vector Q, which determines the spatial resolution of the measurement, so with light elements, for example, those lighter than germanium ($E_K = 11.1$ keV, $Q_{max} = 11.25$ Å^{-1}), other methods must be used: EXAFS, neutron diffraction with isotopic substitution, if suitable isotopes are available, or a combination of two or three techniques. Other advantages of EXAFS are that it is possible to obtain information about three-body correlations[65] and, since the measurements are relatively rapid, structural changes can be studied during rapid variation of temperature or pressure.

The EXAFS signal $\chi(k)$ is defined as the normalized deviation of the absorption coefficient $\mu(k)$ of the sample from its value for an isolated atom $\mu_0(k)$:

$$\chi(k) = \frac{\mu(k) - \mu_0(k)}{\mu_0(k)}.$$

(1.51)

From scattering theory,

$$\chi(k) = \frac{|a(k)|}{k} \int_0^\infty \frac{p(r)}{r^2} e^{-\frac{2r}{\lambda(k)}} \sin[2kr + \varphi(k)] dr,$$

(1.52)

where $|a(k)|$ and $\phi(k)$ are the characteristic back-scattering amplitude and phase shift due to scattering from the neighboring atoms, $\lambda(k)$ is the mean free path of the photoelectron and $p(r)$ is the bond length probability density, proportional to $g(r)$. In the case of small disorder, a Gaussian probability density with variance σ is normally assumed, and Equation 1.52 becomes

$$\chi(k) = -\sum_n \frac{C_1^n(j)}{\left(r_{1j}^n\right)^2} |f_j(k)| \exp\left(-2\left(\sigma_{1j}^n\right)^2 k^2\right) \exp\left(-\frac{2r}{\lambda(k)}\right) \sin\left[2kr_{1j}^n + \varphi_j(k)\right] dr,$$

(1.53)

where, in the notation introduced in Section 1.2.1, the absorbing atom is taken as type 1 and the nth coordination shell is occupied by atoms of type j. For disordered systems like liquids, glasses, and crystalline materials at high temperatures, the assumption of symmetric peaks in $g(r)$ is no longer valid, since the backscattering atoms feel the anharmonicity of the pair potential, and more realistic model functions must be employed.[66]

1.5 Inelastic Scattering

The dynamics of a system can be measured in an inelastic scattering experiment. For the past 50 years, this has been principally the province of neutron scattering, taking advantage of the fact that neutron beam emerging from moderators at reactors or spallation sources have typical energies on the order of 0.025 eV, corresponding to a temperature of about 300 K and comparable with typical energies of collective excitations in solids and liquids.[4,29] With cooled moderators and developments in neutron spectroscopic techniques such as neutron spin-echo spectrometry,[67] the usable energy range has been pushed down to 10^{-6} eV and below, which provides a powerful probe of relaxation processes in complex materials. X-rays start with the disadvantage that the energies must be on the order of 10^4 eV to access the Q values ~10 Å^{-1} of interest for investigations of the structure and dynamics of materials, so that an energy resolution of ~10^{-7} is required to get useful information about collective excitations in solids and liquids. Remarkably, this has been achieved with sophisticated design of energy monochromators and analyzers at the third-generation neutron sources such as the European Synchrotron Radiation Facility (ESRF) in Grenoble, France, the Advanced Photon Source (APS) in Argonne, Illinois, and SPRing-8 in Hyogo Prefecture, Japan. High-resolution IXS techniques have the advantage of overcoming the kinematic limitations affecting many neutron scattering studies and make it possible to study collective excitations in liquids and glasses at low Q.[68] This follows from the requirement that the velocity of the probe in such measurements must be appreciably higher than that of the collective excitation under study.

In inelastic scattering experiments, the energy transfer E—or equivalently the excitation frequency $\omega = E/\hbar$—is measured in addition to the scattering vector $\vec{Q}$. In the neutron case this is given by

$$E = \frac{\hbar^2 k_i^2}{2m_n} - \frac{\hbar^2 k_f^2}{2m_n},\tag{1.54}$$

and in the x-ray case by

$$E = \hbar c k_i - \hbar c k_f.\tag{1.55}$$

To accomplish this measurement with either probe, both the magnitude and direction of both $\vec{k}_i$ and $\vec{k}_f$ must be defined in the design of the scattering apparatus.

The intensity of this scattering process is reduced to a double differential cross section, which for neutron scattering is given by

$$\frac{\mathrm{d}^2\sigma}{\mathrm{d}\Omega\,\mathrm{d}E} = \frac{k_1}{k_0}\left\{ \left|\sum_{a=1}^{n} c_a \bar{b}_a^2 \right| S(Q,E) + \left[\sum_{a=1}^{n} c_a \left(\overline{b_a^2} - \bar{b}_a^2 \right)\right] S^{\mathrm{inc}}(Q,E)\right\}\tag{1.56}$$

where the $S(Q,E)$ and $S^{\mathrm{inc}}(Q,E)$ are the *coherent* and *incoherent partial scattering functions* (sometimes called *dynamical structure factors*).[4,29] Their physical significance can be understood if we make Fourier transforms to (Q,t) space:

$$S(Q,E) = \frac{1}{2\pi\hbar}\int_{-\infty}^{\infty} I(Q,t)e^{-iEt/\hbar}\,\mathrm{d}t,$$

$$\tag{1.57}$$

$$S^{\mathrm{inc}}(Q,E) = \frac{1}{2\pi\hbar}\int_{-\infty}^{\infty} I^s(Q,t)e^{-iEt/\hbar}\,\mathrm{d}t,$$

where $I(Q,t)$ and $I^s(Q,t)$ are called, respectively, the *total* and *self intermediate scattering functions*. Their values at zero time, or alternatively the integrals of $S(Q,E)$ and $S^{inc}(Q,E)$ over the entire energy region, are

$$I(Q,t=0) = \int_{-\infty}^{\infty} S(Q,E)\,dE = S(Q),$$

$$I^s(Q,t=0) = \int_{-\infty}^{\infty} S^{inc}(Q,E)\,dE = 1.$$

(1.58)

Thus, $I(Q,t)$ represents the time development of the instantaneous partial structure factor $S(Q)$ introduced in Section 1.2. On the other hand, $I^s(Q,t)$ is related to the distribution in space that a single particle of type a is likely to occupy after time t. The time dependent quantity thus contains useful information about the trajectories of individual particles, whereas its value at time zero is trivially equal to one: each particle has not had time to move.

For multicomponent systems, $S(Q,E)$ and correspondingly $I(Q,t)$ are weighted averages over the different pairs of atom types, as in the diffraction case, while $S^{inc}(Q,E)$ and $I^s(Q,t)$ are averages over the different individual atom types. Furthermore, in neutron scattering, the relative contributions of coherent and incoherent scattering will depend on the various elements and isotopes in the samples. Most natural elements, as well as ^2H and ^6Li, are mostly coherent scatterers, whereas naturally abundant hydrogen is mostly incoherent and natural lithium and silver, for example, are a mixture of both (see Table 1.1). These facts must be taken into consideration in the interpretation of the scattering data. In IXS, b_a in the above equations and those that follow is replaced by $f_a(Q)$, the atomic form factor for species a. In this case, every atom of a given element scatters identically so the incoherent term does not appear. (In the x-ray field the term *incoherent* is often used instead to denote the Compton scattering, in which an x-ray scatters inelastically from an individual electron, providing information about the momentum distribution of the electrons in the ground state.)

In the case of a solid sample it is convenient to distinguish four dynamical regimes of neutron scattering, illustrated schematically in Figure 1.20 for the case of a disordered system like a liquid or a glass. For convenience we frame the discussion in terms of the weighted average structure factor $S(Q)$ introduced in Section 1.2 and the corresponding weighted average scattering function $S(Q,E)$. Figure 1.20a shows a typical structure factor $S(Q)$ in which we pick out a particular value of Q, say Q_1, and discuss the time development $I(Q_1,t)$ of the correlations contributing to $S(Q_1)$. Figure 1.20b shows various time regimes that may appear in $I(Q_1,t)$, and Figure 1.20c the corresponding features in $S(Q_1,E)$ obtained by the Fourier transform, Equation 1.57.

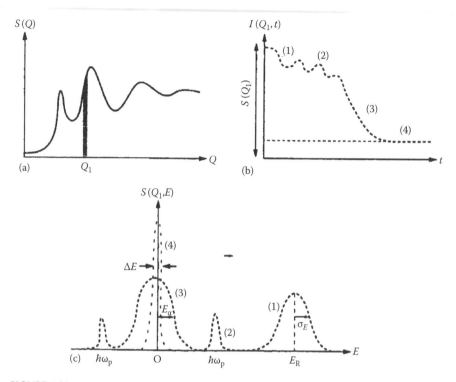

FIGURE 1.20
Schematic illustration of dynamical regimes probed by INS: (a) structure factor $S(Q)$, highlighting a specific scattering vector Q_1; (b) intermediate-scattering function $I(Q_1,t)$; and (c) scattering function $S(Q_1,E)$. The numbers denote the (1) recoil, (2) one-phonon, (3) quasielastic, and (4) elastic scattering regimes. (From Price, D.L., Saboungi, M.L., and Bermejo, F.J., Dynamical aspects of disorder in condensed matter, *Rep. Prog. Phys.*, 66, 413, 2003. With permission.)

1. The conceptually simplest scattering event is one that takes place as if the target nucleus is independent of its neighbors. This is in fact what happens at short times, where $I(Q_1,t)$ falls off from its value at $t=0$, generally with an approximately Gaussian behavior. In the limit of large Q, this *recoil scattering* is the dominant contribution to $S(Q_1,E)$, consisting of a peak on the neutron energy-loss side ($E>0$) centered at the recoil energy $E_R = \hbar^2 Q_1^2 / 2M$ with a shape that reflects the momentum distribution of the system in its ground state. In particular, the variance in energy is related to the mean kinetic energy K: $\sigma_E^2 = K\hbar^2 Q_1^2 / M$.

2. If there are collective excitations with a frequency ω_p, $I(Q_1,t)$ has an oscillatory part and $S(Q_1,E)$ has a peak centered at $\pm\hbar\omega_p$, generally referred to as *one-phonon scattering*. In single crystals the phonons can be labeled by a wave vector $\vec{Q}$ and a branch index j: if the vibrational motion is harmonic, $S(\vec{Q}_1,E)$ has a delta-function form $S_1(\vec{Q}_1)$ $\delta(E \pm \hbar\omega_p)$. In a polycrystalline sample, $S(Q_1,E)$ is an orientational

average over all directions of $\vec{Q}$. In glass, phonons still exist although they can no longer be labeled by a single value of Q.

3. If there are relaxation processes in which the correlations decay at some characteristic rate $\alpha(Q)$ at a given value of Q, $S(Q_1,E)$ has a broadened component centered at $E=0$, called *quasielastic scattering*. Because of the higher energy resolution, quasielastic neutron scattering (QENS) is more commonly used. In the case of an exponential time decay $\exp(-\alpha t)$, $S(Q_1,E)$ has a Lorentzian form proportional to $L_\alpha(Q, E)$, where

$$L_\alpha(Q,E) = \frac{\hbar\alpha(Q)}{E^2 + [\hbar\alpha(Q)]^2}. \tag{1.59}$$

4. If there are structural correlations that last for long times (more precisely, times $t \gg \hbar/\Delta E$ where ΔE is the energy resolution of the experiment), which is the case in a solid where the atoms execute thermal motions about fixed equilibrium positions, $I(Q_1,t)$ contains a nonzero time-independent term. This will give a delta-function $S_{el}(Q_1)$ $\delta(E)$ term in the scattering function, generally referred to as *elastic scattering*.

In a liquid, purely elastic scattering (regime 4) does not exist, and both the quasielastic scattering (regime 3) and collective excitations (regime 2) will generally be heavily damped and only show up as recognizable peaks at sufficiently low Q, sometimes called the *generalized hydrodynamic* regime. In the purely hydrodynamic regime that can be probed by light scattering, the collective excitations represent density fluctuations and are often referred to as Brillouin peaks, while the quasielastic scattering is due to entropy fluctuations and is called the Rayleigh peak. At higher Q the scattering reflects the dynamics of single particles, which evolves continuously into the recoil scattering at very high Q (regime 1).

As stated above, the integral of $S(Q,E)$ over the entire range of E is equal to $S(Q)$. There is also a simple expression for the second energy moment of $S(Q,E)$:

$$\int_{-\infty}^{\infty} S(Q,E)E^2 dE = \frac{\hbar^2 Q^2}{2M}kT, \tag{1.60}$$

which, taken together with Equation 1.58, implies that the standard deviation of the width is given by

$$\sigma(E) = \sqrt{\frac{kT}{2MS(Q)}} \cdot \hbar Q. \tag{1.61}$$

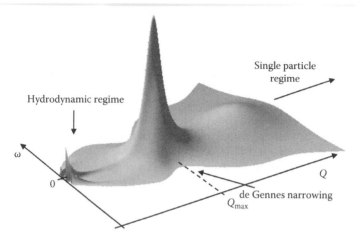

FIGURE 1.21

A typical form of $S(Q,E)$ for a classical liquid, as a function of scattering vector Q and frequency $\omega = E/\hbar$. (From Price, D.L., Experimental techniques, in *High-Temperature Levitated Materials*, Cambridge University Press, Cambridge, U.K., p. 45, 2010. With permission.)

This shows that the distribution in E narrows values of Q where $S(Q)$ has a maximum, a behavior first pointed out by de Gennes.[69] A typical behavior of $S(Q,E)$ for a liquid, illustrating these various regimes, is shown in Figure 1.21.

The expressions given above strictly refer to the coherent scattering function. In regime (1), the incoherent scattering function will have the same form at high Q, while in regime (2) the delta function is replaced by a continuous function of Q describing a density of states. In regimes (3) and (4) the general form will be the same but with a different, generally more slowly varying, Q dependence (Figure 1.20). In an actual condensed system, especially as the complexity increases, there will be coupling between the different types of motion and the simple forms given above must be replaced by more complicated functions of Q and E. Nevertheless, the distinction between the four dynamical regimes will generally be meaningful.

Neutrons from a moderator in a reactor or placed adjacent to a target in a pulsed spallation source have typical energies that correspond to ambient temperature—~25 meV—and wavelengths on the order of 1.8 Å, both of which are well matched to the energy transfers and scattering vectors of interest in the dynamics of condensed matter. Measurement of these requires a definition of the energy, and hence the wave vector, of both incident and scattered beams. For measurements at high E as in regime (1), the higher epithermal neutron intensities at spallation sources are advantageous. Conversely, QENS measurements at high E resolution are generally carried out with neutron beams produced by cold moderators.

1.5.1 Selected Examples

Single-walled carbon nanohorns (SWNH) consist of graphitic structures formed out of a single-walled graphene sheet with an average size of 2–3 nm.[70] They adopt a hornlike shape and aggregate to form flowerlike structures with sizes of about 80–100 nm.[71] These nanostructures exhibit very large surface areas approaching 1500 m^2 g^{-1} and are therefore potential candidates for gas and liquid storage. H$_2$ adsorption isotherms suggested a higher adsorbate density than that of the liquid at 20 K, with values approaching those of solid H$_2$ near its triple point,[72] a behavior attributed to strong quantum effects. Subsequently, Tanaka et al.[73] reported isosteric heats of adsorption for H$_2$ on SWNHs nearly three times larger than as those on single-walled carbon nanotubes, corresponding to binding energies as high as 100–120 meV. This increase in H$_2$ binding energy was attributed to strong solid–fluid interactions at the conical tips. Neutron scattering measurements of H$_2$ adsorption in SWNHs[74] showed that thermal cycling from 10 to 290 K and back to 10 K recovered a spectrum close to that measured after the initial low-temperature loading, indicating that most of the H$_2$ remained firmly attached to these carbon nanostructures.

Fernandez-Alonso et al.[74] investigated the dynamical behavior and molecular bonding leading to this result, with its important implications for the hydrogen economy, using QENS and INS. In addition to the measurements on H$_2$ adsorbed in the SWNHs, they also measured bulk liquid H$_2$ for comparison (Figure 1.22).

Figure 1.22 displays a comparison of QENS spectra for the bulk liquid H$_2$ (a–c) and H$_2$ adsorbed on the SWNHs at three values of scattering vector Q both at 15 K. An obvious difference between the two sets of data is the strictly elastic component in the SWNHs, indicating the presence of species with highly reduced mobility at temperatures and pressures where bulk H$_2$ is liquid. The elastic contribution has an integrated intensity about four times that of the broadened component. The broadened spectra in both cases could be fitted with a Lorentzian function of energy with half width at half maximum (HWHM)

$$\Gamma = \frac{\hbar D Q^2}{1 + D Q^2 \tau},$$

(1.62)

corresponding to a rapid jump diffusion model. At small Q, Γ takes the Fickian diffusion value of $\hbar D Q^2$, where D is the diffusion constant, while at large Q, corresponding to distances that are short on the atomic scale, it tends to the asymptotic value $\hbar / \tau$ where τ is a mean residence time between jumps. The fitted curves in Figure 1.22 correspond to values of $D = 4.3 \times 10^{-5}$ and 9.6×10^{-5} cm^2 s^{-1}, and $\tau = 0.20$ and 5.06 ps, for the bulk liquid H$_2$ and H$_2$ adsorbed on the SWNHs, respectively. Thus, the mobile fraction of H$_2$ in the SWNHs undergoes jump motions that are some 2.3 times faster than those

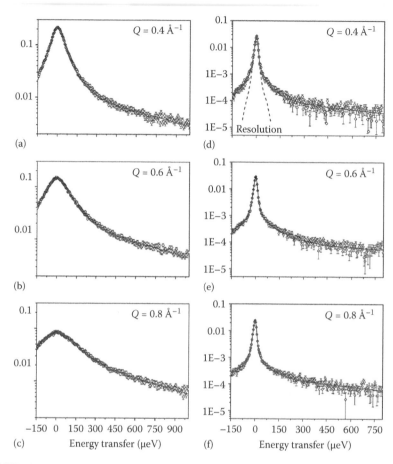

FIGURE 1.22

QENS spectra for (a)–(c) bulk liquid H_2 and (d)–(f) H_2 adsorbed on SWNHs both at $T = 15\,K$. The solid lines are model fits to the data. The narrow line in (d) represents the instrumental energy resolution. (From Fernandez-Alonso, F. et al., *Phys. Rev. Lett.*, 98, 215503, 2007. With permission.)

found in the bulk liquid at the same temperature, whereas the residence time increases by a factor of about 25 due to interaction with the carbon matrix. The falloff of the intensity of the elastic line with increasing Q indicates strong quantum effects that persist at temperatures well above the boiling point of the bulk liquid.

The INS measurements probed the transition between para- and ortho-rotational levels that occurs at 14.7 meV in bulk gaseous H_2. Figure 1.23 shows that in the case of H_2 adsorbed on the SWNHs this is split into three components due to a Stark splitting of the rotational transition arising from the presence of electrostatic interactions between the H_2 and the carbon matrix. The intensity ratio of the Stark-split component relative to the unperturbed central line is roughly 3.3, i.e., about 75% of the molecules are interacting

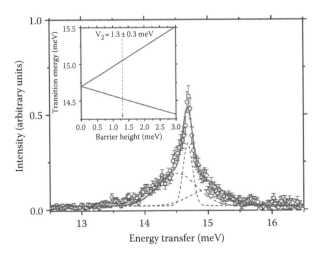

FIGURE 1.23

INS spectra from H_2 adsorbed on SWNHs at 15 K. The dotted lines show the fit of a central component due to free rotation and two side bands caused by Stark splitting. The inset shows a schematic diagram of the Stark energy splittings. (From Fernandez-Alonso, F. et al., *Phys. Rev. Lett.*, 98, 215503, 2007. With permission.)

with the carbon lattice strongly enough to perturb their high-frequency internal rotation.

This ratio agrees with the amount of immobile H_2 extracted from the QENS data, reinforcing the notion that the most energetically favorable SWNH adsorption sites lead to a solid-like H_2 phase characterized by a significant angular anisotropy in the interaction potential. The magnitude of Stark splitting suggests a barrier height for this potential of 1.3 meV. Furthermore, since most of the spectral intensity appears below the free-rotor line, the preferred orientation of the H_2 molecular axis must be parallel to the SWNH surface.

This work showed that both the stochastic and vibrational dynamics of free and bound H_2 species are strongly altered due to their adsorption on the carbon substrate, and also explained the finding of Tanaka et al.[73] of two preferential adsorption sites with significantly different mobilities. It provided important microscopic evidence for the idea that the strong H_2–carbon interactions lead to the firm attachment of a substantial component of the adsorbed hydrogen at temperatures as high as 290 K, with important implications for use of these materials in hydrogen storage.

The last example of this chapter concerns two recent studies[58,59] on dynamical properties of sugar solutions confined in an aqueous silica gel. These new materials, at the borderline between materials science and biology, are particularly interesting because they represent a simplified model of biological membranes. Sugar molecules are well known as stabilizing agents for cell membranes upon dehydration, but the responsible process is still unidentified. The great affinity between sugar and water molecules—evidenced by the large number of hydrogen bonds between these two species—seems to be the cause of this exceptional property. In this work, the authors focused on solutions of glucose and trehalose which are a mono- and a disaccharide, respectively.

QENS is able to selectively probe, at the picosecond timescale, the dynamics of complex organic systems placed under confinement by means of H/D contrast. With this method it is then possible to distinguish the dynamics of solute and solvent and to access the timescales appropriate to the dynamics of each. A desired amount of hydrogenated glucose was dissolved in D_2O in order to reduce the scattering from the solvent and then to enhance that of the solute. Three different samples were prepared: a sugar-free silica gel (A1) and silica gels containing 15 and 30 wt% of glucose (A2 and A3). The $S(Q,E)$ for two of the gels are shown in the left side of Figure 1.24. To obtain information on the dynamics of the glucose molecules, the data from the pure gel (A1) were subtracted from those for the gels containing glucose (A2 and A3). The resulting signals, denoted as A2-A1 and A3-A1, were fitted with a combination of a delta function, two Lorentzian functions, and a sloping background. The Q^2 dependence of the full width at half maximum (FWHM) of the two Lorentzian functions, which is shown in the right side of Figure 1.24, was least squares fitted by

$$FWHM = \alpha_1 + \frac{\beta_1 Q^2}{(1 + \gamma_1 Q^2)},$$ (1.63)

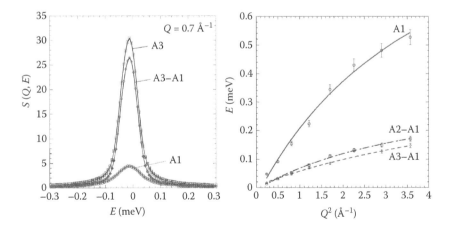

FIGURE 1.24
(Left) Dynamical structure factors $S(Q,E)$ at $Q=0.7\,\text{Å}^{-1}$ for the sugar-free gel (A1) and the gel containing 30 wt% (A3); (Right) FWHM as a function of Q^2 for the sugar-free gel and the gels containing 15 wt% (A2) and 30 wt% (A3) of glucose, (A2–A1) and (A3–A1) denote the resulting signals after subtraction of the scattering from the gel host, corresponding to the water dynamics of the 15 and 30 wt% solutions. The lines represent least-squares fits described in the text.

in the case of the narrower Lorentzian function, and by the following expression in the case of the broader one:

$$FWHM = \alpha_2 + \beta_2 Q^2.$$ (1.64)

According to the model of Teixeira et al.,[75] the translational diffusion constant D and an effective jump distance l can be extracted from the parameters β_1 and γ_1 associated with the rapid jump diffusion model (Equation 1.62) and α_2 can be associated with the relaxation time for free orientational diffusion τ_R. At 300 K, the diffusion rate of confined glucose molecules was found to decrease from 0.74×10^{-5} to 0.46×10^{-5} cm^2 s^{-1} for sugar concentrations going from 15 to 30 wt%, respectively. The same evolution was also observed for confined trehalose molecules, for which the translational diffusion coefficients were found equal to 0.44×10^{-5} and 0.2×10^{-5} cm^2 s^{-1} for the same sugar concentrations. Then, the QENS study on confined solutions showed that the diffusion rate of mono- and disaccharides slows down by a factor close to 2 as the sugar concentration is doubled. A similar behavior was previously observed in bulk solutions,[76] indicating that the confinement in silica gel does not substantially modify the dynamical properties of the sugar molecules. Moreover, the estimate of the pore size diameter of 20 nm obtained from the SANS measurement on the pure gel (A1) provided an upper limit of the length scale at which the dynamics of the glucose molecules are affected by confinement. More interestingly, it appears that these aqueous silica gels containing glucose or trehalose solutions present macroscopically a solid behavior with the microscopic dynamics of the corresponding bulk liquid solutions.

References

1. Alcoutlabi, M. and McKenna, B. 2005. Effects of confinement on material behaviour at the nanometre size scale. *J. Phys.: Condens. Matter* 17:R461–R524.
2. Morineau, D., Xia, Y., and Alba-Simionesco, C. 2002. Finite-size and surface effects on the glass transition of liquid toluene confined in cylindrical mesopores. *J. Chem. Phys.* 117:8966.
3. Zorn, R., Hartmann, L., Frick, B., Richter, D., and Kremer, F. 2002. Inelastic neutron scattering experiments on the dynamics of a glass-forming material in mesoscopic confinement. *J. Non-Cryst. Solids* 307:547.
4. Price, D.L. and Sköld, K. 1986. Introduction to neutron scattering. In *Neutron Scattering, Methods of Experimental Physics*, eds. K. Sköld and D.L. Price, Vol. 23 (Part A), p. 1. New York: Academic Press.
5. Faber, T.E. and Ziman, J.M. 1965. A theory of the electrical properties of liquid metals III. The resistivity of binary alloys. *Philos. Mag.* 11:153.
6. Bhatia, A.B. and Thornton, D.E. 1970. Structural aspects of the electrical resistivity of binary alloys. *Phys. Rev. B* 2:3004–3012.

7. March, N.H. and Tosi, M. 1992. *Atomic Dynamics in Liquids*. Mineola, NY: Dover Publications.
8. Enderby, J.E., North, D., and Egelstaff, P.A. 1966. Partial structure factors of liquid Cu-Sn. *Philos. Mag.* 14:961.
9. Raoux, D. 1993. Differential and partial structure factors by x-ray anomalous wide angle scattering. In: *Methods in the Determination of Partial Structure Factors of Disordered Matter by Neutron and Anomalous X-Ray Diffraction*, eds. J.B. Suck, D. Raoux, P. Chieux, and C. Riekel, p. 130. Singapore: World Scientific.
10. Price, D.L. and Saboungi, M.-L. 1998. Anomalous x-ray scattering from disordered materials. In *Local Structure from Diffraction*, eds. S.J.L. Billinge and M.F. Thorpe, p. 23. New York: Plenum Press.
11. Price, D.L., Saboungi, M.-L., and Barnes, A.C. 1998. Structure of vitreous germania. *Phys. Rev. Lett.* 81:3207–3210.
12. Goldbach, A., Saboungi, M.-L., Iton, L.E., and Price, D.L. 2001. Approach to band-gap alignment in confined semiconductors. *J. Chem. Phys.* 115:11254.
13. Armand, P., Saboungi, M.-L., Price, D.L., Iton, L., Cramer, C., and Grimsditch, M. 1997. Nanoclusters in zeolite. *Phys. Rev. Lett.* 71:2061.
14. Goldbach, A., Saboungi, M.-L., Iton, L., and Price, D.L. 1999. Stabilization of selenium in zeolites: An anomalous x-ray scattering study. *J. Chem. Soc., Chem. Commun.* 997.
15. Edler, K.J., Reynolds, P.A., and White, J.W. 1998. Small-angle neutron scattering studies on the mesoporous molecular sieve MCM-41. *J. Phys. Chem. B* 102:3676–3683.
16. Connolly, J., Singh, M., and Buckley, C.E. 2004. Determination of size and ordering of pores in mesoporous silica using small angle neutron scattering. *Phys. B* 350:224–226.
17. Dore, J.C., Webber, J.B.W., and Strange, J.H. 2004. Characterisation of porous solids using small-angle scattering and NMR cryoporometry. *Colloids Surf. A* 241:191–200.
18. Rouquerol, J., Avnir, D., Fairbridge, C.W., Evertt, D.H., Haynes, J.H., Pernicone, N., Ramsay, J.D., Sing, K.S.W., and Unger, K.K. 1994. Recommendations for the characterization of porous solids. *Pure Appl. Chem.* 66:1739–1758.
19. Windsor, C.G. 1988. An introduction to small-angle neutron scattering. *J. Appl. Cryst.* 21:582–588.
20. Jackson, A.J. 2008. Introduction to Small-Angle Neutron Scattering and Neutron Reflectometry. http://www.ncnr.nist.gov/summerschool/ss08/pdf/SANS_NR_Intro.pdf
21. Kádár, G. and Rosta, L. 2003. *Neutron Scattering—Introductory Course to ECNS'99*. Budapest, Hungary: Prosperitás Press.
22. Kline, S.R. 2000. Fundamentals of Small-Angle Neutron Scattering. NCNR Summer School. http://www.ncnr.nist.gov/programs/sans/pdf/sans_fund.pdf
23. Hammouda, B. 1995. A Tutorial on Small-Angle Neutron Scattering from Polymers. http://www.ncnr.nist.gov/programs/sans/pdf/polymer_tut.pdf
24. Hammouda, B. 2008. Probing Nanoscale Structures—The SANS Toolbox. http://www.ncnr.nist.gov/staff/hammouda/the_SANS_toolbox.pdf
25. Feigin, L.A. and Svergun, D.I. 1987. *Structure Analysis by Small-Angle X-Ray and Neutron Scattering*. New York: Plenum Press.
26. Chen, S.-H. and Lang, T.-S. 1986. *Methods of Experimental Physics—Neutron Scattering*, 23 (Part B). New York: Academic Press.

27. Higgins, J.S. and Benoit, H.C. 1994. *Polymers and Neutron Scattering*. Oxford, U.K.: Clarendon Press.
28. Lindner, P. and Zemb, T. 1991. *Neutron, X-ray and Light Scattering*. Amsterdam, The Netherlands: North-Holland.
29. Squires, G.L. 1978. *Introduction to the Theory of Thermal Neutron Scattering*. New York: Cambridge University Press.
30. Glatter, O. and Kratky, O. 1982. *Small Angle X-Ray Scattering*. New York: Academic Press.
31. Glinka, C. 2000. SANS from Dilute Particle Systems. NCNR Summer School. http://www.ncnr.nist.gov/programs/sans/pdf/SANS_dilute_particles.pdf
32. Kline, S.R. 2006. Reduction and analysis of SANS and USANS data using IGOR Pro. *J. Appl. Cryst.* 39:895–900.
33. Kline, S.R. 2000. SANS from Concentrated Dispersions. NCNR Summer School. http://www.ncnr.nist.gov/programs/sans/pdf/sans_fund.pdf
34. Schmidt, P.W. 1991. Small-angle scattering studies of disordered, porous and fractal systems. *J. Appl. Cryst.* 24:414–435.
35. Teixeira, J. 1988. Small-angle scattering by fractal systems. *J. Appl. Cryst.* 21:781–785.
36. Sinha, S.K. 1989. Scattering from fractal structures. *Physica D* 38:310–314.
37. Michels, A. and Weissmüller, J. 2008. Magnetic-field-dependent small-angle neutron scattering on random anisotropy ferromagnets. *Rep. Prog. Phys.* 71:066501–066538.
38. Neylon, C. 2008. Small angle neutron and X-ray scattering in structural biology: Recent examples from the literature. *Eur. Biophys. J.* 37:531–541.
39. Lipfert, J. and Doniach, S. 2007. Small-angle X-ray scattering from RNA, proteins, and protein complexes. *Annu. Rev. Biophys. Biomol. Struct.* 36:307–327.
40. Wignall, G.D. and Melnichenko, Y.B. 2005. Recent applications of small-angle neutron scattering in strongly interacting soft condensed matter. *Rep. Prog. Phys.* 68:1761–1810.
41. Melnichenko, Y.B. and Wignall, G.D. 2007. Small-angle neutron scattering in materials science: Recent practical applications. *J. Appl. Phys.* 102:021101–021124.
42. Chu, B. and Hsiao, B.S. 2001. Small-angle X-ray scattering of polymers. *Chem. Rev.* 101(6):1727–1761.
43. Fratzl, P. 2003. Small-angle scattering in materials science—A short review of applications in alloys, ceramics, and composite materials. *J. Appl. Cryst.* 36:397–404.
44. Bhattacharyya, S., Lelong, G., and Saboungi, M.-L. 2006. Recent progress in the synthesis and selected applications of MCM-41: A short review. *J. Exp. Nanosci.* 1(3):375–395.
45. Lelong, G., Bhattacharyya, S., Kline, S.R., Cacciaguerra, T., Gonzalez, M.A., and Saboungi, M.-L. 2008. Effect of surfactant concentration on the morphology and texture of MCM-41 materials. *J. Phys. Chem. C* 112:10674–10680.
46. Né, F. and Zemb, T. 2003. Determination of compaction and surface in mesostructured materials using SAXS. *Appl. Cryst.* 36:1013–1018.
47. Bauer, B.J., Hobbie, E.K., and Becker, M.L. 2006. Small-angle neutron scattering from labeled single-wall carbon nanotubes. *Macromolecules* 39:2637–2642.
48. Kainourgiakis, M.E., Steriotis, T.A., Kikkinides, E.S., Charalambopoulou, G.C., Ramsay, J.D.F., and Stubos, A.K. 2005. Combination of small angle neutron scattering data and mesoscopic simulation techniques as a tool for the structural characterization and prediction of properties of bi-phasic media. *Chem. Phys.* 317:298–311.

49. Bras, W. and Ryan, A.J. 1998. Sample environments and techniques combined with small angle X-ray scattering. *Adv. Coll. Interf. Sci.* 75:1–43.
50. Hoinkis, E. 2004. Small-angle scattering studies of adsorption and of capillary condensation in porous solids. *Part. Part. Syst. Charact.* 21:80–100.
51. Dolino, G., Bellet, D., and Faivre, C. 1996. Adsorption strains in porous silicon. *Phys. Rev. B* 54:17919–17929.
52. Zickler, A., Jähnert, S., Funari, S.S., Findenegg, G.H., and Paris, O. 2007. Pore lattice deformation in ordered mesoporous silica studied by in situ small-angle X-ray diffraction. *J. Appl. Cryst.* 40:s522–s526.
53. Isnard, O. 2007. A review of in situ and/or time resolved neutron scattering. *C.R. Physique* 8:789–805.
54. Zholobenko, V.L., Khodakov, A.Y., Impéror-Clerc, M., Durand, D., and Grillo, I. 2008. Initial stages of SBA-15 synthesis: An overview. *Adv. Coll. Interf. Sci.* 142:67–74.
55. Impéror-Clerc, M., Grillo, I., Khodakov, A.Y., Durand, D., and Zholobenko, V.L. 2007. New insights into the initial steps of the formation of SBA-15 materials: An in situ small angle neutron scattering investigation. *Chem. Comm.* 8:834–836.
56. Boué, F., Cousin, F., Gummel, J., Oberdisse, J., Carrot, G., and El Harrak, A. 2007. Small angle scattering from soft matter—Application to complex mixed systems. *C.R. Physique* 8:821–844.
57. Doe, C., Jang, H.-S., Kline, S.R., and Choi, S.-M. 2009. Subdomain structures of lamellar and reverse hexagonal pluronic ternary systems investigated by small angle neutron scattering. *Macromolecules* 42:2645–2650.
58. Lelong, G., Price, D.L., Douy, A., Kline, S., Brady, J.W., and Saboungi, M.-L. 2005. Molecular dynamics of confined glucose solutions. *J. Chem. Phys.* 122:164504–164510.
59. Lelong, G., Price, D.L., Brady, J.W., and Saboungi, M.-L. 2007. Dynamics of trehalose molecules in confined solutions. *J. Chem. Phys.* 127:065102–06518.
60. Porcar, L., Warr, G.G., Hamilton, W.A., and Butler, P.D. 2005. Shear-induced collapse in a lyotropic lamellar phase. *Phys. Rev. Lett.* 95:078302–078305.
61. Liberatore, M.W., Nettesheim, F., Vasquez, P.A., Helgeson, M.E., Wagner, N.J., Kaler, E.W., Cook, L.P., Porcar, L., and Hu, Y.T. 2009. Microstructure and shear rheology of entangled wormlike micelles in solution. *J. Rheol.* 53(2):441–458.
62. Nikitenko, S., Beale, A.M., van der Eerden, Ad M.J., Jacques, S.D.M., Leynaud, O., O'Brien, M.G., Detollenaere, D., Kaptein, R., Weckhuysen, B.M., and Bras, W. 2008. Implementation of a combined SAXS/WAXS/QEXAFS set-up for time-resolved *in situ* experiments. *J. Synchrotron Rad.* 15: 632–640.
63. Meneau, F., Sankar, G., Morgante, N., Winter, R., Catlow, C.R.A., Greaves, G.N., and Thomas, J.M. 2002. Following the formation of nanometer-sized clusters by time-resolved SAXS and EXAFS techniques. *Faraday Discuss.* 122:203–210.
64. Bras, W. 2003. SAXS/WAXS experiments using extreme sample environments. *Nucl. Instrum. Methods Phys. Res. B* 199:90–97.
65. Di Cicco, A., Trapananti, A., Faggioni, S., and Filipponi, A. 2003. Is there icosahedral ordering in liquid and undercooled metals? *Phys. Rev. Lett.* 91:135505–135509.
66. Filipponi, A., Di Cicco, A., and Natoli, C.R. 1995. X-ray-absorption spectroscopy and *n*-body distribution functions in condensed matter. I. Theory. *Phys. Rev. B* 52:15122–15134.
67. Mezei, F. 1972. Neutron spin echo: A new concept in polarized thermal neutron techniques. *Z. Phys.* 255:146.

68. Burkel, E. 1991. *Inelastic Scattering of X-Rays with Very High Energy Resolution.* Berlin, Germany: Springer.
69. de Gennes, P.G. 1959. Liquid dynamics and inelastic scattering of neutrons. *Physica* 25:825–839.
70. Kasuya, D., Yudasaka, M., Takahashi, K., Kokai, F., and Iijima, S. 2002. Selective production of single-wall carbon nanohorn aggregates and their formation mechanism. *J. Phys. Chem. B* 106:4947.
71. Wang, H., Chhowalla, M., Sano, N., Jia, S., and Amaratunga, G.A.J. 2004. Large-scale synthesis of single-walled carbon nanohorns by submerged arc. *Nanotechnology* 15:546.
72. Tanaka, H., Kanoh, H., El-Merraoui, M., Steele, W.A., Yudasaka, M., Iijima, S., and Kaneko, K. 2004. Quantum effects on hydrogen adsorption in internal nanospaces of single-wall carbon nanohorns. *J. Phys. Chem. B* 108:17457.
73. Tanaka, H., Kanoh, H., Yudasaka, M., Iijima, S., and Kaneko, K. 2005. Quantum effects on hydrogen isotope adsorption on single-wall carbon nanohorns. *J. Am. Chem. Soc.* 127:7511.
74. Fernandez-Alonso, F., Bermejo, F.J., Cabrillo, C., Loutfy, R.O., Leon, V., and Saboungi, M.-L. 2007. Nature of the bound states of molecular hydrogen in carbon nanohorns. *Phys. Rev. Lett.* 98:215503.
75. Teixeira, J., Bellissent-Funel, M.C., Chen, S.-H., and Dianoux, A.J. 1985. Experimental determination of the nature of diffusive motions of water molecules at low temperature. *Phys. Rev. A* 31:1913.
76. Lelong, G., Howells, W.S., Brady, J.W., Talon, C., Price, D.L., and Saboungi, M.-L. 2009. Translational and rotational dynamics of monosaccharide. *J. Phys. Chem. B* 113:13079–13085.
77. Price, D.L. 2010. Experimental techniques. In *High-Temperature Levitated Materials*, p. 45. Cambridge, U.K.: Cambridge University Press.

2

Studying Diffusion and Mass Transfer at the Microscale

Christian Chmelik, Douglas M. Ruthven, and Jörg Kärger

CONTENTS

2.1 Introduction

Diffusion, i.e., the irregular movement of the elementary constituents of matter, notably of atoms and molecules, is among the fundamental and universal phenomena in nature [1] and provides the basis for numerous technological processes [1–3]. This is especially true for the nanoporous materials that have been developed and applied in molecular separations [4] and heterogeneous catalysis [5]. The benefit of these materials results from the similarity of their pore sizes and the sizes of the molecules involved in these processes. However, this similarity also inhibits molecular propagation rates and may therefore lead to a reduction in performance.

Sophisticated synthesis methods and post-synthesis modification techniques [6–10] have been developed to provide tailor-made nanoporous materials with pore sizes fine-tuned according to the molecular sizes of the desired key products. For performance enhancement, sophisticated procedures for producing nanoporous materials with hierarchical pore structures are under discussion [11], where the presence of meso- and macropores ("transport" pores) ensures fast matter exchange between the regions of nanoporosity (i.e., the locations of molecular separation and/or conversion) and the surroundings [12–16].

Conventionally, the transport properties of nanoporous materials have been determined by measuring rates of molecular uptake or release on particle ("crystallite") assemblies. Such investigations are referred to as "macroscopic" measurements [17,18]. Over many years, transport inhibition by the intracrystalline pore network was considered to be the rate-determining process in such experiments [19]. Only with the introduction of pulsed field gradient nuclear magnetic resonance (PFG NMR) to diffusion studies in zeolites [20–22], molecular diffusion under confinement could be directly measured and, in many cases, was found to be notably faster than previously assumed. In fact, molecular diffusivities in zeolites are often much closer to those in the pure liquid than to the typical diffusivities in solids [23]. As a "microscopic" technique [17,18], PFG NMR is able to monitor diffusion paths over distances much smaller than the crystal sizes, so that the diffusion data are unaffected by the crystal boundary and by any extracrystalline resistances.

The message of PFG NMR has meanwhile been confirmed in numerous studies, including measurements by quasi-elastic neutron scattering (QENS) [24,25] as another "microscopic" technique as well as by molecular modeling [26–30]. The various extracrystalline resistances to heat and mass transfer, which can limit adsorption/desorption rates, have also been explored in some detail. Therefore, in "macroscopic" diffusion measurements today one should be able to rule out misinterpretations due to such effects. However, although a number of systems show good agreement between macroscopic and microscopic measurements [31–35], there are also several systems for which comparative diffusion studies involving microscopic and macroscopic measurements still yield widely different diffusivities [35–37]. An example of such behavior is shown in Figure 2.1 in which the diffusivities of *n*-hexane in silicalite-1, as measured by several different techniques, are compared.

When there is disagreement, it is generally found that the measured diffusivity increases as the length scale of the measurement decreases. Thus, the diffusivities derived from QENS (for which the displacements are nanometers [25]) tend to slightly exceed the PFG NMR diffusivities (for which the displacements are typically micrometers [38]), while the techniques that follow molecular uptake on an individual crystallite (mesoscopic measurements [17,18]) or with an assemblage of crystals yield the smallest diffusivities. Such a finding has to be related to the existence of a hierarchy of transport resistances acting in addition to the diffusional resistance of the nanopores themselves. If the technique is able to resolve molecular displacements much smaller than the separation between these additional resistances, the vast

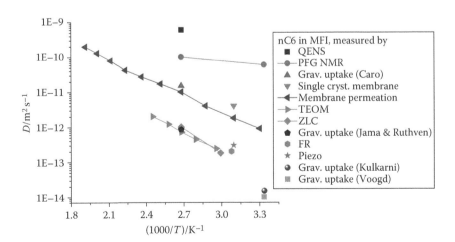

FIGURE 2.1
Arrhenius plot showing a comparison between the diffusivities of *n*-hexane in silicalite-1 measured (after 1989) by various experimental techniques. (Courtesy of S. Brandani.)

majority of the recorded molecules and hence the resulting diffusivities will be unaffected by these resistances. With increasing displacements, however, their influence becomes more and more significant, leading to the observed decrease in the apparent diffusivities that are measured by the "more macroscopic" techniques.

Although internal transport resistances were suggested as a possible explanation for the differences in the measured diffusivities some time ago [33], it was only in 2002 that PFG NMR studies covering a large range of observation times and hence of diffusion path lengths [39,40] provided independent evidence of their existence. More recently [41], the occurrence of stacking faults as directly observed by transmission electron microscopy (TEM) has been correlated with a notable difference between the QENS and PFG NMR diffusivities.

On assessing the transport properties of nanoporous materials, one has to be aware of the complications provided by their real structure as, most likely, a quite general feature. Being able to follow molecular diffusion paths from about 100 nm up to more than tens of micrometers, PFG NMR is able to cover exactly those diffusion path lengths that are critical for the practical application of these materials. By its very nature, PFG NMR operates under equilibrium conditions and provides information about probability distributions of displacements over microscopic distances within macroscopic samples. Very recently, interference microscopy (IFM [42]) and IR microimaging (IRM [43]) have been introduced as complementary techniques that allow the evolution of the transient concentration profiles of the guest molecules to be followed in an individual crystal.

Following an introduction to the principles of measurement and the accessible information in the subsequent Section 2.2, we illustrate the potential of these microscopic techniques for diffusion measurement with examples taken from their application to various types of nanoporous materials, including zeolites (Section 2.3.1), metal organic frameworks (MOFs) (Section 2.3.2), ordered mesoporous materials (Section 2.3.3), and random pore networks (Section 2.3.4).

2.2 Techniques Applied

2.2.1 Pulsed Field Gradient Nuclear Magnetic Resonance

Diffusion measurement by the PFG NMR technique is based on the fact that the resonance frequency of nuclear spins is proportional to the intensity of the applied magnetic field. The principle of the method is illustrated in Figure 2.2. By applying, in addition to a constant magnetic field, a highly inhomogeneous magnetic field (the "gradient pulses") over two short intervals of time, the diffusion path length covered by each individual molecule

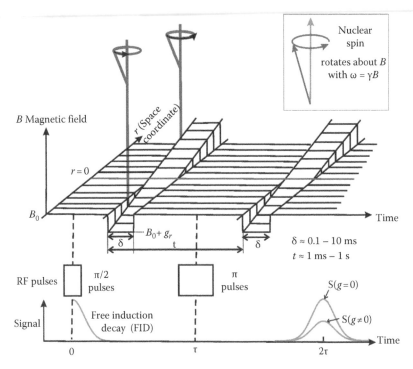

FIGURE 2.2
Basic principle of PFG NMR. In addition to a constant magnetic field, a highly inhomogeneous magnetic field (the "gradient pulses") is applied over two short intervals of time. The diffusion path length covered by each individual molecule in the time interval t between these two gradient pulses is "recorded" by the difference in the respective resonance frequencies, which appear as a difference in the precessional phases accumulated during these pulses. PFG NMR may thus be shown to yield directly the probability distribution $P(x,t)$ that, during time t, an arbitrarily selected molecule within the sample is shifted over a distance x in the direction of the applied field gradient.

in the time interval t between these two gradient pulses is recorded by the difference in the respective resonance frequencies. PFG NMR may, thus, be shown to yield directly the probability distribution $P(x,t)$ that, during time t, an arbitrarily selected molecule within the sample is shifted over a distance x in the direction of the applied field gradient referred to, in Figure 2.2, as the space coordinate [38]. The function $P(x,t)$ is referred to as the mean propagator and is in fact the Fourier transform of the primary experimental data, namely, the attenuation of the NMR signal intensity under the influence of the applied gradient pulses [44,45]. It is the key function for the description of molecular redistribution within a sample and explains the significance of PFG NMR in exploring the molecular dynamics of complex systems [46–48] (see also Figure 2.3).

Although a large spectrum of nuclei, including ^{13}C [49], ^{15}N [50], ^{19}F [51], have been applied for diffusion studies in porous media, the vast majority of

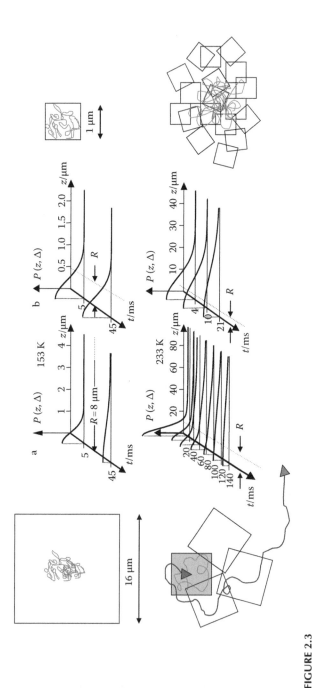

FIGURE 2.3

Mean propagator for ethane self-diffusion in NaCaA zeolite crystallites of different size (mean crystallite diameter of 16 μm [left] and 1 μm [right]) and cartoons of typical diffusion paths under the considered situations. They include the limiting cases of long-range diffusion (bottom right), intracrystalline diffusion (top left), and restricted diffusion (top right). The intermediate case shown on bottom left allows an estimate of surface resistances as considered in Section 2.2.1.4. (From Kärger, J. and Heink, W., *J. Magn. Reson.*, 51, 1, 1983.)

measurements have been performed with protons, the nucleus of hydrogen. Hydrogen is contained in most molecules of technical relevance and offers the best measuring conditions with respect to both signal intensity and spatial resolution [38]. The space and timescales typically covered in proton PFG NMR are in the range of micrometers (for x) and milliseconds till seconds (for t). Both the observation times and the minimum displacements still observable are controlled by the nuclear magnetic relaxation times of the considered spins. Best measuring conditions are provided for guest molecules of high rotational mobility in host systems with low contents of paramagnetic impurities. With sophisticated experimental arrangements [52,53] today, under optimum conditions, field gradient NMR diffusion studies allow the recording of molecular displacements down to the order of 50 nm. This is also the lower limit of displacements accessible with the home-built PFG NMR spectrometer FEGRIS 400 [54,55], which was used in these studies. Recent exploitation of the neutron spin-echo method has enhanced the range of displacements accessible by QENS diffusion measurement to about 10 nm [25]. Complete coverage of the gap in the displacements as accessible by the two "microscopic" techniques of diffusion measurement under (macroscopic) equilibrium may thus turn out to be a realistic goal for the near future.

Starting from the general measuring principle of PFG NMR, we now illustrate those special cases of application, to which we shall refer in the diffusion studies in Sections 2.3.1 through 2.3.4.

2.2.1.1 Normal, Isotropic Diffusion

Normal diffusion (see, e.g., [23,56]) in an isotropic system is described by Fick's first law

$$j = -D \frac{\partial c(x,t)}{\partial x} \tag{2.1}$$

where
  j is the flux
  $\partial c / \partial x$ is the concentration gradient in the (arbitrarily chosen) x direction

Under equilibrium conditions and hence with uniform concentrations, diffusion fluxes can only be observed by labeling. Therefore, under equilibrium, Equation 2.1 has to consider the fluxes and concentrations of correspondingly "labeled" molecules within an "unlabeled" surrounding, and the factor of proportionality, D, is referred to as the tracer diffusivity or self-diffusivity. By distinguishing between "labeled" and "unlabeled" molecules, one may ensure a uniform overall concentration (i.e., concentration sum of the "labeled" and "unlabeled" molecules) and vanishing overall fluxes, due to a mutual compensation of the fluxes by the labeled and unlabeled molecules.

In PFG NMR, the labeling is effected by the resonance frequencies of the nuclear spins. Since the thermal energy exceeds the interaction energy of the nuclear spins with the external magnetic field by orders of magnitude, this "labeling" does not affect the intrinsic system dynamics.

Combination of Equation 2.1 with the requirement of mass conservation, $\partial c/\partial t = -\partial j/\partial x$, yields Fick's second law

$$\frac{\partial c}{\partial t} = D \frac{\partial^2 c}{\partial x^2} \tag{2.2}$$

The probability distribution of molecular displacements, i.e., the propagator $P(x,t)$, as directly accessible by PFG NMR, results as the solution of Equation 2.2 with the initial concentration $c(x,0) = \delta(x)$ and the boundary condition $c(\pm\infty, t) = 0$, yielding the Gaussian

$$P(x,t) = (4\pi Dt)^{-1/2} \exp\left[-\frac{x^2}{(4Dt)}\right] \tag{2.3}$$

The mean squared value $\langle x^2(t) \rangle \left(\equiv \int_{-\infty}^{+\infty} x^2 P(x,t) dx \right)$ of the molecular displacements during the observation time t (the distance between the two gradient pulses) serves as a useful measure of the rate of molecular redistribution. With Equation 2.3, one obtains Einstein's famous diffusion equation

$$\langle x^2(t) \rangle = 2Dt \tag{2.4}$$

Thus, by stating that (1) the propagator obeys a Gaussian dependence and that (2) the mean squared width of this distribution (i.e., the mean square displacement) increases linearly with the observation time as required by the Einstein relation, Equation 2.4, a PFG NMR measurement can provide direct experimental evidence of normal diffusion. In terms of the primary PFG NMR data, this requires that (1) the semilogarithmic plot of signal intensity versus the squared gradient intensity yields a straight line and that (2) the slope of this line increases in proportion to the observation time [38]. It is important to emphasize that this assessment clearly only refers to the space scale of the observation. Both for smaller displacements (e.g., over distances comparable with the pore sizes) and larger displacements, notable deviations may occur.

2.2.1.2 Anisotropic Diffusion

Anisotropic diffusion cannot be described by the simple form of Fick's first law, Equation 2.1, since the diffusion coefficient has to be replaced by a diffusion tensor. Equations as simple as Equation 2.1 will apply only in the directions of the principal axes of the diffusion tensor, with the diffusivities as the principal elements. Analysis of PFG NMR diffusion data is facilitated, however, by the fact that in general a random distribution of all orientations within the PFG NMR sample tube may be assumed. Thus, one may derive well-defined analytical expressions, representing correspondingly weighted superpositions of the propagator (or, correspondingly, of the PFG NMR signal attenuations). They include the main tensor elements as the only free parameters that become accessible by fitting the theoretical expressions to the experimental data [46,47,57]. The fitting procedure is greatly simplified for systems with rotational symmetry [58].

2.2.1.3 Long-Range Diffusion

In PFG NMR studies with beds of nanoporous particles, one may distinguish between two limiting cases. The different situations are illustrated in Figure 2.3. For diffusion path lengths much smaller than the particle sizes, any significant disturbance by the particle boundary and/or the space between the particles may be excluded (upper left case). It appears as if the measurements were performed within particles of infinite size. Implying particle homogeneity, the measurements would reveal the features of normal isotropic or anisotropic diffusion. However, normal diffusion may also be observed under the opposite conditions, namely, for diffusion path lengths notably exceeding the sizes of the individual nanoporous particles (bottom right in Figure 2.3). Provided that (1) the observation time may be chosen to be much larger than the molecular mean lifetime in the individual nanoporous particles and that (2) we confine ourselves to a space scale notably exceeding the particle sizes, again a Gaussian propagator is observed. One may show that now the effective diffusivity $D_{(l.r.)}$ as appearing in Equations 2.2 and 2.4 is approached by the simple product $p_{inter}D_{inter}$ [38,59] of the relative amount of molecules in the space between the particles (the "crystallites," p_{inter}) and their diffusivity (D_{inter}).

At low temperatures, the molecules may not have enough energy to leave the crystallites (upper right case in Figure 2.3). Here, the mean square displacement approaches the crystal radius and becomes independent of t (upper right cartoon)—the characteristic feature of "restricted diffusion."

2.2.1.4 Transition Range: Determination of Surface Resistances

In many cases (and, in particular, at sufficiently high temperatures), the coefficient of long-range diffusion significantly exceeds the intraparticle diffusivity. Under such conditions and for observation times of the order of the mean lifetime of the molecules within the nanoporous particles, the overall propagator may be easily decomposed into two constituents. One of them, the broader one, comprises all those molecules that, during the given observation time t, were able to pass from one particle to another one. The width of the narrow one is of the order of the mean particle size. It comprises those molecules that, during time t, have remained within the same particle. Recording the relative intensity $m(t)$ of the broad constituent as a function of time is referred to as the "NMR tracer desorption technique" [38,59,60]. It has been exploited for the detection and quantification of additional transport resistances ("surface barriers") on the external surface of nanoporous crystallites. For this purpose, one determines the "first moment" $M_1 = \int_0^{\infty} (1 - m(t)) dt$ of the NMR tracer desorption curve, which, for spherical particles of radius R with a finite surface permeability α, results as a sum

$$M_1 = \frac{R^2}{(15D)} + \frac{R}{(3\alpha)} \tag{2.5}$$

of the contributions of diffusion ($M_{1,\text{diff}} = R^2/(15D)$) and of surface permeation ($M_{1,\text{bar}} = R/(3\alpha)$) to overall transport inhibition. The surface permeability is defined by the relation [61]

$$j_{\text{surf}} = \alpha \left(c(0) - c_{\text{eq}} \right) \tag{2.6}$$

as the factor of proportionality between the flux through the surface and the difference between the actual surface concentration ($c(0)$) and the concentration in equilibrium with the external atmosphere (c_{eq}). Following the procedure described in Section 2.2.1.1, one is able to determine the intraparticle diffusivity. Hence, with the microscopically determined value of R and the total moment derived from the NMR tracer desorption curve, one may estimate the surface permeability α. A significant uncertainty (typically a factor of two) in the total moments resulting from NMR tracer desorption experiments severely limits the applicability of this method. Higher accuracies have recently been shown to become accessible by including the information provided by the variation of the width of the narrow propagator constituent with varying observation time [62,63]. However, these options are still far inferior to the options provided by IFM and IRM, which are discussed in Section 2.2.2.

2.2.1.5 Selective Diffusion Measurements

The remarkable potential of PFG NMR includes the option to measure the diffusion of one component in a mixture by using different "probe" nuclei, e.g., ^{13}C [49], ^{15}N [50], and ^{19}F as the probe nuclei [51,64–67]. By applying high-resolution NMR techniques, it is even possible to measure simultaneously the displacement of different molecular species. In this way, the measurement of the diffusivity of both the reactant and product molecules during a chemical reaction has become possible [68–70]. Very recently, the combination of magic angle spinning and PFG NMR [71] has been shown to yield a remarkable improvement in this approach for diffusion studies with porous materials [72–75].

Depending on the given pore structure, molecular diffusion may proceed at very different rates in different regions. Provided that the exchange time between these different regions exceeds the observation time, PFG NMR is able to record the spectra for all these different diffusivities, thus yielding much more information than is provided by a single average overall diffusivity. By variation of the guest pressure over the sample and/or by temperature variation, the weight of these different contributions may be varied. This includes, in particular, the option of a progressive freezing by temperature reduction. Since freezing starts in the largest pores, and the frozen regions do not contribute anymore to the signal measured by PFG NMR [76,77], it is possible to focus progressively on molecular diffusion in the pores with the smallest diameters.

2.2.2 Interference Microscopy and IR Microimaging

2.2.2.1 Principles of Measurement

Both techniques are based on monitoring the evolution of concentration profiles in a particular nanoporous particle/crystallite during molecular uptake or release. Therefore, corresponding devices for sample activation and guest dosing have to be incorporated into the optical arrangement of observation (Figures 2.4 and 2.5). In both techniques, the quantity directly recorded is the concentration integral $\int c(x,t)dx$, with x denoting the coordinate in the observation direction.

In IFM, the determination of the concentration integral is based on the measurement of the phase shift $\Delta\varphi^L$ between the light beams passing the crystal and the corresponding beams through the surrounding atmosphere (see schematic sketch in Figure 2.4d). Changes $\Delta(\Delta\varphi^L)$ in this difference (which appear in corresponding changes of the interference patterns [42,78,79]) are caused by changes Δn of the refractive index which, in turn, are a consequence of the changing concentrations Δc within the crystal. To a first, linear approximation, one, therefore, may note

$$\Delta\left(\Delta\varphi^L(t)\right) \propto \int_0^L \Delta n(t)dx \propto \int_0^L \Delta c(x,y,z,t)dx \qquad (2.7)$$

As indicated by the signs of proportionality, IFM is unable to provide absolute concentration data. Therefore, the resulting concentration profiles are generally plotted in relative units. Information about absolute concentrations can be obtained from conventional (macroscopic) sorption measurement at the relevant guest pressures or by direct comparison with the results of IRM.

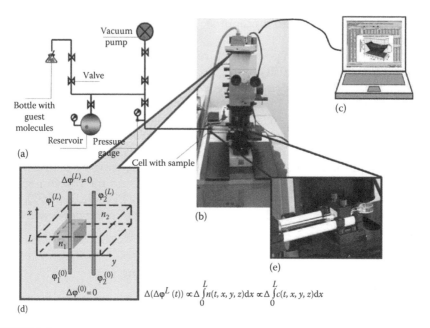

FIGURE 2.4
Experimental setup and basic principle of interference microscopy (IFM) for diffusion measurements in nanoporous systems. (a) Schematic representation of the vacuum system (static)
with the optical cell containing the sample. (b) Interference microscope with CCD camera on
top, directly connected to (c) the computer. (d) Basic principle: changes in the intracrystalline
concentration during diffusion of guest molecules cause changes in the refractive index of the
crystal (n_1) and hence in the phase difference $\Delta\varphi$ of the two beams. Variation of the difference
of the optical path length yields the variation of the intracrystalline concentration by the indicated equation. (e) Close-up view of the optical cell containing the crystal under study.

We have used a Carl-Zeiss JENAPOL interference microscope with a
Mach–Zehnder type interferometer [43]. The concentration integrals are
obtained with a spatial resolution of $\Delta x \times \Delta y = 0.5\,\mu m \times 0.5\,\mu m$. The minimum
temporal separation between two subsequent concentration profiles as pres-
ently accessible is 10 s, but there are no principle limitations toward smaller
time intervals of observation.

IRM is based on the frequency-dependent absorption of infrared light,
corresponding to the special features of the molecules under study (see
Figure 2.5) [80]. We have used a Fourier-Transform IR microscope (Bruker
Hyperion 3000) [43] composed of a spectrometer (Bruker Vertex 80v) [81,82]
and a microscope with a focal plane array (FPA) detector [83] as a device for
microimaging. The FPA detector consists of an array of 128×128 single detec-
tors with a size of $40\,\mu m \times 40\,\mu m$ each. By means of a 15× objective, in the
focal plane, i.e., at the position of the crystal under study, a resolution of up to
$2.7\,\mu m \times 2.7\,\mu m$ is gained. Each single element of the FPA detector records the
IR signal. The intensity of the IR light as a function of the wavelength, i.e., the

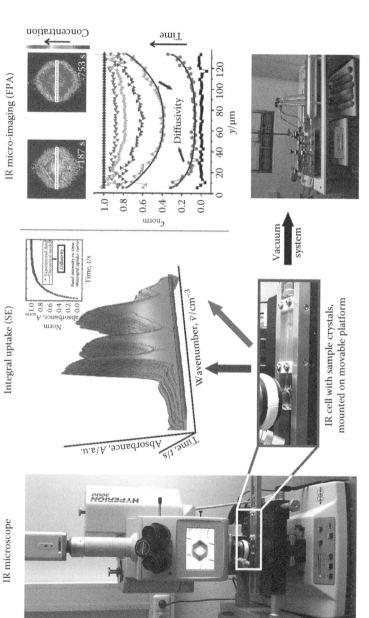

FIGURE 2.5

Experimental setup and basic principle of IR microscopy (IRM). An optical cell is connected to a vacuum system. After mounting it on a movable platform under the microscope, one individual crystal is selected for the measurement. Changes in area under IR bands of the guest species are related to the guest concentration. The spectra can be recorded as signal integrated over the whole crystal (SE, single-element detector) or with a spatial resolution of up to 2.7 μm by the imaging detector (FPA, focal plane array).

transmission spectrum, is determined by means of the spectrometer using Fourier transformation [81]. According to Lambert-Beer's law, the concentration of a particular molecular species is proportional to the intensity of the "absorption band," defined as the negative logarithm of the ratio between the transmission spectrum of the sample over the relevant frequency range and of the corresponding background signal [81,82]. By comparison with a standard, also information about the absolute number of molecules becomes attainable.

2.2.2.2 Attainable Information and Correlation with the PFG NMR Diffusivity Data

While PFG NMR records the rate of molecular redistribution under macroscopic equilibrium, the application of IFM and IRM is based on the observation of changes in concentration and has to operate, therefore, under nonequilibrium conditions. Thus, the diffusivity introduced by Fick's first law (Equation 2.1) refers to directly observable mass transfer. For distinction from the self-(or tracer) diffusivity obtainable at equilibrium, this is referred to as the transport diffusivity for which, in the following, we use the notation D_T.

In addition to studying the overall uptake or release, IRM may also be applied to multicomponent systems. Thus, also the observation of co- and counter-diffusion becomes possible. These options in particular include counter-diffusion studies of differently labeled molecules, e.g., by using deuterated and undeuterated species of the same chemical compound. The thus attainable diffusivities and surface permeabilities are obtained under quasi-equilibrium and correspond to the PFG NMR data, i.e., to the self-diffusivities and (equilibrium) surface permeabilities.

Introducing the "thermodynamic factor" $\partial \ln p / \partial \ln c$, by the relation

$$D_T = D_{T0} \frac{\partial \ln p}{\partial \ln c} \tag{2.8}$$

the transport diffusivity may be expressed as a "corrected" diffusivity D_{T0}. The thermodynamic factor is a measure of the deviation from proportionality between pressure p and the corresponding equilibrium concentration $c(p)$. For Langmuir-type sorption isotherms ("type I" isotherms [4]), which are typical of nanoporous host–guest systems, the thermodynamic factor increases with increasing guest concentration, starting from unity in the limit of sufficiently small concentrations. According to the Maxwell–Stefan model of diffusion [84], the corrected diffusivity may be correlated with the self-diffusivity by the expression [35,85]

$$\frac{1}{D} = \frac{1}{D_{T0}} + \frac{\theta}{\mathcal{D}_{AA}} \tag{2.9}$$

with the pore filling factor, $\theta = c/c_{max}$. $Ð_{AA}$ is referred to as the mutual Maxwell–Stefan diffusivity. Its reciprocal value is a direct measure of the frictional drag of the diffusants with each other (while $1/D_{T0}$ represents the friction of the diffusants with the pore walls). For Langmuir-type isotherms, combination of Equations 2.8 and 2.9 yields the inequality

$$D \leq D_{T0} \leq D_T \qquad (2.10)$$

Such an interdependence may be nicely demonstrated in QENS or IRM experiments [24,80,86], which allow the simultaneous determination of all relevant quantities. Note that for systems where molecular propagation is controlled by the rate of guest jumps between adjacent cages, the mutual interference of the diffusants and, hence, $1/Ð_{AA}$ becomes negligibly small. Under such conditions, according to Equation 2.9, the self- and corrected diffusivities coincide [80].

It is noteworthy that in most recent systematic IRM diffusion studies with ZIF-8, a member of the novel family of MOF-type nanoporous host systems, the thermodynamic factor turned out to become also smaller than one. In such cases, the transport diffusivities were found to be smaller rather than larger than the self-diffusivities [80]! This is exactly the situation predicted by Equation 2.8 if, as in the given case, the self-diffusivities are approached by the corrected diffusivities.

For sufficiently small concentrations, i.e., for $\theta \to 0$ and $\partial \ln p / \partial \ln c$ approaching unity (Henry-type sorption isotherms), with Equations 2.8 and 2.9, relation 2.10 is easily found to become an equality. This means in particular that, at the rather low concentrations typical of high-temperature catalytic reactions, the information provided by the PFG NMR equilibrium measurements (namely, the self-diffusivities D) may be directly used for an assessment of the relevant transport diffusivities D_T.

As with the diffusivities introduced by Equation 2.1, the surface permeabilities (Equation 2.6) may also be different for equilibrium and nonequilibrium conditions. Although well understood in the context of pore diffusion, this point has received less attention in connection with surface resistance, probably because the direct measurement of surface permeability has become possible only recently with the introduction of the techniques presented in this overview.

We focus on the application of IFM and IRM to materials traversed by one- and two-dimensional pore systems. For such systems, by observation perpendicular to these pore systems, the concentration integral (Equation 2.7) degenerates to the simple product of the concentration $c(x, y, z, t)$ and the crystal thickness L since the concentration becomes independent of x. Three-dimensional pore systems require more complicated procedures of analysis as described, e.g., in Refs [78,79,82,87–89].

Following Equations 2.1 and 2.6, in both IFM and IRM, the determination of the diffusivities and surface permeabilities may be based on the measurement of the molecular fluxes. Since fluxes are determined by the number of molecules passing a certain cross section per time interval, they may be easily determined from the difference in area between two subsequently recorded concentration profiles divided by the time interval between their measurement (see Figure 2.6). The transport diffusivity results as a factor of proportionality between the flux (determined in this way) and the slope of the concentration at the considered cross section (Equation 2.1). Correspondingly, for surface permeabilities, one has to correlate the flux entering the whole crystal with the difference between the actual concentration close to the crystal surface and the concentration eventually attained at equilibrium. A summary of alternative approaches to the desired transport parameters and their assessment is given in Ref. [87].

Although based on essentially the same schemes of analysis, with similar comparative advantages and disadvantages, IFM and IRM benefit from their complementarity. While the higher spatial resolution makes IFM particularly useful for application to relatively small crystals, the evidence of IRM significantly benefits from its ability to provide guest-selective measurements. However, even in cases where the small crystal size does not allow spatially resolved IR measurements, focusing on a particular crystal notably enhances the power of evidence of such studies in comparison with conventional uptake and release measurements (see Section 2.1). Concentrating on a particular crystal in immediate contact with the guest atmosphere excludes any disturbance by a "bed" effect. Moreover, for an isolated single crystal, the ratio between the external surface area and the sample volume assumes the largest possible value and exceeds that of particle beds as applied in conventional macroscopic sorption measurements by orders of magnitude. Since it is this ratio that determines the rate of exchange of the sorption heat with the surroundings [35,90] in single-particle uptake and release experiments, "heat" effects are also essentially excluded [91].

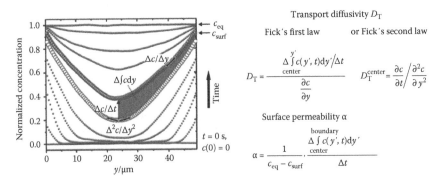

FIGURE 2.6
Data analysis of transient concentration profiles illustrated for the example of methanol uptake in ferrierite (pressure step $0 \rightarrow 5\,$mbar in the gas phase, 298 K). The transport parameters can be calculated directly from the profiles by the indicated relations, as well as by fitting with numerical solutions of Fick's laws.

2.3 Results and Discussion

2.3.1 Zeolites

2.3.1.1 Intracrystalline Diffusion: Zeolites Y and NaX

Zeolite Y is the key-component of fluid catalytic cracking (FCC) catalysts [92]. Although available in nature as large crystals [93], the crystal sizes of laboratory-synthesized zeolites of type Y are typically in the range of 1 μm. For many years, this excluded the option of PFG NMR measurements of intracrystalline diffusion. Only due to recent success in synthesizing Y-type zeolite crystallites of sufficiently large sizes [94] and progress in PFG NMR methodology [54], the measurement of intracrystalline diffusion in zeolite Y has become possible. Ref. [95] provides a survey of the diffusivities, obtained with *n*-octane as a typical guest molecule under FCC conditions with both the ammonium-exchanged form and the zeolites after hydrothermal treatment (zeolite "USY"). In addition to their absolute values (8×10^{-11} m^2 s^{-1} at 213 K and half pore filling), the diffusivities were found to remain essentially unaffected by the hydrothermal treatment. This suggests that the de-alumination initiated by the hydrothermal treatment leads to the formation of isolated mesopores rather than to an interconnected system of transport pores that might give rise to an acceleration of intracrystalline molecular transport. Most importantly, with this finding, one may also exclude that the benefit of hydrothermal treatment is due to an enhancement of the intracrystalline diffusivities in the zeolite Y component of FCC catalysts, as has been suggested in previous publications [96]. In fact, from PFG NMR diffusion measurements, the intracrystalline diffusivities have been shown to have essentially no influence on the overall FCC reaction rates. We shall return to this point in Section 2.3.1.4 where we provide a comparison between the intracrystalline and the "long-range" (in this case: the intra"particle") diffusivities.

These measurements were of course made under "clean" conditions in the laboratory, whereas, under operating conditions in a catalytic cracker, the catalyst is strongly coked, leading to a significant increase in mass transfer resistance. The extent to which conclusions drawn from "clean" measurements remain valid for the real system is therefore subject to some uncertainty. Nevertheless, the major point that the dominant mass transfer resistance is intraparticle diffusion rather than intracrystalline diffusion is probably valid. The improved performance associated with mesopore formation is probably attributable to an increase in the external surface area (on which the larger molecules are cracked) rather than to enhancement of intracrystalline diffusion.

In contrast to zeolite Y, its aluminum-rich isomorph NaX may be synthesized as relatively large crystals (~100 μm). In this way, starting from about 100 nm, PFG NMR is able to cover intracrystalline diffusion paths over more than one order of magnitude, without any essential disturbance by effects due to the crystal surface and/or the intercrystalline space. Measurements of this type can reveal transport resistances with mutual separations comparable to the diffusion path lengths. While, in many cases, this type of measurement revealed linearity between the mean square displacement and the observation time as to be expected from Equation 2.4 for a constant diffusivity D [97,98], there are also examples of notable deviations from linearity [39,40]. Though these deviations are a strong indication of transport resistances, so far any attempt at their direct observation and attribution to certain types of lattice defects have not been successful. A recent study combining QENS and PFG NMR diffusion measurements with TEM [41], with zeolite NaX as a host system and *n*-octane as guest molecules, provided direct evidence of stacking faults (of mirror-twin type on (111)-type planes of the cubic framework), which could be correlated with the fact that the short-range self-diffusivities, as measured by QENS, notably exceeded the PFG NMR values. Since, in these PFG NMR studies, there was no perceptible effect of the observation time (and hence of the displacements) on the PFG NMR diffusivities, the separation between adjacent faults can be estimated to be of the order of or below 100 nm.

2.3.1.2 Transport Resistances at the Crystal Surface: Zeolites NaCaA and Ferrierite

Hydrothermal treatment of NaCaA-type zeolites [99,100] and/or contact with hydrocarbons at elevated temperatures [101] are well known to give rise to the formation of transport resistances notably exceeding the influence of intracrystalline diffusion on the overall uptake or release process of hydrocarbons. In fact, by the application of the NMR tracer desorption technique [60], long-term contact with water vapor or with hydrocarbons at elevated temperatures was found to notably reduce the exchange rate of probe molecules with the surroundings, while their intracrystalline diffusivities remained essentially unaffected. Only very recently [62,63], surface resistances could be identified to exist already on the surface of the as-synthesized zeolite crystallites of type NaCaA, with an effect comparable with the resistance exerted by intracrystalline diffusion. This conclusion resulted from a more detailed analysis of the time dependence of the mean square displacement of those molecules, which, during the "observation time" t, remained confined by one and the same crystal. It does not allow, however, more than an order-of-magnitude estimate.

Notably better conditions for quantitative measurement of surface resistances are provided by IR and IFM. Since they are able to record molecular fluxes, these techniques can determine surface resistances directly (from Equation 2.6). This greatly enhances the accuracy of the measurement. So far, however, this type of measurement has not been possible with zeolites of type LTA. In the cation-containing species (NaCaA), the existence of the bivalent cations leads to adsorption isotherms that are too favorable and therefore make it difficult to achieve reversible adsorption and desorption with a single crystallite. Cation-free zeolites of type LTA [102–104], on the other hand, cannot yet be prepared as single crystals of such homogeneity as is required for IR and IFM.

Much better prospects for quantifying surface resistances by the combined application of IRM and IFM are provided by zeolites of the ferrierite type [42,105,106]. As an example, Figure 2.7a shows the transient concentration profiles during desorption of methanol in the direction of the eight-ring channels of ferrierite by a pressure step from the initial (equilibrium)

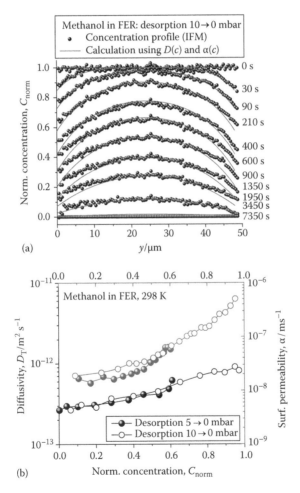

FIGURE 2.7
Concentration profiles (a) and the resulting (transport) diffusivities and surface permeabilities (b) during methanol desorption from 10 to 0 mbar in ferrierite. By comparison with IRM, a (normalized) concentration of 1 is estimated to correspond to about 1.2 mmol g^{-1} [42]. The full lines in (a) show the concentration profiles recalculated on the basis of the diffusivities and permeabilities shown in (b) by the corresponding solutions of Fick's second law.

pressure of 10 mbar to 0 mbar. As explained above, correlating the fluxes (the area between sequential profiles divided by the time interval between them) with either the concentration gradients (via Equation 2.1) or the difference between the actual and the equilibrium boundary concentration (via Equation 2.6) leads to both the intracrystalline diffusivities and the surface permeabilities (Figure 2.7b). For comparison, Figure 2.7b also displays the permeabilities and diffusivities resulting from another desorption experiment, revealing satisfactory agreement. This self-consistency is also reflected by the agreement between the measured concentration profiles and the profiles calculated by solving Fick's second law with the intracrystalline diffusivities and surface permeabilities (formulating the boundary conditions) appearing from the representations in Figure 2.7a.

Plotting the integral of each individual profile as a function of time provides the information attainable by conventional uptake or release experiments. It is noteworthy that, for the host–guest system under consideration, the resulting uptake or release curves provide good examples of "disguised kinetics" [107]. Analyzing the transient sorption curves by conventional models suggests intracrystalline diffusion control but in fact, as shown by interference and IR microscopy, the overall mass transfer rate is also significantly affected by the finite rate of surface permeation.

2.3.1.3 Mesoscopic Observation of Molecular Uptake: Water in Sodalite

Usually, the measurement of intracrystalline transport is facilitated by the application of large crystals. They allow the most accurate observation of intracrystalline concentration profiles and of their variation with time. However, in special cases, the opposite situation may occur. This is particularly true for extremely small molecular mobilities (or surface permeabilities) where, with the dependencies as given by Equation 2.5, large crystal sizes would lead to diverging response times that complicate the observation of transport phenomena. In such cases, one may benefit from the use of small crystallites, even with the loss of the option of intracrystalline spatial resolution.

Such a situation arises in the study of the rate of water diffusion in hydroxy sodalites. In Ref. [108], hydroxy sodalite was suggested as a promising candidate for fabricating membranes for dewatering organic alcohol/water mixtures. From permeation studies, the transport diffusivity of water at room temperature is estimated to be of the order of 10^{-18} m^2 s^{-1} [109]. These extremely low diffusivities reflect the fact that, with a diameter of 0.27 nm, the "windows" between the cages of the sodalite pore structure essentially coincide with the diameters of the water molecules.

In parallel with these studies, IR microscopy was applied to record water uptake and release with pressure steps between 0 and 10 mbar [110]. The sodalite crystals of about 0.5 μm diameter are clumped to conglomerates with a size of 20–200 μm. In view of the extremely low intracrystalline diffusivities, any limitation by long-range diffusion within the conglomerates may be excluded. In the experiments, time constants of 5000 s for uptake and 11,000 s for release were observed. Under the simplifying assumption of spherical particle shapes and constant diffusivity, with Equation 2.5 (considering only the term of diffusion limitation), the respective transport diffusivities are found to be 0.8×10^{-18} and 0.4×10^{-18} m^2 s^{-1}. These data are in good agreement with the estimate based on permeation measurements [109]. The slightly smaller diffusivity results from the larger time constant of desorption. This arises because, for most nanoporous host–guest systems, transport diffusivities increase with increasing loading [35], so that the time constant of molecular uptake has to exceed that of release [106].

2.3.1.4 Long-Range Diffusion: Mass Transfer in Beds of Zeolite NaX and in FCC Catalyst Particles

In general, PFG NMR measurements of diffusion in nanoporous materials are performed with closed sample tubes. Thus, following essentially isosteric measuring conditions, the gas phase pressure over the porous material varies dramatically with varying temperature. This leads, under the measuring regime of long-range diffusion (i.e., by observing mean path lengths notably exceeding the size of the individual nanoporous crystallites), to a decisive change in the mechanism of molecular diffusion through the intercrystalline space with increasing temperature, namely, from Knudsen diffusion (dominating molecule-wall collisions) to bulk diffusion (dominating mutual collisions between the diffusing molecules) [59,111]. In an Arrhenius plot of the diffusivity, the transition between these two mechanisms appears as a distinctive change from a straight line at low temperatures (in the Knudsen regime, the temperature dependence of $D_{(l.r.)} = p_{\text{inter}} D_{\text{inter}}$ is essentially given by p_{inter}, which increases exponentially, in proportion to the gas pressure) to a plateau (since, under the conditions of bulk diffusion, the increase of p_{inter} with further increasing temperature is compensated by the decrease of D_{inter}, which is proportional to $1/p$ and $1/p_{\text{inter}}$) [112].

Estimating the magnitude of p_{inter} from the intercrystalline pore volume and the adsorption isotherm, and of D_{inter} by the gas kinetic approach $\frac{1}{3} \langle v \rangle \lambda$ (with $\langle v \rangle$ the mean velocity and λ the mean distance between successive collisions), one is able to determine a tortuosity factor, which describes the enhancement of the diffusion path in the intercrystalline pore space in comparison with the free gas space (in the regime of bulk diffusion) or with a straight tube (for Knudsen diffusion) [59,112,113]. Application of this procedure to the long-range diffusion of ethane in beds of zeolite NaX [113] yielded tortuosity factors that notably differed between the regimes of bulk and Knudsen diffusion. Since, in both cases, molecular propagation occurs in the same pore space, this result may, at a first glance, be considered to be counterintuitive. It may be rationalized, however, by realizing that the correlation of subsequent molecular displacements may differ. Under confinement by the same pore surface, the probability for an anti-correlation of subsequent displacements under the conditions of Knudsen diffusion may be shown to be notably larger than in the bulk regime [114–116]. Such behavior could be nicely confirmed by a mesoscopic simulation of the ethane diffusivities in beds of NaX zeolite crystals [117].

We now apply the concept of long-range diffusion to catalyst particles that consist of catalytically active zeolite crystallites dispersed in an inert binder. As a typical example, we consider FCC catalysts. The particle sizes are typically between 30 and 100 µm and notably exceed the size of the individual zeolite crystallites with sizes of, in general, around 1 µm. The performance of these catalyst particles depends on the rate of exchange between the active sites within the individual zeolite crystals and the surroundings of the catalyst particle. Ideally, an exploration of the controlling mechanisms should be based on the separate determination of the rate of molecular propagation in both the zeolitic component (intracrystalline diffusion) and through the whole particle (long-range diffusion or intra"particle" diffusion). Exactly this option is provided by the PFG NMR technique.

As an example, Figure 2.8 displays the results of such diffusion studies with *n*-octane as a typical representative of the guest molecules during FCC [118]. The figure also indicates the need for extrapolation from the temperature range of diffusion measurements to the technical conditions of FCC. Interestingly, the activation energy of intraparticle diffusion (43.4 ± 1.3 kJ/mol) is found to be in satisfactory agreement with the isosteric heat of adsorption of *n*-octane in the catalyst (around 40 kJ/mol for the *n*-octane loading of around 0.6 mmol g^{-1} of the zeolite in the catalyst), which has been obtained by measuring the adsorption isotherms of *n*-octane. This suggests that, in complete analogy with the well-known behavior of long-range diffusion in

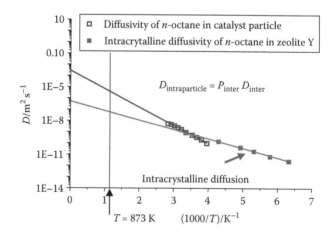

FIGURE 2.8
Arrhenius plot of the coefficients of intraparticle ("long-range") and intracrystalline diffusion of *n*-octane in an FCC catalyst. (From Kortunov, P. et al., *Chem. Mat.*, 17, 2466, 2005.)

zeolite beds in the Knudsen regime [59,112,113], the temperature dependence of the intraparticle diffusivity is determined by that of the fraction of adsorbate molecules in the meso- and macropores of the particles [118]. A more accurate extrapolation procedure is described below.

By inserting the respective sizes and diffusivities into Equation 2.5 (and neglecting any influence by additional surface barriers on the individual crystallites and/or the catalysts particles), the transport resistances (as represented by the corresponding first moment), by diffusion through the zeolite catalysts, may be shown to be notably smaller than the transport resistance for diffusion through the (much larger) FCC catalyst particle. This finding significantly deviates from the general view as resulting from (conventional) macroscopic measurements in which the intracrystalline diffusivities in zeolite Y appear to be far below the intraparticle (i.e., the long-range) diffusivity [119–121]. The result of the PFG NMR studies was nicely confirmed in a series of comparative diffusion–reaction experiments [122,123] in which the zeolitic component has been left completely unaffected while, by deliberately varying the architecture of the transport pores in the binder, the long-range diffusivity was varied by a factor of 2. Most importantly, in these studies, the catalytic performance expressed by the oil-to-gasoline conversion was found to increase with increasing long-range diffusivities.

In Ref. [124], PFG NMR has been applied to measure the tortuosity factor of FCC catalysts. In these studies, PFG NMR is applied to determine the mean square displacement in a given time interval for the free liquid and a sample of the catalyst with liquid-filled pores, so that the tortuosity simply results as the ratio between these quantities which, according to Equation 2.4, also results as the ratio between the corresponding diffusivities. In these studies, the tortuosity factors were found to decrease with decreasing diameters of the meso- and macropores.

With a simple gas-kinetic approach [64,65,112], in Ref. [125], the thus-determined tortuosities were successfully shown to yield excellent agreement between the long-range diffusivities under gas-phase adsorption and model predictions. Representative data for the commercial catalysts under study are shown in Figure 2.9. Consistent tortuosities of 2.0 for catalyst 1 and 2.5 for catalyst 2 were obtained using both water and cyclooctane as the test liquids.

Under transient conditions, the effective macropore diffusivity is given by

$$D_{\text{eff}} = \frac{n_{\text{pore}}}{n_{\text{pore}} + n_{\text{ads}}} \frac{D}{\tau}, \quad \frac{1}{D} = \frac{1}{D_{\text{K}}} + \frac{1}{D_{\text{M}}} \qquad (2.11)$$

If we assume Langmuir equilibrium, the molecular densities in the adsorbed and fluid phases will be related by

$$\frac{n_{\text{ads}}}{n_{\text{pore}}} = \frac{bq_s}{1 + bc} \qquad (2.12)$$

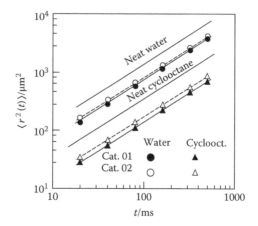

FIGURE 2.9
Mean square displacement as a function of time, determined by PFG NMR, for water and cyclooctane in two samples of commercial catalyst at room temperature. (From Stallmach, F. and Crone, S., *Diffusion Fundamentals*, J. Kärger, F. Grinberg, and P. Heitjans (Eds.), Leipziger Universitätsverlag, Leipzig, Germany, p. 474, 2005.)

Thus, with the tortuosity determined as outlined above, the effective diffusivity can be estimated at any temperature provided the Langmuir parameters are known. The application of this approach is illustrated in Figure 2.10 in which the effective self-diffusivities for ethylene in a porous catalyst, measured over a wide range of temperatures by PFG NMR, are compared with the values predicted from Equations 2.11 and 2.12.

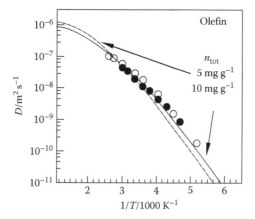

FIGURE 2.10
Experimental effective self-diffusivities for ethylene in a porous catalyst at two different loadings compared with the values predicted from Equations 2.11 and 2.12. (From Stallmach, F. and Crone, S., *Diffusion Fundamentals*, J. Kärger, F. Grinberg, and P. Heitjans (Eds.), Leipziger Universitätsverlag, Leipzig, Germany, p. 474, 2005.)

2.3.2 Metal Organic Frameworks

The MOFs [126,127] offer an important class of new materials, which have been found to provide ideal host systems for the study of molecular transport under a variety of different conditions of molecular confinement. So far, however, this benefit has to be purchased by an enhanced tendency toward lattice instability. The multitude of pore architectures includes the option to vary the dimensionalities of the pore arrangement but, again, this benefit comes at the cost of an enhanced tendency toward lattice instability. With the introduction of the concept of single-file diffusion into molecular dynamics under pore confinement [128,129], one-dimensional pore arrangements are of particular interest [130]. Also, in this respect, the advent of the MOFs offers novel opportunities to which we will refer in the following two sections.

2.3.2.1 Manganese Formate

Figure 2.11 shows a typical representative of the MOF manganese formate crystals under study [131,132]. From Figure 2.11b, the pore system may be seen to be a one-dimensional arrangement of cavities. Figure 2.11c shows transient concentration profiles as observed for methanol uptake by a pressure step from 0 to 10 mbar [133]. Molecular uptake is seen to occur predominantly along the y direction as to be expected from the one-dimensional structure of the pore system. Figure 2.11d represents the profiles within that part of the crystal, where the structure—judging from the perfect and highly symmetric shape of the profiles—appears to be close to perfect. Either via Fick's second law, with the first and second derivatives as indicated in the figure, or by simply applying Fick's first law to the molecular fluxes, as described in Section 2.2.2.2 (see also Ref. [87]), one may determine the transport diffusivities and surface permeabilities shown in Figure 2.11e.

Most remarkably, the transport diffusivities are thus found to be essentially independent of concentration. Though, in general, the corrected diffusivity (Equation 2.8) rather than the transport diffusivity itself is found to remain constant with varying concentration [35], there is no nature-given reason for such a dependency. In fact, the constancy of the transport diffusivity (rather than of the corrected diffusivity), as observed in our studies with this particular host–guest system, would nicely comply with the requirement of a system consisting of well-defined lattice sites, which can only be occupied by one molecule, with hard-core repulsion as the only mechanism of molecular interaction [134–137].

Comparing Figures 2.7 and 2.11, we also note that there appears to be no general correlation between the concentration dependences of diffusion and surface permeation: in ferrierite it is the diffusivity, in manganese formate it is the surface permeability of methanol, which reveals the stronger concentration dependence.

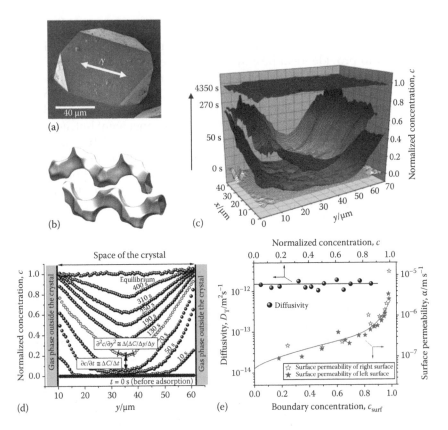

FIGURE 2.11
Methanol uptake by MOF manganese formate by a pressure step from 0 to 10 mbar at 298 K as studied by IFM, with a typical crystal under study (a) and a schematics of the pore structure (b). From the two-dimensional concentration plots (c), best profiles could be determined at $x = 41\,\mu m$ (d), leading to the transport diffusivities and surface permeabilities as shown in (e). (From Kortunov, P. et al., *J. Am. Chem. Soc.*, 129, 8041, 2007.)

2.3.2.2 Zn(tbip)

Zn(tbip) (H$_2$tbip = 5-*tert*-butyl isophthalic acid) [138,139] is a highly stable representative of the family of MOFs. Owing to this stability, it was possible to perform dozens of adsorption and desorption cycles with one and the same crystal, without any perceptible loss in reproducibility. This reproducibility was the main prerequisite for an in-depth investigation of the dependence of the surface resistances on the external experimental conditions, namely, on both the equilibrium concentration (as determined by the external gas phase) and the actual boundary concentration. Figure 2.12a shows the results of this study, with propane as the probe molecule.

From measurements covering a large spectrum of conditions (including adsorption and desorption runs as well as large and small pressure steps),

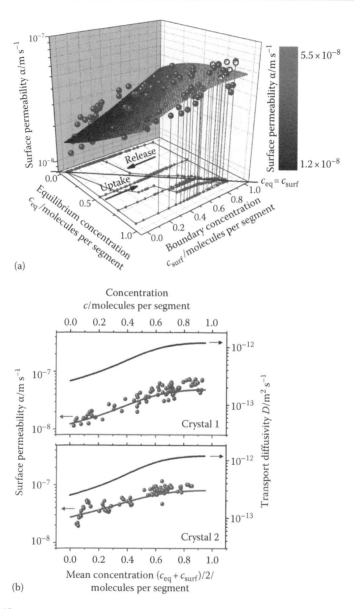

(a)

(b)

FIGURE 2.12

Surface permeability of a MOF Zn(tbip) crystal for propane at room temperature plotted (a) as a function of both the actual concentration at the crystal boundary (c_{surf}) and the concentration in equilibrium with the external gas phase (c_{eq}), and (b) as a function of the mean concentration $(c_{eq} + c_{surf})/2$ (i.e., along the plane in (a)). For comparison, (b) also displays the transport diffusivities and the corresponding data for a second MOF Zn(tbip) crystal. (From Tzoulaki, D. et al., *Angew. Chem. Int. Ed.*, 48, 3525, 2009.)

it was found (Figure 2.12a and [140]) that neither the step size nor the step direction are of significant influence on the resulting surface permeabilities. In all cases, the mean value $(c_{eq} + c_{surf})/2$ of the boundary and equilibrium concentrations was found to essentially determine the rate of surface permeation. The best fit to a dependence $\alpha((c_{eq} + c_{surf})/2)$ is displayed in Figure 2.12a by the plane with shades from black (lowest permeabilities) to gray (highest permeabilities). This result is actually to be expected since surface resistance is in essence equivalent to permeation through a membrane and it has been shown that when the diffusivity is concentration dependent, the permeance is governed (to an excellent approximation) by the mean diffusivity [141,142].

The same dependence is shown by the one-dimensional representation in Figure 2.12b (top). The representation on the bottom of Figure 2.12b shows the permeabilities measured for another crystal. Both representations also show the diffusivities. Comparison of these representations reveals three remarkable results:

1. In contrast to the systems considered so far (see Figures 2.7 and 2.11), for propane in Zn(tbip), the diffusivities and surface permeabilities follow essentially identical concentration dependences. In Ref. [140], this similarity in the concentration dependences of intracrystalline transport diffusion and surface permeation was also observed with ethane and *n*-butane as probe molecules. Most remarkably, coinciding trends in the concentration dependences of surface permeation and intracrystalline diffusion were even observed in tracer exchange experiments between propane and deuterated propane. Under conditions of tracer (i.e., self-) diffusion and tracer permeation (leaving open whether, eventually, also the term self-permeation might synonymously be used), diffusivity and permeability both decrease with increasing loading. This behavior reflects the inhibition of site exchange between neighboring guest molecules (which increases with increasing loading and, under ideal single-file conditions, would reduce the diffusivity to zero [130,139]).

2. The coincidence in the trends of the diffusivities and surface permeabilities for the Zn(tbip) crystals under study is taken as an indication that the finite rate of surface permeation is caused by the total blockage of a large number of channel openings, with only a few of them essentially freely accessible. Otherwise, if there were an additional blockage to essentially all pore openings, one should expect to see differences between the respective concentration dependencies as observed, e.g., for the ferrierite and MOF manganese formate crystals (see Sections 2.3.1.2 and 2.3.2.1).

3. Comparison of crystals 1 and 2 reveals essentially the same intracrystalline diffusivities while there is a notable difference between the surface permeabilities. Since damages to the crystal structure are

more likely to occur close to the surface, rather than in the crystal bulk phase, such behavior may be explained by structural variation between different crystals of the same batch. Most remarkably, however, the surface permeabilities on either side of a given crystal are generally found to agree with each other. This indicates that the crystallization process itself does not give rise to the formation of surface resistances, since then it would be difficult to explain why they differ from crystal to crystal but coincide on either side of a given crystal. The formation of resistances on the surface of a crystal is obviously associated with the surroundings which should, generally, be uniform on either side of a crystal but which, depending on the conditions of storage or pretreatment, may clearly vary within one batch of crystals.

In a test experiment, this symmetry was broken by simply breaking the crystal. To demonstrate the surface structure, Figure 2.13a shows scanning

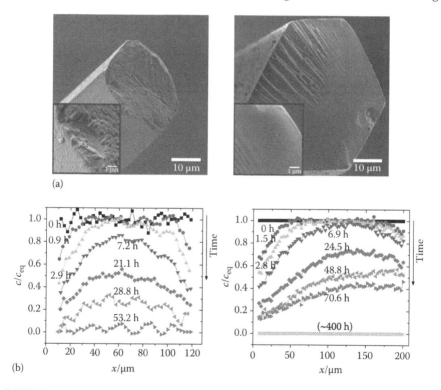

FIGURE 2.13
Scanning electron micrographs of the face of an as-synthesized MOF Zn(tbip) crystal (a) (left) and of the face of a freshly broken crystal (right) and transient concentration profiles during the tracer exchange of propane and deuterated propane within such crystals (b) (left side: as-synthesized; right side: freshly broken crystal, with the fresh face on the left).

electron micrographs of both the genuine external surface (left) and the freshly created surface after crystal breaking (right). It is seen that amorphous deposits, which appear on the outer surface, do not occur on the freshly broken faces. The resulting asymmetry in the surface permeabilities is nicely reproduced in the transient profiles: while in the as-synthesized crystal (left), the profiles are symmetric, uptake through the freshly broken face (left side of right representation) in Figure 2.13b is notably faster, giving rise to a pronounced asymmetry in the concentration profiles.

2.3.3 Ordered Mesoporous Materials

Mesoporous materials of type SBA-15 [143] are nicely suited to demonstrate the ability of PFG NMR to correlate structural features of the host with the patterns of molecular mobility. We demonstrate here two options for such studies. In Section 2.3.3.1, we investigate the interparticle exchange in differently textured SBA-15 samples, namely, in "bundles" and "fibers" [12]. Section 2.3.3.2 is dedicated to the correlation of molecular diffusion and hysteresis, with particular emphasis on the role of metal deposition.

2.3.3.1 Pore Connectivity in Silica SBA-15

In general, interparticle exchange in mesoporous host–guest systems may proceed via both the particle interconnections and the gas phase. PFG NMR offers a simple means for suppressing the latter option [58]. Applying guest molecules in abundance, reduction of the measuring temperature below the bulk freezing point creates a frozen phase in the interparticle space. Then, guest exchange between different particles cannot occur anymore through the free space and has to proceed through particle interconnections or along the particle surface. For this type of studies, nitrobenzene has been found to be a particularly useful probe molecule [77,144–147].

In Ref. [148], by this type of measurement and using the method of analysis presented in Section 2.2.1.2, guest diffusion in SBA-15 particles was shown to be described by a diffusion tensor of rotational symmetry. Figure 2.14 displays the key parameters of these studies, namely, the diffusivities along the axis of rotational symmetry (D_{par}) and the diffusivities in the perpendicular direction (D_{perp}). The data are plotted as a function of the observation time. In addition to these observation times, we have also indicated the root mean square displacements, resulting from them and the respective mean diffusivities via Einstein's relation, Equation 2.4.

Corresponding to their shape, the SBA-15 particles under study are referred to as fibers and bundles, with the respective diffusivities given on top and at the bottom of Figure 2.14. As the most striking feature, the diffusivities are found to decrease with increasing observation times in the fibers (top of Figure 2.14), while they increase in the bundles (bottom of Figure 2.14). Diffusivities decreasing with increasing observation times may be easily

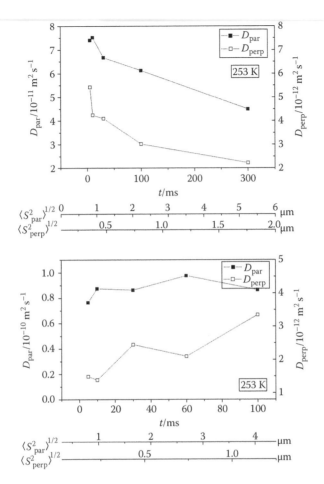

FIGURE 2.14
Self-diffusivities D_{par} and D_{perp} as a function of diffusion (i.e., observation) time for SBA-15 fibers (top) and bundles (bottom). For an illustration of the distances over which these diffusivities have been measured, the abscissa displays the values of the mean-square displacements $\langle s^2_{par}\rangle$ and $\langle s^2_{perp}\rangle$, resulting from an average value of the diffusivities in the respective directions by application of Equation 2.4. (From Naumov, S. et al., *Microporous Mesoporous Mater.*, 110, 37, 2008.)

referred to a hierarchy of transport resistances (just as reported in Refs [39,40], as evidence of the existence of internal barriers in MFI-type zeolite crystallites). In the case of SBA-15 fibers, embedded in the frozen guest phase, one may expect analogous behavior. The probability of encounters with the rigid frozen phase should clearly increase with increasing observation time, progressively inhibiting molecular propagation. The displacements covered (a couple of micrometers in particle longitudinal extension and fractions of micrometers in perpendicular direction) reflect the geometrical situation [12,148]. Correspondingly, the absolute values of the diffusivities in longitudinal direction (D_{par}) exceed those in the perpendicular direction by more than an order of magnitude. The fact that diffusion is not ideally one-dimensional may be easily referred to the microporosity of SBA-15, which ensures matter exchange between adjacent channels. Moreover, transport perpendicular to the main channel direction may quite generally be referred to lattice defects which, though often essentially invisible to conventional methods of structural analysis, from diffusion measurements [40,58,139], may be expected to occur quite commonly in nanoporous materials.

Structural defects may also be considered as the origin of the unusual increase in the diffusivities with increasing observation time as observed with the SBA-15 bundles (bottom of Figure 2.14) [148]. The bundle extensions and hence the space in which the guest molecules may freely diffuse notably exceed the mean diffusion paths so that one should not expect substantial effects of constriction as observed with the fibers. In contrast, on their diffusion paths, the guest molecules might get into regions of even higher mobility with increasing observation times. Such behavior would presumably lead to the observed diffusivity enhancement with increasing observation time.

2.3.3.2 Mesopore Diffusion, Hysteresis, and the Effect of Impregnation

PFG NMR self-diffusion measurements have been recently shown [146,147,149] to reflect very clearly the features of sorption hysteresis in the diffusion patterns of nanoporous materials. It could be demonstrated that by considering either adsorption or desorption, for the same guest pressure in the surrounding atmosphere, one may attain quite different diffusivities corresponding to the differences in the amounts adsorbed on the adsorption and desorption branches of the hysteresis loop. This concept has been used in [150] to explore the structural and dynamic features of mesoporous SBA-15. In these studies, it was of particular interest to learn how these features of the SBA-15 samples were modified after their impregnation with vanadium silicalite-1 nanoparticles. Figure 2.15 provides a survey of the attained diffusivities, with n-hexane used as a probe molecule. The measurements have been performed by stepwise variation of the guest pressure in the surrounding atmosphere at room temperature.

As a general feature, all samples show a hysteresis loop in the measured diffusivities. Quite generally, the diffusivities on the adsorption branch notably exceed those on the desorption branch at the same pressure. To rationalize this tendency, we have to take into account that, with the given mean diffusion path lengths of up to 50 μm, PFG NMR is measuring in the regime of long-range diffusion. This means (see Section 2.2.1.3) that the measured diffusivities are the product of the relative amount of molecules in the free space around the adsorbent particles (p_{inter}) and their diffusivity (D_{inter}). Within the hysteresis loop, on the desorption branch, the total number of molecules adsorbed notably exceeds those on the adsorption branch (when the pores are not yet filled). This means in turn that, during the desorption

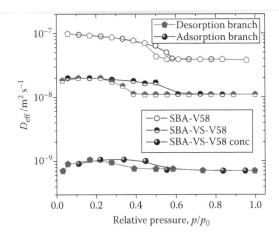

FIGURE 2.15
Mean diffusivity of *n*-hexane in different SBA-15 samples at room temperature, after either pressure increase (adsorption branch) or decrease (desorption branch), as a function of the relative pressure p/p_0. SBA-V58 denotes the starting SBA-15 material; SBA-VS-V58 and SBA-VS-V58 conc denote the sample impregnated with vanadium silicalite-1 at low and high concentrations, respectively. (From Meynen, P. et al., *Microporous Mesoporous Mater.*, 99, 14, 2007.)

branch, the relative number of molecules in the gas phase, i.e., p_{inter}, has to be notably smaller than for the adsorption branch. Since, at identical external pressures, also the diffusivities D_{inter} are identical, the long-range diffusivity $p_{inter} D_{inter}$ during the desorption is expected to be smaller than on the adsorption branch. Exactly this behavior is observed in all measurements.

Following the same reasoning, the decrease of the long-range diffusivities of the SBA-15 samples with increasing amount of vanadium silicalite-1 has to be taken as a surprise. Assuming that the impregnation procedure leaves the extra-particle space unchanged, while the intraparticle pore space is reduced, the relative number of guest molecules in the free space (p_{inter}), and hence the long-range diffusivity $p_{inter} D_{inter}$ should be expected to increase. Obviously, exactly the opposite behavior is observed. One may expect, however, that the particle texture will be notably affected by the impregnation procedure, leading to particle compaction and hence to a reduction of the external space, resulting in a decrease of both p_{inter} and D_{inter}. With such a mechanism, the observed tendency may be reasonably explained.

2.3.4 Random Pore Networks

2.3.4.1 *Diffusion in Mesoporous Aluminosilicate under Freezing Conditions*

Figure 2.16 shows the Arrhenius dependence of the diffusivities of *n*-decane and *n*-heptadecane and of their mixtures in a mesoporous silicate (SiAl15) [151,152]. From transmission electron micrographs [151], the pore network is

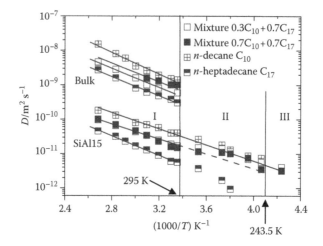

FIGURE 2.16

Diffusion coefficients of *n*-decane and of *n*-heptadecane and of their mixtures confined in the microporous aluminosilicate SiAl15 and in bulk as a function of the reciprocal temperature. The vertical lines indicate the freezing points of the neat liquid. (From Krutyeva, M. et al., *Microporous Mesoporous Mater.*, 120, 104, 2009.)

known to be of worm-like channel structure, with a mean pore diameter of about 2.4 nm. The vertical lines indicate the bulk freezing temperature of the two liquids. On considering the temperature range above the *n*-heptadecane freezing temperature (295 K), the activation energies (i.e., the slopes in the Arrhenius plots) are found to essentially coincide for the free liquid and the liquid under pore confinement. The dramatic difference in the absolute values has to be referred, therefore, to a pronounced effect of tortuosity. As is to be expected, the mixture diffusivities lie between those of the individual components.

Below the bulk freezing point of *n*-heptadecane, one may observe a notable deviation of the *n*-heptadecane diffusivities from the Arrhenius dependence toward smaller diffusivities. This behavior may be attributed to the freezing process. Since freezing starts within the largest pores (see Section 2.2.1.5), an increasing fraction of the molecules contributing to the observed signal and hence to the measured diffusivities will be situated in the smaller pores, with an, accordingly, smaller diffusivity.

It is interesting to note that, by contrast, the diffusivities in the mixture deviate from the Arrhenius dependence toward higher diffusivities. This behavior is related to the fact that in these studies, we dropped the option of selective diffusion measurement. (Due to the similarity in the high-resolution spectra of *n*-decane and *n*-heptadecane, this would have required a much more expensive and time-consuming study with pairs of samples, with one of the components in each pair applied in the per-deuterated form, see Section 2.2.1.5.) Therefore, by similar reasoning as for the explanation of the deviation of the pure *n*-heptadecane sample, we may now state that, with decreasing temperature, an increasing fraction of the *n*-heptadecane molecules will become frozen so that the relative fraction of the more mobile *n*-decane molecules increases with decreasing temperature, leading now to mean diffusivities deviating to values higher than those expected from the Arrhenius dependence.

2.3.4.2 Diffusion in Activated Carbon: Indications of Long-Range and of Micropore Diffusion

Activated carbons represent an important class of nanoporous materials with potentials for the generation of hierarchical pore structures [6]. Discrimination of the governing transport mechanisms is often facilitated by considering the diffusion properties of a whole spectrum of guest molecules. As an example for such a procedure, Table 2.1 provides a comparison of the mean diffusivities of a series of guest molecules within an activated carbon of type LMA12 [153,154] with the corresponding bulk diffusivities. LMA12 has an essentially bimodal pore structure, with a micropore volume of $0.23\,cm^3\,g^{-1}$ and a mesopore volume of $0.48\,cm^3\,g^{-1}$ [153].

Depending on the molecular species under consideration, the presence of the activated carbon may obviously lead to diffusivities that may be either larger or smaller than those for the corresponding liquid phase. For an explanation of this finding, it is helpful to consider the variation of the boiling points, which are also given in the table. The diffusivity in the sorbed state is found to decrease with increasing boiling point, corresponding to a decrease in the gas phase pressure at the (constant) measuring temperature. Exactly such behavior has to be expected under the regime of long-range diffusion, where the effective diffusivity is proportional to the relative

TABLE 2.1

Mean Diffusion Coefficients at Room Temperature for a Series of Molecules under Confinement by the Activated Carbon LMA12 ($D_{sorbate}$) and Comparison with the Corresponding Bulk Data (D_{liquid})

	$D_{sorbate}$ $(10^{-9}\,m^2\,s^{-1})$	D_{liquid} $(10^{-9}\,m^2\,s^{-1})$	$D_{sorbate}/D_{liquid}$	Boiling Point (K)
Acetone	5.0	4.6	1.1	329.7
Ethanol	1.7	1.1	1.5	351.6
Toluene	0.6	2.3	0.26	384.2
n-Decane	0.24	1.4	0.17	447.2
Nitrobenzene	0.09	0.8	0.11	484.2

The ratio of the diffusivities and the boiling temperatures are also given.
Source: Krutyeva, M.F. et al., *Microporous Mesoporous Mater.*, 120, 91, 2009.

amount of molecules in the free space. In view of this tendency, one may conclude that the observed mass transfer corresponds mainly to molecular migration through the largest mesopores and macropores which, obviously, affect the overall mass transfer. One may also conclude that, under such conditions, diffusion in the micropores can only be recorded if the observation times may be kept notably smaller than the mean exchange time between the micropores and the remaining pore network. In this way, with an observation time of 10 ms, for ethanol in LMA12, the diffusivity in the micropores could be determined to be as small as 4×10^{-13} m^2 s^{-1} [153], i.e., by close to four orders of magnitude smaller than the overall diffusivity.

2.4 Conclusion

In parallel with the impressive increase in the spectrum of nanoporous materials over the last couple of years, experimental progress in the PFG technique of NMR and the introduction of IFM and IRM have led to the formation of a most versatile, powerful tool kit to probe these materials with respect to their transport properties. This information is in particular provided for diffusion path lengths from the sub-micrometer range up to fractions of millimeters and covers the range of particular relevance for the many technological applications of these materials in heterogeneous catalysis, adsorption, and separation.

Following a survey of the potential of these measuring techniques, we have provided examples of their application. The selected systems cover a large spectrum of substances, including micro- and mesoporous materials, with both ordered and random pore structures.

It could be demonstrated that the applied techniques of microscopic diffusion measurement were able to provide specific information about each of these materials. So far, however, this information was based on the application of either the PFG NMR method or the microscopic techniques. This situation is caused by the difference in the timescales of these techniques, typically milliseconds in PFG NMR and a few seconds for the microscopic techniques. These limitations are, however, not fundamental and, in the future, further development of the experimental techniques should lead to convergence of the ranges of measurement.

Further progress in the application of the techniques of diffusion measurement to the characterization of porous materials and, eventually, to their optimization with respect to the requirements of their technological use will clearly depend on the intensity by which these novel options are used by the community. We hope that this contribution will have contributed to this aim.

Acknowledgments

The examples of experimental evidence given in this chapter were attained within the frame of the INSIDE-POReS Network of Excellence, in particular in cooperation with Jürgen Caro, Pegie Cool, Marc-Olivier Coppens, Deborah Jones, Freek Kapteijn, Francisco Rodríguez-Reinoso, Michael Stöcker, Doros Theodorou, Etienne F. Vansant, and Jens Weitkamp, and are published as a review work in the Journal Chemical Engineering Technology [155], which we have taken as the basis of this chapter.

References

1. P. Heitjans and J. Kärger. *Diffusion in Condensed Matter: Methods, Materials, Models*. Springer, Berlin, Heidelberg, Germany, 2005.
2. N.Y. Chen, T.F. Degnan, and C.M. Smith. *Molecular Transport and Reaction in Zeolites*. VCH, New York, 1994.
3. D. Ruthven. The technological impact of diffusion in nanopores. In: *Leipzig, Einstein, Diffusion*, J. Kärger (Ed.), Leipziger Universitätsverlag, Leipzig, Germany, 2007, p. 123.
4. F. Schüth, K.S.W. Sing, and J. Weitkamp. *Handbook of Porous Solids*. Wiley-VCH, Weinheim, Germany, 2002.
5. G. Ertl, H. Knözinger, F. Schüth, and J. Weitkamp. *Handbook of Heterogeneous Catalysis*. Wiley-VCH, Weinheim, Germany, 2008.
6. H. Marsh and F. Rodríguez-Reinoso. *Activated Carbon*. Elsevier, Amsterdam, the Netherlands, 2006.
7. M.E. Davis. *Nature* 2002, *417*, 813.
8. A. Corma. Towards a rationalization of zeolite and zeolitic materials synthesis. In: *Recent Advances in the Science and Technology of Zeolites and Related Materials*, E. van Steen, M. Claeys, and L.H. Callanan (Eds.), Elsevier, Amsterdam, the Netherlands, 2004, p. 25.
9. H. Robson and K.P. Lillerud. *Verified Syntheses of Zeolitic Materials*. Elsevier, Amsterdam, the Netherlands, 2001.
10. E.F. Vansant. *Pore Size Engineering in Zeolites*. Wiley, New York, 1990.
11. G. Wang, E. Johannessen, C.R. Kleijn, S.W. de Leeuw, and M.O. Coppens. *Chem. Eng. Sci.* 2007, *62*, 5110.
12. R. Pitchumani, W.J. Li, and M.O. Coppens. *Catalysis Today* 2005, *105*, 618.
13. O. Sel, D. Kuang, M. Thommes, and B. Smarsly. *Langmuir* 2006, *22*, 2311.
14. H. Preising and D. Enke. *Chemie Ingenieur Technik* 2006, *78*, 1333.
15. J.-H. Smatt, C. Weidenthaler, J.B. Rosenholm, and M. Linden. *Chem. Mater.* 2006, *18*, 1443.
16. R. Ryoo, S.H. Joo, and S. Jun. *J. Phys. Chem. B* 1999, *103*, 7743.
17. J. Kärger. *Ind. Eng. Chem. Res.* 2002, *41*, 3335.
18. J. Kärger and D. Freude. *Chem. Eng. Technol.* 2002, *25*, 769.
19. R.M. Barrer. *Advan. Chem. Ser.* 1971, *102*, 1.

20. J. Kärger and J. Caro. *J. Chem. Soc. Faraday Trans. I* 1977, *73*, 1363.
21. J. Kärger and D.M. Ruthven. *J. Chem. Soc. Faraday Trans. I* 1981, *77*, 1485.
22. J. Kärger and H. Pfeifer. *Zeolites* 1987, *7*, 90.
23. H. Mehrer. Introduction and case studies in metals and binary alloys. In: *Diffusion in Condensed Matter: Methods, Materials, Models*, P. Heitjans and J. Kärger (Eds.), Springer, Berlin, Heidelberg, Germany, 2005, p. 3.
24. H. Jobic and D. Theodorou. *Microporous Mesoporous Mater.* 2007, *102*, 21.
25. H. Jobic. Investigation of diffusion in molecular sieves by neutron scattering techniques. In: *Adsorption and Diffusion*, H.G. Karge and J. Weitkamp (Eds.), Springer, Berlin, Heidelberg, Germany, 2008, p. 207.
26. D.N. Theodorou, R.Q. Snurr, and A.T. Bell. Molecular dynamics and diffusion in microporous materials. In: *Comprehensive Supramolecular Chemistry*, G. Alberti and T. Bein (Eds.), Pergamon, Oxford, U.K., 1996, p. 507.
27. S.M. Auerbach. *J. Chem. Phys.* 1997, *106*, 7810.
28. R. Haberlandt and J. Kärger. *Chem. Engin. J.* 1999, *74*, 15.
29. P. Demontis, L.A. Fenu, and G.B. Suffritti. *J. Phys. Chem. B* 2005, *109*, 18081.
30. E. Beerdsen, D. Dubbeldam, and B. Smit. *Phys. Rev. Lett.* 2006, *96*, Art. No. 044501.
31. D.M. Ruthven, *Zeolites* 1993, *13*, 594.
32. P. Grenier, F. Meunier, P.G. Gray, J. Kärger, Z. Xu, and D.M. Ruthven. *Zeolites* 1994, *14*, 242.
33. J. Kärger and D.M. Ruthven. *Zeolites* 1989, *9*, 267.
34. H. Jobic, J. Kärger, C. Krause, S. Brandani, A. Gunadi, A. Methivier, G. Ehlers, B. Farago, W. Haeussler, and D.M. Ruthven. *Adsorption* 2005, *11*, 403.
35. D.M. Ruthven, S. Brandani, and M. Eic. Measurement of diffusion in microporous solids by macroscopic methods. In: *Adsorption and Diffusion*, H.G. Karge and J. Weitkamp (Eds.), Springer, Berlin, Heidelberg, Germany, 2008, p. 45.
36. H. Jobic, W. Schmidt, C. Krause, and J. Kärger. *Microporous Mesoporous Mater.* 2006, *90*, 299.
37. V. Bourdin, S. Brandani, A. Gunadi, H. Jobic, C. Krause, J. Kärger, and W. Schmidt. *Diffusion Fundamentals (Online)* 2005, *2*, 83.
38. J. Kärger. Diffusion measurement by NMR techniques. In: *Adsorption and Diffusion*, H.G. Karge and J. Weitkamp (Eds.), Springer, Berlin, Heidelberg, Germany, 2008, p. 85.
39. S. Vasenkov and J. Kärger. *Microporous Mesoporous Mater.* 2002, *55*, 139.
40. S. Vasenkov, W. Böhlmann, P. Galvosas, O. Geier, H. Liu, and J. Kärger. *J. Phys. Chem. B* 2001, *105*, 5922.
41. A. Feldhoff, J. Caro, H. Jobic, C.B. Krause, P. Galvosas, and J. Kärger. *Chem. Phys. Chem.* 2009, *10*, 2429.
42. J. Kärger, P. Kortunov, S. Vasenkov, L. Heinke, D.B. Shah, R.A. Rakoczy, Y. Traa, and J. Weitkamp. *Angew. Chem. Int. Ed.* 2006, *45*, 7846.
43. L. Heinke, C. Chmelik, P. Kortunov, D.M. Ruthven, D.B. Shah, S. Vasenkov, and J. Kärger. *Chem. Eng. Technol.* 2007, *30*, 995.
44. J. Kärger and W. Heink. *J. Magn. Reson.* 1983, *51*, 1.
45. R.M. Cotts. *Nature* 1991, *351*, 443.
46. J. Kärger, H. Pfeifer, and W. Heink. *Adv. Magn. Reson.* 1988, *12*, 2.
47. P.T. Callaghan. *Principles of NMR Microscopy*. Clarendon Press, Oxford, U.K., 1991.
48. R. Kimmich. *NMR Tomography, Diffusometry, Relaxometry*. Springer, Berlin, Germany, 1997.

49. J. Kärger, H. Pfeifer, F. Stallmach, N.N. Feoktistova, and S.P. Zhdanov. *Zeolites* 1993, *13*, 50.
50. N.K. Bär, P.L. McDaniel, C.G. Coe, G. Seiffert, and J. Kärger. *Zeolites* 1997, *18*, 71.
51. J. Kärger, H. Pfeifer, S. Rudtsch, W. Heink, and U. Gross. *J. Fluorine Chem.* 1988, *39*, 349.
52. P.T. Callaghan, M.E. Komlosh, and M. Nyden. *J. Magn. Reson.* 1998, *133*, 177.
53. I.Y. Chang, F. Fujara, B. Geil, G. Hinze, H. Sillescu, and A. Tolle. *J. Non-Cryst. Solids* 1994, *172*, 674.
54. P. Galvosas, F. Stallmach, G. Seiffert, J. Kärger, U. Kaess, and G. Majer. *J. Magn. Reson.* 2001, *151*, 260.
55. F. Stallmach and P. Galvosas. *Ann. Rep. NMR Spectrosc.* 2007, *61*, 51.
56. J. Kärger, *Leipzig, Einstein, Diffusion*. Leipziger Universitätsverlag, Leipzig, Germany, 2007.
57. U. Hong, J. Kärger, R. Kramer, H. Pfeifer, G. Seiffert, U. Müller, K.K. Unger, H.B. Lück, and T. Ito. *Zeolites* 1991, *11*, 816.
58. F. Stallmach, J. Kärger, C. Krause, M. Jeschke, and U. Oberhagemann. *J. Am. Chem. Soc.* 2000, *122*, 9237.
59. J. Kärger and D.M. Ruthven. *Diffusion in Zeolites and Other Microporous Solids*. Wiley & Sons, New York, 1992.
60. J. Kärger. *AIChE J.* 1982, *28*, 417.
61. J. Crank. *The Mathematics of Diffusion*. Clarendon Press, Oxford, U.K., 1975.
62. M. Krutyeva, X. Yang, S. Vasenkov, and J. Kärger. *J. Magn. Reson.* 2007, *185*, 300.
63. M. Krutyeva, S. Vasenkov, X. Yang, J. Caro, and J. Kärger. *Microporous Mesoporous Mater.* 2007, *104*, 89.
64. F. Rittig, C.G. Coe, and J.M. Zielinski. *J. Am. Chem. Soc.* 2002, *124*, 5264.
65. F. Rittig, C.G. Coe, and J.M. Zielinski. *J. Phys. Chem. B* 2003, *107*, 4560.
66. S.S. Nirvarthi and A.V. McCormick. *J. Phys. Chem.* 1995, *99*, 4661.
67. U. Hong, J. Kärger, and H. Pfeifer. *J. Am. Chem. Soc.* 1991, *113*, 4812.
68. U. Hong, J. Kärger, B. Hunger, N.N. Feoktistova, and S.P. Zhdanov. *J. Catal.* 1992, *137*, 243.
69. H.B. Schwarz, S. Ernst, J. Kärger, B. Knorr, G. Seiffert, R.Q. Snurr, B. Staudte, and J. Weitkamp. *J. Catal.* 1997, *167*, 248.
70. H.B. Schwarz, H. Ernst, S. Ernst, J. Kärger, T. Röser, R.Q. Snurr, and J. Weitkamp. *Appl. Catal. A* 1995, *130*, 227.
71. A. Pampel, J. Kärger, and D. Michel. *Chem. Phys. Lett.* 2003, *379*, 555.
72. A. Pampel, M. Fernandez, D. Freude, and J. Kärger. *Chem. Phys. Lett.* 2005, *407*, 53.
73. A. Pampel, F. Engelke, P. Galvosas, C. Krause, F. Stallmach, D. Michel, and J. Kärger. *Microporous Mesoporous Mater.* 2006, *90*, 271.
74. M. Fernandez, J. Kärger, D. Freude, A. Pampel, J.M. van Baten, and R. Krishna. *Microporous Mesoporous Mater.* 2007, *105*, 124.
75. M. Fernandez, A. Pampel, R. Takahashi, S. Sato, D. Freude, and J. Kärger. *Phys. Chem. Chem. Phys.* 2008, *10*, 4165.
76. J.H. Strange, M. Rahman, and E.G. Smith. *Phys. Rev. Lett.* 1993, *71*, 3589.
77. R. Valiullin and I. Furo, *J. Chem. Phys.* 2002, *116*, 1072.
78. U. Schemmert, J. Kärger, C. Krause, R.A. Rakoczy, and J. Weitkamp. *Europhys. Lett.* 1999, *46*, 204.
79. U. Schemmert, J. Kärger, and J. Weitkamp. *Microporous Mesoporous Mater.* 1999, *32*, 101.

80. C. Chmelik, H. Bux, J. Caro, L. Heinke, F. Hibbe, T. Titze, and J. Kärger. *Phys. Rev. Lett.* 2010, 104, 085902.
81. P.R. Griffiths and J.A. des Hasth. *Fourier Transform Infrared Spectroscopy.* Wiley & Sons, New York, 1986.
82. H.G. Karge and J. Kärger. Application of IR spectroscopy, IR microscopy, and optical interference microscopy to diffusion in zeolites. In: *Adsorption and Diffusion*, H.G. Karge and J. Weitkamp (Eds.), Springer, Berlin, Heidelberg, Germany, 2008, p. 135.
83. Y. Roggo, A. Edmond, P. Chalus, and M. Ulmschneider. *Anal. Chim. Acta.* 2005, 535, 79.
84. R. Krishna. *Chem. Eng. Sci.* 1990, 45, 1779.
85. D. Paschek and R. Krishna. *Phys. Chem. Chem. Phys.* 2001, 3, 3185.
86. H. Jobic, J. Kärger, and M. Bee. *Phys. Rev. Lett.* 1999, 82, 4260.
87. L. Heinke and J. Kärger. *New J. Phys.* 2008, 10, 023035.
88. L. Heinke and J. Kärger. *J. Chem. Phys.* 2009, 130, 044707.
89. L. Heinke, P. Kortunov, D. Tzoulaki, M. Castro, P.A. Wright, and J. Kärger. *Europhys. Lett.* 2008, 81, 26002.
90. D.M. Ruthven and L.K. Lee. *AIChE J.* 1981, 27, 654.
91. L. Heinke, C. Chmelik, P. Kortunov, D.B. Shah, S. Brandani, D.M. Ruthven, and J. Kärger. *Microporous Mesoporous Mater.* 2007, 104, 18.
92. J. Weitkamp and L. Puppe. *Catalysis and Zeolites.* Springer, Berlin, Heidelberg, Germany, 1999.
93. M. Goddard and D.M. Ruthven. *Zeolites* 1986, 6, 445.
94. C. Berger, R. Gläser, R.A. Rakoczy, and J. Weitkamp. *Microporous Mesoporous Mater.* 2005, 83, 333.
95. P. Kortunov, S. Vasenkov, J. Kärger, R. Valiullin, P. Gottschalk, M.F. Elia, M. Perez, M. Stöcker, B. Drescher, G. McElhiney, C. Berger, R. Gläser, and J. Weitkamp. *J. Am. Chem. Soc.* 2005, 127, 13055.
96. M.L. Occelli and P. O'Connor. *Fluid Catalytic Cracking V: Materials and Technological Innovations.* Elsevier, Amsterdam, The Netherlands, 2001.
97. W. Heink, J. Kärger, H. Pfeifer, and F. Stallmach. *J. Am. Chem. Soc.* 1990, 112, 2175.
98. K. Ulrich, D. Freude, P. Galvosas, C. Krause, J. Kärger, J. Caro, P. Poladli, and H. Papp. *Microporous Mesoporous Mater.* 2009, 120, 98.
99. J. Kärger, W. Heink, H. Pfeifer, M. Rauscher, and J. Hoffmann. *Zeolites* 1982, 2, 275.
100. J. Kärger, M. Bülow, B.R. Millward, and J.M. Thomas. *Zeolites* 1986, 6, 146.
101. J. Kärger, H. Pfeifer, R. Richter, H. Fürtig, W. Roscher, and R. Seidel. *AICHE J.* 1988, 34, 1185.
102. A. Corma, F. Rey, J. Rius, M.J. Sabater, and S. Valencia. *Nature* 2004, 431, 287.
103. A. Corma, J. Kärger, and C. Krause. *Diffusion Fundamentals (Online)* 2005, 2, 87.
104. J. Kärger, *Microporous Mesoporous Mater.* 2008, 116, 715.
105. R.A. Rakoczy, Y. Traa, P. Kortunov, S. Vasenkov, J. Kärger, and J. Weitkamp. *Microporous Mesoporous Mater.* 2007, 104, 1195.
106. P. Kortunov, L. Heinke, S. Vasenkov, C. Chmelik, D.B. Shah, J. Kärger, R.A. Rakoczy, Y. Traa, and J. Weitkamp. *J. Phys. Chem. B* 2006, 110, 23821.
107. P. Kortunov, C. Chmelik, J. Kärger, R.A. Rakoczy, D.M. Ruthven, Y. Traa, S. Vasenkov, and J. Weitkamp. *Adsorption* 2005, 11, 235.
108. S. Khajavi, J.C. Jansen, and F. Kapteijn. *J. Membr. Sci.* 2009, 326, 153.
109. S. Khajavi, F. Kapteijn, and J.C. Jansen. *J. Membr. Sci.* 2007, 299, 63.

110. L. Heinke. *Mass Transfer in Nanoporous Materials: A Detailed Analysis of Transient Concentration Profiles.* Leipzig University, Leipzig, Germany, 2008.
111. S. Zschiegner, S. Russ, R. Valiullin, M.-O. Coppens, A.J. Dammers, A. Bunde, and J. Kärger. *Eur. Phys. J.* 2008, 109.
112. J. Kärger and P. Volkmer. *J. Chem. Soc. Faraday Trans. I* 1980, *76*, 1562.
113. S. Vasenkov, O. Geier, and J. Kärger. *Eur. Phys. J. E* 2003, *12*, 35.
114. V.N. Burganos. *J. Chem. Phys.* 1998, *109*, 6772.
115. B. Derjaguin. *C. R. (Dokl.) Acad. Sci. URSS* 1946, *7*, 623.
116. J.M. Zalc, S.C. Reyes, and E. Iglesia. *Chem. Eng. Sci.* 2004, *59*, 2947.
117. G.K. Papadopoulos, D.N. Theodorou, S. Vasenkov, and J. Kärger. *J. Chem. Phys.* 2007, *126*, 094702.
118. P. Kortunov, S. Vasenkov, J. Kärger, M.F. Elia, M. Perez, M. Stöcker, G.K. Papadopoulos, D. Theodorou, B. Drescher, G. McElhiney, B. Bernauer, V. Krystl, M. Kocirik, A. Zikanova, H. Jirglova, C. Berger, R. Gläser, J. Weitkamp, and E.W. Hansen. *Chem. Mater.* 2005, *17*, 2466.
119. A.M. Avila, C.M. Bidabehere, and U. Sedran. *Chem. Eng. J.* 2007, *132*, 67.
120. J. Kärger, *Chem. Eng. J.* 2009, *145*, 522.
121. A.M. Avila, C.M. Bidabehere, and U. Sedran. *Chem. Eng. J.* 2009, *145*, 525.
122. P. Kortunov, S. Vasenkov, J. Kärger, M.F. Elia, M. Perez, M. Stöcker, G.K. Papadopoulos, D. Theodorou, B. Drescher, G. McElhiney, B. Bernauer, V. Krystl, M. Kocirik, A. Zikanova, H. Jirglova, C. Berger, R. Gläser, J. Weitkamp, and E.W. Hansen. *Magn. Reson. Imaging* 2005, *23*, 233.
123. J. Kärger and S. Vasenkov. *Microporous Mesoporous Mater.* 2005, *85*, 195.
124. S. Vasenkov and P. Kortunov. *Diffusion Fundamentals* 2005, *1*, 2.1.
125. F. Stallmach and S. Crone. Analytical model for extrapolation of experimental NMR diffusion studies to reaction conditions for formulated catalyst particles. In: *Diffusion Fundamentals*, J. Kärger, F. Grinberg, and P. Heitjans (Eds.), Leipziger Universitätsverlag, Leipzig, Germany, 2005, p. 474.
126. H. Li, M. Eddaoudi, M. O'Keeffe, and O.M. Yaghi. *Nature* 1999, *402*, 276.
127. S. Kaskel, Porous metal-organic frameworks. In: *Handbook of Porous Solids*, F. Schüth, K.S.W. Sing, and J. Weitkamp (Eds.), Wiley-VCH, Weinheim, Germany, 2002, p. 1190.
128. L. Riekert. *Adv. Catal.* 1970, *21*, 281.
129. J. Kärger, M. Petzold, H. Pfeifer, S. Ernst, and J. Weitkamp. *J. Catal.* 1992, *136*, 283.
130. J. Kärger. Single-file diffusion in zeolites. In: *Molecular Sieves, Science and Technology: Adsorption and Diffusion*, H.G. Karge and J. Weitkamp (Eds.), Springer, Berlin, Germany, 2008, p. 329.
131. D.N. Dybtsev, H. Chun, S.H. Yoon, D. Kim, and K. Kim. *J. Am. Chem. Soc.* 2004, *126*, 32.
132. M. Arnold, P. Kortunov, D.J. Jones, Y. Nedellec, J. Kärger, and J. Caro. *Eur. J. Inorg. Chem.* 2007, 60.
133. P. Kortunov, L. Heinke, M. Arnold, Y. Nedellec, D.J. Jones, J. Caro, and J. Kärger. *J. Am. Chem. Soc.* 2007, *129*, 8041.
134. K.W. Kehr, K. Mussawisade, G.M. Schütz, and T. Wichmann. Diffusion of particles on lattices. In: *Diffusion in Condensed Matter—Methods, Materials, Models*, P. Heitjans and J. Kärger (Eds.), Springer, Berlin, Germany, 2005, p. 975.
135. C. Saravanan, F. Jousse, and S.M. Auerbach. *Phys. Rev. Lett.* 1998, *80*, 5754.
136. F.J. Keil, R. Krishna, and M.O. Coppens. *Rev. Chem. Eng.* 2000, *16*, 71.
137. H. Ramanan, S.M. Auerbach, and M. Tsapatsis. *J. Phys. Chem. B* 2004, *108*, 17171.

138. L. Pan, B. Parker, X. Huang, D.H. Olson, J.-Y. Lee, and J. Li. *J. Am. Chem. Soc.* 2006, *128*, 4180.
139. L. Heinke, D. Tzoulaki, C. Chmelik, F. Hibbe, J. van Baten, H. Lim, J. Li, R. Krishna, and J. Kärger. *Phys. Rev. Lett.* 2009, *102*, 065901.
140. D. Tzoulaki, L. Heinke, J. Li, H. Lim, D. Olson, J. Caro, R. Krishna, C. Chmelik, and J. Kärger. *Angew. Chem.* 2009, *48*, 3525.
141. H.L. Frisch. *J. Phys. Chem.* 1958, *62*, 401.
142. D.M. Ruthven. *Chem. Eng. Sci.* 2007, *62*, 5745.
143. D.Y. Zhao, Q.S. Huo, J.L. Feng, B.F. Chmelka, and G.D. Stucky. *J. Am. Chem. Soc.* 1998, *120*, 6024.
144. R. Valiullin and I. Furo. *Phys. Rev. E.* 2002, *6603*, 1508.
145. J. Kärger, R. Valiullin, and S. Vasenkov. *New J. Phys.* 2005, *7*, 15.
146. R. Valiullin, P. Kortunov, J. Kärger, and V. Timoshenko. *J. Phys. Chem. B* 2005, *109*, 5746.
147. R. Valiullin, S. Naumov, P. Galvosas, J. Kärger, H.-J. Woo, F. Porcheron, and P.A. Monson. *Nature* 2006, *430*, 965.
148. S. Naumov, R. Valiullin, J. Karger, R. Pitchumani, and M.O. Coppens. *Microporous Mesoporous Mater.* 2008, *110*, 37.
149. J. Kärger, F. Stallmach, R. Valiullin, and S. Vasenkov. Diffusion in nanoporous materials. In: *NMR Imaging in Chemical Engineering*, S. Stapf and S.-I. Han (Eds.), Wiley-VCH, Weinheim, Germany, 2006, p. 231.
150. V. Meynen, P. Cool, E.F. Vansant, P. Kortunov, F. Grinberg, J. Kärger, M. Mertens, O.I. Lebedev, and G. Van Tendeloo. *Microporous Mesoporous Mater.* 2007, *99*, 14.
151. J. Roziere, M. Brandhorst, R. Dutartre, M. Jacquin, D. Jones, P. Vitse, and J. Zajac. *Mater. Chem.* 2001, *11*, 3264.
152. M. Krutyeva, F. Grinberg, J. Kärger, and D. Jones, *Microporous Mesoporous Mater.* 2009, *120*, 104.
153. M. Krutyeva, F. Grinberg, F. Futardo, P. Galvosas, J. Kärger, A. Silvestre-Albero, A. Sepúlveda-Escribano, J. Silvestre-Albero, and F. Rodríguez-Reinoso. *Microporous Mesoporous Mater.* 2009, *120*, 91.
154. J.M. Juárez-Galán, A. Silvestre-Albero, J. Silvestre-Albero, and F. Rodríguez-Reinoso. *Microporous Mesoporous Mater.* 2009, *117*, 519.
155. J. Kärger, J. Caro, P. Cool, M.O. Coppens, D. Jones, F. Kapteijn, F. Rodríguez-Reinoso, M. Stöcker, D. Theodorou, E.F. Vansant, and J. Weitkamp. *Chem. Eng. Technol.* 2009, *32*, 1494.

3

Nanoscale Microscopies

Anthony A.G. Tomlinson

CONTENTS

3.1 Introduction

The term "microscopy" immediately brings to our mind the optical or electron microscope. The former, the oldest type of microscope, allows magnifications up to ×10,000, using glass lenses, while the latter, using magnetic fields of special coils to focus electrons, allows magnifications up to ×1,000,000. In the early 1980s, the microscope evolution gave rise to the atomic force microscope, which is capable of magnifying up to $\times 1 \times 10^6$ in horizontal x- and y-planes and a vertical z-plane (overcoming image two-dimensionality). Parallel developments and synergies of instrumentation, image processing and analysis (all), tip technologies (especially AFM), and combined microscopy/spectroscopy (especially combining confocal methodologies) have been breakneck since 2000. Consequently, such a short introduction cannot but be selective, we hope that no injustice has been done to any microscopy, yet the flavor and place of each in nanoscience are apparent.

3.2 Fundamentals

3.2.1 Electron Microscopy

Figure 3.1 compares a light microscope with a transmission electron microscopy (TEM) and scanning electron microscopy (SEM). In a light microscope, a beam is condensed so as to illuminate a particular area of a specimen and then passes through an objective lens to be finally projected for visual observation. By analogy, in a TEM, the "illumination" source is an electron emitter, which is then condensed to fall on the specimen, to be transmitted through a projection lens and ultimately onto an image plane, i.e., a fluorescent screen (the "eye" as it were; in modern TEMs and SEMs: charge-coupled devices [CCDs] that have undergone a veritable revolution since 2000).

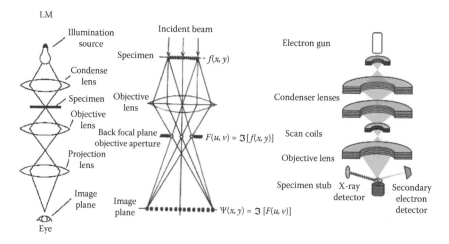

FIGURE 3.1
Comparison between a light microscope (left), a transmission electron microscope (center), and a scanning electron microscope (right). (Amelinckx, S., Van Dyck, D., Van Landuyt, J., and Van Tandeloo, G. (eds.), *Handbook of Microscopy*, VCH, Weinheim, Germany, 1997. Copyright Wiley-VCH Verlag GmbH & Co. KGaA.)

By contrast, as seen in the figure, SEM is not an optical instrument because there are no image-forming lenses; it instead utilizes electron optics (the condenser lenses 1–3) to focus the electron beam onto the specimen under observation. The image is interrogated with a probe-forming signal detection device—and optical analogies are unhelpful.

3.2.1.1 Electron Optics: The Basics

Refraction of a beam (i.e., a wave) occurs when it enters a medium having a different optical density (OD). In light optics, this is achieved when a light wave moves from air into glass. In an electron microscope, however, there is a vacuum that has an OD much lower (1.0) than that of glass. Since the electron beam cannot enter a conventional lens having a different OD, a force must be applied, which also causes the electron "illumination" beam to bend. The force may be applied either electromagnetically or electrostatically. Before describing these in detail, we recall that in light optics, the refractive index (μ) *changes abruptly* at a surface and remains constant between surfaces, which basically allows imaging lenses (i.e., glass) to be constructed. By contrast, in electron optics *changes in refractive index occur gradually* so that rays are continuous curves—not straight lines. Refraction of electrons must be accomplished using fields in space around charged electrodes or solenoids. These fields can assume only certain distributions consistent with field theory.

We begin with the simpler electrostatic lens, the principle of which is shown in Figure 3.2.

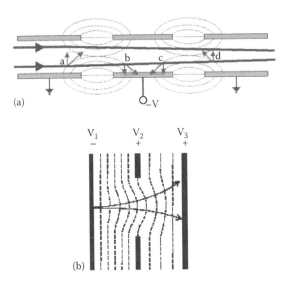

FIGURE 3.2
(a) Basic principle of the electrostatic lens. (b) Beam form. (From Salmeron, M.B., Use of the atomic force microscope to study mechanical properties of lubricant layers, *MRS Bull.*, 18(5), 20, May 1993.)

As illustrated in Figure 3.3, an electromagnetic lens is more complex. The left-hand side shows that it consists of a single wire wrapped many times around circular pole pieces. The latter, made of soft iron, are usually separate from the iron shroud and concentrate the magnetic field of the entire lens into a very small region. Apart from the wire coil and pole pieces, an electromagnetic lens also has plumbing for cooling water, which removes heat generated by the poles.

Due to the magnetic field, the electrons follow a helical trajectory, which converges at a fine focal point after it emerges from the lens (DC-powered magnets behave similar to converging glass lenses). The field strength determines the focal length, which varies with

$$(\text{focal length}) \ f = K\left(\frac{V}{i^2}\right)$$

where
  K is the constant, based on the number of turns of lens coil wire and the geometry of the lens
  V is the accelerating voltage
  i is the milliamps of current that passes through the coil

Focus and magnification of the electron beam are achieved via potentiometer controls, which vary the current to the various lenses.

The magnetic field strength is in turn determined by how much wire is wrapped around the poles and the current passing through it (as in an ordinary electromagnet). The larger the current, the stronger the lens (i.e., α_g in Figure 3.4 increases) and, consequently, the shorter the focal length. Note, however, that some of the electrons are scattered electrons that do not actually strike it but appear merely as dark areas on the final TEM image (Figure 3.4, right-hand side). Worse, some will be only partially scattered and will reach the screen in an anomalous position; i.e., they will give rise to a false signal, hence degrading the image. These forward scattered electrons are eliminated by placing an aperture below the specimen, to give the final basic electromagnetic lens design, shown only schematically in Figure 3.4—a very strong lens with a very short focal length.

Both specimen and chosen apertures are positioned in the electron beam (i.e., inserted into the lens) each on a stage fitted with a rod. Nonmagnetic materials, such as copper, brass, or platinum, are used.

Aperture characteristics and lens defects are, therefore, as important as those of light optics (if not more so). For example, although a small opening has the advantage of stopping scattered electrons and hence enhancing image contrast, it also drastically reduces the project lens α angle, which lowers image resolution (Figures 3.4 and 3.5). Figure 3.5 illustrates how different beam wavelengths will be focused at different positions when the focal length of the lens changes.

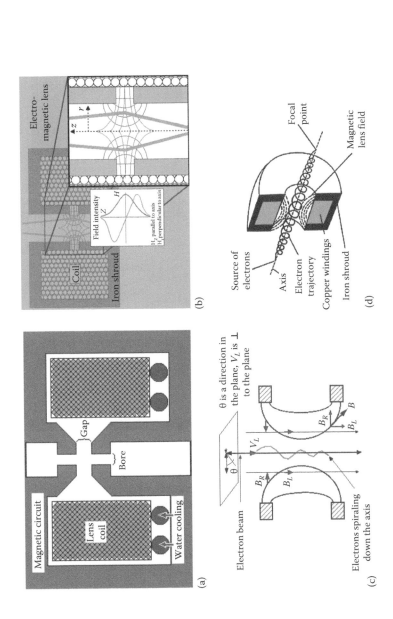

FIGURE 3.3

The electromagnetic lens. (a) Basic structure. (b) Magnetic lines of force between the poles, with inset field intensity // and ⊥ to the beam axis. The B field has two components: B_L = longitudinal component (down the axis) and B_R = radial component (⊥ to the axis). (c) Effect of magnetic field on e-beam. (d) Illustrating the helical trajectory. (Williams, D.B. and Carter, C.B., *Transmission Electron Microscopy: A Textbook for Materials Science*, Plenum Press, New York, 1996. Copyright Wiley-VCH Verlag GmbH & Co. KGaA.)

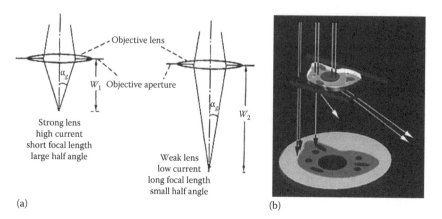

FIGURE 3.4

(a) Strong and weak lenses. (b) Illustrating the effect of scattered light. (From Salmeron, M.B., Use of the atomic force microscope to study mechanical properties of lubricant layers, *MRS Bull.*, 18(5), 20, May 1993.)

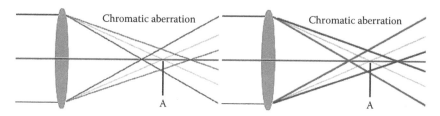

FIGURE 3.5

Illustrating chromatic aberration.

In light optics, this scattering is referred to as chromatic aberration of a lens and it results in fringes around the image because a zone of focus is present. This leads to an important difference between an electron microscope and a light microscope. In light optics, higher-energy wavelengths are of course bent more strongly than lower-energy ones (remember the prism) and therefore have smaller f. In electron optics, the opposite is the case: higher-energy wavelengths are *less* affected than lower-energy ones and give rise to *higher f*. Chromatic aberration can be corrected by simply using a monochromatic beam, which in an electron microscope is achieved by having a very stable electron acceleration voltage. In some electron microscopes on the market, an electromagnetic (i.e., converging) lens is combined with an electrostatic (i.e., diverging) lens to create an achromatic lens. (This is analogous to doublet lenses used in light optics.) There are, however, several other important lens defects to be taken into account before we pass on to the images provided by TEM.

First, electrons can be differentially scattered within the sample itself—some being slowed down more than others; a polychromatic beam is

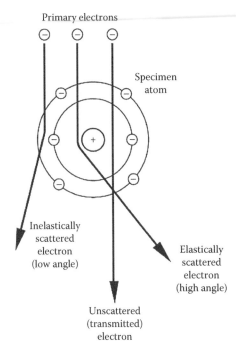

Primary electrons

Specimen atom

Inelastically scattered electron (low angle)

Elastically scattered electron (high angle)

Unscattered (transmitted) electron

FIGURE 3.6
Elastically and inelastically scattered electrons. (Amelinckx, S., Van Dyck, D., Van Landuyt, J., and Van Tandeloo, G. (eds.), *Handbook of Microscopy*, VCH, Weinheim, Germany, 1997. Copyright Wiley-VCH Verlag GmbH & Co. KGaA.)

produced from a monochromatic one. This type of chromatic aberration is more serious at the edges of the lens and is reduced with increase in image contrast, by placing an aperture immediately after the specimen. Spherical aberration also derives from the fact that waves enter and leave the lens field at differing angles arriving at differing f (Figures 3.5 and 3.6).

Again, a small aperture cutting off the worst part of the lens is the best way to reduce this defect. Different means are used when a lens is not completely symmetrical. Focal planes are now produced, which give rise to astigmatic images, i.e., the image is elongated in a plane xy with respect to xz (see Figure 3.7).

A ring of electromagnets (a "stigmator") is positioned around the beam, their strength and polarity being adjusted so that the beam becomes more perfectly circular in cross section. Finally, and especially recalling the importance of apertures, we come to diffraction that can always occur at the edges of an aperture, where it spreads out the beam's focal point instead of concentrating it. Although very important, being the basis of electron diffraction using TEM, for reasons of space this will not be treated here; an excellent introduction can be found in Ref. [2].

We have insisted on electron beam manipulation through lenses and defect corrections not only to bring out analogies with light absorption, but, more importantly, to underline the differences between them. The first is, of course, the source, which is basically a means for providing the electron

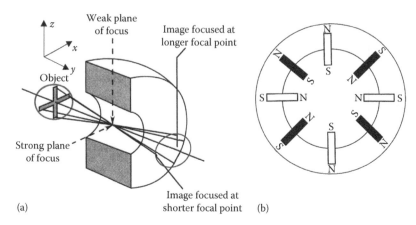

FIGURE 3.7
(a) Origin of astigmatic distortion. (b) Principle of a stigmator. (Amelinckx, S., Van Dyck, D., Van Landuyt, J., and Van Tandeloo, G. (eds.), *Handbook of Microscopy*, VCH, Weinheim, Germany, 1997. Copyright Wiley-VCH Verlag GmbH & Co. KGaA.)

with energy to overcome the work function, defined as the energy (or work) required to eject an electron from a metal surface (Figure 3.8a). For filament sources, increasing the filament current increases the beam current but only up to a saturation point (Figure 3.8b).

Electron sources (usually mounted at the top of EMs) are of many kinds and are summarized in Table 3.1.

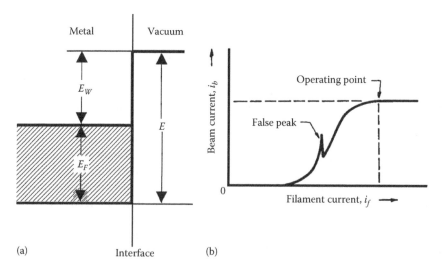

FIGURE 3.8
Principle of electron sources. (a) The work function. (b) Basic operational characteristics of an electron source. (From http://www4.nau.edu/microanalysis/Microprobe-SEM/Instrumentation.html. With permission.)

TABLE 3.1

Electron Sources Used in TEM (See Resources for Detailed Characteristics Including Those for SEM)

Electron Source	Design	Functioning
Thermionic emitters	(a)	Utilize heat to overcome the work function of a material, e.g., tungsten wire bent in a loop.
LaB6 cathodes	(b)	Heat is applied via separate resistance wire or ceramic mounts (the filament current is separate from the heating current).
Electron gun	(c)	A W filament generates an electron cloud which passes through a cathode-anode array to give a highly coherent e-beam
Field emitter	(d)	A single oriented tungsten crystal etched to a fine tip. Depends on the emission of electrons that are stripped from the W atoms by a high electric field
Field emission gun	(e)	Can be "cold" or thermally assisted to overcome the work function; a high-voltage field of 3 keV is required

Most recent electron microscopes no longer have LaB_6-type filaments but field emission sources, which provide a more intense and also more coherent beam. The electron source is configured with heating supply, bias resistor, and anode plate; particularly important is that the source must be centered so that there is no lateral displacement (or tilt) with respect to the anode plates. TEM microscope design configurations may be found in Ref. [3] as well as (animated) tutorials to illustrate operation under Resources.

3.2.1.1.1 Imaging Modes

There are two imaging modes in TEM—bright field and dark field. In the bright-field mode an aperture is placed in the back focal plane of the objective lens, which allows only the direct beam to interact with the specimen, whereas in dark-field mode the imaging beam is a diffracted beam and may be either on- or off-axis (see Figure 3.9).

The imaging modes have been particularly useful in biology for following the surface topology of specimens [4].

3.2.2 High-Resolution TEM

We now turn to high-resolution TEM (HRTEM) [1], where differently than in conventional microscopy, amplitudes (i.e., adsorption by the sample) are not used for image formation. Instead, contrast arises from interference of the electron wave with itself and within the image plane. Since phases of these waves are not directly recordable, in general, the amplitude resulting from this interference is measured. The phase of the electron wave still carries the information about the sample and generates contrast in the image, hence the name *phase-contrast imaging*. This is true, however, only when the sample is so thin that amplitude variations affect the image only slightly (the "weak-phase object approximation," or WPOA). Electron-wave interaction with the crystal structure of the sample is not yet fully understood; fortunately, a qualitative understanding of the interaction suffices. The electron arriving at the sample can be approximated as a plane wave incident on the sample surface. Penetrating the sample, it is then attracted by the positive atomic potentials of the atom cores, channeling along the atom columns of the lattice (s-state model). Simultaneously, interaction between the electron wave in different atom columns leads to Bragg diffraction. Following the interaction with the sample, the exit wave, immediately below the sample $\phi_e(\chi, u)$, as a function of the spatial coordinate χ, is a superposition of a plane wave and myriad diffracted beams having differing in-plane spatial frequencies u (high spatial frequencies correspond to large distances from the optical axis). The phase change of $\phi_e(\chi, u)$ compared with the incident wave now gives peaks at the atom columns. The exit wave finally passes through the microscope's imaging system (usually a CCD). Note that the image recorded is not a direct representation of the crystal structure of the sample; the relationship of the exit wave with the image wave is highly nonlinear and depends on the

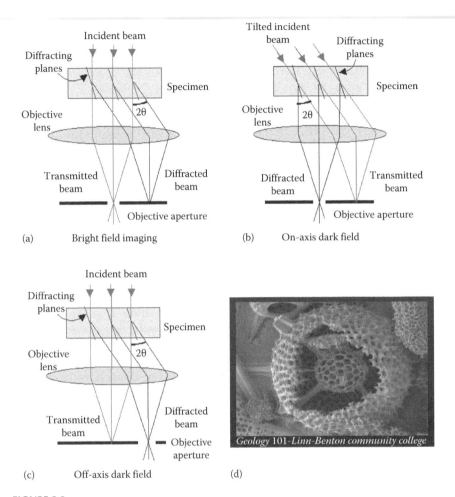

FIGURE 3.9
TEM imaging modes. (a) Bright field imaging. (b) On-axis dark field imaging. (c) Off-axis dark field imaging. (d) A radiolarian taken in BF mode. (Spence, J.C.H., *High-Resolution Electron Microscopy*, 3rd edn., 2003 by permission from Oxford University Press.)

microscope's aberrations. It is described by the *contrast transfer function* (ctf), which assuming that WPOA holds (thin sample) is given by [3]

$$ctv(\upsilon) = A(\upsilon)E(\upsilon)\sin\big(\chi(\upsilon)\big)$$

where
$A(\upsilon)$ is the aperture function
$E(\upsilon)$ describes the wave attenuation for higher spatial frequency υ (this is also referred to as the envelope function)
$\chi(\upsilon)$ is a function of the aberrations of the electron optical system

The last sinusoidal form of ctf determines the sign possessed by components of frequency υ and will enter contrast in the final image. Taking into account spherical aberration to the third order alone and defocus, χ is rotationally symmetric about the microscope optical axis and therefore depends only on $\upsilon = |\upsilon|$ given by

$$\chi(\upsilon) = \frac{\pi}{2} C_s \lambda^3 \upsilon^4 - \pi \Delta f \lambda \upsilon^2$$

where
  C_s is the spherical aberration coefficient
  λ is the electron wavelength
  Δf is the defocus

Defocus is easily controlled and measured to high precision in TEM, so the ctf can be easily altered; differently than in optical microscopy, defocusing increases the precision (and therefore the interpretation of images).

E cuts off beams, scattered above a certain critical angle (given by, e.g., the objective pole piece) and limits attainable resolution. It is the envelope function $E(\upsilon)$ that dampens the signal beams scattered at high angle, thus imposing a maximum to the transmitted spatial frequency. This maximum determines the highest resolution attainable with a microscope (the "information limit") and can be described as the product of single envelopes:

$$E(\upsilon) = E_s(\upsilon)E_c(\upsilon)E_d(\upsilon)E_v(\upsilon)E_D(\upsilon)$$

where
  $E_s(\upsilon)$ is the angular spread of source
  $E_c(\upsilon)$ is the chromatic aberration
  $E_d(\upsilon)$ is the specimen drift
  $E_v(\upsilon)$ is the specimen vibration
  $E_D(\upsilon)$ is the detector

Specimen drift and vibration are minimized easily in a suitable working environment. Instead, the spherical aberration C_s limits spatial coherence and defines $E_s(\upsilon)$ and the chromatic aberration, together with current and voltage instabilities, which define the temporal coherence in $E_c(\upsilon)$.

Taking the probe to have a Gaussian electron density distribution, the spatial envelope function is given by

$$E_s(\upsilon) = \exp\left[-\left(\frac{\pi\alpha}{\lambda}\right)^2\left(\frac{\delta\chi(\upsilon)}{\delta\upsilon}\right)^2\right] = \exp\left[-\left(\frac{\pi\alpha}{\lambda}\right)^2\left(C_s\lambda^3\upsilon^3 + \Delta f \lambda \upsilon\right)^2\right]$$

$$E_c(\upsilon) = \exp\left[-\frac{1}{2}(\pi\lambda\delta)^2\upsilon^4\right]$$

where δ is the focus spread due to the chromatic aberration C_c

$$\delta = C_c\sqrt{4\left(\frac{\Delta I_{obj}}{I_{obj}}\right)^2 + \left(\frac{\Delta E}{V_{acc}}\right)^2 + \left(\frac{\Delta V_{acc}}{V_{acc}}\right)^2}$$

The terms $\Delta I_{obj}/I_{obj}$ and $\Delta V_{acc}/V_{acc}$ represent instabilities in the currents in the objective lens and the high-voltage supply of the electron gun. $\Delta E/V_{acc}$ is the energy spread. Defocusing is thus an important part of any TEM "sitting." Choosing the right defocus value Δf, $\chi(\upsilon)$ flattens, creating a wide band, where low spatial frequencies are transferred into image intensity with a similar phase.

For example, in Gaussian focus, the defocus is set to zero (the sample is in focus). In Scherzer defocus, terms in υ^4 with the parabolic term $\Delta f\upsilon^2$ of $\chi(\upsilon)$—ADD.

However, until the 1990s the problem of spherical aberration (briefly mentioned above) remained unsolved and resolution remained low. This problem has recently been solved by a combination of instrumental modifications (introducing entrance and exit hexapoles and substituting the TEM lenses by "transfer doublet lenses," see Refs [1,4] for details) and imaging and filtering techniques. Since noise filtering is particularly important for materials science problems, we consider it [5] in some detail.

3.2.2.1 Noise Filtering

The FT (F_0) of an observed image may be expressed as the sum of a signal F_c originating from the periodic part (the crystals) and a broad background F_b from the nonperiodic part (the amorphous materials), and thus $F_0 = F_c + F_b$. The signal F_c is confined to diffraction-like spots and is usually strong, while the background F_b distributes more evenly. Thus, if the image contains a single domain of a crystal, it is easy to select F_c by using a mask that is made up of a set of periodic holes. However, this simple periodic masking does not work for a nonideal crystal.

The Wiener filter uses an adaptive mask M_w, which is obtained by seeking a solution that minimizes the summed square difference between the signal F_c and its estimate $\hat{F}_w$:

$$\sum\left|\hat{F}_w - F_c\right|^2 = \sum\left|M_w F_0 - F_c\right|^2 \Rightarrow \text{ minimum}$$

where we approximate the estimate with a product of a mask M_w and the FT of an observed image: $M_w F_0$.

When we assume that the signal and noise are uncorrelated, the appropriate solution for M_w is given by

$$M_w = \frac{|F_c|^2}{|F_0|^2} \approx \frac{\left(|F_0|^2 - |F_b|^2\right)}{|F_0|^2}$$

where we use $|F_0|^2 \approx (|F_c|^2 + |F_b|^2)$.

Thus, we have an applicable form of the Wiener filter given by

$$\hat{F}_w = M_w F_0 \approx \frac{|F_0|^2 - |\hat{F}_b|^2}{|F_0|^2} F_0 = \left[1 - \frac{|\hat{F}_b|^2}{|F_0|^2} \right] F_0$$

where $|\hat{F}_b|^2$ is an estimate of a nonperiodic background. Here, M_w, and thus $\hat{F}_w$, is set to zero, if $|F_0| - |\hat{F}_b| \leq 0$, in other words, when the estimated background is higher than the observed value [5].

3.2.2.2 Variants

The past 10 years has seen a blooming of TEM variations, especially instrumental. These include scanning TEM (STEM) in which the scan coils in some TEMs allow them to be used in scanning. A conical electron beam is focused through the specimen by lens in front of it. Images are formed by scanning the probe across the specimen (by double-deflection shift coils) and detecting the transmitted electrons. These are either on the optic axis, forming a bright-field image, or have been scattered to high angles, to form a dark-field image [6].

Turning to SEM, this is used mainly for solid specimens and has a much lower resolution than TEM, ca 10 nm (unless field emission SEM is used). However, the ca 20,000 magnification gives images a high depth of field, and often a 3D effect, despite much less image processing than TEM. A raster scans the specimen (as in STM and AFM) with a primary electron beam, and the secondary electrons coming from the specimen are detected (x-rays produced by the specimen can also be detected) [7].* There are considerable differences in sample preparation between TEM and SEM. For HRTEM, the requirements of very thin specimens, often the need for tomography to obtain sections extensive use of filters, etc., is a science in itself (a figure is often quoted of only ca 1 cm^3 of material(s) having been investigated since its invention). Conversely, for SEM, specimens are usually plated with a fine coating

---

* The most useful single general reference for SEM: good on specimen preparation.

of metal or englobed in a plastic, i.e., sample preparation is much simpler (see Resources) explaining the ubiquitous use of SEM in biology and medicine.

3.2.3 Image Processing and Analysis

All modern TEMs, SEMs, and for that matter SPMs usually are provided by suppliers with image analysis systems, making an overall scheme impossible. However, the major stages in all image processing and analysis can perhaps be summarized as in Figure 3.10.

The stages can be summarized as follows [4,8]:

- Delimit the area of interest and establish basic mathematical transformation (FT, etc.) over that area.
- Merge data (more important in SEM—to increase focus depth—and confocal microscopy—to obtain 3D images).
- Repeat measurements (if necessary) to eliminate artifacts.

3.2.4 Scanning Probe Microscopies

When a fine tip is brought sufficiently close (a few Å) to a sample surface, the wave functions of the electrons of the tip and those of the surface overlap. By applying a voltage between the tip and the sample, a tunneling current is established through the vacuum barrier, as illustrated in Figure 3.11.

The intensity of this current is a measure of the overlap between these two wave functions and depends very strongly on the tip–sample distance. By analogy with the 1D tunneling problem, it is possible to estimate the tunneling current I and its dependence on the tip–sample distance d. In the limit of low temperature and low bias voltage, V, one gets

$$I = V \cdot C(V)^{\exp}(-2k_0 s)$$

where
$C(V)$ is the factor accounting for deviation from Ohm's law
k_0 is the inverse decay length of electrons in a vacuum

$$k_0 = \left[\left(\frac{2m}{h^2} \right) f \right]^{1/2}$$

where
f is the work function
m is the mass of the electron

In the above equation, f is the height of the potential barrier associated with the vacuum gap. For a typical value $f = 4\text{eV}$, the previous equation

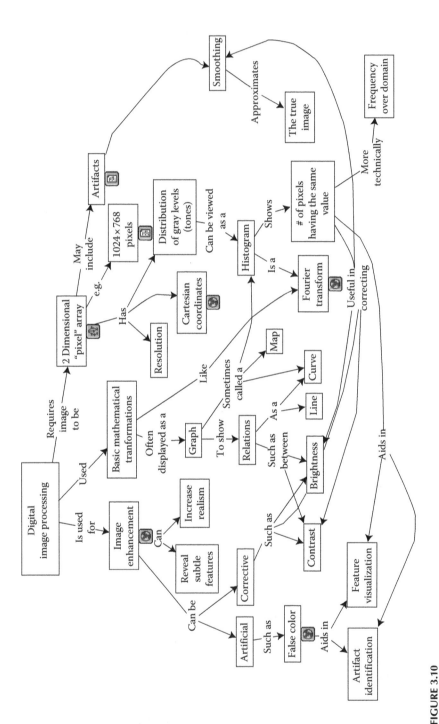

FIGURE 3.10
Basic stages of image processing and analysis. (With permission from CRC Press and J. Wiley & Sons.)

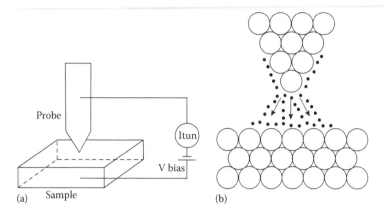

FIGURE 3.11
(a) The general principle of STM. (b) Illustrates the fact that it depends on interaction between tip and substrate orbitals. (From Magonov, S.N., Surface characterization of materials at ambient conditions by scanning tunneling microscopy (STM) and atomic force microscopy (AFM), *Appl. Spectrosc. Rev.*, 28(1-2), 1, 1993.)

predicts that the tunneling decreases by an order of magnitude when the gap is increased by only 1 Å. This extremely strong dependence on distance is the basis of the STM.

This can be schematized as shown in Figure 3.12. "Scanning" derives from the fact that the tip is moved backward and forward (or "rastered") systematically over the surface to give a map. After scanning over the surface, the image obtained consists of a $Z(X, Y)$ map of the Z position of the tip versus its lateral position (X, Y). The simplest scheme for plotting an image is to draw the Z movement of the tip on a recorder table while it is scanned along a series of parallel lines of the surface.

Historically, the basic current was used in most studies. In this mode, a feedback circuit measures the tunneling current I and adjusts the Z position

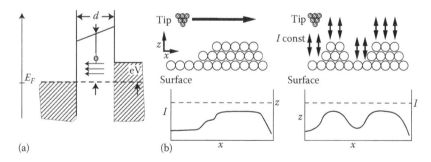

FIGURE 3.12
Left: Energy levels involved in STM; Right: Modes of operation of STM. (a) constant current mode; (b) constant height mode. (From Roth, E., Humbert, A., and Salvan, F., (eds.), Scanning tunneling microscopy and spectroscopy, in *Proceedings of the EUROANALYSIS VI Conference—Reviews on Analytical Chemistry*, pp. 93–116.)

of the tip (normal to the mean surface) continuously in order to keep this current constant and equal to a reference value, I_{ref}. This is referred to as the *constant current mode* and is shown in Figure 3.12a. A raster scan is taken in the x-direction and the images collected (below in Figure 3.12b) reflect an electron tunneling current map (left ordinate) for x, y positions for this fixed height above the conductive surface (right ordinate). Alternatively, the tip can be scanned over the surface at nearly constant z position, while the current is monitored, which is referred to as the *constant height mode* (see Figure 3.12a). The tip is moved up and down in the z direction to match the measured tunneling current to that set by the user. Repeating this procedure at each x, y position produces a topographic map of the surface consisting of the tip height z (left ordinate) required to equalize the measured and set tunneling currents I (right ordinate).

Images can also be displayed in many other ways. For example, the darkness at each point (X, Y) in the image can be determined by the height producing a so-called grayscale image. In this case, the feedback network keeps the average current constant, and the rapid variations in current as the tip passes over the surface now feature as atoms plotted versus the scan position (X, Y). The image again consists of multiple scans displaced laterally from each other in the Y direction (i.e., perpendicular to the scan direction X).

The basic current mode has the advantage that it can be used especially to track surfaces that are not atomically flat, whereas the constant height mode allows for much faster imaging of atomically flat surfaces.

3.2.4.1 Atomic Force Microscopy

The interatomic force between point-like objects or atoms is usually described by the Lennard–Jones potential:

$$U(r) = -U_0 \left[\left(\frac{r_0}{z} \right)^{12} - \left(\frac{r_0}{z} \right)^6 \right]$$

where
  z is the distance between atoms
  U_0 and r_0 are the energy and the distance between atoms in equilibrium, respectively

Interatomic force interactions change their character from attractive to repulsive when the distance between atoms becomes smaller than r_0. When the outermost atoms of the probing tip come close to the atoms of a sample, an overlap of electronic clouds causes the appearance of a strong repulsive force, and that force is proportional to the total electron density. Dispersive long-range (van der Waals) forces are responsible for the attractive part of the L–J

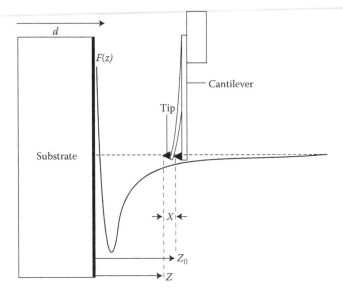

FIGURE 3.13
Principle of AFM. (From Salmeron, M.B., Use of the atomic force microscope to study mechanical properties of lubricant layers, *MRS Bull.*, 20, May 1993.)

potential, and attractive forces can also be induced by Coulomb and dipole interactions.

The force, calculated from the deflections of a cantilever, is plotted against the displacement (distance) of a piezo-drive working in a cyclic regime. It is first moved up vertically to bring the sample into contact with the tip, and then moved to withdraw the sample from the tip. During this engagement and when the tip is close to the sample, an attractive force is measured. The tip attraction to the surface is enhanced by any adsorbed thin liquid layers on the approaching surface. On further motion, the apex of the tip comes into contact with the surface and the net interaction then becomes repulsive (point 0,0 in Figure 3.13). The situation is different when the sample is moved in the opposite direction.

During withdrawal of the sample, the tip is stuck to the sample when the latter is situated at an even higher vertical position than that at which the tip was pulled into the surface during the approach. The difference in the two situations produces a well-defined hysteresis in the force versus distance curve, also shown in Figure 3.13. A pronounced pull-out force—which can be used practically, as discussed below—is detected before the tip and the sample is disconnected. Imaging of surfaces can be realized at any point along the A–A line in the force curve by pre-adjustment of the photodetector signal, which can be changed during scanning (Figure 3.14).

Most modern AFMs are configured to have integrated optical spectroscopy capabilities; an example is illustrated in Figure 3.15.

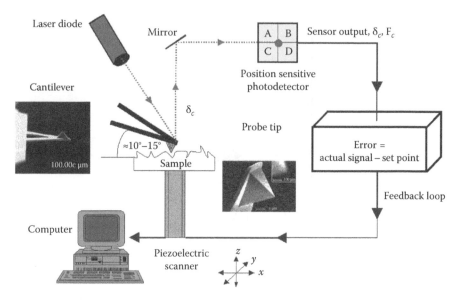

FIGURE 3.14
Typical AFM instrument layout. (From Spence, J.C.H., *High-Resolution Electron Microscopy*, 3rd edn., Oxford University Press, New York, 2003.)

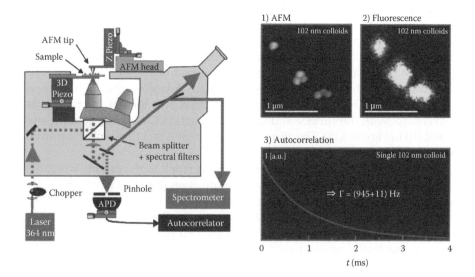

FIGURE 3.15
A combined optical microscope/AFM. (Poole, K., *Imaging Microscopy*, 2007, 2008, 8, 43. Copyright Wiley-VCH Verlag GmbH & Co. KGaA. With permission.)

3.2.4.2 Confocal Microscopy (LSCM—Laser Scanning Confocal Microscopy)

The term "confocal" means "having the same focus" that is accomplished by focusing a condenser lens to the same focal plane as an objective lens. Laser light is passed through an excitation filter, some of which is then passed through an objective and the rest through a dichroic reflector, so that both arrive at the specimen. The united beam is then passed through an emission filter (Figures 3.16 through 3.18) to be collected by a PMT [12].

Only a very small part of the specimen is illuminated at the same time, differently than for the much wider area covered by a conventional fluorescence microscope. Compared with the latter, confocal microscopy

1. Increases the effective resolution
2. Improves the signal/noise (S/N) ratio
3. Allows z-axis scanning
4. Reduces blurring of the image caused by light scattering
5. Gives depth perception in z-sectioned images
6. Allows electronic magnification of the image
7. Allows clear probing of thick specimens

Modern confocal microscopes use the objective lens for focusing both the light source and the image and are therefore "confocal" by default because this is by necessity the same focal plane.

A focus motor moves the stage in precise increments; multiple images from consecutive planes are scanned individually.

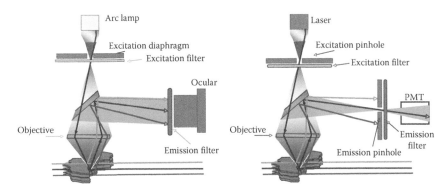

FIGURE 3.16
The basic principle of confocal microscopy. (From Rajwa, B., *The Principles of Confocal Microscopy*, ppt file Lect-7-Rajwa.ppt, Lecture 6. http://www.cyto.purdue.edu/flowcyt/educate/pptslide. htm.)

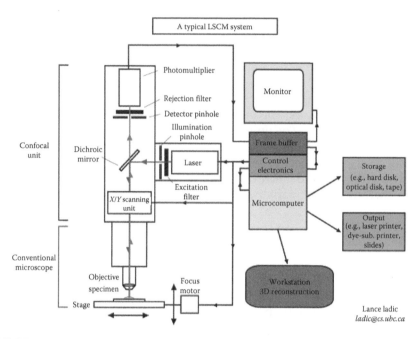

FIGURE 3.17

A typical LSCM system. (From Sheppard, C.J.R. et al., *Confocal Microscopy*, BIOS Scientific Publishers (Microscopy handbooks series) 1997; Wilson, T. (ed.), *Confocal Microscopy*, Academic Press, 1990.)

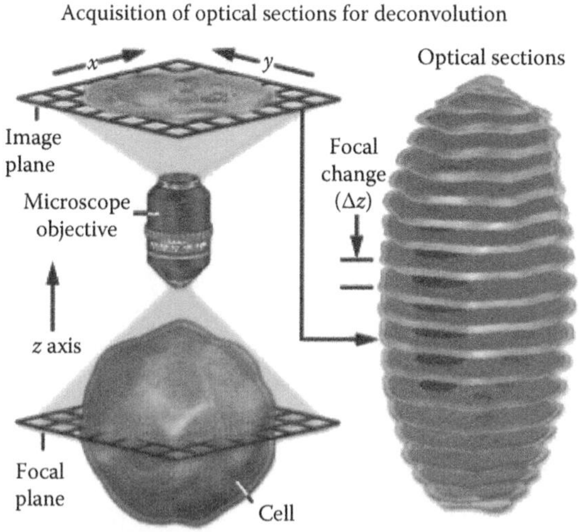

FIGURE 3.18

Optical sectioning (i.e., focusing at differing focal lengths) in confocal microscopy. (From Sheppard, C.J.R. et al., *Confocal Microscopy*, BIOS Scientific Publishers (Microscopy handbooks series) 1997; Wilson, T. (ed.), *Confocal Microscopy*, Academic Press, 1990.)

Image processing in confocal microscopy differs somewhat from that in the other SPMs because it is an optical technique, so that analogous processing as in optics can be used.

Thus, optical sections are readily obtained for deconvolution (one of the reasons why this microscopy is so useful in medicine). A confocal (like a deconvolution system) can collect images in a z-stack by repeatedly sampling the specimen at different focal planes, as illustrated in Figure 3.18.

3.3 Selected Examples

3.3.1 Environmental TEM of Cu/ZnO Catalysts for MeOH Synthesis

Studying chemical processes at gas–solid interfaces in situ with atomic resolution, long a challenge, is now being rapidly met. The reason why applying HRTEM to such studies is so difficult is that the HRTEM image resolution is degraded due to (a) scattering of the electron beam by gas atoms (see Figure 3.6) and (b) specimen drift due to heating. To minimize both, the gas atoms in the microscope must be minimized either by limiting the pressure or by confining the gas along the path of the beam into a layer that is as thin as possible. The two methods for achieving this, differentially pumped vacuum systems and windowed cells, although providing environmental TEM images having lattice fringes with spacing <0.2–0.3 nm bar and T several hundreds of degree celcius, do so at pressures to ca 10×10^{-3}, i.e., far from catalytic reaction conditions. A recent nanoreactor, a microelectromechanical (MEMS) system, shown in Figure 3.21, instead, allows imaging of a Cu/ZnO catalyst for MeOH synthesis using H_2 [14]. Direct observation of Cu

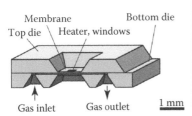
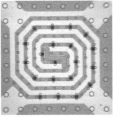

FIGURE 3.19

Illustration of the nanoreactor device. (a) Schematic cross section of the nanoreactor. (b) Optical close-up of the nanoreactor membrane. The bright spiral is the Pt heater. The small ovaloids are the electron-transparent windows. The circles are the SiO_2 spacers that define the minimum height of the gas channel. (c) A low-magnification TEM image of a pair of superimposed 10 nm thick windows. Their alignment creates a highly electron-transparent (bright) square through which imaging can be performed. (From *Ultramicroscopy*, 108, Creemer, J.F., Helveg, S., Hoveling, G.H., Ullmann, S., Molenbroek, A.M., Sarro, P.M., and Zandbergen, H.W., Atomic-scale electron microscopy at ambient pressure, 993–998, Copyright 2008, with permission from Elsevier.)

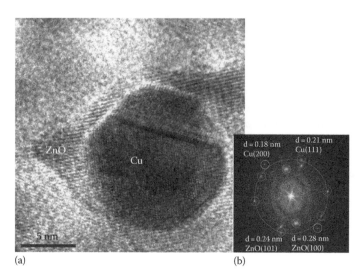

(a)   (b)

FIGURE 3.20

A representative HRTEM image of the Cu/ZnO catalyst during exposure to 1.2 bar hydrogen at 500°C. (a) The image displays lattice fringes of a twinned Cu nanocrystal and of the ZnO support. (b) A Fourier transform of (a). The bright dots represent sets of lattice fringes. Their lattice spacing corresponds to the distance to the origin and reveals the crystallographic identity. The large circle corresponds to a spacing of 0.21 nm. The smallest, resolved lattice spacing is 0.18 nm. (From *Appl. Catal. A General*, 348, Kurr, P. et al., 153, Copyright 2008, with permission from Elsevier.)

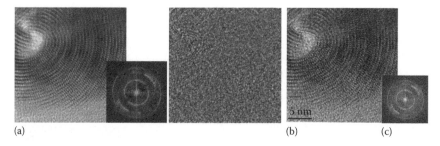

(a)   (b)   (c)

FIGURE 3.21

Above: Processing of the image in b (c is the FT). Right: Mathematical subtraction of the processed image from the original image in b using 2DLWF. The original image with 512 × 512 pixels was divided into partial images with 64 × 64 pixels. (From Kogure, T., Eilers, P.H.C., and Ishizuka, K., Application of optimum HRTEM noise filters in mineralogy and related sciences, *Microsc. Anal.*, 22(6), S11–S14 (EU), 2008.)

nanocrystal growth on a subnanosecond timescale during heating to 500°C at 1.2 bar H_2 is possible.

3.3.2 HRTEM of Chrysotile

For inorganics, overlap of noisy contrast from amorphous material reduces resolution in the signal deriving from the periodic surface structure. It may arise from various sources and depends on the preparation technique(s) used to obtain the periodic (or, for that matter, complex nanostructured) material. Noisy contrast can also derive from layers damaged by ion beams during sample preparation or—more commonly, as in the example here from mineralogy—from damage during HRTEM observation itself. One such example is chrysotile, $(Mg_3Si_2O_5(OH)_4)$, a component of asbestos and a natural multiwall nanotube. The structure consists of layers rolled up to accommodate misfit of lateral dimensions between the tetrahedral and octahedral sheets.

3.3.2.1 Use of Wiener filters

In Ref. [15], to estimate the background contribution IFbI in Fourier space, the authors first assumed that the contribution varies slowly. First, radial background was estimated by taking an average from a periodic structure after excluding strong intensities corresponding to diffraction (giving an "RWF"). Then, they took into account the fact that the radial background will network when structure information appears at the same distance from the origin (again, in Fourier space). To do this, they utilized a P-spline fitting to finally estimate a smoothed 2D background in Fourier space (2DWF). They allowed, finally, for the fact that when a periodic structure is small in size and its orientation changes locally the background estimated from the FT of the whole image may not be adequate. This was done by estimating the 2D background in Fourier space, *locally*, for each small image area given, by dividing a whole image (2DLWF). The resultant is shown in Figure 3.21.

(Compare the resolution of the chrysotile image with those in the pioneering account of HRTEM in layered materials of Ref. [16].)

3.3.3 Nanostructures of Porous Silicon

The next example is a study of anodization of Czochralski-grown *n*- and *p*-doped Si<111> and Si<100> by anodic etching in HF solution to give porous Si. Under various anodization conditions, as defined in Figure 3.22, it was found that

1. Anodization in the boundary transition region gave rise to materials showing high attenuation of x-rays (<111> and <100> absorptions for Si<111> and Si<100>, respectively). No materials obtained by anodization outside this region gave x-ray attenuation.

2. Formation depends crucially on electrochemical parameters and anodization times and the highest attenuation is shown by Si<111>.

An AFM investigation then showed that the x-ray attenuation correlates with formation of "pillar" (more correctly, from the section analysis of Figure 3.23, "cone-like") structures.

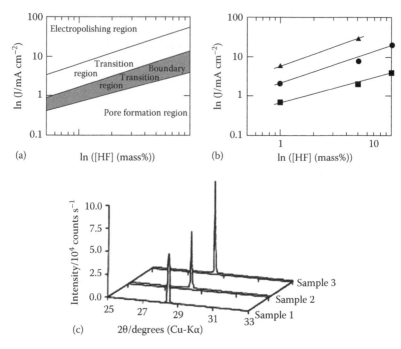

(a)   ln ([HF] (mass%))   (b)   ln ([HF] (mass%))

(c)   2θ/degrees (Cu-Kα)

FIGURE 3.22

(a) Definition of anodization conditions. (b) ln J versus ln [HF] plots for various samples and conditions. (▲) n-Si<111> 4min/5mAcm-1; (•) 10min/2mAcm-1; (■) s p-Si<111> 10min/5mAcm-1. (c) X-ray surface diffraction pattern of Si <111>. (From Lazarouk, S.K. and Tomlinson, A.A.G., Formation of pillared arrays by anodization of silicon in the boundary transition region: An AFM and XRD study, *J. Mater. Chem.*, 7, 667–671, 1997.)

These cone-like arrays are clearly arranged in long trenches and grooves. The more trench-like arrays give relatively flat-topped artifacts, whereas those with more V-shaped grooves have roughly pyramidal structures. The final effect is of arrays of both slanted and almost vertical pseudo-planes as in the clinographic view above. The heights are very variable for both Si<111> and Si<100>, but are never higher than ca 50nm and there is a considerable variation in width, which ranges between 10 and 20nm. The AFM of a sample cut perpendicular to the <111> face demonstrates how powerful the technique is for measuring porosity.

3.4 Future Challenges and Expected Breakthroughs

The speed of change in all microscopies for nanoscience, from light and electron sources through the minor industry surrounding scanning probe

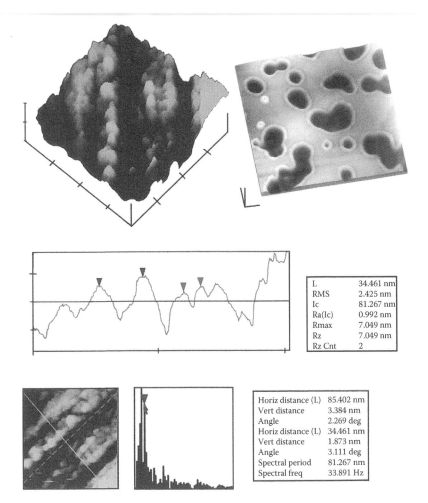

FIGURE 3.23
Typical AFM micrograph (dimensions in nm) of the pillared *n*-Si <111> surface in clinographic projection to underline surface trenches (a) and the same sample, cut perpendicular to the <111> face (b). (c) Section analysis along the white line of the projection. (From Lazarouk, S.K. and Tomlinson, A.A.G., Formation of pillared arrays by anodization of silicon in the boundary transition region: An AFM and XRD study, *J. Mater. Chem.*, 7, 667–671, 1997.)

microscopies (nanopositioning tools, tip technology, etc.) makes identification of expected breakthroughs hazardous. Nevertheless, some possible challenges do appear to be discernible. To begin with, the aim of any structural tool is to reach the highest resolution possible for an atom ordering. For an electron beam, through $\lambda = h/mv$ (v = velocity; e of $1\,\text{MeV} > \lambda = 0.00123\,\text{nm}$), this means approaching the theoretical limit of resolution:

$$= 0.0006\ \text{nm}$$

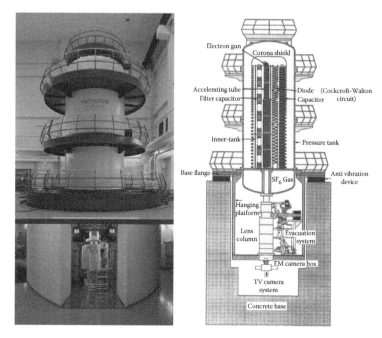

FIGURE 3.24
The Hitachi HU 3000 3MeV TEM; the most powerful TEM. (With permission from Hitachi Corp., http://www.uhvem.osaka-u.ac.jp/en/features.html)

The best resolutions have been achieved with a 1 MeV field emission TEM; the advantages of high-voltage operation being: a) increased specimen penetration, the accelerating voltage increases the ability of the beam to penetrate also increases, b) increased depth of information, the great depth of information allows an entire thick section to be viewed at once. It also allows for stereo pairs to be created from thick sections, and c) reduced beam damage, the increased speed of the electrons actually decreases the likelihood of an inelastic collision. However, even higher-voltage instruments are available such as the 5 MeV Hitachi, shown in Figure 3.24.

On the smaller scale and again for electron beams, the recently announced e-beam on a chip (while also neatly underlying synergy between current nanotechnology and instrumentation) deserves mention. This depends upon vertically aligning carbon nanotubes (CNTs) with precise placement and step repeatability, achieved using an automated thermal and plasma enhanced CVD process.

Figure 3.25 shows a schematic provided by the inventors [19] who are probably right in seeing it as a paradigm shift on considering possible applications: e-beam source for TEMs and SEMs (miniature TEMs?), gas ionization, and an x-ray tool [19]. Given the mass production capabilities, flat panel displays may prove the ultimate application. This is the second innovation involving aligned CNTs with mass production potential, the other—in an

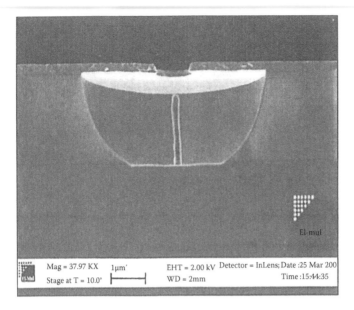

Mag = 37.97 KX   1µm'   EHT = 2.00 kV   Detector = InLens; Date :25 Mar 200
Stage at T = 10.0°   ⊢————⊣   WD = 2mm   Time :15:44:35

FIGURE 3.25
The e-beam on a chip electron source. (From El-Mul, http://www.el-mul.com/inr6873.
html?id=237&sid=275&pid=4037)

apparently unrelated field (again said with extreme caution)—acoustics. Aligned CNTs show thermoacoustic properties, which make them suitable as loudspeaker replacements [20]. It may be possible to combine AFM with acoustic spectroscopy (sometime ago a not very successful adjunct to optical spectroscopy for interrogating the visible region). Returning to TEMs, the major obstacle to higher resolution has been removed by the development of a spherical aberration corrector.

The major remaining challenges in SPMs, especially that of being able to interrogate each surface feature in, e.g., a film or, eventually, a pore separately, are being rapidly met by combined techniques, such as those in Figure 3.26.

For combined AFM-IR-ATR, the sample is deposited at the surface of a single reflection hemispherical ATR crystal mounted at the bottom of an adapted AFM liquid cell and is penetrated by the exponentially decaying evanescent field. A top–down AFM scanner is aligned to the center of the ATR crystal on top of the sample mount.

Raman/AFM and IR/AFM are already established, the third, scanning microwave microscopy (SMM) is not at first sight a spectroscopic technique, although its modification for performing microwave spectroscopy (of, e.g., gaseous species entrapped in porous matrices) seems feasible. The SMM mode (this is another modular attachment to an existing AFM instrument) permits measurement of complex impedance (resistance and reactance), capacitance, and dopant density, together with topography.

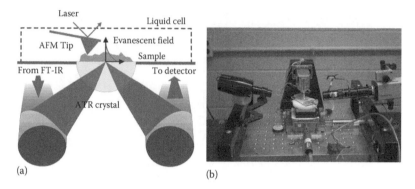

(a)   (b)

FIGURE 3.26
(a) Principle of combined FTIR-AFM; (b) the Agilent Technologies scanning microwave AFM microscope (components in the schematics are not to scale). (From Brucherseifer, M. et al., *Anal. Chem.*, 79(22), 8803–8806, 2007. With permission.)

3.4.1 Remote Control Operation

Finally, nanoscale microscopies are already moving into the era of remote control operation. The first scientific instrument to be operated remotely was a Kratos instrument. This was soon followed by FT-IR spectrometers and other spectroscopic probe techniques, possibly as "fallout" from the parallel technology for remote control developed for operating LANS (Local area Networks) in air pollution control (Figure 3.27).

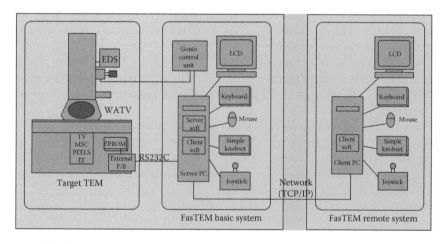

FIGURE 3.27
The Jeol JEM-2010 FasTEM, a typical TEM for telemicroscopy. (With permission from Jeol Corp.)

For electron microscopy, an example of interfacing and local instrument modifications (in the case of a Kratos EM-1500 with an electron energy of 1.5 MeV) can be found in Ref. [22].*

3.5 Conclusions

As microscope resolution has increased this century, so has the possibility of manipulating molecules, even to routine extension of "writing with atoms" and investigating surface chemistry itself (with SPM tips, especially in AFM). Further, the development of filters (in HRTEM) to advances in image processing and analysis, especially incorporation of deconvolution techniques (in confocal and SNOM microscopies, in particular), has added to the armory of surface and interface probing/manipulation. Double and even multiple microscope/spectroscopy techniques are also currently making routine not only topography and molecular structure to (in selected cases) the atomic level but also the chemical nature of the surface species being interrogated. *In situ* techniques are developing in parallel, so a surface can now be probed dynamically as it functions. Microscopy is rapidly becoming an indispensable tool and guide for "tailoring" new materials, and the time when it will also be so for "designing in" required physical and chemical properties "nanoscale up" cannot be far off.

3.6 Questions on Basic Issues

6.1 What is the difference between SEM and TEM as regards the nanomaterials sample used and how does the kind(s) of topology information differ between the two?

6.2 What innovations in image processing techniques have allowed such an increase in resolution (especially of layered materials) in HRTM over the last 15 years? Detail one.

6.3 Describe how you would choose between AFM and STM to probe your nanoporous material?

6.4 You need to study your nanoporous material first in a gas atmosphere then, in a second series of (sorption) experiments, in liquids. Which techniques would you choose and why, and how would a "sitting" proceed?

---

* For a general overview of telemicroscopy.

6.5   "Confocal microscopy is of little utility for investigating nanoporous materials." Is this true? (remember sectioning and the problems of sampling powder materials, remembering that some porous materials can be obtained in platelet form).

References

1. S. Amelinckx, D. Van Dyck, J. Van Landuyt, and G. Van Tandeloo (eds.), *Handbook of Microscopy*, VCH, Weinheim, Germany, 1997.
2. J.M. Cowley (ed.), *Electron Diffraction Techniques*, Oxford University Press, New York, 1993.
3. D.B. Williams and C.B. Carter, *Transmission Electron Microscopy: A Textbook for Materials Science*, Plenum Press, New York, 1996.
4. J.R. Harris, The future of transmission electron microscopy (TEM) in biology and medicine, *Micron*, 2000, *31*, 1.
5. L.D. Marks, Wiener-filter enhancement of noisy HREM images, *Ultramicroscopy*, 1996, *62*, 43; R. Kilaas, Optimal and near-optimal filters in high-resolution electron microscopy *J. Microsc.*, 1998, *190*, 45.
6. P.L. Gai, *Microscopy and Microanalysis*, 2006, *12*, 48; an in-situ STEM application.
7. SEM: M.T. Postek, K.S. Howard, A.H. Johnson, and K.L. McMichael, *Scanning Electron Microscopy. A Student Handbook*, Ladd Industries, Kettering, OH.
8. *The Image Processing Handbook* (2nd edn.), CRC Press, Boca Raton, FL, 1995.
9. R.J. Colton et al. (eds.), *Procedures in Scanning Probe Microscopies*, J. Wiley and Sons, New York, 1998.
10. A. De Stefanis and A.A.G. Tomlinson, *Scanning Probe Microscopies*, TTP Press, Zurich, 2001, which also contains an AFM "sitting."
11. AFM instrument.
12. K. Poole, *Imaging and Microscopy*, 2007, 2008, *8*, 43, for a more advanced version suited for biology.
13. C.J.R. Sheppard, D.M. Hotton, and D. Shotton, *Confocal Microscopy*, Microscopy Handbooks Series, BIOS Scientific Publishers, Oxford, U.K., 1997; T. Wilson (ed.), *Confocal Microscopy*, Academic Press, London, U.K., 1990 (still good on imaging); N.S. Claxton, T.J. Fellers, and M.W. Davidson, *Laser Scanning Confocal Microscopy*, Theory Series Handbook of Olympus Confocal at www.olympusconfocal.com
14. J.F. Creemer, S. Helveg, G.H. Hoveling, S. Ullmann, A.M. Molenbroek, P.M. Sarro, and H.W. Zandbergen, Atomic-scale electron microscopy at ambient pressure, *Ultramicroscopy*, August 2008, *108*(9), 993–998.
15. K. Ishizuka, P.H.C. Eilers and T. Kogure, Optimal noise filters in high-resolution electron microscopy, *Microscopy Today*, 2007, *15*(5), 16–20; Paul H.C. Eilers, Iain D. Currie, Maria Durbán, Fast and compact smoothing on large multidimensional grids, *Computational Statistics & Data Analysis*, 10 January 2006 *50*(1), 61–76.
16. J.M. Thomas, New ways of characterizing layered silicates and their intercalates, *Phil. Trans. R. Soc. Lond. A*, June 14, 1984, *311*, 271–285.

17. S. Lazarouk and A.A.G. Tomlinson, *J. Mater. Chem.*, 1997, *7*, 667. An entire book dedicated to the AFM of membranes: K.C. Khulbe, C.Y. Feng, and T. Matsuura, *Synthetic Polymer Membranes. Characterization by Atomic Force Microscopy*, Springer Laboratory Series, 2008, has appeared, much of it concerned with problems of measuring nanoporosity.
18. www.hitachi.co.jp
19. www.el-Mul.com
20. Lin Xiao, Zhuo Chen, Chen Feng, Liang Liu, Zai-Qiao Bai, Yang Wang, Li Qian, Yuying Zhang, Qunqing Li, Kaili Jiang, and Shoushan Fan, Flexible, stretchable, transparent carbon nanotube thin film loudspeakers, *Nano Letters*, 2008, *8*(12), 4539–4545.
21. Jobin-Yvonne (left); M. Brucherseifer, C. Kranz, and B. Mizaikoff, *Anal. Chem.*, 2007, *79*, 8803 (center); Agilent Technologies (right).
22. G. Fan, P. Mercurio, S. Young, and M.H. Ellisman, *Ultramicroscopy*, 1993, *52*, 499.

Resources

- TEM, HRTEM, and SEM tutorials
- www.Rodenburg.org, a good beginners guide to TEM*
- ucsd.edu/web-course:includes an animated TEM
- acad.udayton.edu for SEM
- Scanning Probe Microscopies:
- AFM tutorial: www.jpk.com
- Confocal microscopy and SNOM SNOM A. Criscenti, *Phys. Stat. Solidi*, 2008, *5*, 2615; a good overview
- Image Processing: Free image analysis programs can be downloaded from the National Institutes of Health (N.I.H., U.S.A.):
- rsb.info.gov/nih-image (NIH Image-Mac;rsb-info.nih.gov/ij (ImageJ-PC).

---

* Overall, the monthly (free) *Microscopy and Analysis*, Wiley & Sons: www.microscopy-analysis.com, contains good short articles on all microscopies. Also useful are the company updates on all new products in the field; many companies also list short-course materials. Finally, a long-forgotten "bible" of surface physics: *Handbook of Surface Metrology*, by D.J. Whitehouse, Institute of Physics, 1994, remains an excellent introduction to the physics and metrology of nonflat surfaces.

4

Calorimetric Techniques

Philip L. Llewellyn

CONTENTS

Several calorimetric methods can be used for the characterization of adsorbent surfaces and of adsorption phenomena. The texture of solids, that is to say the extent of surface area and pore size distribution, can be characterized by thermoporometry and immersion calorimetry. An advantage of the calorimetric approach over more standard methods lies in the possibility to characterize a more realistic surface area in the case of microporous solids. The probing of the chemical nature of surface can also be attained by immersion calorimetry and by adsorption calorimetry.

The following paragraphs will briefly describe these methods and will highlight several results that have been obtained.

4.1 Immersion Calorimetry

Immersion calorimetry is a simple method that can lead to information about the surface area as well as the surface chemistry of a solid. The surface area of a solid can be obtained by the immersion of the solid into a nonporous liquid or by using the modified Harkins and Jura method [1,2]. In this method, the solid is pre-recovered with a film of liquid prior to immersion into the same liquid.

An estimation of a micropore size distribution can be obtained by the immersion into liquids of different molecular dimensions. Finally, the surface chemistry of a solid can be followed by the immersion into various polar liquids.

The experimental procedure is relatively straightforward. The sample cell is blown from glass with a brittle point added to the bottom of the bulb and the top open. After the sample is placed into the cell, it can be outgassed under vacuum in a standard manner or by using sample controlled thermal analysis [3,4]. The top of the cell is then closed and is attached to a glass rod that is found inside the immersion cane. The immersion fluid is added to a tube at the bottom of this cane. This tube can then be screwed on to complete the immersion cane. The whole system is placed into the calorimeter. After thermal equilibrium is attained, the glass rod is pressed down so that the brittle point on the sample cell is broken. This allows the intrusion of the immersion liquid to the sample and a heat effect is measured (Figure 4.1).

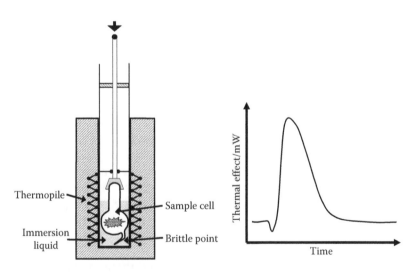

FIGURE 4.1
Schematic diagram of the setup used for immersion calorimetry (left) and the thermal effect measured during an immersion experiment (right).

This heat effect, schematized in Figure 4.1, shows an initial dip before the peak. This endothermic dip corresponds to the initial vaporization of the immersion liquid as it enters the cell. The exothermic peak then corresponds to the wetting of the sample, as well as to effects due to the breaking of the brittle point and the compression—depression of the immersion fluid inside and outside the cell. These different effects can be summarized by the following terms [2]:

$$Q_{exp} = \underset{\text{Energy of immersion}}{\Delta_{imm}U} + \underset{\text{Breaking of brittle point}}{W_b} + \underset{\text{Compression of vapor in bulb}(v)}{\int_0^{V-v} p\,dv}$$

$$+ \underset{\text{Liquefaction of liquid in bulb}(v)\text{ and vaporization of liquid outside bulb}(V)}{\frac{\Delta_{liq}h}{RT}\left[(p-p)^0 V + p^0 v\right]}$$

An estimation of the terms other than the energy of immersion can be obtained by a series of blank experiments with sample cells of various sizes. A plot of the heat effect measured as a function of the volume of the sample cell should then give a slope which is proportional to the heat of vaporization of the immersion liquid.

The heat of immersion that is measured depends on several factors:

- The *surface area of the solid*. For solids of identical surface chemical nature, the immersion energy is proportional to the surface area. A means to overcome the problem of comparing solids with different surface chemistries is to use the modified Harkins and Jura method, which is described in Section 4.1.1.

- The *chemical nature of the surface*. For a given liquid, the immersion energy depends on the chemical nature of the surface. For example, if the liquid is polar, the immersion energy increases with the polarity of the surface chemical functions. An application of such a study is to follow the influence of a treatment (thermal treatment, grafting, etc.) on the nature and density of surface functions.

- The *chemical nature of the liquid*. For a given nonporous surface, the immersion enthalpy depends on the chemical nature of the immersion liquid. Here, an application can be the evaluation of the average dipolar moment of surface sites by the immersion of the solid in liquids of increasing polarity. This leads to an analysis of the hydrophilic or hydrophobic character of a surface.

- The *porosity*. If the solid is microporous, the molecules of the liquid may be too large to penetrate into all the pores. In this case it is interesting to carry out a series of immersion experiments with molecules of different sizes but similar in chemical nature. In this case a micropore size distribution can be obtained and, in some cases, it is possible to follow the kinetics of wetting or pore filling.

4.1.1 Immersion Calorimetry for the Estimation of the Surface Area of a Solid: The Modified Harkins and Jura Method

As mentioned earlier, the surface area of a solid is proportional to the immersion energy that is released on wetting. In many cases, it is possible to immerge the solid into a nonpolar liquid such as hexane. The heat effect measured can then be compared with that obtained with that of a reference solid.

In some cases, though, the surface chemistry of the solid can still play a role on the heat effects measured. One procedure to overcome this problem is known as the modified Harkins and Jura method [2], which is schematized in Figure 4.2.

In this method, after outgassing, the sample is equilibrated at a given relative pressure of the immersion liquid. The surface is thus pre-covered with a liquid film prior to the immersion experiment proper. Thus the heat effect that is measured is simply related to the disappearance of the liquid–vapor interface and is given by the expression [5]

$$\Delta_{imm} H = A \left(\gamma_{lv} - \left\{ \frac{\partial \gamma_{lv}}{\partial T} \right\}_A \right)$$

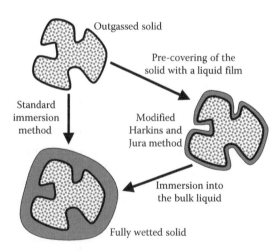

FIGURE 4.2
Schematic diagram comparing the standard immersion method with the modified Harkins and Jura method.

where

$\Delta_{imm}H$ is the immersion enthalpy
A is the area of solid
γ_{lv} is the liquid–vapor surface tension of the wetting liquid
T is the temperature

The question arises as to the thickness of the pre-covered film that should be used. If too thick, there is a risk of pore filling and that the film surface is not representative of the solid covered. Partyka et al. [2] studied this problem in the case of the immersion of a silica sample into water. The sample was pre-covered with water at various relative pressures before immersion experiments. The results thus obtained were then compared with the water adsorption isotherm (Figure 4.3).

Figure 4.3 shows that the immersion energy decreases for samples pre-covered at low relative pressures. From a relative pressure of around 0.4, however, the immersion energy remains constant. This relative pressure corresponds to the formation of two layers of water on the surface of the silica. This study thus implies that if pore filling occurs before the formation of two adsorbed layers, then the Harkins and Jura method will not give valid results. This is typically the case for microporous samples.

The comparison of the results obtained with the Harkins and Jura method with those obtained using the BET (Branauer, Emmet, and Teller) method from adsorption experiments is given in Table 4.1. It can be seen that, in many cases, the agreement is good between the two methods.

An advantage of the Harkins and Jura method is that there is no assumption about the size of the wetting molecule as there is with the BET method.

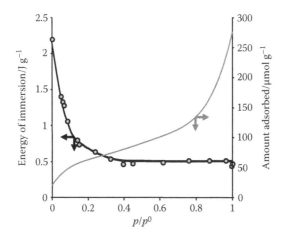

FIGURE 4.3
Immersion energy of water on a silica sample pre-covered with water at various relative pressures and corresponding water adsorption isotherm [2].

TABLE 4.1

Comparison of Specific Surface Areas
Obtained from Adsorption Experiments (a_{BET})
and from Immersion Calorimetry ($a_{Harkins-Jura}$)

Sample	a_{BET}/m^2 g^{-1}	$a_{Harkins-Jura}$/m^2 g^{-1}
Quartz	4.2	4.2
Silica	129	140
Alumina	81	100
Titania	57	63
Gibbsite	24	27
Kaolinite	19.3	19.4
Zinc oxide	2.9	3.1
Gallium hydroxide	21	21.3

Source: Partyka, S. et al., *J. Colloid Interface Sci.*, 68, 21, 1979.

Such experiments require that the solid in question should be fully wetted. However, as noted above, the method is limited to samples that are essentially nonporous. It should be noted that, experimentally, the Harkins and Jura method is quite painstaking.

4.1.2 Immersion Calorimetry for an Evaluation of the Chemical Nature of the Surface

An evaluation of the surface chemistry of samples can be carried out by comparing results with a sample containing few surface chemical sites. This is highlighted in Figure 4.4 in which the immersion energy of titania is compared with that obtained for a graphon sample as a function of the dipole moment of the immersion liquid used [6].

For a sample that has few surface chemical sites, such as the graphon sample, a change in the polarity of the immersion liquid shows little variation in the energy of immersion. On the other hand, it can be seen that the interaction of the increasing dipole moment of the immersion liquid with the titania leads to a slope, which is a function of the surface chemistry of the solid.

Immersion calorimetry can be used to follow the variation of surface chemistry as a function of various thermal treatments. Figure 4.5 shows the variation of immersion energy as a function of the thermal pretreatment temperature in the case of a silica sample [7]. It can be seen that there is little variation in immersion energy for temperatures below 500°C; however, the thermal treatment of silica samples above 500°C leads to a progressive dehydroxylation, which results in a decrease in immersion energy.

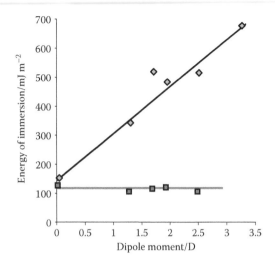

FIGURE 4.4
Immersion energy of titania (diamonds) and graphon (squares) as a function of the dipole moment of the immersion liquid. (Drawn from data in Chessick, J.J. and Zettlemoyer, A.C., *Adv. Catal.*, 11, 263, 1959.)

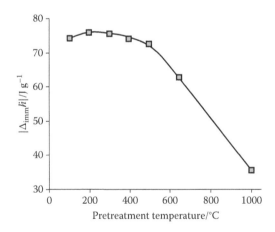

FIGURE 4.5
Variation of immersion energy for precipitated silica as a function of its pre-treatment temperature [7].

4.1.3 Immersion Calorimetry for the Evaluation of Microporosity

It is possible to estimate the micropore size distribution of heterogeneous samples such as carbons via the use of different immersion liquids of different size and similar chemical nature. This is demonstrated in several studies such as those in Refs. [8–10].

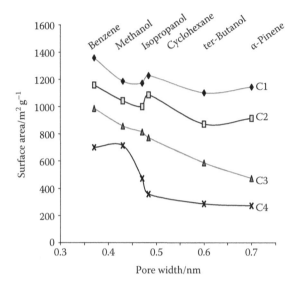

FIGURE 4.6
Accessible surface area of a series of charcoal samples as a function of pore width calculated from the molecular dimensions of different immersion liquids [8].

Figure 4.6 shows the surface area of a series of carbon samples as a function of the pore width [8]. This pore width is calculated from the molecular dimensions of the various immersion liquids used. It can be seen that a distinct cutoff point is observed for sample C4 just before 0.5 nm, which indicates the narrow pore size distribution for this sample. C3, however, seems to show a rather large pore size distribution in the micropore range whereas the other two samples seem to consist of pores of larger dimensions.

An advantage of the use of a calorimetric method for the estimation of micropore surface areas lies in the fact that the energy released on immersion is directly proportional to the interacting surface [11]. This is an advantage with respect to the BET method, which considers the molecular cross section as is schematized in Figure 4.7. Figure 4.7 shows that the BET method, in which the number of molecules that form a statistical monolayer is multiplied by the cross-sectional area of the molecules, can lead to an underestimation and overestimation of ultramicropore and supermicropore surface areas respectively. This is not the case for immersion calorimetry, in which the immersion energy is proportional to the area of interaction with the surface. Thus, for example, the immersion energy of a molecule inside a cylindrical ultramicropore is 3.6 times greater than on an open surface.

To access the micropores, relatively small molecules should be used. This leads to limits for studies at room temperature where mostly alkanes are

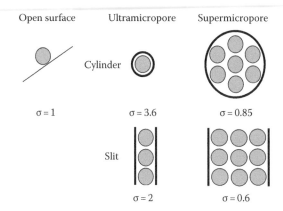

FIGURE 4.7
Diagram highlighting the problems that can be encountered when one considers the molecular cross section "σ" in the estimation of ultra- and supermicropore surface areas. The cases shown are for cylindrical and slit-shaped pores.

used. In such cases, the use of immersion into liquid argon would seem to be advantageous [12–14]. This can be attained using the low temperature calorimeter described in Figure 4.12 and the use of a nonpolar liquid such as argon. In such studies an estimation of the surface areas of microporous carbon and silica samples allows one to go one step further in the determination of the internal surface area of micropores. Liquid nitrogen and liquid argon provide a very similar areal enthalpy of immersion for carbons: for instance, 165 and 160 mJ m^{-2} in nitrogen and argon, respectively, if the surface area of the reference material is measured by the BET method with nitrogen at 77 K [14]. The enthalpies of immersion obtained with silica samples into liquid nitrogen are systematically higher than into liquid argon, but this should not influence the derivation of the "immersion surface area" provided the reference sample is correctly selected. Such results rule out a simplifying assumption initially made by the pioneers of the method, Chessick et al. [12] and Taylor [13], that the areal enthalpy of immersion is somewhat independent of the chemical nature of the adsorbent. This means that a calibration with a nonporous sample is needed for each type of surface. This is not too much of a problem since the duration of a complete calorimetric experiment (after preliminary weighing and outgassing of the sample) is approximately 2 h.

4.1.4 Intrusion of Non-Wetting Fluids

In the case of non-wetting systems, pressure is required for intrusion of the fluid into the pore to occur. This is the case for mercury porosimetry, for example. There is growing interest in highly hydrophobic systems in which water intrusion occurs under pressure. Such systems can be used for shock absorbers or for storing energy. There have been several studies of water intrusion into hydrophobic systems such as pure silica zeolites and grafted mesoporous silicas [15–17].

In such systems the increase in pressure is accompanied by a decrease in volume of the system corresponding to the intrusion of water into the porosity. In this case, the pore size is calculated via the Washburn equation:

$$p = -\frac{2\gamma_{lg}}{r}\cos\theta$$

where
  p is the pressure
  γ_{lg} is the interfacial tension
  r is the radius of curvature
  θ is the contact angle

An estimation of the contact angle thus has to be made. An advantage of combining such a system with a simultaneous calorimetric reading is that the contact angle can be calculated by a combination of the Washburn equation and the following expression [18]:

$$dU = \left(T\frac{\partial(\gamma_{lv}\cos\theta)}{\partial T} - \gamma_{lv}\cos\theta \right)dA$$

where
  U is the internal energy
  T is the temperature
  A is the surface area

An example of the results that can be obtained for the intrusion of water into porous silica is shown in Figure 4.8. From the upper curves and applying the relevant formulas, it is possible to obtain the pore size distribution in the lower curve. The contact angle calculated during this experiment was $126° \pm 5°$.

4.2 Thermoporometry

Thermoporometry is a simple method to obtain the pore size distribution of mesoporous solids. It is based on the phenomenon that liquids will freeze at lower temperatures in confined media than in the bulk.

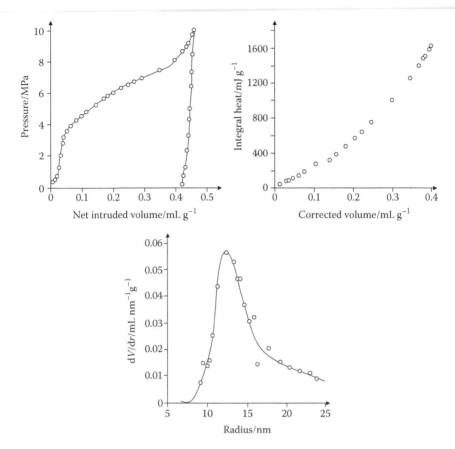

FIGURE 4.8
Results obtained for the intrusion of water into a polymer-grafted silica [18]: upper left pressure/volume curve, upper right heat/volume curve, lower curve—differential pore size distribution calculated from the upper curves.

Experimentally, this phase transition can be followed by differential scanning calorimetry (DSC). A typical experimental protocol (Figure 4.9) begins with the immersion of the solid with a slight excess of liquid before being placed into the DSC. A temperature program starts with a quench to low temperature (typically −80°C for water) before gradual heating (0.5°C–1°C min⁻¹) to just below the bulk melting temperature. In this part of the procedure, the fluid is initially frozen both inside and outside the pores. On heating to just below the bulk melting temperature, only the fluid inside the pores melts leaving the solid outside the pores. The temperature is then lowered again at the same rate on heating for resolidifying the fluid inside the pores. Such a procedure avoids any large deviation from equilibrium and avoids undercooling of the liquid. In some cases the temperatures of liquefaction and solidification differ. This has been

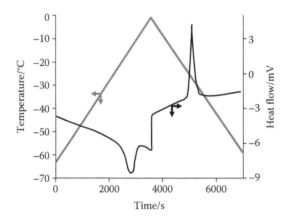

FIGURE 4.9
Heating program and heat flow observed during a thermoporometry experiment.

ascribed to pore shape, as well as to differences in mechanisms on heating and cooling.

From a phenomenological point of view, there is still discussion as to the mechanism of solidification inside the pores. This is schematized in Figure 4.10 in which the mechanisms of interfacial advancement and in situ nucleation are shown. The interfacial advancement mechanism was first described by Everett as the plastic ice model [19]. A further point that can be seen in the figure is that a layer of fluid remains on the pore walls which does not freeze on cooling. It is generally accepted that this nonfreezable film consists of two molecular layers.

The depression in freezing point is described by the equation put forward by Kubelka [20]:

$$\Delta T = T - T_0 = \frac{2\gamma_{sl}MT_0}{R\rho_1\Delta H_s}$$

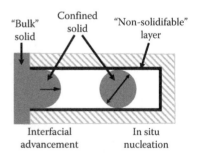

FIGURE 4.10
Diagram showing the different mechanisms of solidification inside a pore. Left: the penetration of a solid interface from the bulk solid outside the pore. Right: nucleation of the solid inside the pore.

where

T and T_0 are the temperatures of freezing in the pore and in the bulk respectively

γ_{sl} is the surface tension

M is the molar mass

R is the radius of curvature

ρ_l is the liquid density

ΔH_s is the enthalpy of solidification

Of these parameters, γ_{sl}, ρ_l, and ΔH_s all vary with temperature. The heat flow on solidification can be given by [21]:

$$Q = T_e \left\{ \Delta S_{s0} + \int_{T_0}^{T_e} \frac{Cp_l - Cp_s}{T} dT + 2\frac{v_l}{R_p - t} \frac{d\gamma_{ls}}{dT} + \left(\frac{\partial v_l}{\partial T}\right)_P (P_l - P_0) + \left(\frac{\partial v_s}{\partial T}\right)_P (P_0 - P_s) \right\}$$

where

T_e is the equilibrium temperature

S_{s0} is the entropy of solidification

Cp_l and Cp_s are the specific heats at constant pressure of the liquid and solid respectively

v_l and v_s are the volumes of the liquid and solid, respectively

t is the thickness of the nonfreezable layer

P_0, P_l, and P_s are the initial pressure and pressures within the liquid and solid phases respectively

This expression can be used to calculate the volume of the pores.

More simple, however, is the calculation of the pore size, but, for this, the variation in the parameters with temperature as well as the thickness of the nonfreezable layer must be considered. For characterization purposes, Quinson and Brun developed simple formulae which model the different variations in temperature and integrate the nonfreezable film thickness [21]. For water the pore radius R_p is given by the following expression:

$$R_p = \frac{-64.67}{\Delta T} + 0.57$$

whereas for benzene the expression used is

$$R_p = \frac{-131.6}{\Delta T} + 0.54$$

For two MCM-41 samples, Figure 4.11 shows the results that can be obtained using thermoporometry. These results compare well with those obtained via other methods.

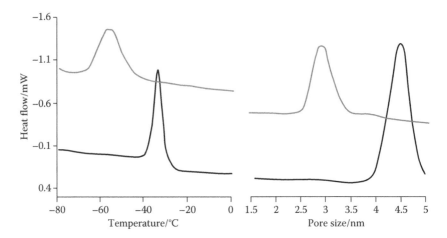

FIGURE 4.11
Thermoporometry on water: experimental cooling curves (left) and pore size calculation (right) obtained with two MCM-41 samples.

There are several advantages of using thermoporometry with respect to adsorption–desorption experiments. Firstly, the experiment is relatively rapid compared to a full adsorption–desorption isotherm. Furthermore, it is possible to analyze relatively fragile samples such as polymers, which can undergo degradation on outgassing or at liquid nitrogen temperature. Finally, it is possible to analyze samples in their application media if used in the liquid phase.

4.3 Isothermal Adsorption Microcalorimetry

Different calorimetric methods are available for the study of adsorption phenomena. While *adsiabatic* calorimetry is more suitable for the determination of heat capacity, both *isoperibol* and *diathermal* calorimetries have been used to follow adsorption phenomena. Isoperibol calorimetry, where no special connection is made between the sample temperature and that of the surroundings, was the first to be applied to adsorption. The experimental setups and the experiments themselves are complicated and comparison of results obtained using this methodology with those obtained by standard, isothermal measurements is difficult. Diathermal calorimeters, where the sample temperature follows that of the surroundings, are most suited to follow gas adsorption phenomena. The isothermal conditions of adsorption manometry can be reproduced so that the thermal events that occur during adsorption

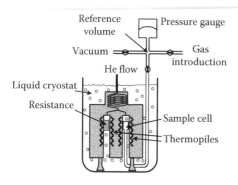

FIGURE 4.12
Schematic representation of a calorimetric setup used for adsorption experiments.

can be measured. The examples given below were obtained under these diathermal or *quasi-isothermal* conditions.

For the characterization of adsorbents, low temperature measurements, in the range of 77 K are generally used. However, data obtained in the temperature region from 25°C to 100°C can often be of interest, for example in studies related to gas storage and gas separation. Specific instruments are required for each of these temperature domains.

An example of a diathermal calorimeter [22] used for low temperature adsorption studies is shown in Figure 4.12. This apparatus is formed of three main parts: the dosing apparatus, the sample cell, and the calorimeter. The calorimeter is a Tian-Calvet apparatus where two thermopiles are mounted in electrical opposition. The calorimeter is placed upside down, like a diving bell in the cryogenic fluid. Each thermopile houses around 1000 thermocouples that provide an overall sensitivity of around 5 mJ. The electrical resistance is housed in the reference thermopile, allowing calibration via the Joule effect. The calorimeter is maintained in a liquid nitrogen (or argon) cryostat containing around 1000 L of cryogenic fluid. A small flow of helium is maintained through the thermopiles. This helium flow maintains the calorimeter under isothermal conditions, as well as allowing a good thermal contact between the sample and the thermopile.

There are, however, two different methods of adsorbate introduction. The first, and most common, is to inject discrete quantities of adsorbtive to the adsorbent. Each introduction of adsorbate to the sample is accompanied by an exothermic thermal effect, until equilibrium is attained. This peak in the curve of energy with time has to be integrated to give an integral (or pseudo-differential) molar enthalpy of adsorption for each dose.

The calorimetric cell (including the relevant amounts of adsorbent and gas phase) is considered as an open system. In this procedure, as well as in the quasi-equilibrium procedure of gas introduction (following section), it is important to consider that the gas is introduced reversibly. However, to calculate the differential enthalpy of adsorption via the discontinuous procedure, one must introduce quantities dn small enough for a given pressure increase dp.

Under these conditions it is possible to determine the differential enthalpy of adsorption $\Delta_{ads}\dot{h}$, via the following expression:

$$\Delta_{ads}\dot{h} = \left(\frac{dQ_{rev}}{dn^a}\right)_T + V_c\left(\frac{dp}{dn^a}\right)_T$$

where
   dQ_{rev} is the heat reversibly exchanged with the surroundings at temperature T, as measured by the calorimeter
   dn^a is the amount adsorbed after introduction of the gas dose
   dp is the increase in pressure
   V_c is the dead space volume of the sample cell within the calorimeter itself (thermopile)

For the observation of subtle adsorption phenomena, such as adsorbate phase changes, an increased resolution in both the isotherm and differential enthalpy curves is required. It would be possible to introduce very small doses of gas to increase the number of points taken. This is both time consuming and may lead to the summation of a number of errors. However, a continuous introduction of gas leads to an infinite resolution in both curves.

Figure 4.13 highlights how this resolution can be interesting. The peak A in the full line would be indicative of an adsorbate phase change, which would go unnoticed using the discontinuous procedure of gas introduction.

In what has been termed the "continuous flow" procedure, the adsorbate is introduced to the system at a defined rate, slow enough that the adsorbate–adsorbent system can be considered to be essentially at equilibrium at all times [23–25]. To verify this equilibrium it is possible to stop the adsorbate flow and verify that the heat signal stops (Figure 4.14).

In this "quasi-equilibrium" state, the quantity of adsorbate admitted to the system Δn can be replaced in adsorption calculations by the rate of gas flow dn/dt. The calorimeter under these conditions measures a heat flow ϕ.

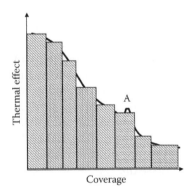

FIGURE 4.13
Comparison of the results obtained using either the discontinuous (bars) or continuous (full line) procedure of gas introduction. The peak A corresponds to an adsorbate phase change which is overlooked when using the discontinuous procedure of adsorptive introduction.

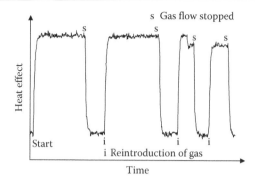

FIGURE 4.14

Adsorption of krypton on silicalite-1. The flow of gas to the sample is stopped periodically to verify the adsorbate–adsorbent equilibrium.

The use of a sonic nozzle allows the gas flow to the sample, $f = dn/dt$, to be kept constant. A rate of adsorption, f^a, can therefore be calculated using the following expression:

$$f^a = \frac{dn^a}{dt} = f - \frac{1}{R}\left(\frac{V_d}{T_d} - \frac{V_c}{T_c}\right)\frac{dp}{dt}$$

Here, V_d and V_c are the volumes of the dosing system and that "accessible" to the calorimeter at temperatures T_d and T_c. The corresponding heat flow ϕ can be given by

$$\phi = \frac{dQ_{rev}}{dt} = \frac{dQ_{rev}}{dn^a} \cdot \frac{dn^a}{dt} = f^a\left(\frac{dQ_{rev}}{dn^a}\right)_T$$

Combining the last two expressions leads to

$$\Delta_{ads}\dot{h} = \left(\frac{dQ_{rev}}{dn^a}\right) + V_c\left(\frac{dp}{dn^a}\right) = \frac{\phi}{f^\sigma} + V_c\frac{dp}{dt}\frac{dt}{dn^a}$$

$$\Delta_{ads}\dot{h} = \frac{1}{f^a}\left(\phi + V_c\frac{dp}{dt}\right)$$

Blank experiments can lead to an estimation of $V_c(dp/dt)$. This term is large during horizontal parts of the isotherm. The error in the estimation of the differential enthalpy thus becomes large. However, in these regions of multilayer adsorption, an estimation of the enthalpy is less interesting and can be readily obtained using the isosteric method.

For micropore filling, or during capillary condensation, the term $V_c(dp/dt)$ is minimal. Effectively, during such phenomena, the increase in pressure with time is small. Furthermore, almost all of the flow of gas to the sample is adsorbed making, $f \approx f^a$. In such cases, the last equation can be simplified to

$$\Delta_{ads}\dot{h} \approx \frac{\phi}{f}$$

Thus if the rate of gas flow, f, is constant, a direct measurement of $\Delta_{ads}\dot{h}$ with the amount adsorbed is recorded [22,25].

An example of the results that can be obtained using combined adsorption manometry/calorimetry is shown in Figure 4.15. This figure represents the direct signals of pressure and heat flow as a function of time, recorded during the adsorption of nitrogen onto a well-organized graphite sample [24,26].

This diagram highlights several points relative to the measurement of differential enthalpies of adsorption using the continuous procedure of gas introduction. It can be seen that the initial introduction of gas, up to 1.5 h, leads to only a slight increase in the pressure signal. This corresponds to a relatively strong signal in the heat flow curve that is the result of monolayer adsorption on a highly organized homogeneous surface. The point "P" corresponds to a small step in the pressure signal and a large peak in the heat flow signal. This phase transition corresponds to the completion of the monolayer in epitaxy with the highly organized substrate [24,26]. At point "s," however, the flow of gas is stopped in order to check the equilibrium. It can be seen that the pressure signal does not change and the heat flow signal decreases to the baseline within the response time of the calorimeter.

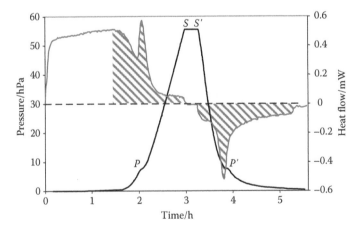

FIGURE 4.15
Plot of the signals of heat flow and pressure obtained during the adsorption of nitrogen on graphite at 77.4 K. (Adapted from Rouquerol, J. et al., *J. Chem. Soc. Faraday Trans. 1*, 73, 306, 1977.)

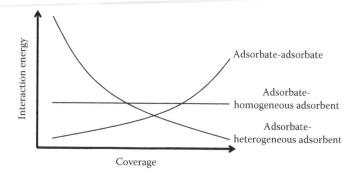

FIGURE 4.16
Hypothetical breakdown of calorimetric curves due to various interactions in play during the adsorption of simple gases at low temperature.

These two points allow the conclusion of a quasi-equilibrium state. At point "s," the vacuum line is opened to desorb the nitrogen and check the reversibility of the system. Note that this is one of the requirements for the above-mentioned calculations. It can be seen that at "P," an effect similar to that produced on adsorption occurs. This and the fact that the two hatched areas are equivalent show the reversibility of this system.

As shown above, the differential enthalpy curves obtained using such adsorption microcalorimetric experiments is a global effect that includes both adsorbate–adsorbent and adsorbate–adsorbate interactions. Various adsorbate filling mechanisms and phase transitions can be highlighted, as well as any structural changes of the adsorbent.

In general, though, the calorimetric curve highlights three different types of behavior as schematized in Figure 4.16. In each system, an increase in the amount of gas adsorbed on a sample leads to an increase in the interactions between the adsorbate molecules. Concerning the adsorbate–adsorbent interactions, the interaction of an adsorbate molecule with an energetically homogeneous surface will give rise to a constant signal.

Finally, in most cases, the adsorbent is energetically heterogeneous due to a pore size distribution and/or a varying surface chemistry (defects, cations, etc.). Initially, one would expect relatively strong interactions between the adsorbing molecules and the surface. The extent of these interactions will then decrease as these specific sites are occupied. Thus, for energetically heterogeneous adsorbents, a gradual decrease in the calorimetric signal is observed. However, each differential enthalpy curve varies and is a composite of varying percentages of each type of interaction.

Both Kiselev [27] and Sing [28] have put forward classifications of differential enthalpy curves. Figure 4.17 shows hypothetical differential enthalpy of adsorption curves which would correspond to the International Union of Pure and Applied Chemistry (IUPAC) [29] classification of adsorption isotherms.

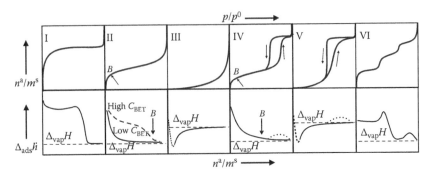

FIGURE 4.17
Six IUPAC classified isotherms (upper row) and corresponding hypothetical differential
enthalpy of adsorption curves (lower row).

For nonporous and macroporous ($d_p > 50\,nm$) solids which give rise to Type
II isotherms, the differential enthalpy curve invariably decreases rapidly to
the enthalpy of vaporization ($\Delta_{vap}H$) of the gas. In several cases where there
exist many specific sites on these materials, this decrease in the curve is less
marked. These differences would seem to correspond to different C values
derived from the BET equation.

Mesoporous materials ($2 < d_p < 50\,nm$) which normally give rise to Type IV
isotherms also give rise to differential enthalpy curves which decrease to
the enthalpy of vaporization ($\Delta_{vap}H$) of the gas under investigation. For solids with a very narrow pore size distribution (MCM-41 type materials, for
example) a slight increase in calorimetric signal of around 0.5–1 kJ·mol⁻¹ is
observed during the capillary condensation step [30].

Systems that give rise to Type III or Type IV isotherms are indicative of
very weak adsorbate–adsorbent interactions. For these systems, the differential enthalpy of adsorption is initially below that of the enthalpy of vaporization of the gas. In such cases, it would seem that entropy effects drive the
adsorption process.

Type VI isotherms are typical for very homogeneous two-dimensional
solids such as graphite. Each step corresponds to the edification of a different adsorbate layer. The differential enthalpy curve is relatively constant for
the initial monolayer coverage. The completion of this monolayer results in
a distinct peak in the differential enthalpy curve which corresponds to the
formation of an epitaxal layer of adsorbate (see, Figure 4.15). It is noteworthy that this two-dimensional disorder–order transition was first observed
by microcalorimetry [26] before being characterized by neutron diffraction
methods.

Finally, the filling of micropores ($d_p < 2\,nm$) is characterized by Type I
isotherms. The initial uptake is characterized by a very small increase in
pressure and is the result of enhanced interactions. Such cases are ideal for

microcalorimetric studies as the technique is at its most sensitive. The differential enthalpy of adsorption curves are typically elevated throughout the pore filling process.

4.3.1 Adsorption Calorimetry for the Characterization of Heterogeneous Adsorbents

Most adsorbents that are encountered can be considered energetically heterogeneous, whether from a textural point of view (pore size distribution, etc.) or from the point of view of their surface chemistry.

A typical example is that of porous silica gel. The pore size distribution is relatively large and the surface contains hydroxyl groups of different energies. Figure 4.18 shows the isotherms and differential enthalpy curves for the adsorption of argon and nitrogen on a microporous silica gel. Both the differential enthalpy curves obtained with argon and nitrogen decrease continually with relative coverage. This is characteristic of large electrical homogeneity and is due to both the surface chemistry and the pore size distribution.

Often the comparison between the results obtained with argon and nitrogen can give an idea of the relative importance of the inhomogeneity due to the pore size distribution and surface chemistry. Indeed, argon being a spherical and nonpolar molecule interacts only very weakly with surface chemical species such as hydroxyl groups. The behavior thus observed is essentially due to the textural properties (pore geometry, pore size distribution, etc.). Nitrogen, however, has a permanent quadrupole which is able to interact with any specific surface groups. The behavior thus observed corresponds to any specific interactions with the surface as well as any interaction due to the textural nature of the sample.

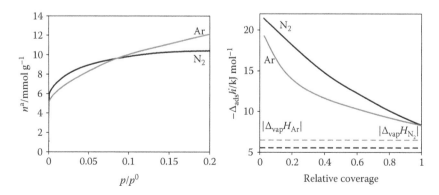

FIGURE 4.18
Isotherms (left) and corresponding differential enthalpies (right) at 77.4 K for nitrogen and argon adsorbed onto a microporous silica gel [31].

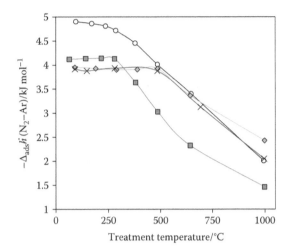

FIGURE 4.19
The difference in the enthalpy of adsorption at zero coverage of argon and nitrogen for different silica samples (A pyrogenic silica, G silica gel, P precipitated silica, S Stöber silica) as a function of thermal treatment [7,32].

The difference in the differential enthalpies of argon and nitrogen can be taken as an indication of the extent of the interactions due to the surface chemistry of the adsorbent under investigation.

An example of the information that can be obtained from the difference in energy of adsorption of argon and nitrogen can be seen in Figure 4.19. Here the difference in energy at zero coverage is plotted for a series of silica samples that had been pretreated to different temperatures. This plot gives an insight into the hydroxyl content on the surface of the silica under investigation. It can be seen that the surface hydroxyl groups start to be transformed into siloxane bridges from around 300°C for the precipitated silica, whereas these groups would seem to be more stable on pyrogenic and Stöber silicas [32].

Porous carbons often have a rather large pore size distribution and as such the differential enthalpy of adsorption curves also suggests energetic heterogeneity. However, these curves often have more features than those for silicas. The adsorption of simple gases onto microporous active carbons generally leads to calorimetric curves containing three different regions during the filling of the micropores. An example is shown in Figure 4.20 for the adsorption of nitrogen and argon onto an activated carbon at 77.4 K. The three regions are clearly shown: the first region, AB, decreases before a second, more horizontal region, BC; the third region, CD, again shows a marked decrease toward the enthalpy of liquefaction of the gas under consideration.

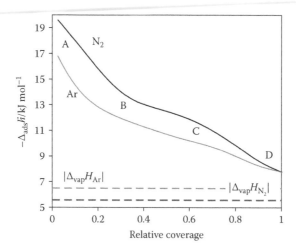

FIGURE 4.20
Differential enthalpies at 77.4 K for nitrogen and argon adsorbed onto an activated carbon.

A number of authors have noted and discussed such phenomena with varying interpretations [33,34]. It would seem that the following conclusions are generally drawn:

- Region AB is characteristic of interactions between the adsorbate and an energetically heterogeneous adsorbent (Figure 4.20). If one considers that the two-dimensional graphite surface is energetically homogeneous, an explanation of the observed heterogeneity has to be found. Such heterogeneity can arise from defects or impurities, as well as from a distribution in micropore size. Although the first two possibilities can be eliminated in some cases, the nature of the preparation and activation of such materials makes a certain pore size distribution inevitable. One can therefore assume that the smallest micropores (or ultramicropores) are filled in this initial region AB.

- Region BC, however, corresponds to a more homogeneous phenomenon and to an enthalpy of adsorption not far from that for the adsorption on a perfect two-dimensional surface ($\approx$14 kJ mol^{-1}). Furthermore, simulation studies [34] have shown that for the adsorption of nitrogen in larger micropores (or supermicropores), above 0.7 nm in diameter, a two-step process may occur. The first step would seem to correspond to the coverage of the pore walls, whereas the second step is the filling of the void space. It would thus seem possible that the region BC corresponds to the coverage of the pore walls. The fact that this step is not completely horizontal, in

comparison to the adsorption on a two-dimensional graphite surface, may be due to curvature effects within the micropores.

- Taking into account the above-mentioned hypothesis, it would seem
  that the region CD corresponds to the completion of the filling of the
  larger micropores.

Clay samples in general can also be considered relatively heterogeneous.
However, often the differential energy curves indicate both heterogeneous
and relatively homogeneous regions. One example is that of kaolinite, which
has a 1:1 sheet structure with a layer repeat distance of 0.72 nm. This is
approximately the distance of the sheets themselves, which means that there
is insufficient space to accommodate intercalated molecules such as water.
The isotherms obtained for such materials are of Type II, which are typical
for nonporous or macroporous materials.

The differential enthalpy curves for argon and nitrogen (Figure 4.21) [35]
show two main regions. The first, AB, corresponds to the adsorption on
defect sites and the adsorption on lateral facets of the materials. These high
energy domains provoke an enhanced interaction with the nitrogen quadrupole. The second region, CD, corresponds to the adsorption on more energetically homogeneous basal planes. Such calorimetric measurements are thus
a simple means to estimate the proportion of lateral and basal planes of such
materials as well as the effect of grinding.

The palygorskites are fibrous clay minerals. Attapulgite and sepiolite are two members of this family which both contain structural micropores. Their structures comprise talc-like layers arranged quincuncially,

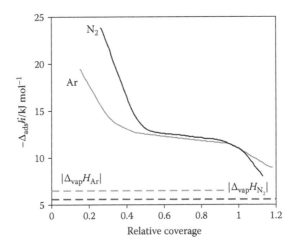

FIGURE 4.21
Enthalpies of adsorption with respect to relative coverage at 77.4 K for nitrogen and argon on
kaolinite. (After Cases, J.M. et al., *Clay Minerals*, 21, 55, 1986. With permission.)

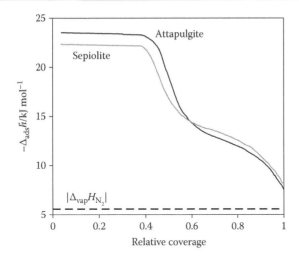

FIGURE 4.22
Enthalpies of adsorption with respect to relative coverage at 77.4 K for nitrogen on attapulgite and sepiolite.

forming microporous channels of rectangular cross section parallel to the longitudinal axis of the crystals [36]. While attapulgite has a pore section of $0.37 \times 0.64 \, nm^2$, the section of sepiolite is of $0.67 \times 1.34 \, nm^2$. The pores contain $Mg(OH_2)_2$ groups situated in the structural micropore walls.

The differential enthalpy curves obtained for the adsorption of nitrogen on sepiolite and attapulgite at 77.4 K are shown in Figure 4.22 [36,37]. Two separate domains can be observed for each of these curves. The first domain up to a relative coverage of 0.4 (Figure 4.22) is quasi-horizontal, which is characteristic of adsorption in highly homogeneous regions. This would seem to correspond to the adsorption within the intrafibrous micropores containing the $Mg(OH_2)_2$ groups.

The second domain corresponds to the end of any micropore filling in the adsorption isotherms. The differential enthalpy curves in this region, in the relative coverage range (0.6–0.9) (Figure 4.22), correspond to a decrease toward the enthalpy of liquefaction. This is characteristic of more energetically heterogeneous regions. It would seem that such regions are found between the fibers. Thus this would seem to correspond to adsorption in these interfibrous micropores.

4.3.2 Adsorption Calorimetry for the Characterization of Homogeneous Adsorbents

As has been seen for the case of attapulgite and sepiolate clays, the presence of energetically homogeneous regions within an adsorbent material will often lead to a near horizontal region in the differential energies of adsorption versus coverage curves. A typical homogeneous nonporous solid is graphite. Indeed the hexagonally ordered carbon of graphite can be considered as a model surface. For the adsorption of nitrogen on graphite, the coverage of the surface is accompanied by a slight increase in the differential energy of adsorption (compare Figure 4.15) which would translate into the addition of homogeneous gas–solid interactions and increasing gas–gas interactions.

Fullerene nanotubes can be considered as a homogeneous porous carbon material. An example of adsorption measurements is shown in Figure 4.23 [38]. Here, two main regions can be distinguished. It is well known that such nanotubes are closed at each end, hence blocking any inherent microporosity. Moreover, these nanotubes arrange themselves into bundles with a porosity of around 0.3 nm between the fibers. This latter porosity should thus be inaccessible.

The first step (AB, Figure 4.23) may be explained by the filling of a small percentage of *unblocked* nanotubes. According to the preparation mode, the quantity of unblocked pores can be in the region of 20%. The second region, BC, would thus seem to be the formation of a monolayer on the external surface of these nanotubes.

From the applications point of view, zeolites and related materials (aluminophosphates, gallophosphates, etc.) are interesting materials. The synthesis of such materials can be adjusted to give a wide range of crystal structures, and an almost infinite variety of chemical compositions make it possible to tailor-make samples for specific applications.

From a fundamental point of view, the regular pore systems can be indexed by x-ray diffraction and the structure can be elucidated using Riedvield

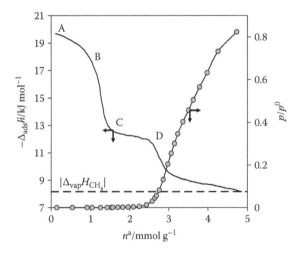

FIGURE 4.23

Enthalpies of adsorption and relative pressures as a function of quantity adsorbed at 77.4 K for methane on carbon nanotubes [38].

refinement-type methods, for example. The zeolite family of materials is thus ideal for the understanding of adsorption phenomena. It is the knowledge gained by such studies, using thermodynamic methods (manometry, calorimetry, etc.), complemented by structural methods (neutron diffraction, x-ray scattering, etc.), which permits, by analogy, the interpretation of adsorption phenomena in more disordered systems. Simulation studies are essential to complete the fundamental understanding.

However, the regularity of the channel systems can lead to distinct adsorption phenomena due to the confinement in such homogeneous systems. Nevertheless, a fine example of the adsorption in an energetically homogeneous solid is given in Figure 4.24.

The behavior shown during the adsorption of methane on silicalite at 77.4 K (Figure 4.24) can be considered almost model. The quasi-horizontal calorimetric signal, corresponding to the entire micropore filling region, would seem to be the result of adsorbent–adsorbate interactions only. One would expect a certain contribution due to adsorbate–adsorbate interactions; however, this would seem to be minimal due to the reduced possibility of such interactions taking place in such a quasi-one-dimensional pore system.

The chemistry of such systems can also be probed by adsorption calorimetry and the adsorption enthalpies at zero coverage can be compared as shown in Figure 4.25.

The adsorption of various simple gases was studied on a series of hydrogen-exchanged ZSM-5 zeolites with different Si/Al ratios. The enthalpies at zero coverage are plotted as a function of the Si/Al ratio in Figure 4.25. It can be seen that for nonspecific gases, such as methane, argon, and krypton, there is little change in adsorption energy. For nitrogen (quadrupole)

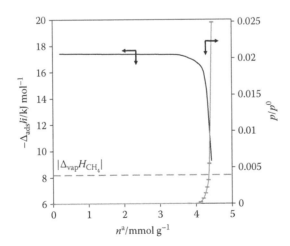

FIGURE 4.24
Enthalpies of adsorption and relative pressure as a function of quantity adsorbed at 77.4 K for methane on silicalite-I [39].

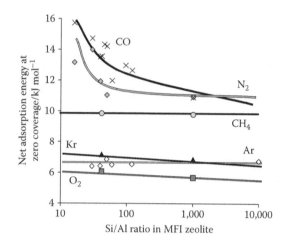

FIGURE 4.25
Adsorption energies for different probe molecules as a function of the Si/Al ratio for various HZSM-5 zeolites.

and carbon monoxide (dipole) with a permanent electric moment, a distinct variation in interaction energy is observed. Indeed, an increase in Si/Al ratio results in a decrease in the content of the compensation cation (H^+). The cations within the structure can be considered specific adsorption sites.

4.3.3 Adsorption Calorimetry for the Detection of Adsorbate Phase Transitions

The adsorbed gas can sometimes be highly influenced by the homogeneity of the surface with which it is in contact, for example for nitrogen adsorption on graphite (Figure 4.15) in which the nitrogen forms an epitaxial film on completion of the surface monolayer. Even more surprising phenomena can occur during adsorption on some zeolites and aluminophosphates, for example silicalite and $AlPO_4$-5.

Figure 4.26 highlights the rather interesting adsorption behavior that can be observed for the adsorption of nitrogen on silicalite at 77 K. The adsorption isotherm exhibits two substeps, α and β. The initial pore filling, however, results in a differential curve which is not completely horizontal. An initial decrease would seem to indicate an enhanced interaction, possibly with defect sites. This curve then increases again, which seems to be characteristic of increasing adsorbate–adsorbate interactions.

The substeps in the isotherm correspond to marked differences in the differential enthalpy curves. Although this second substep, β, was observed in the isotherm prior to any microcalorimetric measurements [40,41], the first substep, α, was initially observed in the calorimetric curve [42]. A complementary study by neutron diffraction was carried out on this system [42].

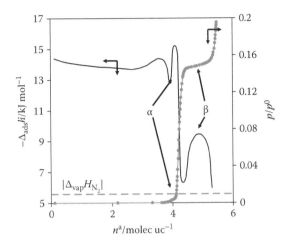

FIGURE 4.26
Differential enthalpies of adsorption and relative pressure as a function of quantity adsorbed at 77.4 K for nitrogen on silicalite [42].

The first substep, α, was concluded to result from an ordering of the adsorbate from a fluid phase to a network fluid. The second substep, β, would seem to correspond to an adsorbate phase transition similar to that previously observed for argon [39], characterized by a network fluid to a "solid-like" adsorbate phase.

Much research has been done about the synthesis of zeolite-like materials with framework species other than silica and alumina. The first family of materials that resulted from this research were the aluminophosphate molecular sieves. Thus $AlPO_4$-5 5 [43] has a unidirectional pore system consisting of parallel circular channels of 0.73 nm diameter. $AlPO_4$-5 has a framework, which, like silicalite-I, is theoretically globally electrically neutral, although the pore openings are slightly larger than those of the MFI-type zeolites. These characteristics make $AlPO_4$-5 an excellent structure for fundamental adsorption studies.

For $AlPO_4$-5, the adsorption isotherms of argon and nitrogen traced up to a relative pressure of 0.2 are indistinguishable (Figure 4.27). Methane adsorbs significantly less, suggesting a different pore filling mechanism.

For the adsorption of methane, an exothermic peak (noted α in Figure 4.27) is observed in the differential enthalpy curve, which would seem to correspond to an energetic term $\approx RT$. This could indicate both a variation of mobility and a variation of the adsorbed methane phase. A complementary neutron diffraction study [45] indicated that the behavior of the methane adsorbed phase is unusual. The methane appears to undergo a transition between two solid-like phases. The first "solid-like" phase corresponds to the adsorption of four molecules per unit cell, while the second phase corresponds to an increase in the amount adsorbed to six molecules per unit cell.

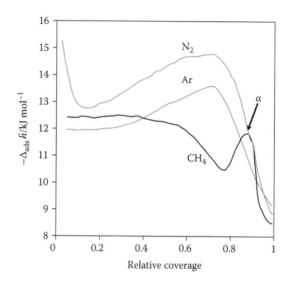

FIGURE 4.27
Differential enthalpies at 77.4 K for nitrogen, argon, and methane adsorbed on $AlPO_4$-5 [44].

This could be the result of a favorable dimensional compatibility between the methane molecule and AlPO4-5 micropore, permitting, from a spatial point of view, the appearance of two relatively dense phases. This hypothesis is supported by simulation studies [46].

4.3.4 Adsorption Calorimetry to Follow the Various Stages of Micropore Filling

For some systems, the different stages of micropore filling can be followed by adsorption calorimetry, for example for large-pore zeolites a two-step filling process can be observed.

The 5A zeolites consist of regularly spaced spherical cages of 1.14 nm diameter. These cages are linked to each other by six circular windows of about 0.42 nm diameter. The negatively charged silico-aluminate framework requires compensation cations. For zeolite 5A, these exchangeable cations are generally a mixture of calcium and sodium.

The 13X zeolite, or faujasite, has very similar primary building blocks to the 5A zeolites. For 13X, however, the spherical cages are 1.4 nm in diameter and they are linked to each other by four circular windows of about 0.74 nm diameter. The exchangeable cations are generally sodium (NaX).

A study of the adsorption of nitrogen at 77 K on 5A and 13X zeolites using quasi-equilibrium, isothermal, adsorption microcalorimetry experiments at 77K [47] detected a step in the differential enthalpies of adsorption, toward the end of micropore filling (Figure 4.28). At the time, this was interpreted as a consequence of specific adsorbate–adsorbate interactions. Recently,

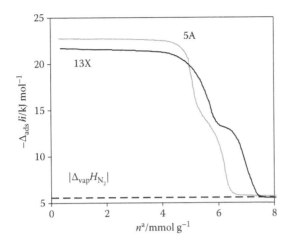

FIGURE 4.28

Differential enthalpies of adsorption and relative pressure as a function of quantity adsorbed at 77.4 K for nitrogen on 5A and 13X zeolites [47].

however, in the light of other microcalorimetry studies, this change in signal has been interpreted as a possible phase change within the cavities [48]. This latter study detected the same phenomenon for a number of other probe molecules, including argon, methane, and carbon monoxide. A second explanation could simply be that the cages are filled in two steps: near the walls and prior to complete filling.

For adsorption on $AlPO_4$-11 an unusual pore filling mechanism can be observed using adsorption calorimetry. $AlPO_4$-11 has a rectilinear pore system similar to that of $AlPO_4$-5. However, the cross section of the pores is elliptical ($0.39 \times 0.63 \, nm^2$). The adsorption of a number of probe molecules occurs in two distinct steps as indicated in Figure 4.29 [48,49]. From the form of the differential enthalpy curve, each step would seem to correspond to a relatively homogeneous filling process. Complementary neutron diffraction experiments showed that the coefficient of diffusion for the first step is around 1/10 of that in the second step [48,49]. This is contrary to the expected behavior where the coefficient of diffusion decreases with increasing adsorbate loading.

In this example, however, the adsorbate is initially situated in the most curved part of the elliptical pore. The curvature acts as a very strong adsorption site, but about 50% of the porosity remains unfilled. In the second adsorption step, the initial adsorbate molecules have to be dislodged to allow the complete filling of the micropores.

4.3.5 Adsorption Calorimetry and Capillary Condensation

As seen above, adsorption calorimetry is of value for examining specific interactions with a surface, or for microporous samples. In general, capillary

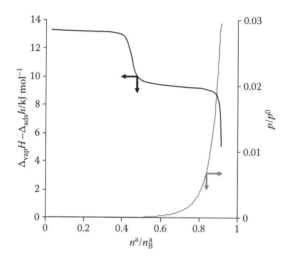

FIGURE 4.29
Differential enthalpy (black) and isotherm (grey) for the adsorption of carbon monoxide on
AlPO$_4$-11 at 77 K.

condensation in mesoporous samples is associated with the enthalpy of liquefaction of the adsorbate fluid. For very well-ordered mesoporous samples,
such as MCM-41, a small increase in the enthalpy of adsorption is observed
during capillary condensation with respect to the liquefaction energy [30,50].

This is shown in Figure 4.30 for methane adsorption on an MCM-41
sample of 4 nm pore diameter. The initial adsorption leads to an enthalpy

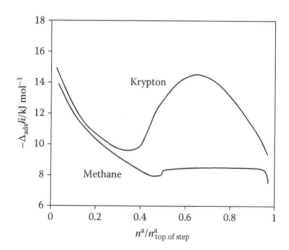

FIGURE 4.30
Differential enthalpy of adsorption as a function of coverage for methane and krypton on
MCM-41 at 77 K.

curve typical for silica (compare with Figure 4.30). The capillary condensation within the mesopores is accompanied by a slight increase of around 0.5–1 kJ mol^{-1}.

Krypton adsorption on MCM-41 at 77 K shows a similar behavior to that of methane. However, the mesopore filling process is accompanied by a surprisingly strong heat effect [30,50]. Complementary experiments suggest that the krypton is adsorbed within the mesopores in a solid-like state. Indeed for samples of such pore size, krypton within the mesopores undergoes a liquid–solid transition at around 83 K [51].

References

1. W. D. Harkins and G. Jura, *J. Am. Chem. Soc.*, 66 (1944) 1362.
2. S. Partyka, F. Rouquerol, and J. Rouquerol, *J. Colloid Interface Sci.*, 68 (1979) 21.
3. J. Rouquerol, *Thermochim. Acta*, 144 (1989) 209.
4. O. Toft Sorensen and J. Rouquerol, eds., *Controlled Thermal Analysis: Origin, Goals, Multiple Forms, Applications and Future*, Kluwer Acad. Publishers, Dordrecht, the Netherlands (2003).
5. W. D. Harkins, *The Physical Chemistry of Surface Films*, Reinhold, New York (1952) p. 275.
6. J. J. Chessick and A. C. Zettlemoyer, *Adv. Catal.*, 11 (1959) 263.
7. Y. Grillet and P. L. Llewellyn, in *The Surface Chemistry of Silica*, A. P. Legrand, ed., Wiley, Chichester, U.K. (1998) Chap 2, p. 23.
8. R. Denoyel, J. Fernandez-Collinas, Y. Grillet, and J. Rouquerol, *Langmuir*, 9 (1993) 515.
9. H. F. Stoeckli, P. Rubstein, and L. Ballerini, *Carbon*, 28 (1990) 907.
10. C. G. de Salazar, A. Sepùlveda-Escribano, and F. Rodrìguez-Reinoso, *Stud. Surf. Sci. Catal.*, 128 (2000) 303.
11. D. H. Everett and J. C. Powl, *J. Chem. Soc. Faraday Trans. 1*, 72(3) (1976) 619.
12. J. J. Chessick, G. J. Young, and A. C. Zettlemoyer, *Trans. Faraday Soc.*, 50 (1954) 587.
13. J. A. G. Taylor, *Chem. Ind.* 49, (1965) 2003.
14. J. Rouquerol, P. Llewellyn, R. Navarrete, F. Rouquerol, and R. Denoyel, *Stud. Surf. Sci. Catal.*, 144 (2002) 171.
15. A. Y. Fadeev and V. Eroshenko, *J. Colloid Interface Sci.*, 187 (1997) 275.
16. V. Eroshenko, R. C. Regis, M. Soulard, and J. Patarin, *J. Am. Chem. Soc.*, 123 (2001) 8129.
17. V. Eroshenko, R. C. Regis, M. Soulard, and J. Patarin, *C. R. Phys.*, 3 (2002) 111.
18. F. Gomez, R. Denoyel, and J. Rouquerol, *Langmuir*, 16 (2000) 4374.
19. D. H. Everett, *Trans. Faraday Soc.*, 57 (1961) 1541.
20. P. Kubelka, *Z. Elekt. Ang. Phys. Chem.*, 38 (1932) 611.
21. M. Brun, A. Lallemand, J. F. Quinson, and C. Eyraud, *Thermochim. Acta*, 21 (1977) 59.
22. J. Rouquerol, in *Thermochimie*, Colloques Internationaux du CNRS, No.201, CNRS Ed., Paris, France (1972) p. 537.

23. C. Letoquart, F. Rouquerol, and J. Rouquerol, *J. Chim. Phys.*, 70 (1973) 559.
24. Y. Grillet, F. Rouquerol, and Jean Rouquerol, *J. Chim. Phys.*, 7–8 (1977) 778.
25. P. Llewellyn and G. Maurin, *C. R. Chimie*, 8(3–4) (2005) 283–302.
26. J. Rouquerol, S. Partyka, and F. Rouquerol, *J. Chem. Soc. Faraday Trans. 1*, 73 (1977) 306.
27. A. V. Kiselev, *Doklady Nauk USSR*, 233 (1977) 1122.
28. K. S. W. Sing, in *Thermochimie*, Colloques Internationaux du CNRS, No.201, CNRS Ed., Paris, France (1972), p. 537.
29. K. S. W. Sing, D. H. Everett, R. A. W. Haul, L. Moscou, R. A. Pierotti, J. Rouquerol, and T. Siemieniewska, *Pure Appl. Chem.*, 57 (1985) 603.
30. P. L. Llewellyn, Y. Grillet, J. Rouquerol, C. Martin, and J.-P. Coulomb, *Surf. Sci.*, 352–354 (1996) 468.
31. D. Atkinson, P. J. M. Carrott, Y. Grillet, J. Rouquerol, and K. S. W. Sing, in *Proceedings of 2nd International Conference on Fundamentals of Adsorption*, A. I. Liapis, ed., Engineering Foundation, NY (1987) p. 89.
32. A. P. Legrand, H. Hommel, A. Tuel, A. Vidal, H. Balard, E. Papier, P. Levitz et al., *Adv. Colloid Interface Sci.*, 33 (1990) 91.
33. F. Rouquerol, J. Rouquerol, and K.S.W. Sing, *Adsorption by Powders and Porous Solids: Principles, Methodology and Applications*, Academic Press, London (1999).
34. P. Brauer, H. R. Poosch, M. V. Szombathely, M. Heuchel, and M. Jarioniec in *Proceedings of 4th International Conference on Fundamentals of Adsorption*, M. Suzuki, ed., Kodansha, Tokyo, Japan (1993) p. 67.
35. J. M. Cases, P. Cunin, Y. Grillet, C. Poinsignon, and J. Yvon, *Clay Miner.*, 21 (1986) 55.
36. Y. Grillet, J. M. Cases, M. François, J. Rouquerol, and J. E. Poirier, *Clays Clay Miner.*, 36 (1988) 233.
37. J. M. Cases, Y. Grillet, M. François, L. Michot, F. Villieras, and J. Yvon, *Clays Clay Miner.*, 39 (1991) 191.
38. M. Muris, N. Dufau, M. Bienfait, N. Dupont-Pavlovsky, Y. Grillet, and J. P. Palmary, *Langmuir*, 16 (2000) 7019.
39. P. L. Llewellyn, J.-P. Coulomb, Y. Grillet, J. Patarin, H. Lauter, H. Reichert, and J. Rouquerol, *Langmuir*, 9 (1993) 1846.
40. P. J. M. Carrott and K. S. W. Sing, *Chem. Ind.*, 22 (1986) 786.
41. U. Müller and K. K. Unger, *Fortschr. Mineral.*, 64 (1986) 128.
42. P. L. Llewellyn, J.-P. Coulomb, Y. Grillet, J. Patarin, G. André, and J. Rouquerol, *Langmuir*, 9 (1993) 1852.
43. S. T. Wilson, B. M. Lok, C. A. Messina, T. R. Cannan, and E. M. Flanigen, *J. Am. Chem. Soc.*, 104 (1982) 1146.
44. Y. Grillet, P. L. Llewellyn, N. Tosi-Pellenq, and J. Rouquerol, in *Proceedings of 4th International Conference on Fundamentals of Adsorption*, M. Suzuki, ed., Kodansha, Tokyo, Japan (1993) p. 235.
45. C. Martin, N. Tosi-Pellenq, J. Patarin, and J.-P. Coulomb, *Langmuir*, 14 (1998) 1774.
46. V. Lachet, A. Boutin, R. J. M. Pellenq, D. Nicholson, and A. H. Fuchs, *J. Phys. Chem.*, 100 (1996) 9006.
47. F. Rouquérol, S. Partyka, and J. Rouquérol, in *Thermochimie*, CNRS Ed., Paris, France (1972) p. 547.
48. N. Dufau, P. L. Llewellyn, C. Martin, J.-P. Coulomb, and Y. Grillet, in *Proceedings of Fundamentals of Adsorption VI*, F. Meunier, ed., Elsevier, Paris, France (1999) p. 63.

49. N. Dufau, N. Floquet, J.-P. Coulomb, P. Llewellyn, and J. Rouquerol, *Stud. Surf. Sci. Catal.*, 135 (2001) 2824.

50. P. L. Llewellyn, C. Sauerland, C. Martin, Y. Grillet, J.-P. Coulomb, F. Rouquerol, and J. Rouquerol, in *Proceedings of Characterisation of Porous Solids IV*, Royal Society of Chemistry, Cambridge, MA (1997) p. 111.

51. J.-P. Coulomb, Y. Grillet, P. L. Llewellyn, C. Martin, and G. André, in *Proceedings of Fundamentals of Adsorption VI*, F. Meunier, ed., Elsevier, Paris, France (1999) p. 147.

5

Combination of In Situ and Ex Situ Techniques for Monitoring and Controlling the Evolution of Nanostructure of Nanoporous Materials

G.N. Karanikolos, F.K. Katsaros, G.E. Romanos,
K.L. Stefanopoulos, and N.K. Kanellopoulos

CONTENTS

5.1 Introduction

Nanoporous materials play an important role in chemical processing, as in many cases they can successfully replace traditional, pollution-prone, and energy-consuming separation, catalytic, and other major processes. These materials are abundantly used as membranes, sorbents, and many other systems such as photocatalytic supports, drug delivery systems, etc., and form the basis of innovative technologies, involving hydrogen storage, high-temperature molecular sieve membrane separations, (hydrogen production, carbon dioxide capture, olefin/paraffin separation, supercritical FCC catalysis, etc.), mainly due to their unique structural or surface physicochemical properties, which can, to an extent, be tailored to meet the specific needs of each process.

Any equilibrium or dynamic process that takes place within the nanopores of a solid is strongly influenced by the topology and the geometrical disorder of the pore matrix. Despite the significant progress in the development of characterization techniques, the complete characterization of the extremely complex nanostructure and the internal surface properties of nanoporous materials still remains a difficult and frequently controversial problem, even if the equilibrium and transport mechanisms themselves are quite simple and well defined. This is mainly due to the great difficulty in accurately representing the complex morphology of the pore matrix. To this end, the application of combined techniques aided by advanced model analysis is of major importance as it is the most powerful tool currently available. On the other hand, no matter how thorough and complete the characterization, it is often inadequate if it is not related to the process under consideration, since one of the most important parameters in any application is the material's ability to retain its properties over a certain period of time. The "changes" induced on materials during their utilization in specific applications are highly relevant and crucial for the economic viability of many applications, such as catalytic, separation, and other processes. In this context, the establishment of innovative combinations of "in situ" and "ex situ" techniques is of great importance. These advanced methodologies will serve as a tool in order to monitor and control the evolution of nanostructure and other properties, which are highly relevant and crucial for the economic viability of a variety of processes involving nanoporous solids. The objective is the establishment of an innovative scheme of combinations of characterization techniques aided by advanced simulation techniques in order to optimize processes, spanning from the synthesis of materials (where particle size control, crystallization kinetics, etc. play a significant role in the product) to processes involving sorbents, membranes, catalysts, and other novel nanomaterials.

The combination of in situ and ex situ techniques aided by appropriate predictive models could become a powerful nanostructure tailoring tool at the nanoscale level aiming at the development of next generation of nanoporous membranes, sorbent, and catalysts by

1. Providing a significant breakthrough in production methodologies replacing empirical procedures, improving standardization and reproducibility while significantly lowering production costs.

2. Optimizing and prolonging the period of optimum performance of nanoporous materials, a very crucial factor for the economic viability of many industrial processes.

3. Determining/predicting the nanoporous materials' performance in the innovative pre-designed realistic process environment with real feed conditions, leading to the development of "smart engineering processes" in several sectors, such as high temperature hydrogen separation membranes, high pressure carbon dioxide separation nanoporous membranes, advanced hybrid membrane/sorbent/catalyst systems for volatile organic compounds (VOCs) removal or oxidative destruction, etc.

This chapter focuses on the applications for the production and characterization of novel nanoporous membranes, carbon nanotube (CNT) membranes, and catalysts.

5.2 The Combination of In Situ and Ex Situ Techniques as a Characterization Tool of Nanoporous Materials

5.2.1 The Development of Models for the Theoretical Analysis of the Combinations of Relative Permeability and Sorption Measurements

The study of the relation between the gas transport properties of porous solids and their pore structure is of great importance for the effective characterization and design of the porous membranes, adsorbents, and catalysts widely used in the chemical, biomedical, and other applications. The performance in these applications depends markedly on their structural characteristics. One relatively easy practical way of obtaining structural information about such materials is through their interaction with fluids.

Fatt (1956a–c) was the first to point out that a network of tubes can effectively be applied to represent multiphase flow in porous media. In his pioneer studies, both uniform and nonuniform networks of capillaries—with certain pore size distributions—were used for the elucidation and the analysis of liquid permeability, liquid–liquid displacement, and relative permeability in oil recovery systems.

For the case of mesoporous and nanoporous materials, static methods involving the analysis of gas and vapor sorption, or liquid (notably mercury) penetration, isotherms are the best known and most widely used on a routine basis. Such porous solids are commonly characterized by their pore size distribution of the pores, $f(\rho)$, and the simplest and industrially most widely employed methods for the determination of $f(\rho)$ involve the analysis of nitrogen sorption and mercury penetration isotherms (Gregg and Sing 1982).

An important point to note, however, is that both static and dynamic methods of characterization of porous solids are based on suitable interpretation of the relevant data. Hence, the structural information derived therefore is model dependent. A second noteworthy point is that current routine applications of the aforesaid methods rely on rather simple models of pore structure. In more realistic model studies of gas or gas and surface transport in porous adsorbents used by Nicholson and Petropoulos (1971, 1973), the porous medium was represented much more realistically as a network of capillaries of randomly varying radius.

In this approach, the analysis of the sorption isotherms is combined with the analysis of the gas relative permeability (or permoporometry), P_R, defined as the permeability of porous medium to a given fluid in the presence of another fluid normalized by the permeability at zero content of foreign sorbate. In the relative permeability, theoretical studies can be simulated by means of a capillary network, consisting of a regular cubic array of nodes joined together by cylindrical tubes. The network is characterized by two variables, the degree of connectivity (measured as the number of capillaries meeting at each junction, n_T) and the distribution of capillary radii, $f(r)$, assumed to be adequately long in order to avoid end effects. In Figure 5.1, the relative permeability curves of three networks with connectivity $n_T = 2$ (serial), $n_T = 4$, and $n_T = \infty$ (parallel model) are depicted at different vapor pressures.

The dependence of the variation of P_R on the pore volume fraction occupied by the sorbed vapor, v_s, was demonstrated by means of numerical calculations for different $f(r)$ (for the simple limiting cases of parallel $n_T = \infty$), where the diminishing of gas flux, which is called percolation threshold, occurs when the larger pore in the parallel array is blocked by the sorbed vapor at $v_s = 1$. On

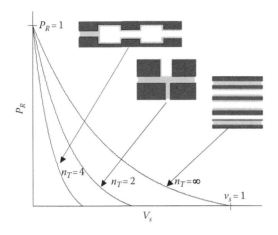

FIGURE 5.1
Relative permeability curves depicting the increase of the percolation threshold with increasing the network's connectivity.

the contrary, in the case of serial arrangement of the pores, it is obvious that the diminishing of gas flux occurs much earlier, when the smaller pore in the serial array is blocked by the sorbed vapor (Kanellopoulos and Petrou 1988).

For the case of the intermediate connectivities, 3D computer models (Figure 5.2) can be applied for the representation of the relative permeability.

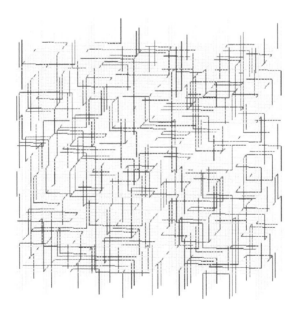

FIGURE 5.2
Three-dimensional computer models representing pore network. (From *Adv. Colloid Interface Sci.*, 76/77, Tzevelekos, K.P., Kikkinides, E.S., Stubos, A.K., Kainourgiakis, M.E., Kanellopoulous, N.K., On the possibility of characterizing mesopourous materials by permeability measurements of condensable vapours: theory and experiments, 373, Copyright 1998, with permission from Elsevier.)

In the latter case, the network is represented by a regular array of nodes joined together by a constant number, n_T, of cylindrical tubes.

The distribution function of capillary radii $f(r)$ is defined in the range $r_a < r < r_b$ subject to the normalizing condition (Fatt 1956c):

$$\int_{r_\alpha}^{r_b} f(r)\,dr = 1.$$

It is convenient to normalize r with respect to the middle radius, $r_m = (r_a + r_b)/2$. Thus, we have

$$\rho = \frac{r}{r_m}; \quad \rho_\alpha = \frac{r_a}{r_m}; \quad \rho_b = \frac{r_b}{r_m}, \quad \sigma = \rho_b - 1 = 1 - \rho_\alpha,$$

$$f(\rho)d\rho = f(r)dr; \quad \rho_\alpha \leq \rho \leq \rho_b.$$

Assuming the thickness, t, to be independent of ρ, t is given by (Kanellopoulos and Petrou 1988; Petropoulos et al. 1989)

$$t = \frac{c_2^{1/3}}{r_m\left[\ln(p/p_0)\right]^{1/3}},$$

where $c_2 = 0.218\,\text{nm}^3$ for N_2 at 77 K. Capillaries with radii below the critical radius, ρ_K, are filled with condensate, as follows, where ρ_K is given as a function of p/p_0 by

$$\rho_K = \frac{c_1}{r_m\,\ln(p/p_0)} + t,$$

where $c_1 = 0.477\,\text{nm}$ for N_2 at 77 K.

The gas flux, J_{ij}, for a cylindrical tube connecting two nodes, i and j, at pressures p_{gi} and, p_{gj}, respectively, in the Knudsen flow regime of interest here is given by

$$J_{ij} = \left(\frac{2\pi r_m^3 \bar{\upsilon}}{3h}\right)\rho_{cij}^3 (\rho_{gi} - \rho_{gj}),$$

where

$\rho_{cij} = \rho_{ij} - t$ is the open core radius of a capillary, partly filled by adsorbate of thickness t

$\bar{\upsilon}$ is the mean gas molecular speed

h is the length of the tube, assumed to be the same for all the network tubes

For continuity of flow, at each node i, we have $\sum_{j=1}^{n_T} J_{ij} = 0$.

The set of the simultaneous linear equations for all nodes i has been solved by the Gauss–Seidel iteration method and the total flux, J, across the network has been obtained for a given pressure drop across the network. The above computation has been repeated for values of p/p_0 between zero and unity and

$$P_R = \frac{J(p/p_0)}{J(0)}$$

has been determined as a function of the sorbate content.

Alternatively, the effective medium approximation (EMA) method can be employed in order to compute the required $P_R(v_g)$ curves in a much simpler manner. By the method above, the actual network is replaced by an effective one of uniform radius (Figures 5.3 and 5.4).

The network permeability is obtained by requiring the same flux to result from both the effective network and the actual partially filled network of

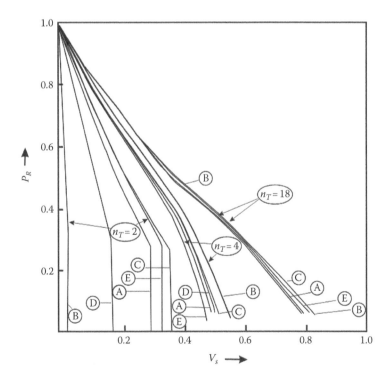

FIGURE 5.3

P_R curves as in Figure 5.1 for different distributions and connectivities. (From *J. Membr. Sci.*, 37(1), Kanellopoulos, N.K., and Petrou, J.K., Relative gas permeability of capillary networks with various size distributions, 1–12, Copyright 1988, with permission from Elsevier.)

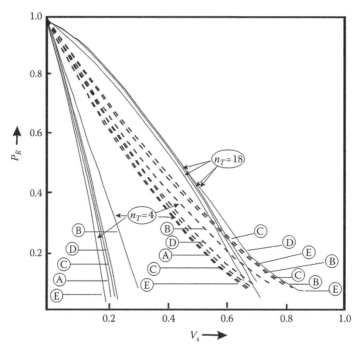

FIGURE 5.4
P_R curves as in Figure 5.3 for the hypothetical cases of pure adsorption (- - -) and pure condensation. (From *J. Membr. Sci.*, 37(1), Kanellopoulos, N.K., and Petrou, J.K., Relative gas permeability of capillary networks with various size distributions, 1–12, Copyright 1988, with permission
from Elsevier.)

radius of the pore size distribution $f(\rho)$. This requires the solution of the
following integral equation, using a false position algorithm:

$$\frac{\left(1-\int_{\rho K}^{\rho_b} \rho^n f(\rho)d\rho\right)}{((n_T/2)-1)} + \int_{\rho K}^{\rho_b} \frac{[P_m-(\rho-t)^3]f(\rho)d\rho}{(\rho-t)^3+((n_T/2)-1)P_m} = 0.$$

In addition, by suitable use of EMA, the relative permeability of a stochastic
network consisting of idealized pores of randomly varying radius, which
are progressively constricted and/or blocked, can be expressed analytically
as a function of the relevant microstructural network parameters (i.e., as an
explicit function of the two network parameters, namely n_T and the pore size
distribution $f(\rho)$, which are functions of the suitable moments of the radius
distribution and network connectivity) to a useful degree of approximation.
The aforesaid treatment can (1) provide a much deeper insight into the effect
and relative importance of individual structural parameters; (2) permit a
greatly simplified method for estimating the network connectivity, n_T; and (3)

check the self-consistency of experimental pore size distributions obtained by conventional sorption isotherm analysis.

In EMA, the actual network is replaced by an "effective" alternate network of uniform radius. The "effective" radius is estimated by requiring equal flux to result from both the effective and the actual networks at the same pressure difference. Upon adsorption the pores follow a probability distribution function $f(x)$ with $x = r - t$ (r is the pore radius and t is the thickness of the adsorbed layer). When condensation occurs, all the pores with radii smaller than a critical value x_k ($x_k = r_k - t$) are blocked (filled with condensed vapor) so that the probability distribution function is $f_c(x) = f_a \cdot \delta(x) + f(x_k \leq x \leq \infty)$ where $\delta(x)$ is the Dirac function, and $f_a = \int_0^{x_k} f_c(x)dx$ is the fraction of the blocked pores. From EMA (Petropoulos et al. 1989),

$$\int_0^\infty \frac{(P_m - c \cdot x^3) \cdot f_c(x)dx}{c \cdot x^3 + v P_m} = 0,$$

where
$v = (n_T/2) - 1$, n_T being connectivity ($2 \leq n_T \leq \infty$)
P_m is the permeability
$c \cdot x^3$ is the conductance (in the Knudsen regime)

By moment expansion of the aforementioned EMA equation the relative permeability can be related explicitly to the relevant microstructural network parameters, namely the suitable moments of the pore size distribution and the network connectivity. The relative permeability, P_R, is then given as

$$P_R = \frac{\bar{x}_c^3 \cdot (1 + \tilde{a}_1) \cdot f_b \cdot (\tilde{a}_{M1} + \tilde{a}_{M2})}{\bar{r}_c^3 \cdot (1 + a_1) \cdot \tilde{a}_M} \cdot \left(\frac{f_b - \lambda}{1 - \lambda} \right),$$

where the relevant parameters (a and λ) are functions of n_T and the moments of psd defined in Petropoulos et al. (1989) and Steriotis et al. (1995). f_b is the fraction of conducting pores ($f_b = 1 - f_a$), $\bar{x}_c = \int_{x_k}^\infty x \cdot (f(x)/f_b)dx$ is the average radius of the open pores, and $\bar{r}_c = \int_0^\infty r \cdot f(r)dr$ is the average radius of the network.

A typical application of this model is the calculation of the pore connectivity, as in the case of the mesoporous alumina membrane. Figure 5.5 illustrates the helium P_R versus the amount of the adsorbed water.

It becomes clear that when the amount adsorbed is between 0% and 30% of the total pore volume there is a sharp decrease in P_R, while for adsorbed fractions over 30% the drop in P_R is smoother and approaches 0 before saturation (at $V_s/V_t \sim 0.83$). At this point (percolation threshold) the last open continuous path of the membrane system is blocked and no flow through the system occurs. Figure 5.5 also shows the EMA model P_R curves for $n_T = 2$, 8, and 18.

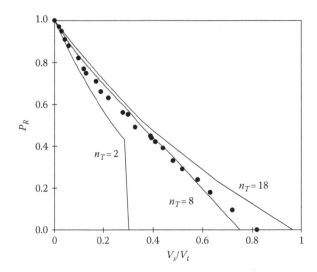

FIGURE 5.5
Prediction of relative permeability curves (lines) based on EMA for a ceramic membrane (alumina) using different connectivity. Circles: experimental points.

By using the best-fitting simulation curve the actual pore connectivity ($n_T=8$ for alumina) can be estimated (Steriotis et al. 1995).

We may also note that an EMA model has recently been developed for the condensed vapor differential permeability (Kikkinides et al. 1997). It is shown that these measurements are very sensitive to the structural characteristics of the membrane (Kainourgiakis et al. 1996), and a complementary dynamic method for the determination of n_T can be based thereon.

Finally, 2D and 3D networks, composed of convergent–divergent flow channels have been employed for the simulation of the relative permeability behavior of a random packing of equal spheres. The relative permeability is calculated as a function of the amount of the sorbed vapor volume, which is the summation of the volume due to multilayer adsorption and capillary condensation and the sizes of the flow channels are determined on the basis of the structural characteristics of the sphere pack (Sasloglou et al. 2001).

The relative permeability is calculated as a function of the amount of the sorbed vapor volume, which is the summation of the volume due to multilayer adsorption and capillary condensation, and the sizes of the flow channels are determined on the basis of the structural characteristics of the sphere pack.

In the current model, the primary nonporous particles of some adsorbents, such as alumina, have been simulated as hard spheres of equal size. Initially, the spheres are placed in a cubic box in regular arrangement at the desired porosity, and subsequently, they are moved one by one, until satisfactory randomness is achieved. It is postulated that equilibrium is achieved when the pair radial distribution function becomes constant.

Subsequently, the sphere packing is divided into Voronoi tetrahedral (Sasloglou et al. 2000). Each tetrahedron has four spheres, one at each vertex, and the cavity of the interior of each tetrahedron simulates the pore space or cavity. Each cavity is connected with the four neighboring cavities via triangular windows. The characteristic radius of a cavity r_c is defined as the radius of a sphere with a volume equal to the volume of the tetrahedron minus the volume occupied by the spherical particles (equivalent sphere). Similarly, the characteristic radius of a window, r_w, is equal to the radius of an "equivalent circle," the area of which is equal to the area of the triangle minus that occupied by the spherical particles.

As the pressure increases on the adsorption curve, cavity i (window i,j) is filled progressively with condensed vapor. It is assumed that the cavity fills with condensate, when the average of the six pendular ring radii (denoted as r_2 in Figure 5.6) exceeds the characteristic radius of the cavity i (rc_{ti}), as reduced in size by the adsorbed layer and pendular ring. Similarly, the window fills with condensate when the average of the radii of the three pendular rings exceeds the characteristic radius of a window i,j rw_{tij}, as reduced in size by the presence of sorbate. On the desorption branch, cavity i is allowed to empty when at least one of the neighboring cavities j is empty and the corresponding window becomes unstable (its radius exceeds the Kelvin radius).

Because of the complexity of the produced network between the random packed spheres, the actual complex in geometrical shape flow channel connecting the adjoining cavities i,j is simplified by a sequence of two conical tubes.

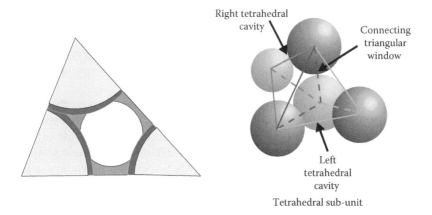

FIGURE 5.6
Pore network model made up of nonporous spherical particles packing in which networks of tetrahedral cavities are interconnected through triangular windows.

The method of modeling vapor sorption in random packed spheres in different porosities is successfully employed for the simulation of the relative gas permeability.

A simplified presentation of the actual flow channel configuration between cavities is employed for the representation of the complex pore geometry. There is a satisfactory agreement with experimental results for the case of CCl_4, and there is some deviation from the experimental results of C_5H_{12}. Furthermore, the predicted hysteresis in the relative permeability of the simulation is larger than that of the experimental results.

The major advantage of the realistic 3D random sphere model is that it takes into account the constrictions' effects of random spheres packed systems, which can be very important in many practical applications, for example, deactivation of catalysts by deposition in the constrictions, membrane separation effected by condensation of the heavy components in the constrictions, and blocking the flow of the light components.

5.2.2 Development of In Situ Adsorption and Neutron Diffraction Techniques for the Structural Characterization of Confined Phases

Sorption of fluids on nanoporous solids is very important in a series of applications such as catalysis, H_2 and natural gas upgrade and storage (chromatographic pressure swing adsorption and membrane), separations, biological and geological processes, etc. (Matranga et al. 1982; Noh et al. 1987; Ruthven et al. 1994). These processes are extremely complex from a fundamental point of view as the properties of sorbed fluids are different from the bulk due to confinement. This is mainly attributed to the combination of solid–fluid interactions and the finite pore sizes, both of which can alter the structural and dynamic properties of the confined fluid, and thus, strongly influence its phase behavior. For instance, confined geometries play significant role on first-order phase transitions and also in the glass formation (Gelb et al. 1999). For the above reasons, there is an increasing interest in the experimental study of fluids confined in nanoporous materials (Frick et al. 2005), including phase transitions (Christenson 2001) and dynamical properties of liquids in nanometer-sized porous materials (Frick et al. 2003).

Adsorption measurements (Gregg and Sing 1982) are abundantly used for the study of pore-confined fluids; however, such methods can only reveal the statistical "ensemble" macroscopic properties and do not provide information on their molecular structure. On the other hand, small-angle scattering (SAS) of x-rays or neutrons techniques (SAXS, SANS) are, nowadays, widely used for their structural characterization, while (x-ray or neutron) diffraction is proved to be an essential tool for the study of the molecular structure and organization of either bulk fluids or confined liquids (Baker et al. 1997; Morishige and Nobuoka 1997; Morishige and Kawano 1999, 2000a,b; Alba-Simionesco et al. 2003; Knorr et al. 2003; Morishige and Iwasaki 2003;

Morineau et al. 2004; Floquet et al. 2005; Morishige and Uematsu 2005; Wallacher et al. 2005; Liu et al. 2006a). Furthermore, adsorption in conjunction with in situ SAS or diffraction can contribute to an in-depth investigation of the phenomenon and resolve some of the implicated *"mysteries."* The enhanced information obtained by such combined methods has motivated the development of several (usually low-pressure) adsorption in situ scattering setups and cells (Li et al. 1994; Mitropoulos et al. 1995; Dolino et al. 1996; Hoinkis 1996; Ramsay and Kallus 2001; Smarsly et al. 2001; Albouy and Ayral 2002; Hofmann et al. 2005; Zickler et al. 2006; Scheiber et al. 2007; Jähnert et al. 2009).

In the following, two cases of structural investigation of confined carbon dioxide in two nanostructured materials are reviewed. Carbon dioxide was the adsorbate of choice since it is generally considered one of the most interesting candidates to study confinement effects because of its linear shape and quadrupole moment (both producing orientational correlations); furthermore, bulk CO_2 has been extensively studied experimentally and theoretically at both subcritical and supercritical thermodynamic states.

5.2.2.1 Structural Study of Confined CO_2 in a Microporous Carbon by In Situ Neutron Diffraction

In general, the supercritical fluids are gaseous compounds, which are compressed to a pressure higher than the critical pressure (P_c) above their critical temperature (T_c) and have nowadays been attracting scientific as well as industrial interest because of their unusual physical properties, since their density can vary significantly even with a slight change of the temperature and pressure. Based on these properties, supercritical fluids are widely used in industry as solvents for extraction and for chemical and biochemical reactions. In most cases carbon dioxide, the simplest liquid composed of linear molecules, is the solvent of choice because it is environmentally benign, abundant, inexpensive, nonflammable, nontoxic, and its supercritical state is easily accessible with a $T_c = 304.1$ K and $P_c = 73.8$ bar.

In the present case, in situ neutron diffraction measurements of adsorbed CO_2 on a microporous carbon sample have been performed along an isotherm slightly above the critical point ($T = 308$ K) and at a range of pressures below the critical one (10–60 bar) (Katsaros et al. 2000; Steriotis et al. 2002b, 2004). The experiment carried out at instrument E2, Helmholtz Zentrum, Berlin. A novel high-pressure apparatus (up to 80 bar) has been constructed for this purpose. By assuming that the interference between adsorbed molecules and carbon walls is not predominant (Ohkubo et al. 1999), the difference of diffraction intensities between the CO_2-loaded and unloaded sample was assigned to the confined CO_2 molecules and, thus, the diffraction patterns of the adsorbed carbon dioxide were deduced by subtracting the diffraction intensities between the CO_2-loaded and unloaded sample.

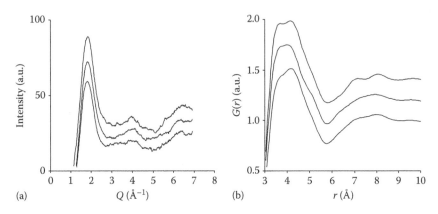

FIGURE 5.7

Confined CO_2 at 308 K and pressures 10, 35, and 60 bar (from bottom to top) in a microporous carbon: (a) Experimental neutron scattering intensities. (b) Intermolecular part of the total radial distribution; the curves are shifted by 0.2 for clarity.

Figure 5.7a shows the corrected neutron diffraction patterns of CO_2 confined in the sample at pressures 10, 35, and 60 bar, respectively.

The spectra clearly exhibit a peak at $Q = 1.8\,\text{Å}^{-1}$ corresponding to the most probable distance between nearest-neighbor molecules. Furthermore, the existence of this peak implies that, at a temperature above the critical point and pressures well below the critical one, adsorbed CO_2 is in *a densified state*, which can be compared to a *very dense supercritical fluid*, or even *bulk liquid*. The case is additionally supported by the appearance of a maximum at about $4\,\text{Å}^{-1}$. This feature has been observed as a tiny bump for dense carbon dioxide thermodynamic states, such as bulk liquid close to the triple point or supercritical fluid under very high pressures. The observed densification can be explained by considering that within micropores, which are no more than a few molecular diameters wide, the potential fields from neighboring walls overlap and produce a very deep potential wells. As a result, the interaction energy between the gas molecules and the adsorbent matrix is increased and adsorption is enhanced while the adsorbed molecules within the confined pore space are arranged in very dense states to compensate the lack of free volume. Additionally, the main peak possesses a degree of asymmetry, which can be attributed to intermolecular orientational correlations between neighboring CO_2 molecules arising mainly from electrical quadrupolar interactions. Finally, at the vicinity of high Q, a maximum is reached at approximately $6.4\,\text{Å}^{-1}$ due to intramolecular correlations.

Figure 5.7b presents the intermolecular parts of the total radial distribution functions for CO_2 confined in the carbon sample along the 308 K isotherm. It appears that in all the pressures examined, the radial distribution functions exhibit a first-neighbor peak around $4\,\text{Å}$, followed by a strongly damped oscillation and a second-neighbor peak at about $8\,\text{Å}$. One can even observe that, at the low-pressure state (10 bar) and clearly at the high-pressure one (60 bar), the main peak reveals the presence of two structures centered at 3.7 and 4.15 Å, respectively. Literature data on bulk supercritical CO_2 (Adya and Wormald 1991) have revealed identical structures, which were attributed to the C–O and O–O interacting distances between orientationally correlated molecules (3.7 Å) and the C–C distance between orientationally uncorrelated molecules (4.15 Å). It, then, appears that the CO_2 molecules are losing their orientational freedom and are preferentially arranging in a T-shape or herringbone configuration (Cipriani et al. 2001), i.e., in a way that C–O and O–O distances are different than the C–C one. Similar correlations have also been observed for liquid as well as for dense supercritical states (van Tricht 1984; Chiappini et al. 1996; Cipriani et al. 2001). In addition, one can notice a split of the second-neighbor peak at $P = 10$ bar, which is more pronounced at $P = 60$ bar implying that the orientational correlations mentioned can expand to the second-neighbor peak, as also observed at a previous neutron diffraction study of bulk high-pressure supercritical CO_2 (Bausenwein et al. 1992). On the other hand, no split of the first- and second-neighbor peaks occurs implying a different configurational arrangement of the confined molecules at 30 bar. However, since the overall experimental resolution is 0.9 Å, the discussion on the split curves of $G(r)$ cannot be completely justified as it pertains to distances smaller than our resolution. Finally, the different configurational arrangement of the confined molecules at 30 bar is further supported by the appearance of a shoulder at 5 Å (within our resolution), since molecular dynamics simulations performed close to the triple point showed that this structure is again related to C–O and O–O correlations (De Santis et al. 1987).

It is worth mentioning that CO_2 permeability measurements through a microporous carbon membrane have revealed a maximum at 308 K and a mean pressure of about 35 bar (Katsaros et al. 1997), also implying a transition in the adsorbate/permeating phase. Recent pertinent grand canonical Monte Carlo (GCMC) simulations have provided structural details of CO_2 adsorbed in microporous carbons (Steriotis et al. 2002a). Following a detailed examination of the local density profiles and angular distributions of molecular axes at different pressures within pores of different size at 308 K, an orientational transition of molecules confined in 1.15 nm slit graphitic pores between 30 and 40 bar was observed. In specific, in such a pore two distinct sub-layers form, one being closer to the wall with molecules lying almost parallel to it and a second one with molecules oriented at a 45° angle with the pore wall (i.e., herringbone structure). As the pressure increases up to 30 bar the density of both sub-layers increases proportionally, while between 30 and 40 bar, the molecules tend to reorient themselves by turning their molecular axes away from the pore wall in a way that the second sub-layer is enriched at the expense of the first one (Figure 5.8). Beyond 40 bar, a proportional increase of density all over the pore with increasing pressure is again observed, in good agreement with the deduced radial distribution functions.

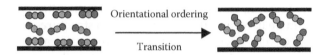

FIGURE 5.8
Orientational transition of CO_2 molecules confined in slit graphitic pores.

5.2.2.2 Structural Study of Confined CO_2 in a Mesoporous Silica by In Situ Neutron Diffraction

In this case, in situ neutron diffraction measurements of adsorbed carbon dioxide have been performed on a model mesoporous material (purely siliceous MCM-41) along an isotherm (253 K) at pressures varying between 0 and 18 bar (subcritical states) (Steriotis et al. 2008). The diffraction pattern of bulk liquid CO_2 ($T = 253$ K, $P = 19.7$ bar) has also been measured. For this purpose, a novel high-pressure adsorption apparatus has been constructed (Figure 5.9). The apparatus is capable of operating at temperatures from 4 K and pressures from high vacuum up to 150 bar, allowing, thus, the expansion of the technique in a broad thermodynamic range.

The experiment was carried out at GEM instrument at ISIS pulsed neutron and muon source (Rutherford Appleton Laboratory, U.K.). GEM is the most advanced materials diffractometer in the world because its detector array has a very large area and a very wide range in scattering angles, corresponding

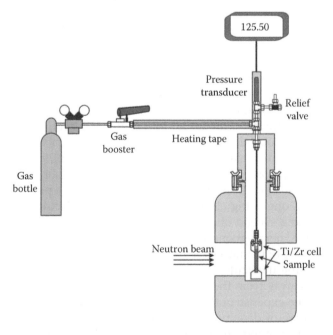

FIGURE 5.9
High-pressure adsorption apparatus for in situ diffraction measurements.

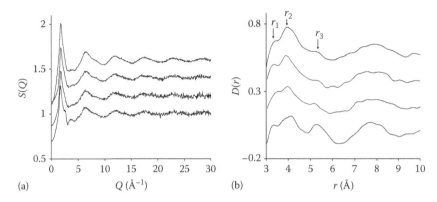

FIGURE 5.10
Confined CO_2 at 253 K and pressures 5, 10, and 18 bar (from bottom to top) in MCM-41; bulk liquid CO_2 at 253 K and 19.7 bar (top): (a) The total-scattering structure factors shifted by 0.2 for clarity. (b) The differential correlation functions are shifted by 0.2 for clarity.

to a Q range varying between 0.02 and 40 Å^{-1}. As a result, the real-space resolution of the Fourier-transformed data is about 0.36 Å.

The structure factor of bulk liquid CO_2 is shown in Figure 5.10a (top). The intermolecular structure peak is located at about 1.79 Å^{-1}, while a long range oscillation corresponding to intramolecular correlations is observed at larger Q values. The main diffraction peak has a minimum at $Q \sim 3$ Å^{-1} followed by a tiny bump near 4 Å^{-1}.

The main peak possesses a degree of asymmetry, which can be attributed to intermolecular orientational correlations between neighboring CO_2 molecules arising mainly from electrical quadrupolar interactions (Chiappini et al. 1996). The structure factors of the confined carbon dioxide phases are also presented in Figure 5.10a. An alternative experimental approach was applied in order to minimize the contribution of the cross correlation term, on the one hand, and, on the other, to eliminate the possible contribution of the adsorbed monolayer structure. This was carried out in an attempt to obtain a clearer picture of "true" confinement effect, i.e., the CO_2 molecular structure only in the "core" of the pores. The correction was implemented by simply using, for all corrections, the experimental differential cross section of the matrix loaded with CO_2 at 5 bar (monolayer coverage) instead of the dry matrix differential cross section. This approach is valid under the assumptions that the correlations between the core fluid and the adsorbed film or the matrix are negligible and that the molecular structure of both multilayers and condensate are the same. The correction was applied to the confined states at pressures 10 and 18 bar, respectively. The patterns clearly exhibit a peak corresponding to the most probable distance between nearest neighbor molecules, while the long-range oscillations at larger Q values observed for the bulk liquid are also present. The peak position shifts slightly to the higher Q region with increasing pressure and reaches that of the bulk liquid at $P = 18$ bar. As mentioned above, at these stages capillary condensation has taken place, the pores are completely filled and the density of the adsorbate is similar to that of the bulk liquid. As in the case of the bulk liquid, the main peaks of the confined CO_2 have a minimum at about $Q = 3\,\text{Å}^{-1}$ followed by a tiny bump near $4\,\text{Å}^{-1}$. The structure factor of the adsorbed CO_2 at the monolayer coverage ($P = 5$ bar), however, exhibits additionally a shoulder on the high Q side of the peak ($Q \approx 2.7\,\text{Å}^{-1}$). The disappearance of the shoulder from the other confined states strongly suggests that its origin is due to the film structure (and possibly cross correlation terms).

Morineau et al. (2004) performed at various temperatures neutron diffraction measurements of methanol confined in MCM-41 samples of different pore sizes, after ensuring that a complete filling has been achieved. In their study, the intrinsic correlations between methanol and silica were modeled, on the basis of canonical Monte Carlo simulations in a cylindrical pore of diameter $24\,\text{Å}$. Their calculated structure factor due to the methanol-matrix cross-term exhibits a double oscillation peak within the Q range of the main diffraction peak of methanol. Our experiment confirms this finding with the presence of the shoulder. Their computational result is, thus, in agreement with our experimental procedure for correcting directly the structure factor of the confined phase, after taking into account the interactions of the CO_2 film with the silica pore walls (either as cross-terms or as film structure). Obviously, this procedure is feasible only when an experimental measurement of the adsorbed film has been attained.

Figure 5.10b illustrates the total differential correlation function, $D(r)$, for both the bulk and the confined CO_2 at 5, 10, and 18 bar, respectively. For the last two pressures (10 and 18 bar), $D(r)$ was calculated after applying the aforementioned adsorbed film correction. The intermolecular part of the correlation functions shows two broad features centered at ~4 and ~8 Å, arising respectively, from the first- and second-neighbor interactions. The first-neighbor peaks reveal three structures located at about $r_1 = 3.3\,\text{Å}$, $r_2 = 4\,\text{Å}$, and $r_3 = 5.2\,\text{Å}$, respectively, in agreement with literature data on bulk CO_2 (Adya and Wormald 1991). Molecular dynamics simulations have revealed that the structure at r_1 is attributed mainly to O–O as well as to C–O pair correlations. Moreover, both C–C and C–O correlations give a positive contribution at r_2, while the structure at r_3 arises from bumps present in the partial C–O and O–O atom–atom pair correlation functions, when quadrupole forces are included in the simulations (De Santis et al. 1987; Chiappini et al. 1996). Based on the deduced $D(r)$ functions, the molecular arrangement in all the confined states seems to be quite similar with that of the bulk liquid; however, certain differences can be observed. First of all, a decrease of the intensity, mainly in the intramolecular region, has been observed (not shown) and this can be explained on the basis of "excluded volume" effects (Morineau and Alba-Simionesco 2003). In addition, the $D(r)$ function that corresponds to the formation of the film ($P = 5$ bar) shows clearly that the structure r_3 tends to become a prominent feature. Based on simulation results, this feature is proved to be strongly potential dependent, becoming evident only when electric quadrupolar interactions are taken into account and can be ascribed to a sharpening of the O–O correlations, arising from increased orientational ordering of the fluid (Chiappini et al. 1996). This seems to be a plausible explanation, as the adsorbed monolayer is subject to enhanced Lennard–Jones and coulombic interactions with the surface atoms. Based on this approach, the silica surface seems to somehow organize the adsorbed film not only in certain "binding" sites but also in a more "oriented" manner in order to counterbalance the strong adsorption potential with the finite size of the surface. One may also mention that the first-neighbor peak distance appears larger than in the bulk phase or the other sorbed states, while a further splitting of the first-neighbor peak can be observed.

In conclusion, the structure factors of CO_2 together with the total differential correlation functions suggest that the confined fluid has, at all confined states, examined *liquid-like properties*; however, several subtle differences, pointing to stronger orientational correlations inside the pores, were observed. These differences are attributed to either *pore wall–fluid interactions* (adsorbed film) *or the confinement of the fluid* (when pores are filled), combined with the relatively *large quadrupole moment* of carbon dioxide.

5.3 Combination of In Situ and Ex Situ Techniques for Monitoring and Controlling the Synthesis of Nanoporous Membranes

The active surface of gas separating membranes usually consists of a thin layer of nanosized amorphous oxide particles, or densely developed zeolitic crystals or graphitic carbon material, and the separation is based on either molecular sieving or adsorption/condensation mechanisms.

A serious obstacle preventing the understanding of the structure of microporous membranes (ceramic, carbon, zeolite, etc.) is the lack of a characterization technique that would yield detailed information as to the exact size and size distribution of the "active" pores.

The difficulty in developing such a technique is due to the complex relationship between structure and properties in complex porous networks (e.g., carbon membranes).

For instance, the existence of a small number of open path mesopores is enough to degrade the separation properties of a membrane, even though in such cases the information obtained by gas sorption experiments does not include the identification of mesopores, as their population does not reach a detectable "critical mass."

With the combination, however, of different techniques such as permeability, sorption, and relative permeability together with molecular simulation techniques (GCMC), such detailed information about the pore structure, as well as its relationship with membrane properties could be obtained.

Still, due to the arduous nature of some of the abovementioned techniques, the number of such studies is very limited in the literature (Sedigh et al. 2000; Nguyen et al. 2003; Lagorsse et al. 2004). In the following paragraphs, emphasis is given to the use of a combination of these characterization techniques for the elucidation of the complex pore morphology of ultramicroporous carbon hollow fiber membranes with asymmetric structure and its correlation with their physicochemical properties.

5.3.1 Low-Cost Asymmetric Carbon Hollow Fibers

The first microporous carbon membranes were produced by Barrer in the 1950s and 1960s by compressing high-surface-area carbon powders at very high pressures. The resulting porous plugs had pores of 5–30 Å diameter and were used to study diffusion of gases and vapors. Nowadays, the carbon membrane research is primarily focused on the development of carbon hollow fibers, because of the high surface area that can be achieved through their compaction into membrane modules.

Numerous synthetic precursors have been used to form carbon hollow fiber membranes, such as polyimide and its derivatives, polyacrylonitrile (PAN), phenolic resin, polyfurfuryl alcohol (PFA), polyvinylidene chloride–acrylate terpolymer (PVDC–AC), phenol, formaldehyde, and cellulose. Among them, the polyimides are of the most promising materials as candidates for the production of polymeric membranes that can be further used for the development of stable nanoporous carbon membranes with satisfactory permeability and selectivity properties.

The polymeric hollow fiber precursors are usually prepared by the dry/wet phase inversion process in a spinning setup. The spinning dope consists of a solution of the polymer of appropriate w/w%. The dry/wet spinning machine is presented in Figure 5.11. The extruded fibers pass first through an air gap of varying length before entering the coagulation bath, which is filled with a polymer non-solvent at room temperature.

The nascent fibers are oriented by means of two guiding wheels and pulled by a third wheel into a collecting reservoir. The as developed polymer hollow fibers are further carbonized in inert gas atmosphere at temperatures up to 1173–1300 K.

The correlation between pyrolysis conditions and permeability and selectivity properties of the prepared carbon hollow fiber membranes is one of the most important research topics of various groups. Many types of carbon hollow fiber membranes have been reported in literature and were produced from a variety of precursor polymers by means of different pyrolysis conditions (Tanihara et al. 1999; Ismail and David 2001). Favvas and coworkers prepared carbon hollow fiber membranes by pyrolysis of polyimides. The

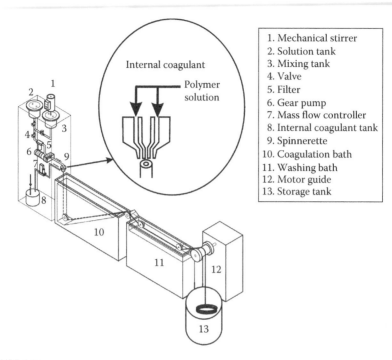

FIGURE 5.11
Diagram of dry/wet spinning setup.

developed fibers exhibited high permselectivity values for H_2/CH_4, O_2/N_2, and CO_2/N_2 couples (Favvas et al. 2007, 2008). The feasibility to control the micropore structure of asymmetric carbon hollow fiber membranes through the optimization of the dry/wet phase inversion conditions and of the subsequent pyrolytic conditions, constitute them as promising materials for application in high temperature gas separation industrial processes such as H_2 enrichment or CO_2 removal.

5.3.1.1 Combination of Ex Situ Techniques: Characterization of Asymmetric Carbon Hollow Fibers

5.3.1.1.1 SEM Analysis

The morphological characteristics of the asymmetric copolyimide precursor and the produced carbon hollow fiber membranes are primarily investigated using SEM analysis (Figure 5.12). The SEM analysis is sufficient for the characterization of the bulk structure of asymmetric porous fibers and can be applied as a useful tool in order to study the effects of several pyrolytic treatments on the bulk morphology of polymeric precursors.

However, there are several issues related to morphology, still beyond the capabilities of the SEM method. Investigating the pore structure of the outer

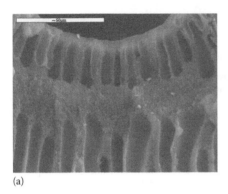
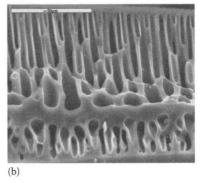

(a)   (b)

FIGURE 5.12
(a) SEM micrograph of the precursor cross section. (b) Cross layer structure of the carbon membrane.

and inner dense layers with SEM is impossible, especially when the pore size resides in the micropore range. The same holds for non-accessible layers embedded into the bulk of the asymmetric fibers. The occurrence of such interface layers usually reflects the boundary region between the opposite streams of the non-solvent, diffusing simultaneously from the inner (bore liquid) and outer (coagulation bath) surface of the fiber.

5.3.1.1.2 Permeability Dependence on Temperature

In order to reach to conclusions about the mean pore size of at least one of the dense layers (outer, interface, and inner) of the asymmetric carbon membranes, the permeance dependence on temperature can be examined, in comparison with relevant studies often encountered in the literature for MFI zeolite membranes (Bernal et al. 2002). In MFI and generally in microporous membranes, the permeance increases with increasing temperature, goes through a maximum and a subsequent minimum, and then increases again (Figure 5.13). The maximum is attributed to the simultaneous occurrence of

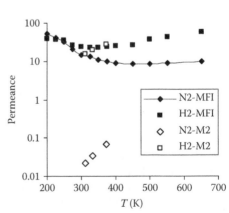

FIGURE 5.13
Permeance evolution (mol/m²/s/Pa) vs. temperature, MFI permeance × 10^8, M2 permeance × 10^{10}.

two competitive mechanisms, namely equilibrium adsorption and activated surface diffusion.

Specifically, the mass transport in micropores takes place by molecules hopping between adsorption sites. This is an activated process and the diffusivity increases with temperature. On the other hand, the amount adsorbed, or else the occupancy degree, decreases with temperature.

Eventually, when the decline in occupancy prevails, permeance decreases. Going further up in temperature, the local minimum indicates the point where adsorption effects become negligible and the molecules inside the pores retain their gaseous character and diffuse from site to site by overcoming an energy barrier, which is imposed by the micropore structure (translational diffusion). As depicted in Figure 5.13, "activated diffusion" (translational diffusion) is the prevailing mechanism for H_2 and N_2 mass transport through a carbon membrane (M2) produced by pyrolysis of a co-polyimide precursor in nitrogen atmosphere up to 1173 K. Specifically, activated diffusion starts to occur at a much lower temperature than that usually observed for MFI membranes, indicating that the mean pore size of at least one of the dense layers is smaller than 0.55 nm, which is the standard size for the zeolite MFI channels.

5.3.1.1.3 *Permeability Dependence on Pressure*

In the cases where the Knudsen type dependence of permeance with temperature ($\propto T^{0.5}$) is observed, we have a first indication of the coexistence of some larger pores (mesopores) within the pore structure of all the dense layers. In order to determine the size of these mesopores, the permeance dependence on pressure can be investigated and the results interpreted in accordance to the Weber's general equation of flux (Weber 1954).

Additionally, relative permeability experiments of an organic vapor/inert gas system (e.g., benzene/Helium) can be applied as a supplementary and more accurate method for the determination of the mesopore size. Concerning the permeance dependence on pressure and according to the theory, Knudsen diffusion (molecular streaming) is predominant when the radius, r, of the pore is much smaller than the mean free path, λ, of the permeating molecule. As the mean pressure increases (λ decreases), the flow regime changes from Knudsen to Poisseuille (viscous flow). Therefore, the permeance evolution versus pressure is represented by a curve, which reaches a minimum in the intermediate region between molecular streaming and viscous flow, and approaches asymptotically to a straight line. The minimum of the Weber's equation occurs at $r/\lambda \sim 0.05$. As an example, the high-pressure permeance results of carbon asymmetric hollow fiber membranes pyrolyzed under different conditions are depicted in Figure 5.14. As it is clear, the transition from Knudsen flow to Poiseuille flow occurs at very low pressures (0.05 MPa) for a carbon membrane pyrolyzed in argon atmosphere at 1300 K and further activated with CO_2 (M3), while for membranes M1 and M2 pyrolyzed in argon at 1300 K and nitrogen at 1173 K, respectively, the Knudsen mechanism persists up to 0.5 MPa.

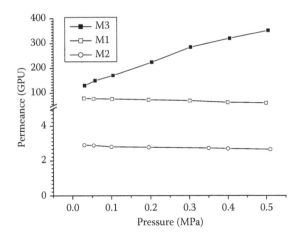

FIGURE 5.14
Pressure dependence of He permeance at 373 K for carbon hollow fiber.

By calculating the mean free path of helium at 373 K, at pressures of 0.05 and 0.5 MPa and based on the aforementioned Weber's treatment, one can conclude that the dense layers of membrane M3 must also contain pores with size above 34 nm, whereas for membranes M1 and M2, the coexisting mesopores must be smaller than 3.5 nm.

5.3.1.1.4 Relative Permeability

Figure 5.15 depicts in comparison the relative permeability curves (Steriotis et al. 1995) of membranes M1 and M3. It is evident that the transport of

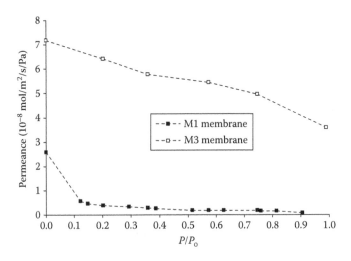

FIGURE 5.15
He/Benzene relative permeability at 307 K (P/P_o is the relative pressure of benzene).

helium through a large portion of the coexisting mesopores of membrane M1 was blocked, due to capillary condensation of benzene vapor at a relative pressure (p/p_0) of 0.3.

The Kelvin radius calculated at this relative pressure is 0.8 nm. Thereby, taking into account the thickness of the adsorbed *t*-layer (Lippens and De Boer 1965), a pore diameter of 2.5 nm can be defined, in good agreement with the prediction of the high-pressure experiments (less than 3.5 nm). A noteworthy issue is that the relative permeability experiment further evidenced the side presence of a population of pores with a diameter of the order of 200 nm. This size can be similarly calculated at the relative pressure of 0.9, where complete blockage of the helium permeance occurs. For the activated membrane M3, the helium permeation had not been blocked at identical relative pressure, indicating the side presence of layer pores, which are greater than 200 nm.

In the analysis up to now, we have discussed the potentiality of gas permeability results, that, when appropriately interpreted with the established theories of mass transport, can provide sufficient information about the pore structure of the several dense layers existing in asymmetric membranes. However, limitations of the gas permeability method are mainly related to its inability to distinguish between the transport characteristics of each layer separately as well as to define the relative population of the several classes of pores (micropores and mesopores) that may constitute each layer's pore structure. Thereby, permeability results proved inadequate for the characterization of the micropore structure of the dense layers of asymmetric carbon membranes because of the side presence of mesopores. In this case, the gas adsorption measurements combined with GCMC simulations (Samios et al. 1997, 2000) can provide complementary information on the micropore structure.

5.3.1.1.5 Micropore Size Distribution via GCMC Simulation

As an example, the GCMC method originally developed by Nicholson et al. (Cracknell and Nicholson 1994; Samios et al. 1997) and further improved by several researchers (Jagiello and Thommes 2004; Konstantakou et al. 2006) can be employed for the simulation of H_2 and CO_2 adsorption (Figures 5.16 and 5.17) in idealized graphite slit-like pores. Details of the method can be found in the recent literature (Konstantakou et al. 2006).

Pore size distributions are obtained (Figures 5.18 and 5.19) after solving the adsorption integral equation and using either the H_2 at 77 K or both H_2 and CO_2 experimental data through a methodology lately developed by our group (Konstantakou et al. 2007). According to this data the micropore structure of the above-described carbon hollow fiber membranes consists mainly of pores with size below 1.8 nm.

When the experimental data from CO_2 adsorption are added in the pore size calculations, the existence of larger pores, also classified as micropores, is revealed. More specifically, by comparing between M1, M3, and M2 in

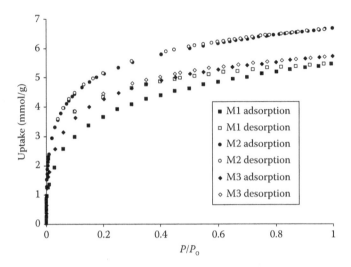

FIGURE 5.16
Low-pressure H_2 adsorption/desorption at 77 K for M1, M2, and M3 membranes.

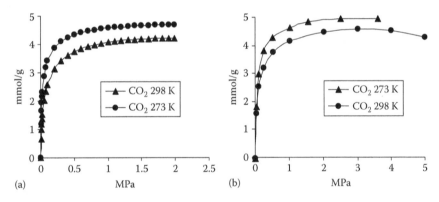

FIGURE 5.17
CO_2 adsorption isotherms: (a) membrane M2 and (b) membrane M3.

Figure 5.18 and between M1 and M2 in Figure 5.19, we can conclude that no micropores above 1.7 nm exist in the fibers carbonized at the higher temperature.

On the contrary, the fiber carbonized at the lower temperature M2 is characterized by the coexistence of larger micropores (1.8–2.6 nm), which however, do not degrade its gas separation performance due to their very small population (Figure 5.19).

In this chapter, a complete analytical methodology, consisting of a combination of advanced ex situ techniques, is described as a tool for the detailed characterization of the pore structure of asymmetric carbon hollow fiber membranes. The same procedure can by applied in all kinds of asymmetric

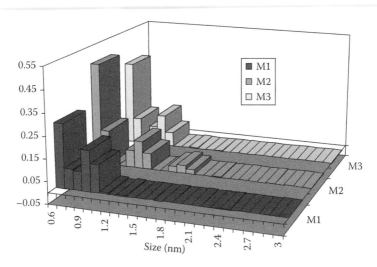

FIGURE 5.18
Pore size distributions for M1, M2, and M3 based on experimental isotherms of H_2 adsorption at 77 K with GCMC method.

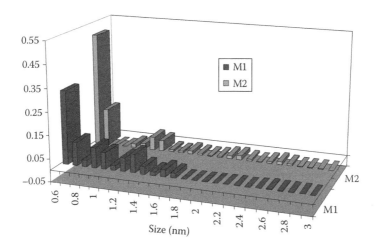

FIGURE 5.19
Pore size distributions for M1 and M2 membranes based on experimental isotherms of H_2 adsorption at 77 K and CO_2 adsorption at 273 and 278 K with GCMC method.

membranes. SEM micrographs of the cross-sectional areas characterize the asymmetric morphology (outer, inner, and intermediate dense layers), of the fibers. Single-phase permeability of several gas molecules at different temperatures are involved as a tool to elucidate the dense (micropore) structure of the layers and reveal the possible coexistence of mesopores. The larger size of these mesopores is defined by high-pressure permeability experiments whereas relative permeability gives an insight into the relative population of these mesopores. Finally, adsorption measurements in combination with GCMC simulation provide sufficient information of the micropore structure.

5.3.2 Combinations of In Situ Permeability and Ex Situ Methods for Monitoring and Controlling the Deposition of SiO_2 within the Separating Layer of Nanoporous Sol–Gel Membranes

Molecular scale separations in aqueous solutions using membranes have been possible in 1963 with the breakthrough technology of asymmetric polymeric membranes by Loeb and Sourirajan (1963).

Extension of the membrane application to gas separation is being hindered by the simple fact that in desalination the main challenge is to separate the permeating water molecule, which is 100% less in diameter that diameter of the salt, whereas the sieving requirements for most of the important gas phase separations are much more challenging, for example, for the CO_2 capture the CO_2/N_2 molecular size difference is less than 0.02 nm and for the methane partial oxidation the CO_2/CH_4 difference is 0.06 nm, whereas the kinetic diameter differences are smaller in the cases of CO_2/CO or ethane/ethylene separation. This requires that new tools should be developed enabling the production of thousands of square meters of molecular sieve membranes with accuracy and repeatability of the order of 0.01 nm.

Since top-down sol–gel techniques for the production of nanopores are capable of tailoring the nanostructure with the accuracy only of the order of 0.2 nm (limited by the minimum particle size), bottom-up nanotechnology techniques need to be developed based on real-time characterization tools, with sensitivity of the order of 0.01 nm during the size tailoring of nanopores.

Nanoengineering by self-replicating nanorobots has been set forth by Drexler, but as has been stated by Nobel Prize winner Smalley, it is not possible due to the "fat finger" problem (Figure 5.20a). In the recent literature (Labropoulos et al. 2008), an alternative approach to the self-replicating nanorobot is proposed, and has been demonstrated that the high temperature permeability measurements with different size probe molecules, during the chemical vapor deposition (Figure 5.20b) inside the pores of nanofiltration membranes, can become an effective tool for monitoring and controlling the minute nanopore size changes of the order of 0.01 nm.

The sensitivity of the method is demonstrated by the comparison of the permeation measurements of the probe molecules N_2 and O_2 during the blocking of the nanopores in a chemical vapor deposition process (Figure 5.20c and d). The permeability is reduced and when it reaches the diameter of N_2 (or O_2), a sudden increase in permeability is measured demonstrating that the in situ permeability method is sensitive enough to distinguish the sizes of N_2 (0.364 nm) and O_2 (0.346 nm) (Labropoulos et al. 2009).

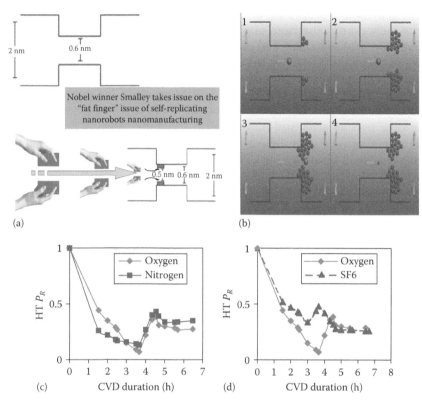

FIGURE 5.20
In order to measure the evolution of pore size, different size probe molecules are used as probe molecules with different kinetic diameters ($N_2 = 0.364$ nm, $O_2 = 0.346$ nm, $SF_6 = 0.55$ nm).

5.3.3 Carbon Nanotube-Based Porous Composites and Combinations of In Situ and Ex Situ Methods for Monitoring and Controlling Their Growth Evolution

Following the initial discovery of multiwall (Iijima 1991; Ebbesen et al. 1992) and single wall (Bethune et al. 1993; Iijima and Ichihashi 1993) carbon nanotubes (CNTs), investigation on these materials is being continuously intensified as novel applications of vital importance to humanity come into light, such as in environment, energy, and health. Unique properties of CNTs including high surface area, resistance to strong acids and bases, high mechanical flexibility and strength, and high electric conductivity (de Lucas et al. 2005; Chakrabarti et al. 2007), which arise from their atomic structures, number of walls, diameters, and lengths, result in many potential applications such as hydrogen storage (Liu et al. 1999), catalyst supports (Toebes et al. 2004), selective adsorption agents (Fujiwara et al. 2001), composite materials (Hinds et al. 2004), nanoelectronic and nanomechanical devices (Tans et al. 1998), and field emission devices (Fan et al. 1999).

The remarkable properties of CNTs have inspired combination of nano-tubes with porous matrices targeting new generation of composites with unique physical, chemical, and electronic characteristics (Vermisoglou et al. 2008; Veziri et al. 2008, 2009). In such a composite, CNTs form an integrated, multidirectional network and thereby contribute to the through-the-thick-ness strength of the material (Baker 1989). Moreover, CNT porous compos-ites coupling the exciting characteristics of CNTs to the large surface area, extensive porous structure and density, and rapid adsorption capability that several porous materials exhibit offer great potential of the generated composites in applications such as energy storage (Emmenegger et al. 2003), hydrogen adsorption (Dillon et al. 1997; Challet et al. 2004), water desalina-tion (Zhang et al. 2006), and pollution control (Dusenbury and Cannon 1996).

Controlled synthesis of CNTs with predictable properties essentially requires fine tuning of the atomic arrangement along the tubes. The three main techniques currently used for CNT synthesis are arc-discharge, laser ablation, and chemical vapor deposition (CVD). The first two operate at extremely high temperatures (thousands of °C) and involve carbon vapor-ization from solid state precursors to provide the carbon source needed for nanotube growth, while CVD utilizes hydrocarbon gases as sources for car-bon atoms and metal catalyst particles as "seeds" for nanotube growth. The advantages of CVD for CNT growth include lower operation temperatures (typically 700°C–900°C), which drastically enhances viability for process up scaling, and smaller amount of byproducts (Dai 2002). Moreover, CVD can yield the formation of various nanostructures, such as straight, bent, and helical, with much greater lengths, as well as synthesis of aligned carbon nanomaterials in high yields (de Lucas et al. 2005).

Controlling the morphology and composition of the catalytic particles dur-ing CVD growth of CNTs is of crucial importance since it strongly affects nanotube characteristics such as thickness, uniformity, and yield. In princi-ple, finely dispersed, nanometer-sized metal particle catalysts that preserve their morphology at the high CVD processing temperatures are required for the growth of CNTs. Recent studies have focused on precisely controlling catalyst particle sizes and uniformity on a variety of support materials (Ago et al. 2005; Ramesh et al. 2005; Ning et al. 2007). However, as the size of the metal particles decreases down to the nanometer scale the ratio of surface area to volume increases considerably. As a result, these particles exhibit extremely high surface energy due to large number of unsaturated surface bonds and tend to agglomerate into bigger and usually irregular metal enti-ties on the support. This phenomenon is more pronounced at elevated tem-peratures. To alleviate this challenge, porous materials have been proposed as supports to prevent catalyst particle coalescence during high-temperature CVD treatment. A porous support exhibiting noncontinuous surface can contribute significantly to particle stabilization by limiting sintering and produce a fine dispersion of well-defined particles, which is desired for controlled growth of CNTs. In addition, the high surface area of the porous support, in contrast to dense substrates, can drastically increase the number of catalytic particles, thus, increasing nucleation sites, which is advantageous to high yield synthesis of CNTs (Ago et al. 2005). A variety of porous supports have been used for catalytic CNT growth that include magnesia (Li et al. 2004), alumina (Peigney et al. 2001), and zeolites (Hiraoka et al. 2003). In addition to the characteristics of the catalyst particles, the interaction of the particles with the support plays a vital role in controlling the growth and quality of the resulting CNTs (Veziri et al. 2009). The support material significantly influences the catalytic activity of metal catalysts and can enhance nanotube yield (Ago et al. 2004).

Another highly sought-after application of CNTs in the field of porous composites is membranes consisting of CNT pores. Fabrication of such membranes has been motivated by theoretical studies indicating that these materials, when appropriately engineered, can exhibit high permeability (Sholl and Johnson 2006). Specifically, recent molecular dynamic models have predicted macroscopic diffusivities of small molecules through CNT membranes that are orders of magnitude higher than in zeolites or any other nanoporous material with comparable pore sizes (Skoulidas et al. 2002; Sokhan et al. 2002; Jakobtorweihen et al. 2005; Chen et al. 2006; Gusev and Guseva 2007). This high flux is attributed to the inherent smoothness of the nanotube inner surfaces that provide nearly frictionless molecular transport (Sokhan et al. 2002; Whitby and Quirke 2007). In addition to high flux, selective action of CNT membranes can be achieved by molecular sieving (Miller et al. 2001) or preferential adsorption. In a recent example for CH_4/H_2 mixtures, atomistic simulations revealed that preferential adsorption of CH_4 onto the nanotube surface favors selective CH_4 transport, in contrast to single Knudsen transport, thus offering separation capability (Chen and Sholl 2006; Skoulidas et al. 2006). Interestingly, sorption separation efficiency can be augmented by attaching functional molecules onto the nanotube internal surface, throughout the length or at the tube ends, which can trigger specific chemical interactions between one or some of the components of the mixture and the functionalized CNT surfaces (Majumder et al. 2005, 2007). In addition to separation applications, aligned CNTs have been proposed as excellent field emitters due to their high aspect ratio that results in a large field enhancement factor producing large current densities at low electric fields (Chhowalla et al. 2001). Such systems can be potentially integrated with signal processing circuits and be part of lab-on-a-chip devices. Vertically aligned multiwall CNTs have been successfully used for detection of blood cholesterol exhibiting distinct electrochemical peaks and high signal-to-noise ratio (Roy et al. 2006). CNT arrays loaded with catalyst particles are also promising as functional reactor elements for catalyzing electrochemical reactions for energy production and storage (Che 1998b; Dai et al. 2002; Du et al. 2005).

Experimentally, CNT membranes have been prepared by first synthesizing individual, aligned CNTs by CVD and then embed them in a silicon nitride matrix (Holt et al. 2006) and in a polymeric matrix (Hinds et al. 2004).

Polymer-coated CNTs were also grown on macroporous alumina substrates by CVD in an attempt to produce supported membranes that can be suitable for upscaling CNT-based separation processes (Mi et al. 2005, 2007). CNTs prepared by laser ablation were aligned by self-assembly (Shimoda et al. 2002a,b) to form a continuous film, and more recently arc discharge-prepared CNTs were oriented by shear flow and subsequently coated by polysulfone in order to seal the structure and provide mechanical strength (Kim et al. 2007). CVD is the method of choice for growth of CNT arrays, yet the critical step that yields tube alignment when no template is involved is the catalyst deposition on the substrate. Laser etching (Terrones et al. 1997), magnetron sputtering (Ren et al. 1998; Choi et al. 2000) electron beam evaporation (Fan et al. 1999), pulse-current electrochemical deposition (Tu et al. 2002), sol–gel processing (Pan et al. 2003), colloidal crystal lithography (Ryu et al. 2007), and confinement in micelles (Liu et al. 2006b) are some of the methods employed and carefully engineered to pattern and control the deposition and morphology of catalyst (Ni, Fe, Co) films and nanoparticles, so that well-defined CNT arrays are formed upon growth. High degree of alignment has been achieved by the aforementioned techniques, yet size uniformity of the resulting nanotubes relies heavily upon the size distribution of the particles, since each particle catalyzes the growth of one nanotube, whose diameter is almost equal to that of the corresponding particle. The templated method, introduced by Martin (1994), tends to eliminate this dependency on the catalyst size distribution since it employs uniform domains, such as membrane pores, to confine nanotube growth inside them, thus, the size-determining factor is the diameter of the domains. Parthasarathy et al. (1995) were the first to use anodized alumina (AAO) templates, followed by Kyotani et al. (1996, 2002) in an attempt to grow CNTs inside the parallel, monodisperse, and cylindrical pores of the template. Che et al. (1998a) synthesized aligned CNTs by CVD using similar templates after introducing Ni, Fe, or Co catalyst into the pores by impregnation. Various research groups adopted the technique in order to study flow phenomena through the CNTs (Miller et al. 2001; Miller and Martin 2002; Kim et al. 2004; Rossi et al. 2004), functionalize them (Kim et al. 2005; Korneva et al. 2005), and evaluate the separation performance of the CNT-modified membranes (Vermisoglou 2008).

Internal nanotube configuration plays a crucial role in most of the applications of CNT-based porous materials; thus, a detailed picture of the overall internal characteristics is critically important, as it determines the efficiency of the resulting nanostructures and affects the decision-making process toward their optimization. Electron microscopy and other on-spot ex situ techniques, although they can provide important information for the inside morphology of the structures, their in-principle extremely localized character does not allow for acquiring a representative view of the overall material. The assumption for uniformity is, in most cases, too optimistic; thus, generalization of microscopy data is often misleading. There is, therefore, an urgent need for employing generic "inside-tubes" techniques in order complementarily to electron microscopy and other localized methods, to provide detailed information of the internal structure variations of CNTs, and possibly other ordered nanomaterials exhibiting hollow morphology and/or interparticle porosity, throughout their dimensions. Ideally, these techniques would operate in situ while growth evolves, so that the process and conditions are manipulated online to meet certain product requirements. Taking advantage of such a toolset, optimization can be performed after a minimal amount of experiments in order to customize CNT internal surface characteristics and provide uniform and functional nanotube arrays optimized for specific applications.

5.3.3.1 CNT Growth on Porous Supports with In Situ Gravimetric Monitoring

In this example, CVD growth of CNTs took place on porous supports of various structural and morphological characteristics, such as activated carbon, with continuous monitoring of the growth evolution through gravimetric analysis, which was achieved by a microbalance system connected online to the reaction zone. The pore structure and density, and the high surface area of the porous support affected metal-dispersion properties, which are of central importance in the catalytic growth of CNTs. CNT growth by CVD involves a large number of tunable parameters that can have an influence on the nature, dimensions, and structural characteristics of the final material, and a statistical study with a screening of all experimental variables would normally be needed for optimization of the process (Porro et al. 2006). However, the in situ gravimetric technique used to monitor the weight change of the samples during carbon deposition was proven to be a valuable tool in simplifying the optimization process, offering control of the growth and of the experimental parameters involved during deposition, and allowing immediate response by the user through manipulation of the growth conditions (CVD duration, temperature, flow rates), based on the gravimetric curve, in order to acquire the optimal product.

The growth/monitoring approach is shown schematically in Figure 5.21. It mainly consists of a flow system with differential mass flow controllers for delivery of carrier gases and precursors, an online microbalance head, and a high temperature furnace that is connected to the microbalance (Veziri et al. 2008). Evolution of CNT growth was monitored in situ by the weight change of the sample during C_2H_2 decomposition expressed by the equation

$$w(\%) = \left[\frac{(m_t - m_o)}{m_o} \right] \times 100,$$

where m_o is the initial weight of the nickel-impregnated activated carbon before introducing the C_2H_2 into the reactor.

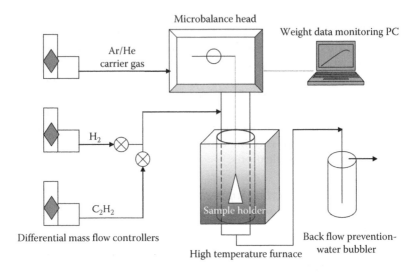

FIGURE 5.21
Diagram of the setup for monitoring CNTs growth. (From *Microp. Mesop. Mat.*, 110(1), Veziri, C.M., Pilatos, G., Karanikolos, G.N., Labropoulos, A., Kordatos, K., Kasselouri-Rigopoulou, V., and Kanellopoulos, N.K. Growth and optimization of carbon nanotubes in activated carbon by catalytic chemical vapor deposition, 41–50, Copyright 2008, with permission from Elsevier.)

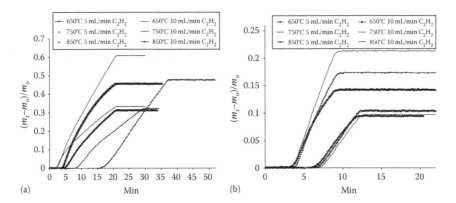

FIGURE 5.22
Online weight change of activated carbon samples during CNT growth as a function of time for (a) 20 min and (b) 5 min CVD. (From *Microp. Mesop. Mat.*, 110(1), Veziri, C.M., Pilatos, G., Karanikolos, G.N., Labropoulos, A., Kordatos, K., Kasselouri-Rigopoulou, V., and Kanellopoulos, N.K., Growth and optimization of carbon nanotubes in activated carbon by catalytic chemical vapor deposition, 41–50, Copyright 2008, with permission from Elsevier.)

Figure 5.22 shows examples of online weight change curves as a function of time offering useful conclusions about CNT growth behavior. Growth rates, as calculated from the curve slopes, increase when temperature increases from 650°C to 750°C, while from 750°C to 850°C there is no significant change. However, given that weight change, as calculated from the in situ microbalance, corresponds to both CNT and amorphous carbon formation and that at temperatures above 800°C there is a dramatic increase in amorphous carbon deposits as also confirmed by SEM imaging, it is concluded that the net growth rate corresponding to CNTs decreases for temperatures higher than 800°C, especially at prolonged reaction times and higher precursor flow rates.

Independent of the CVD parameters tested, considerable growth takes place during the first 5 min of the reaction, which then continuously drops as reaction proceeds. The CNT growth rate drop is even more pronounced at longer times, given that the calculated weight gain corresponds to both CNT and amorphous carbon formation and the amount of amorphous carbon deposits increases with time, as confirmed by SEM images. Therefore, limiting the reaction time to 5 min has the twin advantage of achieving the highest growth rate combined with the minimization of carbonaceous particle formation. The fact that CNT growth increases linearly reaching a maximum after a short period of time and then starts dropping can be attributed to the fast deactivation of the catalytic metal particles from which CNTs nucleate. This deactivation may be due to particle overcoating with carbon or conversion of the metal into a metal carbide or other non-catalytic form (e.g., by absorption of too much carbon into the interior of the particle) (Cassell et al. 1999; Bronikowski 2006). This mechanism is supported by the observation that the growth rate drop at longer times is higher when the carbon feedstock flow rate increases, as calculated by the online gravimetric curves.

CNT growth rate and morphology are also affected by the flow rate of the carbon precursor. The in situ weight change plots shown in Figure 5.22 reveal that the higher the C_2H_2 flow the higher the growth rate is, given that the rest of the reaction conditions are constant. Doubling the C_2H_2 flow rate, for example, from 5 to 10 mL/min, keeping reaction temperature and time constant at 750°C and 5 min, respectively, brings about a 1.1 mg/min increase in the weight gain rate. At higher precursor flows the reaction onset is shifted to shorter times. The weight change plots show that when a 5 mL/min flow rate is used, CNT growth starts approximately 6.5 min after the C_2H_2 is introduced into the CVD chamber, while this time is reduced to 3.5 min when a flow of 10 mL/min is applied.

In conclusion, based on the online weight change data, CNT growth rate increases linearly with temperature in the range of 600°C–800°C and then starts dropping for temperatures over 800°C. Long reaction times broaden the CNT size distribution and increase the amorphous carbon deposits. In addition, after the first 5 min of the reaction, the growth rate appears to decrease as the carbonaceous particles cover the surface of the substrate, especially at high precursor flow rates. A reaction temperature of 750°C with 5 min duration and a precursor flow rate of 10 mL/min appear to constitute the optimal conditions for CNT growth in activated carbon, since they produce a high yield of uniformly sized CNTs, while they minimize amorphous carbon formation. The in situ gravimetric technique, through appropriate recognition of the weight transitions taking place online, can offer useful monitoring capabilities, for example, indicating amorphous carbon formation

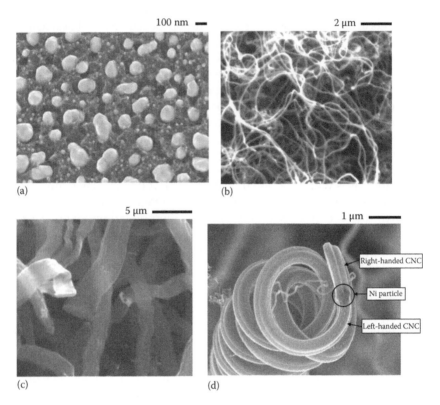

FIGURE 5.23
Polymorphism of carbon nanostructures grown on porous supports. (a) Pre-growth morphology of catalyst deposition, and carbon with (b) fibrous, (c) belt, and (d) coil morphology. The online gravimetric approach can be potentially useful in monitoring possible transitions from one structure to the other during growth. (From *Microp. Mesop. Mat.*, 110(1), Veziri, C.M., Pilatos, G., Karanikolos, G.N., Labropoulos, A., Kordatos, K., Kasselouri-Rigopoulou, V., and Kanellopoulos, N.K., Growth and optimization of carbon nanotubes in activated carbon by catalytic chemical vapor deposition, 41–50, Copyright 2008, with permission from Elsevier.)

and distinguishing the deposition of different structures (e.g., from tubes to coils) (Veziri et al. 2009), due to growth rate variations from one structure to another (Figure 5.23), thus allowing the user to manipulate the process in order to acquire the desired material characteristics.

5.3.3.2 Growth of CNT Membranes Monitored by Online Differential Pressure Measurements and Combination of Sorption and Permeability Techniques

In this section, the growth of aligned CNTs inside the 1D, straight pores of unidimensional-pore templates, such as anodized alumina, is used as synthesis example. A specially designed CVD system was used, in which ethylene was forced to flow through the pores of the template, thus assisting nanotube formation and alignment throughout the template thickness. Online measurement of the differential pressure between the two sides of the alumina template allows continuous monitoring of the evolution of nanotube growth. The internal pore structure of CNT membranes was investigated in detail by employing a combinational toolset consisting of adsorption and permeability-through-the-CNT-pores techniques. The CNT arrays were grown using a suspension of surfactant-coated iron oxide (Fe_3O_4) nanoparticles as catalyst, finely dispersed along the AAO pores by surface pretreatment of the template with NH_4NO_3. Consequently, long nanotube arrays with a diameter comparable to that of the template pores and length that spans the complete thickness of the template were achieved. Combinations of sorption and permeability techniques including *n*-hexane and water adsorption, nitrogen permeability, and relative permeability were employed to evaluate in detail the pore structure of the resulting CNT membranes and provide useful insights into nanotube internal surface and hollow space configuration.

Figure 5.24 shows the online differential pressure curve of the growth. The curve can serve as a first indicator of important transitions taking place, as growth evolves by monitoring pore blocking due to carbon deposits, thus allowing the user to intervene at any stage by terminating the process or manipulating the growth parameters. For instance, the slope changes are indicative of different growth phases, such as carbon source decomposition, nanotube formation, and excessive amorphous carbon deposition, and allow for process optimization by appropriately adjusting the conditions (gas flow rate, reaction duration, etc.).

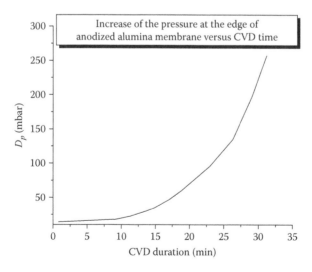

FIGURE 5.24
Diagram that illustrates the differential pressure increase at the two sides of the membrane vs. CVD time.

The effectiveness of the parallel and monodisperse AAO pores as templates for growth of CNT arrays, which has been well demonstrated in the literature, is partially attributed to the smoothness and uniformity of the internal surface of the AAO pores, and to the good adherence between CNT external and AAO internal surfaces, as confirmed by adsorption measurements.

Template surface functionalization by NH_4NO_3 that ensured uniform iron oxide catalyst particle dispersion throughout the pore length, and carbon precursor introduction in a flow-through-the-pores mode during CVD played a crucial role in producing well-aligned and monodisperse CNTs, which are intrinsically open-ended as the iron carbide particles are embedded in the CNT walls, thus leaving the tube interior directly accessible (Pilatos et al. 2010) (Figures 5.25 and 5.26).

Water adsorption studies indicated that the synthesized CNT composite exhibits hydrophilic character that renders the material of particular importance in biological applications.

The high *n*-hexane adsorbed layer thickness at low relative pressures of the resulting CNT membrane and the gradual rise of the adsorption isotherm and corresponding *t*-curve at high relative pressures reveal the existence of

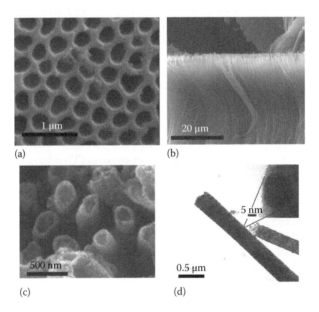

(a)   (b)

(c)   (d)

FIGURE 5.25
Morphology of the synthesized CNT arrays: (a) SEM image of the AAO template, (b) cross-sectional SEM image of aligned CNT arrays after template removal, (c) top-view SEM image of the template-free CNTs, and (d) TEM image of CNTs showing iron carbide particles encapsulated by graphitic layers. (Pilatos, G., Vermisoglou, E.C., Romanos, G.E., Karanikolos, G.N., Boukos, N., Likodimos, V. and Kanellopoulos, N.K., A closer look inside nanotubes: Pore structure evaluation of anodized alumina templated carbon nanotube membranes through adsorption and permeability studies, *Adv. Funct. Mater.*, 2010, 20, 2500–2510. Copyright Wiley-VCH Verlag GmbH & Co. KGaA.)

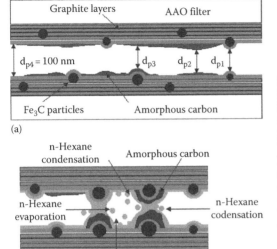

(a)

(b)

FIGURE 5.26
Proposed morphology of the CNT internal surface and hollow space. (Pilatos, G., Vermisoglou, E.C., Romanos, G.E., Karanikolos, G.N., Boukos, N., Likodimos, V. and Kanellopoulos, N.K., A closer look inside nanotubes: Pore structure evaluation of anodized alumina templated carbon nanotube membranes through adsorption and permeability studies, *Adv. Funct. Mater.*, 2010, 20, 2500–2510. Copyright Wiley-VCH Verlag GmbH & Co. KGaA.)

rough internal surfaces that increase the number of adsorption sites, and indicate that gradual condensation of *n*-hexane vapor takes place into the CNT bore space caused by internal diameter alterations. However, the lack of intense hysteresis imply that the internal constrictions that alter nanotube bore diameter are small in number and randomly distributed approximately in the middle of the nanotube length (Figure 5.27).

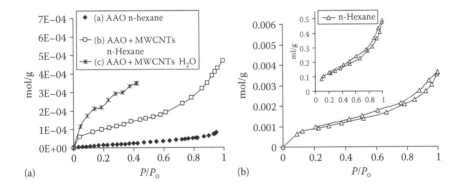

(a)

(b)

FIGURE 5.27
(a) Adsorption isotherms of the AAO and AAO+MWCNTs membranes at 307 K measured by the gravimetric method. (b) *n*-Hexane adsorption isotherm of the template-free MWCNTs at 307 K. The inset is the adsorption isotherm of the MWCNTs expressed in liquid volume of adsorbed substance per gram of solid. (Pilatos, G., Vermisoglou, E.C., Romanos, G.E., Karanikolos, G.N., Boukos, N., Likodimos, V. and Kanellopoulos, N.K., A closer look inside nanotubes: Pore structure evaluation of anodized alumina templated carbon nanotube membranes through adsorption and permeability studies, *Adv. Funct. Mater.*, 2010, 20, 2500–2510. Copyright Wiley-VCH Verlag GmbH & Co. KGaA.)

These constrictions are attributed to a small amount of iron carbide particles that are protruded toward the center of the tubes and/or covered by a relatively thick graphitic layer, as well as to amorphous carbon deposits, the existence of which was confirmed by Raman spectroscopy. The internal dimensions of the CNTs were estimated by various techniques: SEM analysis revealed an internal radius of 50 nm, while single phase permeability experiments, which take into account internal surface morphology alterations throughout the tube length, gave an average overall value of 40 nm. The small variation between the two aforementioned values indicate that the produced CNTs exhibit well-defined cylindrical morphology and monodispersity, though it is also indicative of the presence of the internal constrictions that occupy a small portion of the overall bore volume. Relative permeability confirmed that these constrictions are few in number; yet, they can create necks inside the tubes that can locally reduce the hollow diameter of the tubes to as low as 10 nm (Figure 5.28).

These localized morphological details can be of crucial importance especially in flow applications, yet electron microscopy and other typical on-spot analytical methods are difficult to detect. Taking advantage of the combinational toolset described above and the respective findings, systematic optimization by process manipulation can be performed in order to customize CNT internal surface characteristics and provide uniform and functional nanotube arrays optimized for specific applications.

To tailor CNT size to subnanometer thickness, aluminophosphate templates consisting of 1D channels with nominal diameters ranging from 0.5 to 1.2 nm can be used. *c*-Oriented $AlPO_4$-5 (AFI-type structure) films, for example, can be prepared by seeded growth, which consists of seeding

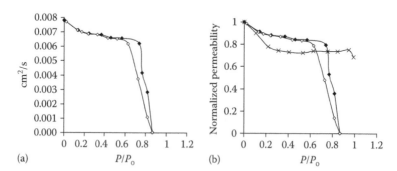

(a)   (b)

FIGURE 5.28

(a) Relative permeability curves of the AAO+MWNTs membrane, closed symbols = adsorption, open symbols = desorption; (b) normalized relative permeability curves of the AAO+MWNTs membrane in comparison with the curve of an AAO membrane with pores of 100 nm in diameter (x symbols). (Pilatos, G., Vermisoglou, E.C., Romanos, G.E., Karanikolos, G.N., Boukos, N., Likodimos, V. and Kanellopoulos, N.K., A closer look inside nanotubes: Pore structure evaluation of anodized alumina templated carbon nanotube membranes through adsorption and permeability studies, *Adv. Funct. Mater.*, 2010, 20, 2500–2510. Copyright Wiley-VCH Verlag GmbH & Co. KGaA.)

the surface of appropriately functionalized substrates with a monolayer of $AlPO_4$ nanoparticles (seeds) followed by sequential hydrothermal growth, under conditions that favor the desirable preferred orientation to yield a continuous and oriented film (Karanikolos et al. 2007, 2008; Veziri et al. 2010). Growth behavior, in this case, can be controlled by appropriate manipulation of the precursor mixture and hydrothermal conditions. Altering the water content can affect growth direction, which can be continuously tuned between in-plane and *c*-out-of-plane (Figure 5.29a through d). Employing precrystallization of the precursor mixture under proper conditions can yield a controllable and uniform growth of prism-shaped thin oriented crystals (Figure 5.29e). Given the large number of parameters that can be involved in such process and affect the quality of the final film, a pseudo in situ environment can be established to provide a full-scale monitoring of the material throughout the growth. The approach is based on process differentiation into multiple substages and preparation of snap-shot samples taken at specific short time intervals during synthesis. For example, the same experiment is initiated at the same time in a large number of identical reactors (cells) that operate under the same set of conditions and parameters (temperature, pressure, species concentrations), and sequentially the reaction is terminated in each one of the samples at predefined times so that the corresponding products are characterized ex situ (e.g., by electron microscopy, XRD, Raman, nitrogen porosimetry). Integration of all the data will give a representative picture of the evolution of all the major characteristics of the material.

SWCNT growth can be performed in the optimized films by pyrolysis of the structure directing agent (SDA) inside the channels (Tang et al. 1998; Wang et al. 2000a) and may be assisted by external flow of carbon precursors and catalysts (Figure 5.30). In situ differential permeability, in this case, can be applied to monitor film growth, defect formation/repair, and CNT growth evolution. The prepared membranes consisting of ultrasmall diameter CNT pores can be ideal to experimentally test high-flux transport, which is anticipated to be comparable to the flow rate of water in biological protein membranes (aquaporin) through ordering of water molecules in the channel.

5.3.3.3 CNT Growth and Monitoring by In Situ Electron Microscopy

Recent advances in special and temporal resolution in situ electron microscopy provide an extremely effective monitoring at atomic level of the evolution of nanostructure of the catalyst during nanotube growth and in situ determination of physical, electronic, and mechanical properties of the synthesized nanotubes. This approach can not only shed light on the formation mechanism and provide the properties of an individual CNT but can also give the structure of the nanotube through electron imaging and diffraction, providing an ideal technique for understanding the property–structure relationship of well-defined nanotubes (Wang et al. 2000b). In one

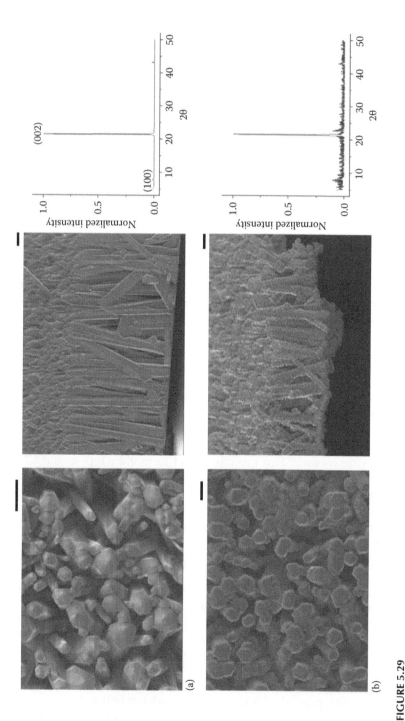

FIGURE 5.29
c-Oriented aluminophosphate AFI films with various morphological characteristics grown by sequential hydrothermal growth on seeded substrates. (a) Columnar AlPO$_4$–5 crystals by secondary growth from dilute precursor mixtures that favor *c*-out-of-plane growth.

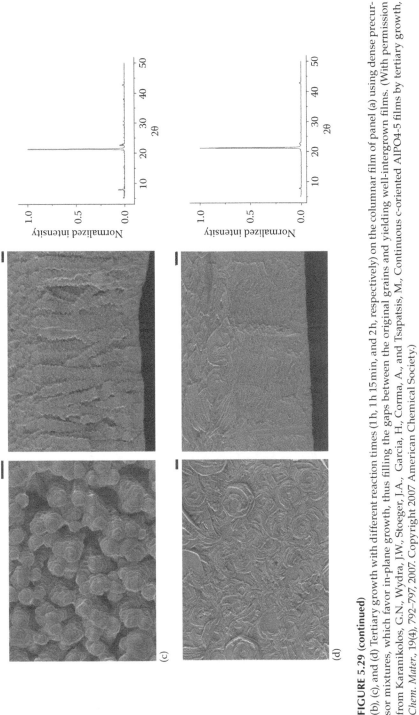

FIGURE 5.29 (continued)

(b), (c), and (d) Tertiary growth with different reaction times (1h, 1h 15min, and 2h, respectively) on the columnar film of panel (a) using dense precursor mixtures, which favor in-plane growth, thus filling the gaps between the original grains and yielding well-intergrown films. (With permission from Karanikolos, G.N., Wydra, J.W., Stoeger, J.A., Garcia, H., Corma, A., and Tsapatsis, M., Continuous c-oriented AlPO4-5 films by tertiary growth, *Chem. Mater.*, 19(4), 792–797, 2007. Copyright 2007 American Chemical Society.)

(continued)

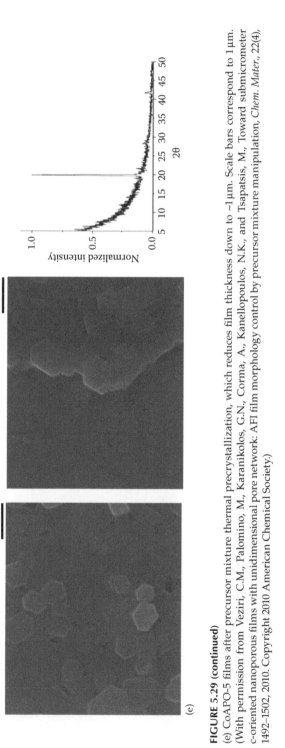

FIGURE 5.29 (continued)

(e) CoAPO-5 films after precursor mixture thermal precrystallization, which reduces film thickness down to ~1 μm. Scale bars correspond to 1 μm. (With permission from Veziri, C.M., Palomino, M., Karanikolos, G.N., Corma, A., Kanellopoulos, N.K., and Tsapatsis, M., Toward submicrometer c-oriented nanoporous films with unidimensional pore network: AFI film morphology control by precursor mixture manipulation, *Chem. Mater.*, 22(4), 1492–1502, 2010. Copyright 2010 American Chemical Society.)

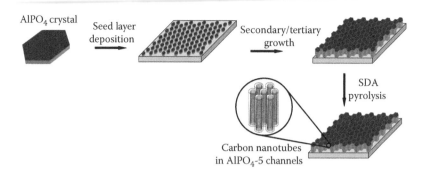

FIGURE 5.30
Schematic of $AlPO_4$-5 thin film fabrication and CNT growth by SDA pyrolysis.

of the first examples of in situ growth, CNT formation was initiated by an electron beam that is carried out in a TEM chamber compatible with in situ and real-time observation of the growth (Yasuda et al. 2002). Such an observation showed that the growth occurs in two steps: rapid formation of rod-like carbon (first step) and slow graphitization of its wall (second step), while the hollow inside the CNT is not formed in the first step but develops during the second. Mechanical properties of CNTs have been measured in situ by a hybrid nanorobotic manipulation system inside a SEM and a TEM (Nakajima et al. 2006). The SEM manipulators have been constructed with 8 degrees of freedom (DOFs) with three units for effective TEM sample preparation. The TEM manipulator consists of a 3-DOF manipulator actuated with four multilayer piezoelectric actuators and a 3-DOF passively driven sample stage. To show the effectiveness of the system, the Young's modulus of a CNT was measured to be 1.23 TPa inside the TEM, after being premanipulated inside the SEM. Electron field emission characteristics of individual MWCNTs were investigated in situ inside a TEM (Jin et al. 2005). For a single MWCNT, it was found that while field-emission can hardly occur from the side of the nanotube, a curved nanotube may result in finite side emission and the best emission geometry is the top emission geometry. Current–voltage (I–V) measurements made at different vacuum conditions and voltage sweeps emphasized the importance of the adsorbates on the electron field emission of MWCNTs. For a contaminated MWCNT, although the field emission current was reduced, the stability of its emission was improved. A current of up to several tens of µA was observed for a single MWCNT, but it was found that long-time emission usually results in drastic structure damage that may lead to sudden emission failure.

Sharma et al. (2009) recently reported important findings concerning CVD growth of nanotubes on catalyst particles that had undergone site-specific deposition using electron beam induced decomposition (EBID). These particles provided an undisrupted view for observing the structural transformations that occur during catalytic CVD of CNTs. High-resolution lattice

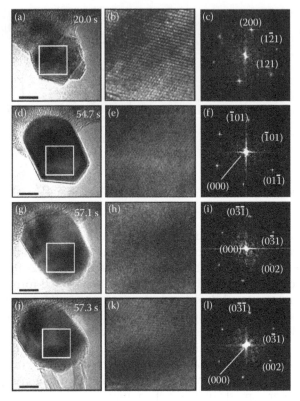

FIGURE 5.31
Evolution of catalyst particle transformation and CNT growth. The relative time lapsed in seconds is given in the upper left corner. The scale bar is 5 nm. (With permission form Sharma, R., Moore, E., Rez, P., and Treacy, M.M.J., Site-specific fabrication of Fe particles for carbon nanotube growth. *Nano Lett.*, 9(2), 689–694, 2009. Copyright 2009 American Chemical Society.)

images revealed both stable and unstable intermediate phases formed during the catalytic growth of CNTs (Figure 5.31).

CNTs nucleate and grow from Fe_3C particles formed during the heating of iron oxide nanoparticles in acetylene via reduction of iron oxide. Although it was generally assumed that cementite phase may be an intermediate step for CNT growth, these findings indicated that cementite is quite stable and its decomposition is not necessary for CNT nucleation and growth, as the enclosed particles after growth retain the cementite structure. The stability of the Fe_3C particles after CNT growth was confirmed by recent developments of Fe_3C-enclosing CNTs with high cementite content by templated growth and evaluation of their magnetic properties owing to the Fe_3C presence (Vermisoglou et al. 2010). Parametric studies by altering growth conditions inside the TEM have been carried out to determine the effect of temperature and pressure on the growth of CNTs by the Ni-catalyzed CVD of acetylene (Sharma et al. 2007). These experiments showed that straight, single-wall, CNTs tend to form at higher temperatures and lower pressures while bent, zigzag, multiwall CNTs form at lower temperatures and higher pressures. These highly innovative measurements can be potentially combined with other in situ techniques, such as gravimetric and permeability, operating under the same conditions, to provide superior monitoring and control over growth and properties of CNTs and other nanostructures.

References

Adya, A. K. and C. J. Wormald. 1991. First direct observation of orientationally correlated molecules in $CO_2(l)$ by neutron diffraction. *Mol Phys* 74: 735–746.

Ago, H., K. Nakamura, N. Uehara, and M. Tsuji. 2004. Roles of metal-support interaction in growth of single- and double-walled carbon nanotubes studied with diameter-controlled iron particles supported on MgO. *J Phys Chem B* 108(49): 18908–18915.

Ago, H., S. Imamura, T. Okazaki, T. Saitoj, M. Yumura, and M. Tsuji. 2005. CVD growth of single-walled carbon nanotubes with narrow diameter distribution over Fe/MgO catalyst and their fluorescence spectroscopy. *J Phys Chem B* 109(20): 10035–10041

Alba-Simionesco, C., G. Dosseh, E. Dumont, B. Frick, B. Geil, D. Morineau, V. Teboul, and Y. Xia. 2003. Confinement of molecular liquids: Consequences on thermodynamic, static and dynamical properties of benzene and toluene. *Eur Phys J E* 12: 19–28.

Albouy, P.-A. and A. Ayral. 2002. Coupling x-ray scattering and nitrogen adsorption: An Interesting approach for the characterization of ordered mesoporous materials. Application to hexagonal silica. *Chem Mater* 14: 3391–3397.

Baker, J. M., J. C. Dore, and P. Behrens. 1997. Nucleation of ice in confined geometry. *J Phys Chem B* 101: 6226–6229.

Baker, R. T. K. 1989. Catalytic growth of carbon filaments. *Carbon* 27(3): 315–323.

Bausenwein, T., H. Bertagnolli, D. Gutwerk, K. Toumldheide, and P. Chieux. 1992. The structure of fluid carbon dioxide by neutron diffraction at high pressure and by RISM calculations. *Mol Phys* 76: 127–141.

Bernal M. P., J. Coronas, M. Menéndez, and J. Santamaría. 2002. Characterization of zeolite membranes by temperature programmed permeation and step desorption. *J Memb Sci* 195: 125–138.

Bethune, D. S., C. H. Kiang, M. S. de Vires, G. Gorman, R. Savoy, J. Vazquez, and R. Beyers. 1993. Cobalt-catalysed growth of carbon nanotubes with single-atomic-layer walls. *Nature* 363: 605–607.

Bronikowski, M. J. 2006. CVD growth of carbon nanotube bundle arrays. *Carbon* 44(13): 2822–2832.

Cassell, A. M., J. A. Raymakers, J. Kong, and H. J. Dai. 1999. Large scale CVD synthesis of single-walled carbon nanotubes. *J Phys Chem B* 103(31): 6484–6492.

Chakrabarti, S., H. Kume, L. J. Pan, T. Nagasaka, and Y. Nakayama. 2007. Number of walls controlled synthesis of millimeter-long vertically aligned brushlike carbon nanotubes. *J Phys Chem C* 111(5): 1929–1934.

Challet, S., P. Azais, R. J. M. Pellenq, O. Isnard, J. L. Soubeyroux, and L. Duclaux. 2004. Hydrogen adsorption in microporous alkali-doped carbons (activated carbon and single wall nanotubes). *J Phys Chem Solids* 65(2–3): 541–544.

Che, G., B. B. Lakshmi, C. R. Martin, E. R. Fisher, and R. S. Ruoff. 1998a. Chemical vapor deposition based synthesis of carbon nanotubes and nanofibers using a template method. *Chem Mater* 10(1): 260–267.

Che, G., B. B. Lakshmi, E. R. Fisher, and C. R. Martin. 1998b. Carbon nanotubule membranes for electrochemical energy storage and production. *Nature* 393(6683): 346–349.

Chen, H. and D. S. Sholl. 2006. Predictions of selectivity and flux for CH_4/H_2 separations using single walled carbon nanotubes as membranes. *J Memb Sci* 269(1–2): 152–160.

Chen, H., J. K. Johnson, and D. S. Sholl. 2006. Transport diffusion of gases is rapid in flexible carbon nanotubes. *J Phys Chem B* 110(5): 1971–1975.

Chhowalla, M., K. B. K. Teo, C. Ducati, N. L. Rupesinghe, G. A. J. Amaratunga, A. C. Ferrari, D. Roy, J. Robertson, and W. I. Milne. 2001. Growth process conditions of vertically aligned carbon nanotubes using plasma enhanced chemical vapor deposition. *J Appl Phys* 90(10): 5308–5317.

Chiappini, S., M. Nardone, F. P. Ricci, and M. C. Bellissent-Funel. 1996. Neutron diffraction measurements on high pressure supercritical CO_2. *Mol Phys* 89: 975–987.

Choi, Y. C., Y. M. Shin, Y. H. Lee, B. S. Lee, G.-S. Park, W. B. Choi, N. S. Lee, and J. M. Kim. 2000. Controlling the diameter, growth rate, and density of vertically aligned carbon nanotubes synthesized by microwave plasma-enhanced chemical vapor deposition. *Appl Phys Lett* 76(17): 367–2369.

Christenson, H. K. 2001. Confinement effects on freezing and melting. *J Phys Condens Matter* 13: R95–R133.

Cipriani, P., M. Nardone, F. P. Ricci, and M. A. Ricci. 2001. Orientational correlations in liquid and supercritical CO_2: Neutron diffraction experiments and molecular dynamics simulations. *Mol Phys* 99: 301–308.

Cracknell, R. F. and D. Nicholson. 1994. Grand canonical Monte Carlo study of Lennard-Jones mixtures in slit pores. Part 3—Mixtures of two molecular fluids: Ethane and propane. *J Chem Soc Faraday Trans* 90: 1487.

Dai, G.-P., C. Liu, M. Liu, M.-Z. Wang, and H.-M.Cheng. 2002. Electrochemical hydrogen storage behavior of ropes of aligned single-walled carbon nanotubes. *Nano Lett* 2(5): 503–506.

Dai, H. J. 2002. Carbon nanotubes: Synthesis, integration, and properties. *Acc Chem Res* 35: 1035–1044.

de Lucas, A., A. Garrido, P. Sanchez, A. Romero, and J. L. Valverde. 2005. Growth of carbon nanofibers from Ni/Y zeolite based catalysts: Effects of Ni introduction method, reaction temperature, and reaction gas composition. *Ind Eng Chem Res* 44(22): 8225–8236.

De Santis, A., R. Frattini, D. Gazzillo, and M. Sampoli. 1987. The potential model dependence of the neutron radial and partial distribution functions for liquid CO_2. *Mol Phys* 60: 21–31.

Dillon, A. C., K. M. Jones, T. A. Bekkedahl, C. H. Kiang, D. S. Bethune, and M. J. Heben. 1997. Storage of hydrogen in single-walled carbon nanotubes. *Nature* 386(6623): 377–379.

Dolino, G., D. Bellet, and C. Faivre. 1996. Adsorption strains in porous silicon. *Phys Rev B* 54: 17919–17929.

Du, C., J. Yeh, and N. Pan. 2005. High power density supercapacitors using locally aligned carbon nanotube electrodes. *Nanotechnology* 16(4): 350.

Dusenbury, J. S. and F. S. Cannon. 1996. Advanced oxidant reactivity pertaining to granular activated carbon beds for air pollution control. *Carbon* 34(12): 1577–1589.

Ebbesen, T. W. and P. M. Ajayan. 1992. Large-scale synthesis of carbon nanotubes. *Nature* 358: 220–222.

Emmenegger, C., P. Mauron, P. Sudan, P. Wenger, V. Hermann, R. Gallay, and A. Zuttel. 2003. Investigation of electrochemical double-layer (ECDL) capacitors electrodes based on carbon nanotubes and activated carbon materials. *J Power Sources* 124(1): 321–329.

Fan, S., M. G. Chapline, N. R. Franklin, T. W. Tombler, A. M. Cassell, and H. Dai. 1999. Self-oriented regular arrays of carbon nanotubes and their field emission properties. *Science* 283(5401): 512–514.

Fatt, I. 1956. (a) The network model of porous media I. Capillary pressure characteristics. *Trans AIME* 207:144–159. (b) The network model of porous media II. Dynamic properties of a single size tube network. *Trans AIME* 207: 160–163. (c) The network model of porous media III. Dynamic properties of networks with tube radius distribution. *Trans AIME* 207: 164–181.

Favvas, E. P., G. C. Kapantaidakis, J. W. Nolan, A. Ch. Mitropoulos, and N. K. Kanellopoulos. 2007. Preparation, characterization and gas permeation properties of carbon hollow fiber membranes based on Matrimid® 5218 precursor. *J Mater Process Technol* 186:102–110.

Favvas, E. P., E. P. Kouvelos, G. E. Romanos, G. I. Pilatos, A. Ch. Mitropoulos, and N. K. Kanellopoulos. 2008. Characterization of highly selective microporous carbon hollow fiber membranes prepared from a commercial co-polyimide precursor. *J Porous Mater* 15(6): 625–633.

Floquet, N., J. P. Coulomb, P. Llewellyn, G. André, and R. Kahn. 2005. Growth mode of hydrogen in mesoporous MCM-41. Adsorption and neutron scattering studies. *Adsorption* 11: 679–684.

Frick, B., R. Zorn, and H. Büttner (eds.). 2005. Proceedings of the International Workshop on Dynamics in Confinement. *J Phys IV* 10: Pr7 3–343.

Frick, B., M. Koza, and R. Zorn (eds.). 2003. Proceedings of the Second International Workshop on Dynamics in Confinement. *Eur. Phys. JE* 12: 5–194.

Fujiwara, A., K. Ishii, H. Suematsu, H. Kataura, Y. Maniwa, S. Suzuki, and Y. Achiba. 2001. Gas adsorption in the inside and outside of single-walled carbon nanotubes. *Chem Phys Lett* 336(3–4): 205–211.

Gelb, L. D., K. E. Gubbins, R. Radhakrishnan, and M. Sliwinska-Bartkowiak. 1999. Phase separation in confined systems. *Rep Prog Phys* 62: 1573–1659.

Gregg, S. J. and K. S. W. Sing. 1982. *Adsorption, Surface Area and Porosity*, 2nd ed., Academic Press, London, U.K.

Gusev, A. A. and O. Guseva. 2007. Rapid mass transport in mixed matrix nanotube/polymer membranes. *Adv Mater* 19(18): 2672–2676.

Hinds, B. J., N. Chopra, T. Rantell, R. Andrews, V. Gavalas, and L. G. Bachas. 2004. Aligned multiwalled carbon nanotube membranes. *Science* 303(5654): 62–65.

Hiraoka, T., T. Kawakubo, J. Kimura, R. Taniguchi, A. Okamoto, T. Okazaki, T. Sugai, Y. Ozeki, M. Yoshikawa, and H. Shinohara. 2003. Selective synthesis of double-wall carbon nanotubes by CCVD of acetylene using zeolite supports. *Chem Phys Lett* 382: 679–685.

Hofmann, T., D. Wallacher, P. Huber, R. Birringer, K. Knorr, A. Schreiber, and G. H. Findenegg. 2005. Small-angle x-ray diffraction of Kr in mesoporous silica: Effects of microporosity and surface roughnes. *Phys Rev B* 72: 064122(1–7).

Hoinkis, E. 1996. Small angle neutron scattering study of C_6D_6 condensation in a mesoporous glass. *Langmuir* 12: 4299–4302.

Holt, J. K., H. G. Park, Y. Wang, M. Stadermann, A. B. Artyukhin, C. P. Grigoropoulos, A. Noy, and O. Bakajin. 2006. Fast mass transport through sub-2-nanometer carbon nanotubes. *Science* 31(5776): 1034–1037.

Iijima, S. 1991. Helical microtubules of graphitic carbon. *Nature* 354: 56–58.

Iijima, S. and T. Ichihashi. 1993. Single-shell carbon nanotubes of 1-nm diameter. *Nature* 363: 603–605.

Ismail, A. F. and L. I. B. David. 2001. A review on the latest development of carbon membranes for gas separation. *J Memb Sci* 193: 1–18.

Jagiello, J. and M. Thommes. 2004. Comparison of DFT characterization methods based on N_2, Ar, CO_2 and H_2 adsorption applied to carbons with various pore size distributions. *Carbon* 42: 1227–1232.

Jähnert, S., D. Müter, J. Prass, G. A. Zickler, O. Paris, and G. H. Findenegg. 2009. Pore structure and fluid sorption in ordered mesoporous silica. I. Experimental study by in situ small-angle x-ray scattering. *J Phys Chem C* 113: 15201–15210.

Jakobtorweihen, S., M. G. Verbeek, C. P. Lowe, F. J. Keil, and B. Smit. 2005. Understanding the loading dependence of self-diffusion in carbon nanotubes. *Phys Rev Lett* 95(4): 044501.

Jin, C., J. Wang, M. Wang, J. Su, and L.-M.Peng. 2005. In-situ studies of electron field emission of single carbon nanotubes inside the TEM. *Carbon* 43(5): 1026–1031.

Kainourgiakis, M. E., A. K. Stubos, N. D. Konstantinou, N. K. Kanellopoulos, and V. Milisic. 1996. A network model for the permeability of condensable vapours through mesoporous media. *J Memb Sci* 114(2): 215–225.

Kanellopoulos, N. K. and J. K. Petrou. 1988. Relative gas permeability of capillary networks with various size distributions. *J Memb Sci* 37(1): 1–12.

Karanikolos, G. N., J. W. Wydra, J. A. Stoeger, H. Garcia, A. Corma, and M. Tsapatsis. 2007. Continuous c-oriented AlPO4-5 films by tertiary growth. *Chem Mater* 19(4): 792–797.

Karanikolos, G. N., H. Garcia, A. Corma, and M. Tsapatsis. 2008. Growth of AlPO4-5 and CoAPO-5 films from amorphous seeds. *Microporous Mesoporous Mater* 115(1–2): 11–22.

Katsaros, F. K., Th. A. Steriotis, A. K. Stubos, A. Ch. Mitropoulos, N. K. Kanellopoulos, and S. Tennison. 1997. High pressure gas permeability of microporous carbon membrane. *Microporous Mater* 8: 171–176.

Katsaros, F. K., T. A. Steriotis, K. L. Stefanopoulos, N. K. Kanellopoulos, A. Ch. Mitropoulos, M. Meissner, and A. Hoser. 2000. Neutron diffraction study of adsorbed CO_2 on a carbon membrane. *Physica B* 276: 901–902.

Kikkinides, E. S., K. P. Tzevelekos, A. K. Stubos, M. E. Kainourgiakis, and N. K. Kanellopoulos. 1997. Application of effective medium approximation for the determination of the permeability of condensable vapours through mesoporous media. *Chem Eng Sci* 52(16): 2837–2844.

Kim, B. M., S. Sinha, and H. H. Bau. 2004. Optical microscope study of liquid transport in carbon nanotubes. *Nano Lett* 4(11): 2203–2208.

Kim, B. M., S. Qian, and H. H. Bau. 2005. Filling carbon nanotubes with particles. *Nano Lett* 5(5): 873–878.

Kim, S., L. Chen, J. K. Johnson, and E. Marand. 2007. Polysulfone and functionalized carbon nanotube mixed matrix membranes for gas separation: Theory and experiment. *J Memb Sci* 294(1–2): 147–158.

Knorr, K., D. Wallacher, P. Huber, V. Soprunyuk, and R. Ackermann. 2003. Are solidified fillings of mesopores basically bulk-like except for the geometric confinement? *Eur Phys J E* 12: 51–56.

Konstantakou, M., S. Samios, Th. A. Steriotis, M. Kainourgiakis, G. K. Papadopoulos, E. S. Kikkinides, and A. K. Stubos. 2006. Determination of pore size distribution in microporous carbons based on CO_2 and H_2 sorption data. *Stud Surf Sci Catal* 160: 543–550.

Konstantakou, M., Th.A. Steriotis, G. K. Papadopoulos, M. Kainourgiakis, E. S. Kikkinides, and A. K. Stubos. 2007. Characterization of nanoporous carbons by combining CO_2 and H_2 sorption data with the Monte Carlo simulations. *Appl Surf Sci* 253: 5715–5720.

Korneva, G., H. Ye, Y. Gogotsi, D. Halverson, G.Friedman, J.-C.Bradley, and K. G. Kornev. 2005. Carbon nanotubes loaded with magnetic particles. *Nano Lett* 5(5): 879–884.

Kyotani, T., L.-F. Tsai, and A. Tomita. 1996. Preparation of ultrafine carbon tubes in nanochannels of an anodic aluminum oxide film. *Chem Mater* 8(8): 2109–2113.

Kyotani, T., L.-F. Tsai, and A. Tomita. 2002. Formation of ultrafine carbon tubes by using an anodic aluminum oxide film as a template. *Chem Mater* 7(8):1427–1428.

Labropoulos, A. I., G. E. Romanos, N. Kakizis, G. I. Pilatos, E. P. Favvas, and N. K. Kanellopoulos. 2008. Investigating the evolution of N_2 transport mechanism during the cyclic CVD post-treatment of silica membranes. *Microporous Mesoporous Mater* 110: 11.

Labropoulos, A. I., G. E. Romanos, G. N. Karanikolos, F. K. Katsaros, N. K. Kakizis, and N. K. Kanellopoulos. 2009. Comparative study of the rate and locality of silica deposition during the CVD treatment of porous membranes with TEOS and TMOS. *Microporous Mesoporous Mater* 120: 177–185.

Lagorsse, S., F. D. Magalhães, and A. Mendes. 2004. Carbon molecular sieve membranes Sorption, kinetic and structural characterization. *J Memb Sci* 241: 275–287.

Li, J.-C., D. K. Ross, L. D. Howe, K. L. Stefanopoulos, J. P. A. Fairclough, R. Heenan, and K. Ibel. 1994. Small-angle neutron-scattering studies of the fractal-like network formed during desorption and adsorption of water in porous materials. *Phys Rev B* 49: 5911–5917.

Li, Y., X. B. Zhang, L. H. Shen, J. H. Luo, X. Y. Tao, F. Liu, G. L. Xu, Y. W. Wang, H. J. Geise, and G. Van Tendeloo. 2004. Controlling the diameters in large-scale synthesis of single-walled carbon nanotubes by catalytic decomposition of CH4. *Chem Phys Lett* 398(1–3): 276–282.

Lippens, B. C. and J. H. De Boer. 1965. Studies on pore systems. V. The t method. *J Catal* 4: 319–323.

Liu, C., Y. Y. Fan, M. Liu, H. T. Cong, H. M. Cheng, and M. S. Dresselhaus. 1999. Hydrogen storage in single-walled carbon nanotubes at room temperature. *Science* 286(5442): 1127–1129.

Liu, E., J. C. Dore, J. B. W. Webber, D. Khushalani, S. Jähnert, G. H. Findenegg, and T. Hansen. 2006a. Neutron diffraction and NMR relaxation studies of structural variation and phase transformations for water/ice in SBA-15 silica: I. The overfilled case. *J Phys Condens Matter* 18: 10009–10028.

Liu, X., T. P. Bigioni, Y. Xu, A. M. Cassell, and B. A. Cruden. 2006b. Vertically aligned dense carbon nanotube growth with diameter control by block copolymer micelle catalyst templates. *J Phys Chem B* 110(41): 20102–20106.

Loeb, S. and S. Sourirajan. 1963. Sea water demineralization by means of an osmotic membrane, in saline water conversion II. R. F. Gould (ed.), *Advances in Chemistry Series Number 38*, American Chemical Society, Washington, DC, pp. 117–132.

Majumder, M., N. Chopra, and B. J. Hinds. 2005. Effect of tip functionalization on transport through vertically oriented carbon nanotube membranes. *J Am Chem Soc* 127(25): 9062–9070.

Majumder, M., X. Zhan, R. Andrews, and B. J. Hinds. 2007. Voltage gated carbon nanotube membranes. *Langmuir* 23(16): 8624–8631.

Martin, C. R. 1994. Nanomaterials: A membrane-based synthetic approach. *Science* 266(5193): 1961–1966.

Matranga, K. R., A. L. Myers, and E. D. Glandt. 1982. Storage of natural gas by adsorption on activated carbon. *Chem Eng Sci* 47: 1569–1579.

Mi, W., J. Y. S. Lin, Y. Li, and B. Zhang. 2005. Synthesis of vertically aligned carbon nanotube films on macroporous alumina substrates. *Microporous Mesoporous Mater* 81(1–3): 185–189.

Mi, W., Y. S. Lin, and Y. Li. 2007. Vertically aligned carbon nanotube membranes on macroporous alumina supports. *J Memb Sci* 304(1–2): 1–7.

Miller, S. A. and C. R. Martin. 2002. Controlling the rate and direction of electroosmotic flow in template-prepared carbon nanotube membranes. *J Electroanal Chem* 522(1): 66–69.

Miller, S. A., V. Y. Young, and C. R. Martin. 2001. Electroosmotic flow in template-prepared carbon nanotube membranes. *J Am Chem Soc* 123(49): 12335–12342.

Mitropoulos, A. Ch., J. M. Haynes, R. M. Richardson, and N. K. Kanellopoulos. 1995. Characterization of porous glass by adsorption of dibromomethane in conjunction with small-angle x-ray scattering. *Phys Rev B* 52: 10035–10042.

Morineau, D. and C. Alba-Simionesco. 2003. Liquids in confined geometry: How to connect changes in the structure factor to modifications of local order. *J Chem Phys* 118: 9389–9400.

Morineau, D., R. Guégan, Y. Xia, and C. Alba-Simionesco. 2004. Structure of liquid and glassy methanol confined in cylindrical pores. *J Chem Phys* 121: 1466–1473.

Morishige, K. and H. Iwasaki. 2003. X-ray study of freezing and melting of water confined within SBA-15. *Langmuir* 19: 2808–2811.

Morishige, K. and K. Kawano. 1999. Freezing and melting of methyl chloride in a single cylindrical pore: Anomalous pore-size dependence of phase-transition temperature. *J Phys Chem B* 103: 7906–7910.

Morishige, K. and K. Kawano. 2000. (a) Freezing and melting of methanol in a single cylindrical pore: Dynamical supercooling and vitrification of methanol. *J Chem Phys* 112: 11023–11029. (b) Freezing and melting of nitrogen, carbon monoxide, and krypton in a single cylindrical pore. *J Phys Chem B* 104: 2894–2900.

Morishige, K. and K. Nobuoka. 1997. X-ray diffraction studies of freezing and melting of water confined in a mesoporous adsorbent (MCM-41). *J Chem Phys* 107: 6965–6969.

Morishige, K. and H. Uematsu. 2005. The proper structure of cubic ice confined in mesopores. *J Chem Phys* 122: 044711(1–4).

Nakajima, M., F. Arai, and T. Fukuda. 2006. In situ measurement of Young's modulus of carbon nanotubes inside a TEM through a hybrid nanorobotic manipulation system. *IEEE Trans Nanotechnol* 5(3): 243–248.

Nguyen, C., D. D. Do, K. Haraya, and K. Wangd. 2003. The structural characterization of carbon molecular sieve membrane (CMSM) via gas adsorption. *J Memb Sci* 220: 177–182.

Nicholson, D. and J. H. Petropoulos. 1971. Capillary models for porous media: III. Two-phase flow in a three-dimensional network with Gaussian radius distribution. *J Phys D Appl Phys* 4(2): 81–189

Nicholson, D. and J. H. Petropoulos. 1973. Capillary models for porous media: IV. Flow properties of parallel and serial capillary models with various radius distributions. *J Phys D Appl Phys* 6(14): 1737–1744.

Ning, G. Q., Y. Liu, F. Wei, Q. Wen, and G. H. Luo. 2007. Porous and lamella-like Fe/MgO catalysts prepared under hydrothermal conditions for high-yield synthesis of double-walled carbon nanotubes. *J Phys Chem C* 111(5): 1969–1975.

Noh, J. S., R. K. Argaval, and J. A. Schwarz. 1987. Hydrogen storage systems using activated carbon. *Int J Hydrogen Energy* 12: 693–700.

Ohkubo, T., T. Iiyama, K. Nishikawa, T. Suzuki, and K. Kaneko. 1999. Pore-width dependent ordering of C_2H_5OH molecules confined in graphitic slit nanospaces. *J Phys Chem B* 103(11): 1859–1863.

Pan, Z., H. Zhu, Z. Zhang, H.-J. Im, S. Dai, D. B. Beach, and D. H. Lowndes. 2003. Patterned growth of vertically aligned carbon nanotubes on pre-patterned iron/silica substrates prepared by sol–gel and shadow masking. *J Phys Chem B* 107(6): 1338–1344.

Parthasarathy, R. V., K. L. N. Pan, and C. R. Martin. 1995. Template synthesis of graphitic nanotubules. *Adv Mater* 7(11): 896–897.

Peigney, A., P. Coquay, E. Flahaut, R. E. Vandenberghe, E. De Grave, and C. Laurent. 2001. A study of the formation of single- and double-walled carbon nanotubes by a CVD method. *J Phys Chem B* 105(40): 9699–9710.

Petropoulos, J. H., J. K. Petrou, and N. K. Kanellopoulos. 1989. Explicit relation between relative permeability and structural parameters in stochastic pore networks. *Chem Eng Sci* 44(12): 2967–2977.

Pilatos, G., E. C. Vermisoglou, G. E. Romanos, G. N. Karanikolos, N. Boukos, V. Likodimos, and N. K. Kanellopoulos. 2010. A closer look inside nanotubes: Pore structure evaluation of anodized alumina templated carbon nanotube membranes through adsorption and permeability studies. *Adv. Funct. Mater.* 20: 2500–2510.

Porro, S., S. Musso, M. Giorcelli, A. Chiodoni, and A. Tagliaferro. 2006. Optimization of a thermal-CVD system for carbon nanotube growth. *Phys E* 37(1–2): 16–20.

Ramesh, P., T. Okazaki, R. Taniguchi, J. Kimura, T. Sugai, K. Sato, Y. Ozeki, and H. Shinohara. 2005. Selective chemical vapor deposition synthesis of double-wall carbon nanotubes on mesoporous silica. *J Phys Chem B* 109(3): 1141–1147.

Ramsay, J. D. F. and S. Kallus. 2001. Characterization of mesoporous silicas by in situ small angle neutron scattering during isothermal gas adsorption. *J Non-Cryst Solids* 285: 142–147.

Ren, Z. F., Z. P. Huang, J. W. Xu, J. H. Wang, P. Bush, M. P. Siegal, and P. N. Provencio. 1998. Synthesis of large arrays of well-aligned carbon nanotubes on glass. *Science* 282(5391): 1105–1107.

Rossi, M. P., H. Ye, Y. Gogotsi, S. Babu, P. Ndungu, and J.-C. Bradley. 2004. Environmental scanning electron microscopy study of water in carbon nanopipes. *Nano Lett* 4(5): 989–993.

Roy, S., H. Vedala, and W. Choi. 2007. Vertically aligned carbon nanotube probes for monitoring blood cholesterol. *Nanotechnology* 17(4): S14.

Ruthven, D. M., F. Shamasuzzaman, and K. S. Knaebel. 1994. *Pressure Swing Adsorption*. Wiley-VCH, New York.

Ryu, K., A. Badmaev, L.Gomez, F. Ishikawa, B. Lei, and C. Zhou. 2007. Synthesis of aligned single-walled nanotubes using catalysts defined by nanosphere lithography. *J Am Chem Soc* 129(33): 10104–10105.

Samios, S., A. K. Stubos, N. K. Kanellopoulos, R. F. Cracknell, G. K. Papadopoulos, and D. Nicholson. 1997. Determination of micropore size distribution from grand canonical Monte Carlo simulations and experimental CO2 isotherm data. *Langmuir* 13: 2795–2802.

Samios, S., A. K. Stubos, G. K. Papadopoulos, N. K. Kanellopoulos, and F. Rigas. 2000. The structure of adsorbed CO2 in slitlike micropores at low and high temperature and the resulting micropore size distribution based on GCMC simulations. *J Colloid Interface Sci* 224: 272–290.

Sasloglou, S. A., J. K. Petrou, N. K. Kanellopoulos, and G. P. Androutsopoulos. 2000. Realistic random sphere pack model for the prediction of sorption isotherms. *Microporous Mesoporous Mater* 39(3): 477–483.

Sasloglou, S. A., J. K. Petrou, N. K. Kanellopoulos, and G. P. Androutsopoulos. 2001. Realistic random sphere pack model for the prediction of relative permeability curves. *Microporous Mesoporous Mater* 47(1): 97–103.

Scheiber, A., I. Ketelsen, G. H. Findenegg, and E. Hoinkis. 2007. Thickness of adsorbed nitrogen films in SBA-15 silica from small-angle neutron diffraction. *Stud Surf Sci Catal* 160: 17–24.

Sedigh, M. G., M. Jahangiri, P. K. T. Liu, M. Sahimi, and Th. T. Tsotsis. 2000. Structural characterization of polyetherimide-based carbon molecular sieve membranes. *AIChE* 46: 2245–2255.

Sharma, P., P. Rez, M. Brown, G. Du, and M. M. J. Treacy. 2007. Dynamic observations of the effect of pressure and temperature conditions on the selective synthesis of carbon nanotubes. *Nanotechnology* 18(12): 125602.

Sharma, R., E. Moore, P. Rez, and M. M. J. Treacy. 2009. Site-specific fabrication of Fe particles for carbon nanotube growth. *Nano Lett* 9(2): 689–694.

Shimoda, H., S. J. Oh, H. Z. Geng, R. J. Walter, X. B. Zhang, L. E. McNeil, and O. Zhou. 2002a. Self-assembly of carbon nanotubes. *Adv Mater* 14(12): 899–901.

Shimoda, H., L. Fleming, K. Horton, and O. Zhou. 2002b. Formation of macroscopically ordered carbon nanotube membranes by self-assembly. *Phys B* 323(1–4): 135–136.

Sholl, D. S. and J. K. Johnson. 2006. Making high-flux membranes with carbon nanotubes. *Science* 312(5776): 1003–1004.

Skoulidas, A. I., D. M. Ackerman, J. K. Johnson, and D. S. Sholl. 2002. Rapid Transport of Gases in Carbon Nanotubes. *Phys Rev Lett* 89(18): 1859011–1859014.

Skoulidas, A. I., D. S. Sholl, and J. K. Johnson. 2006. Adsorption and diffusion of carbon dioxide and nitrogen through single-walled carbon nanotube membranes. *J Chem Phys* 124(5): 054708–054707

Smarsly, B., C. Göltner, M. Antonietti, W. Ruland, and E. Hoinkis. 2001. SANS investigation of nitrogen sorption in porous silica. *J Phys Chem B* 105: 831–840.

Sokhan, V. P., D. Nicholson, and N. Quirke. 2002. Fluid flow in nanopores: Accurate boundary conditions for carbon nanotubes. *J Chem Phys* 117(18): 8531–8539.

Steriotis, Th. A., F. K. Katsaros, A. Mitropoulos, A. K. Stubos, and N. K. Kanellopoulos. 1995. Characterisation of porous solids by simplified gas relative permeability measurements. *J Porous Mater* 2: 73–77.

Steriotis, Th. A., G. K. Papadopoulos, A. K. Stubos, and N. K. Kanellopoulos. 2002a. A Monte Carlo study on the structure of carbon dioxide adsorbed in microporous carbons. *Stud Surf Sci Catal* 144: 545–552.

Steriotis, Th. A., K. L. Stefanopoulos, A. Ch. Mitropoulos, N. K. Kanellopoulos, A. Hoser, and M. Hofmann. 2002b. Structural studies of supercritical CO_2 in confined space. *Appl Phys A* 74: 1333–1335.

Steriotis, Th. A., K. L. Stefanopoulos, N. K. Kanellopoulos, A. Ch. Mitropoulos, and A. Hoser. 2004. The structure of adsorbed CO_2 in carbon nanopores: A neutron diffraction study. *Coll Surf A* 241: 239–244.

Steriotis, Th. A., K. L. Stefanopoulos, F. K. Katsaros, R. Gläser, A. C. Hannon, and J. D. F. Ramsay. 2008. In situ neutron diffraction study of adsorbed carbon dioxide in a nanoporous material: Monitoring the adsorption mechanism and the structural characteristics of the confined phase. *Phys Rev B*, 78: 115424(1–10).

Tang, Z. K., H. D. Sun, J. Wang, J. Chen, and G. Li. 1998. Mono-sized single-wall carbon nanotubes formed in channels of AlPO4-5 single crystal. *Appl Phys Lett* 73(16): 2287–2289.

Tanihara, N., H. Shimazaki, Y. Hirayama, S. Nakanishi, T. Yoshinaga, and Y. Kusuki. 1999. Gas permeation properties of asymmetric carbon hollow fiber membranes prepared from asymmetric polyimide hollow fiber. *J Memb Sci* 160: 179–186.

Tans, S. J., A. R. M. Verschueren, and C. Dekker. 1998. Room-temperature transistor based on a single carbon nanotube. *Nature* 393(6680): 49–52.

Terrones, M., N. Grobert, J. Olivares, J. P. Zhang, H. Terrones, K. Kordatos, W. K. Hsu, J. P. Hare, P.D. Townsend, K. Prassides, A. K. Cheetham, H. W. Kroto, and D. R. M. Walton. 1997. Controlled production of aligned-nanotube bundles. *Nature* 388(6637): 52–55.

Toebes, M. L., Y. H. Zhang, J. Hajek, T. A. Nijhuis, J. H. Bitter, A. J. van Dillen, D. Y. Murzin, D. C. Koningsberger, and K. P. de Jong. 2004. Support effects in the hydrogenation of cinnamaldehyde over carbon nanofiber-supported platinum catalysts: Characterization and catalysis. *J Catal* 226(1): 215–225.

Tu, Y., Z. P. Huang, D. Z. Wang, J. G. Wen, and Z. F. Ren. 2002. Growth of aligned carbon nanotubes with controlled site density. *Appl Phys Lett* 80(21): 4018–4020.

Tzevelekos, K. P., E. S. Kikkinides, A. K. Stubos, M. E. Kainourgiakis, and N. K. Kanellopoulos. 1998. On the possibility of characterising mesoporous materials by permeability measurements of condensable vapours: Theory and experiments. *Adv. Colloid Interface Sci.* 76–77: 373–388.

van Tricht, J. B., H. Fredrikze, and J. van der Laan. 1984. Neutron diffraction study of liquid carbon dioxide at two thermodynamic states. *Mol Phys* 52: 115–127.

Vermisoglou, E. C., G. Pilatos, G. E. Romanos, G. N. Karanikolos, N. Boukos, K. Mertis, N. Kakizis, and N. K. Kanellopoulos. 2008. Synthesis and characterisation of carbon nanotube modified anodised alumina membranes. *Microporous Mesoporous Mater* 110(1): 25–36.

Vermisoglou, E. C., G. N. Karanikolos, G. Pilatos, E. Devlin, G. E. Romanos, C. M. Veziri, and N. K. Kanellopoulos. 2010. Aligned carbon nanotubes with ferromagnetic behavior. *Adv Mat* 22(4): 473–477.

Veziri, C. M., G. Pilatos, G. N. Karanikolos, A. Labropoulos, K. Kordatos, V. Kasselouri-Rigopoulou, and N. K. Kanellopouios. 2008. Growth and optimization of carbon nanotubes in activated carbon by catalytic chemical vapor deposition. *Microporous Mesoporous Mater* 110(1): 41–50.

Veziri, C. M., G. N. Karanikolos, G. Pilatos, E. C. Vermisoglou, K. Giannakopoulos, C. Stogios, and N. K. Kanellopoulos. 2009. Growth and morphology manipulation of carbon nanostructures on porous supports. *Carbon* 47(9): 2161–2173.

Veziri, C. M., M. Palomino, G. N. Karanikolos, A. Corma, N. K. Kanellopoulos, and M. Tsapatsis. 2010. Toward submicrometer c-oriented nanoporous films with unidimensional pore network: AFI film morphology control by precursor mixture manipulation. *Chem Mater* 22(4): 1492–1502.

Wallacher, D., M. Rheinstaedter, T. Hansen, and K. Knorr. 2005. Neutron diffraction study of he solidified in a mesoporous glass. *J Low Temp Phys* 138: 1013–1024.

Wang, N., Z. K. Tang, G. D. Li, and J. S. Chen. 2000a. Materials science—Single-walled 4 angstrom carbon nanotube arrays. *Nature* 408(6808): 50–51.

Wang, Z. L., P. Poncharal, and W. A. de Heer. 2000b. Measuring physical and mechanical properties of individual carbon nanotubes by in situ TEM. *J Phys Chem Solids* 61(7): 1025–1030.

Weber, S. 1954. Kgl. Danske Videnskab. *Selskab Mat Fys Medd* 28: 2.

Whitby, M. and N. Quirke. 2007. Fluid flow in carbon nanotubes and nanopipes. *Nat Nanotechnol* 2(2): 87–94.

Yasuda, A., N. Kawase, and W. Mizutani. 2002. Carbon-nanotube formation mechanism based on in situ TEM observations. *J Phys Chem B* 106(51): 13294–13298.

Zhang, D. S., L. Y. Shi, J. H. Fang, and K. Dai. 2006. Removal of NaCl from saltwater solution using carbon nanotubes/activated carbon composite electrode. *Mater Lett* 60(3): 360–363.

Zickler, G. A., S. Jähnert, W. Wagermaier, S. S. Funari, G. H. Findenegg, and O. Paris. 2006. Physisorbed films in periodic mesoporous silica studied by in situ synchrotron small-angle diffraction. *Phys Rev B* 73: 184109(1–10).

Part II

Fundamentals, Recent Advances, and Expected Developments of Simulation Methods

6

Mesoscopic Methods

P.M. Adler and J.-F. Thovert

CONTENTS

6.1 Introduction

Porous media are important in many industries such as the oil industry, chemical engineering, underground water management, storage of nuclear wastes, and CO_2 sequestration, to name a few. The determination of the macroscopic properties of the related media is a subject of great importance.

The major purpose of this paper is to review some of the recent advances of our group about the generation of porous media and the study of a series of phenomena in such media. The reader may consult [1] to appreciate some of the progress that has been done in the last 10 years. This paper is by no means an exhaustive review, but such a review can be started by browsing our various contributions and the references therein.

This paper is organized as follows. The numerical generation of porous media is addressed first because of its fundamental importance. Section 6.2 reviews the major tools that we developed. The first one is the standard reconstruction, which is based on the measurement of porosity and of the correlation function of the media on thin sections. This now elementary technique can be extended in different ways by using several random fields. Vugular media, media with several solid phases, and heterogeneous media, to name a few, can be generated by such adequate extensions.

Grain reconstruction, which is based on the random insertion of polydisperse spheres, is recalled in Section 6.2.2. It gives excellent results in terms of the properties of the generated media.

Such a procedure provides a transition to the generation of packings of star particles, which is described in Section 6.2.3. Recent progress on polydisperse packings of spheres and of concave particles obtained by sequential deposition are surveyed.

When a porous medium is generated, it is important to measure its geometrical characteristics. Our techniques are briefly illustrated in Section 6.2.4 for packings of polydisperse spheres.

The rest of the chapter is devoted to the study of various phenomena in the simulated media. The most important phenomena are conduction and permeability and they are addressed in Section 6.3. Illustrations are given for packings of polydisperse spheres.

Some other phenomena, which have been recently studied, are gathered in Section 6.4. The properties of the apparent diffusion coefficient in a motionless two-phase configuration are summarized in Section 6.4.1. The mechanical properties of random porous media have been revisited with an enormously increased computing power; the methodology and the results are given in Section 6.4.2. Of course, wave velocities in dry media can be deduced from the previous results. This is extended to media saturated by one phase in Section 6.4.3. Finally, this short review is ended by some concluding remarks in Section 6.5.

6.2 Geometry

The geometry of porous media is important due to the fact that it controls many important properties. The most conspicuous examples are the macroscopic conductivity and the permeability.

A historical review of the experimental studies of the geometry can be found in [1] and is not repeated here. These studies have been revolutionarized by a new tool called microtomography, which enables to determine the inner structure of rocks with a resolution better than 1 μm. Such a resolution is sufficient for sandstones, but not for most carbonates.

A picture of historical interest is the Fontainebleau sandstone displayed in Figure 6.1, which was the first published image of this sort [2] with a simultaneous determination of the macroscopic properties. The precise determination of the porous structures has considerably facilitated the derivation of their macroscopic properties, but the need for their generation has not decreased. In the rest of this section, some recent progress is reviewed.

6.2.1 Reconstructed Porous Media

This name was first introduced in [3,4]. It is defined as the numerical simulation of porous media with given statistical properties. Several such processes are briefly recalled.

6.2.1.1 The Standard Reconstruction of Porous Media

The reconstruction process of a 3D random porous medium with a given porosity ε and a given correlation function $R_Z(\mathbf{u})$ was devised by

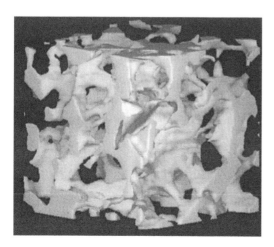

FIGURE 6.1
View of the pore space of Fontainebleau sandstones obtained by CMT [2]. The pores are shown as white and the solid as black; i.e., the pores are opaque and the rock transparent. (From Spanne, P. et al., *Phys. Rev. Lett.*, 73, 2001, 1994.)

Adler et al. [5]; on the average, the resulting medium is homogeneous and isotropic. Spatially, periodic and isotropic samples are obtained in the following way.

The statistical properties of the medium are measured on thin sections. The phase function $Z(\mathbf{x})$ is introduced

$$Z(\mathbf{x}) = \begin{cases} 1, & \text{if } \mathbf{x} \text{ belongs to the pore space} \\ 0, & \text{otherwise} \end{cases}, \qquad (6.1)$$

where $\mathbf{x}$ denotes the position in the space with respect to an arbitrary origin.

The porosity ε and the correlation function $R_Z(\mathbf{u})$ can be defined by the statistical averages (which are denoted by an overbar)

$$\varepsilon = \overline{Z(\mathbf{x})}, \qquad (6.2a)$$

$$R_Z(\mathbf{u}) = \frac{\overline{[Z(\mathbf{x}) - \varepsilon][Z(\mathbf{x} + \mathbf{u}) - \varepsilon]}}{\varepsilon - \varepsilon^2}, \qquad (6.2b)$$

where $\mathbf{u}$ is the lag. For isotropic porous media, R_Z only depends on $u = |\mathbf{u}|$.

A random and discrete field $Z(\mathbf{x})$ can be derived from a Gaussian field $X(\mathbf{x})$, which is first correlated and then thresholded.

A Gaussian and uncorrelated field $X(\mathbf{x})$ can be generated at the center of each elementary cube. The random variables $X(\mathbf{x})$ are assumed to be normally distributed with a mean equal to 0, a variance equal to 1, and they are independent. A Gaussian field $Y(\mathbf{x})$ with a given correlation function $R_Y(\mathbf{u})$ can be derived from the field $X(\mathbf{x})$ using a Fourier transform technique. $R_Y(\mathbf{u})$ can also be derived from $R_Z(\mathbf{u})$ measured on thin sections [5]. In this chapter, the correlation function $R_Y(\mathbf{u})$ is always chosen to be

$$R_Y(\mathbf{u}) = e^{-(\pi u/l)^2}, \qquad (6.3)$$

where l is the correlation length. Finally, a discrete field $Z(\mathbf{x})$ is obtained by application of a threshold, which is such that the average value of $Z(\mathbf{x})$ is equal to ε.

An example of the reconstructed medium is displayed in Figure 6.2a with $l = 24a$ where a is the size of the elementary cubes. Such media are called unimodal reconstructed media; they are characterized by a single correlation length.

6.2.1.2 Reconstruction of Porous Media by Two Fields

Vugular media possess several sizes of pores. More precisely, the small-scale pores are characterized by a porosity ε_p and a correlation length l_p while the large-scale pores (also called vugs) are characterized by a porosity ε_v and a correlation length l_v.

(a)   (b)   (c)

FIGURE 6.2
Consolidated porous media. The content of the unit cell is shown for a unimodal reconstructed porous medium with $\varepsilon = 0.2$ (a), the bimodal (or vugular) porous media PV (b), and NPV (c).

Such media can be constructed in two steps [6]. First, a field $Z_p(\mathbf{x})$ is generated with a porosity ε_p and a correlation length l_p, which corresponds to the small-scale pores. Then, another field $Z_v(\mathbf{x})$, which corresponds to the vugs, is generated; $Z_v(\mathbf{x})$ is statistically independent of $Z_p(\mathbf{x})$, with a porosity ε_v and a correlation length l_v. The bimodal porous medium is obtained by the superposition of the two pore systems

$$Z(\mathbf{x}) = Z_p(\mathbf{x}) + Z_v(\mathbf{x}) - Z_p(\mathbf{x})Z_v(\mathbf{x}). \tag{6.4}$$

The total porosity ε is derived by averaging the previous formula; since the two fields $Z_p(\mathbf{x})$ and $Z_v(\mathbf{x})$ are independent, one obtains

$$\varepsilon = \varepsilon_p + \varepsilon_v - \varepsilon_p\varepsilon_v. \tag{6.5}$$

Other statistical properties have been derived by Moctezuma-Berthier et al. [6]. Figure 6.2b and c provides examples of porous media that were generated by this technique with $l_p = 5a$, $l_v = 30a$. Figure 6.2b represents a sample with percolating vugs (PV) and Figure 6.2c with non-percolating vugs (NPV) though the total porosity is approximately the same and equal to 0.36. Such media are also called bimodal reconstructed media.

6.2.1.3 Extension of the Standard Reconstruction Method

The advantage of this procedure is its relative simplicity and the essential feature that the porosity and the correlation function can be measured on thin sections in order to reproduce the structure of a given medium.

It can be extended in several ways, one of them being already described in the previous section. Other extensions were done for several solid phases [7] or for heterogeneous media [8].

However, this method has some drawbacks and one of them is the fact that the percolation threshold is around 0.1 [8], which is still too high when compared with the threshold of real porous media.

6.2.2 Grain Reconstruction of Porous Media

Thovert et al. [9] presented a reconstruction technique, which introduces an underlying granular structure, but only makes use of geometrical parameters that can be measured on images of real samples. From this standpoint, the methodological approach is the same as for the correlation technique; all geometrical quantities are measured and a medium is generated with the same statistical characteristics. This technique is based on a Poissonian penetrable sphere model, conditioned by the experimental solid size distribution. Thus, the size distribution of the solid phase should be quantified in the first place, and a sizing technique that provides the required information is introduced.

These techniques were applied in [9] to the analysis of a low-porosity Fontainebleau sandstone sample, based on a high-resolution 3D digital image obtained by x-ray computed microtomography (CMT). First, the geometry and the transport properties of the real sample were thoroughly characterized. Then, the same analysis was repeated on a numerically reconstructed sample, which allows a direct assessment of the merit of the reconstruction algorithm, with respect to a variety of geometrical and transport-related criteria. Special emphasis is put on the quantification of local variability in the real material, and on its rendering in the reconstructed one.

6.2.3 Packings of Star Particles

Grain packings have given rise to a considerable interest for a long time, as a model for various types of porous media such as geological materials like soils. A considerable literature has accumulated over the years on this topic (see the references in [10,11]). Two major kinds of methods are used to generate random packings, namely, random sequential addition and collective rearrangement. The two methods can mimic the physical processes, which generate real particle packings, such as deposition by gravity for the first one while the second one is closer to generation of packings by shaking. The first method usually generates packings with a larger porosity than collective rearrangement.

Random sequential addition has been successfully used to pack star particles. These particles may have any size and shape, provided that they can be described in a spherical polar coordinates system (r, θ, ϕ) attached to it by a single valued function $\rho(\theta,\phi)$. The inner volume of the particle is defined by

$$r \leq \rho(\theta, \phi). \qquad (6.6)$$

Obviously, any convex particle shape can be described by (6.6).

Our random packings result from the successive deposition of grains in a "gravitational" field. The grains are introduced at a random location above the bed already in place, and fall until they reach a local minimum of their potential energy. Sometimes, a dynamic language is used, but the reader should not be misled, since the Newton's laws of motion are never solved. During their fall, any displacement and rotation, which contribute to lower their barycenter are allowed.

As a general rule, a mobile particle is allowed to slip freely on the bed surface as long as the elevation of the barycenter can be diminished. Moreover, each elementary displacement of a grain is independent of its previous position and orientation increments. An adjustable parameter favors either translation or rotation of the particle, when both motions could lower its elevation. Finally, the interactions are reduced to steric exclusion. A variant of this rule has been devised to simulate short-range attractive forces, which could create permanent links between grains. After contact, a settling grain can be allowed to rotate around the contact point without slip (but the contact may move if the grain rolls on the bed). For instance, for parallelepipedic grains, if a vertex comes in contact with an underlying plane solid surface, the particle would rotate until one of its edges and eventually one of its faces becomes tangent to this surface.

The position of a particle is represented by the location $\mathbf{r} = (x,y,z)$ of its barycenter, and by a set ω of three angles that give the orientation of the particle with respect to the coordinate system. The z-axis is oriented upward. The grains are deposited in a square vertical box, with a flat bottom at $z = 0$, and periodicity conditions along the x and y directions in order to avoid the well-known hard-wall effects.

6.2.3.1 Polydisperse Packings

Two major types of packings are considered. First, bidisperse packings of spheres of radii r and R (with $r \leq R$) are generated. The two radii are generally chosen in such a way that their average is equal to 1 in order to keep the same precision in the packing generation. The ratio R/r is denoted by ρ and the volume fraction of large grains by f_R, respectively. More precisely,

$$R = \frac{2\rho}{1+\rho}, \quad r = \frac{2}{1+\rho}. \tag{6.7}$$

However, when $\rho \geq 5$ and $f_R \leq 0.7$, the number of small grains, which results from (6.7) becomes numerically overwhelming. Thus, in such cases, r was kept equal to 2/5, with $R = 2\,\rho/5$.

Second, polydisperse spheres with a lognormal distribution of the radius R are randomly packed. More precisely, the grain size distributions in number $f(R)$ and in volume $v(R)$ are given by

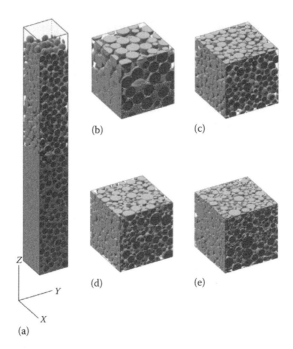

FIGURE 6.3
Polydisperse packings of spheres. Illustration of the segregation effect for a bidisperse packing with a ratio $\rho = 8$ and $f_R = 0.85$ (a). (b) is an enlarged view of the middle part of (a). Lognormal packings with $\sigma = 0.2$ (c), 0.4 (d), and 0.6 (e).

$$v(R) = \frac{1}{\sqrt{2\pi}\sigma R}\exp\left[\frac{(\ln R - \mu)^2}{2\sigma^2}\right], \quad f(R) = \frac{\langle R^3 \rangle}{\sqrt{2\pi}\sigma R^4}\exp\left[\frac{(\ln R - \mu)^2}{2\sigma^2}\right] \quad (6.8)$$

where
   μ is the average of the logarithm of the radius
   σ is its standard deviation

Experimentally, real packings are mostly characterized by $v(R)$, but numerical packings are more easily generated with $f(R)$. In order to compare the various results, μ is always equal to 1.

   Examples of these packings are displayed in Figure 6.3, where the segregation effect is also illustrated.

6.2.3.2 Nonconvex Star Particles

It should be noted that most works were devoted to random packings of spheres, whether they are mono or polydisperse and that the studies devoted to nonspherical particles are much less common and are mostly

experimental. Since packings of spherical and ellipsoidal particles are easily generated, new grain shapes are based on these forms. Therefore, all the
particles are composed by a sphere and each new form of grain is obtained
by addition to this sphere of one or more identical prolate ellipsoids with the
same center. The positions of the ellipsoids are chosen in order to produce
grains, which are relatively regular and isotropic. However, two anisotropic
types of grains will also be considered.

The grains are defined by the distance R between a point $\mathbf{X}_G(X_G, Y_G, Z_G)$
called the center and its surface. Often $\mathbf{X}_G$ is the center of gravity of the particle. Let us consider each particle in its local coordinates with the origin at
$\mathbf{X}_G$. The distance R is determined in the polar coordinates for each orientation (Θ, ϕ); Θ is the inclination with respect to the z-axis and ϕ is the rotation
of the x-axis around the z-axis (Figure 6.4a). Then, the surface of the particle
is represented as

$$r = R(\theta, \phi). \tag{6.9a}$$

For a spherical grain, the function $R(\Theta, \phi)$ is simply equal to the radius of the
sphere R.

For an ellipsoidal grain with semiaxes (R_x, R_y, R_z) aligned with the axes of
coordinates, it can be written as

$$R_{ell}(\theta, \phi) = \left\{ \frac{\cos^2 \theta}{R_z^2} + \sin^2 \theta \left[\frac{\cos^2 \phi}{R_x^2} + \frac{\cos^2 \phi}{R_y^2} \right] \right\}^{-1/2}. \tag{6.9b}$$

In all cases, two semiaxes of each ellipsoid are identical and smaller than the
third one. Therefore, the orientation of the ellipsoid is fully described by the
orientation of its major axis.

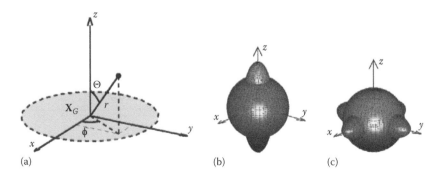

(a)   (b)   (c)

FIGURE 6.4
(a) The particle definition in the polar coordinates $(\theta; \phi)$. Examples of particles: (b) one-ellipsoid
particle and (c) two-ellipsoid particle.

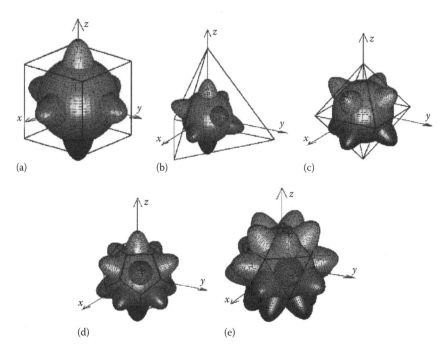

FIGURE 6.5
Regular particles: (a) C-shaped (cube), (b) T-shaped (two tetrahedra), (c) O-shaped (octahedron), (d) D-shaped (dodecahedron), and (e) I-shaped (icosahedron).

Since all the grains are made of one sphere and a few ellipsoids, the corresponding functions $R(\Theta,\phi)$ are expressed in terms of R_{sph} and $R_{ell}(\Theta,\phi)$. More precisely, $R(\Theta,\phi)$ can be expressed as

$$R(\theta,\phi) = \max\left\{R_{sph}, R_{ell}^{(j)}(\theta,\phi)\right\}, \quad j = 1, \dots N_{ell}, \quad (6.10)$$

where
N_{ell} is the number of ellipsoids added to the sphere in the particle
$R_{ell}^{(j)}$ is deduced from (6.9b) by an appropriate rotation

Examples of such particles are given in Figures 6.4 and 6.5.

6.2.4 Geometric Properties of Polydisperse Packings

Many geometric properties can be measured on numerical samples of porous media starting with porosity. Much more can be done such as the extraction of the graph of the pore space, the statistical characterization of the underlying equivalent capillary network, and the pore size distribution spectrum. Determinations of these quantities for reconstructed media can be found in [12].

Let us illustrate the numerical determination of some geometrical properties on the polydisperse packings of spheres. Let V_p and S_p denote the pore volume and the surface of the solid interface, respectively. According to standard definitions [3], the hydraulic radius R_H is defined as the ratio of the total pore volume to its surface area

$$R_H = \frac{V_p}{S_p}. \tag{6.11}$$

An equivalent radius $\tilde{R}$ can be defined as the inverse of the harmonic mean radius

$$\tilde{R} = \left[\int \frac{v(R)\mathrm{d}R}{R} \right]^{-1}, \tag{6.12a}$$

where $v(R)$ is the probability density in volume (cf. (6.8)). $\tilde{R}$ has interesting properties, since it can also be written as

$$\tilde{R} = \frac{\int R^3 f(R)\mathrm{d}R}{\int R^2 f(R)\mathrm{d}R}, \tag{6.12b}$$

and as

$$\tilde{R} = \frac{\int 4\pi R^2 f(R) R\,\mathrm{d}R}{\int 4\pi R^2 f(R)\mathrm{d}R}. \tag{6.12c}$$

In the last expression, the average is weighted by the sphere surfaces.

Another property is worth mentioning. Whatever the radius distribution, the following relation holds for spheres

$$R_H = \frac{\varepsilon}{3(1-\varepsilon)} \tilde{R}. \tag{6.13}$$

Therefore, two sphere packings with the same porosity and the same $\tilde{R}$ have the same hydraulic radius and the same surface area of the pores whatever the radius distribution.

These quantities can be determined on the packings generated by the algorithm, since the coordinates of the sphere centers and the sphere radii are recorded. The spheres whose centers are located between two horizontal planes are known; the total solid volume and the total solid interface can be readily deduced.

Another possible way consists in discretizing the spheres and then performing the previous measurements on the discretized array. More precisely, space is discretized into $N_{cx} \times N_{cy} \times N_{cz}$ elementary cubes of size a; generally, $N_{cx} = N_{cy}$. Whenever the center of an elementary cube falls in the solid (resp. fluid) phase, the whole cube is considered as filled with solid (resp. fluid). Note that the transport calculations are done on these discretized arrays.

It is crucial to realize that volumes are relatively insensitive to the way they are measured, but this is not the case for surfaces where a systematic bias is present. More precisely, numerical evaluations of isotropic surfaces are overestimated by a factor 3/2 by the discrete numerical scheme used here. However, this effect is partly compensated by the nonzero area of the grain contacts in the discretized representation, which lowers the wetted area.

The porosity of bidisperse packings is displayed in Figure 6.6. First, it should be noted that the porosity of monodisperse packings ε_m is given by

$$\varepsilon_m \approx 0.41, \tag{6.14}$$

in agreement with Coelho et al. [10].

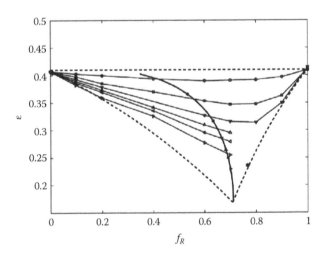

FIGURE 6.6
The porosity of the bidisperse packings as a function of the volume fraction of the larger grains. From top to bottom, data are for $\rho = 2$ (○), 3 (□), 4 (▽), 5 (△), 6 (◁), and 8 (▷). The two lower dotted lines correspond to (6.16) and (6.17). The thick solid line corresponds to the variations of the minimal porosity for various values of ρ provided by (6.18) and (6.19); the dots on this curve provide the value of ρ. The star corresponds to the bottom layer of a bed with $\rho = 8$ and $f_R \geq 0.8$ (see Figure 6.3).

For bidisperse packings, the general physical effect is clear. If ρ is close to 1, the porosity variations are very small; in the opposite case, they may be quite large with a minimum that is located around $f_R = 0.70$; f_R is recalled to be the volume fraction of large grains. The general shape of these curves is in agreement with the data that can be found in the literature (see [13] for instance); a more precise comparison is done at the end of this section. The curves are seen to be incomplete in Figure 6.6. This is due to the so-called segregation effect, which occurs for $\rho \geq 5$. The small particles tend to accumulate at the bottom of the packing, where they fill entirely the void space between the large particles. Some attention was paid to this effect in [11] to which the interested reader is referred.

The value f_{R_0} of f_R in the lower part of the packing can be estimated in the following approximate way in a binary mixture with an extreme size ratio $\rho \ll 1$. For an overall volume equal to 1, the large particles fill a volume $1 - \varepsilon_m$. The small particles fill the complement to 1, i.e., ε_m. The volume of the small particles is equal to $\varepsilon_m(1 - \varepsilon_m)$. Therefore, f_{R_0} can be expressed as

$$f_{R_0} = \frac{1}{1 + \varepsilon_m}, \tag{6.15}$$

which is about 0.71 for $\varepsilon_m = 0.41$. This situation corresponds to the maximal density for a binary mixture. The porosity in the lower part of the packing is simply ε_m^2. These simple properties are seen to be independent of the size ratio ρ provided that it is small.

One can easily derive the shapes of the envelopes, which are seen in Figure 6.6. If the proportion of large grains is smaller than f_{R_0}, they are dispersed within a bed of small ones and the porosity is

$$\varepsilon = \frac{(1 - f_R)\varepsilon_m}{1 - f_R\varepsilon_m} \geq \varepsilon_m^2. \tag{6.16}$$

Conversely, if the population of large grains is larger, the amount of small ones is insufficient to fill the macropores. The porosity can be formally expressed as

$$\varepsilon = \frac{f_R - 1 + \varepsilon_m}{f_R}. \tag{6.17}$$

These two curves are illustrated in Figure 6.6.

Finally, Yu [14] summarized previous experimental results and gave the variations of the minimum of porosity for each radius ratio ρ if $\rho \leq 0.741$

$$\varepsilon_{\min} = \varepsilon_m - \varepsilon_m(1 - \varepsilon_m)(1 - 2.35\rho + 1.35\rho^2). \tag{6.18}$$

The other situation $\rho > 0.741$ is not used in this chapter. The fractional volume v_R for which this minimum occurs is given by

$$v_R = \frac{1-\rho^2}{1+\varepsilon_m}. \tag{6.19}$$

This quantity is displayed in Figure 6.6. It does not agree well with our numerical data because the numerical packings are loose, while the experimental packings are closely packed. Our data correspond to packings generated under the action of gravity without any further rearrangement in order to decrease the porosity. It is useful to notice that for large values of ρ when segregation occurs, some numerical experiments were performed; packings of large particles were built first and the small particles were dropped later; results were the same as when large and small particles were dropped together.

6.2.5 Mixed Structures

The combination of the previous generating tools may yield a very large number of structures.

For instance, in order to simulate a sand bed partially plugged by calcite, one can superpose a random packing of spheres and a calcite field obtained by the standard reconstruction algorithm. The resulting media are displayed in Figure 6.7. The radii of the spheres range from 4 to 16 (in arbitrary units) with a Gaussian distribution; the mean and the standard deviation are 10 and 2.5, respectively. The packing porosities are about 40%.

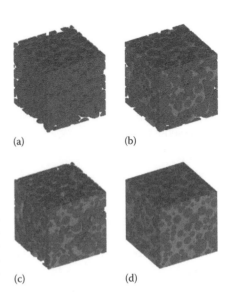

(a)   (b)

(c)   (d)

FIGURE 6.7
Samples of consolidated grain packings. Dark grey corresponds to quartz and light grey to calcite. (a) Quartz parking without calcite filling. (b) Quartz packing with $S_c = 0.3$. (c) Quartz packing with $S_c = 0.7$. (d) Quartz packing with $S_c = 1$.

6.3 Conductivity and Permeability

The governing equations and their methods of solution are briefly recalled in this section. In all cases, the macroscopic coefficients are deduced by integrating the local fields, obtained by solving the transport equations at the pore scale. The solutions of these equations are illustrated for random packed beds.

6.3.1 Conduction

The electric terminology is used here, but the following developments are also valid for thermal conduction and for diffusion of Brownian particles, whose size is small with respect to a typical size of the medium. The local flux $\mathbf{q}$ is equal to

$$\mathbf{q} = -\Sigma_0 \nabla \psi, \tag{6.20}$$

where Σ_0 is the fluid conductivity. Electrical and thermal conductions are both governed by a Laplace equation ([15] and [3], where additional details are given), which corresponds to the conservation of the local electrical flux

$$\nabla^2 \psi = 0, \tag{6.21}$$

where ψ is the local electrical potential, together with the no-flux boundary condition at the wall S_p when the solid phase is assumed to be insulating

$$\mathbf{n} \cdot \nabla \psi = 0, \quad \text{on } S_p, \tag{6.22}$$

where $\mathbf{n}$ is the unit vector normal to S_p.
$\nabla \psi$ is assumed to be spatially periodic with a period aN_c in the three directions of space. In addition, either the macroscopic potential gradient or the average electrical flux

$$\overline{\mathbf{q}} = \frac{1}{\tau_0} \int_{\partial \tau_0} \mathbf{R} \mathbf{q} \cdot ds \tag{6.23}$$

is specified.

These two quantities are related by the symmetric positive definite conductivity tensor Σ

$$\bar{q} = -\Sigma \cdot \overline{\nabla \psi}, \tag{6.24}$$

which depends only upon the geometry of the medium.

On the average, for an isotropic random medium, Σ is a spherical tensor equal to $\Sigma.\mathbf{I}$. For deposited packings, the x- and y-directions play equivalent roles, but one may expect a different behavior along the z-axis. In the following, for the sake of simplicity, Σ denotes the average of the conductivities along the x- and y-axes, which were indeed always found equal within statistical fluctuations.

The Neumann problem (6.21 through 6.23) is solved via a second-order finite-difference formulation. A conjugate-gradient method turned out to be very effective for the problem at hand, primarily because it is better suited to vectorial programming than implicit relaxation schemes.

The formation factor F is generally defined as the inverse of the dimensionless macroscopic conductivity

$$F = \frac{\Sigma_0}{\Sigma}. \tag{6.25}$$

The length scale Λ defined by Johnson et al. [16] can be used in order to characterize porous media.

Λ is essentially a volume-to-surface pore ratio with a measure weighted by the local value of the electric field $\mathbf{E(x)}$ in a conduction process

$$\Lambda = 2 \frac{\int_{V_p} |\mathbf{E(x)}|^2 \, dv}{\int_{S_p} |\mathbf{E(x)}|^2 \, ds}. \tag{6.26}$$

This quantity can be calculated when the Laplace equation (6.21) is solved in order to derive Σ.

Systematic calculations were performed in [11] for the polydisperse packings, described in Section 6.2.3.1. Precise estimations of the macroscopic conductivity were obtained by means of extrapolations to infinite discretization of the porous media. Results are shown in Figure 6.8 for both types of packings.

Overall correlations for conductivity of polydisperse packings were derived in [11]. The first one has the form of a classical Archie's law [20]

$$\Sigma_\infty = 0.808\varepsilon^{1.2}. \tag{6.27}$$

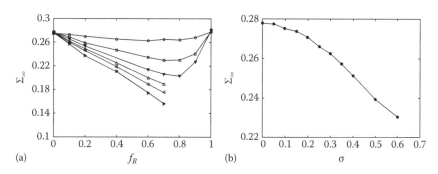

FIGURE 6.8
The extrapolated conductivity Σ_∞. Bidisperse packings (a) Σ_∞ as a function of the volume fraction of the larger grains f_R; data are for $\rho=2$ (o), 3 (□), 4 (▽), 5 (△), 6 (◁), and 8 (▷). Lognormal packings (b) Σ_∞ as a function of the standard deviation σ.

The length scale Λ was also systematically evaluated. The following expression provides an estimate of Λ_∞ within a few percent for all the grain size distributions considered in [11]

$$\frac{\Lambda_\infty}{\tilde{R}} = \frac{2\varepsilon}{9(1-\varepsilon)}(3.5-1.6\varepsilon). \tag{6.28}$$

6.3.2 Stokes Flow

The low Reynolds number flow of an incompressible Newtonian fluid is governed by the usual Stokes equations

$$\nabla p = \mu_f \nabla^2 \mathbf{v}, \quad \nabla \cdot \mathbf{v} = 0, \tag{6.29}$$

where $\mathbf{v}$, p, and μ_f are the velocity, pressure, and viscosity of the fluid, respectively.

In general, $\mathbf{v}$ satisfies the no-slip condition at the wall

$$\mathbf{v} = 0 \quad \text{on } S_p. \tag{6.30a}$$

Because of the spatial periodicity of the medium, it can be shown ([17] and [14]) that $\mathbf{v}$ possesses the following property:

$\mathbf{v}$ is spatially periodic along the three directions of space. (6.30b)

This system of equations and conditions applies locally at each point **R** of the interstitial fluid. In addition, it is assumed that either the seepage velocity vector is specified, i.e.,

$$\bar{\mathbf{v}} = \frac{1}{\tau_0} \int_{\partial \tau_0} \mathbf{R}\mathbf{v} \cdot d\mathbf{s} = \text{a prescribed constant vector,} \qquad (6.31a)$$

or else that the macroscopic pressure gradient is specified,

$$\overline{\nabla p} = \text{a prescribed constant vector.} \qquad (6.31b)$$

Note that (6.31a) is easily derived from the identity valid for an incompressible fluid $\mathbf{v} = \nabla \cdot (\mathbf{R}\mathbf{v})$ (cf. [3]). Since the system (6.29 through 6.31b) is linear, it can be shown that $\bar{\mathbf{v}}$ is a linear function of $\bar{p}$.

These two quantities are related by the permeability tensor **K** such that

$$\bar{\mathbf{v}} = -\frac{1}{\mu_f} \mathbf{K} \cdot \overline{\nabla p}. \qquad (6.32)$$

Here, **K** is a symmetric tensor that is positive definite. It depends only on the geometry of the system and thus can be simplified when the porous medium possesses geometric symmetries. Its diagonal component K_{xx} was calculated by imposing $\bar{p}$ along the x-axis. It is simply denoted by K in the following.

The numerical method that is used here is a second-order finite-difference scheme identical to the one first described by Lemaître and Adler [17]. In order to cope with the continuity equation, the so-called artificial compressibility method was applied with a staggered marker-and-cell (MAC) mesh [18].

The length scale Λ was proved to be very useful in many ways. As suggested by Johnson et al. [16], Pengra et al. [19] showed that the permeability K can be determined via Λ

$$K = \frac{\Lambda^2}{8F}, \qquad (6.33)$$

where F is the formation factor. This was recently confirmed by systematic calculations of Valfouskaya et al. [21] who computed K, F, and Λ for a variety of porous media. Moreover, Valfouskaya et al. [21] showed that Λ is indeed approximately equal to twice the inverse of the surface-to-volume ratio

$$\Lambda \approx 2\frac{V_p}{S_p}. \qquad (6.34)$$

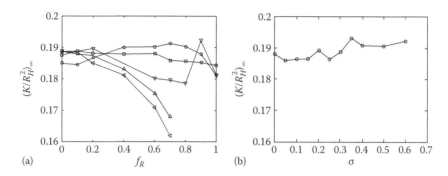

FIGURE 6.9
The dimensionless extrapolated permeability $(K/R_H^2)_\infty$ as a function of the volume fraction of the larger grains f_R for all the bidisperse packings (a) and as a function of the standard deviation σ for the lognormal packings (b). Data in (a) are for $\rho = 2$ (o), 3 (□), 4 (▽), 5 (△), and 6 (◁).

Extrapolation to an infinite discretization was made for permeability by means of the classical Carman–Kozeny equation [22]

$$K = \frac{\varepsilon R_H^2}{k}, \tag{6.35a}$$

where k is called the Kozeny constant; it is supposed to be of the order of a few units and almost constant for a large variety of conditions

$$K_{cal} = \frac{\varepsilon R_H^2}{k_{cal}}. \tag{6.35b}$$

Various representations were proposed in [11] to gather the results. The simplest one is the following. $(K/R_H^2)_\infty$ is displayed in Figure 6.9 for the two types of studied packings as a function of the geometric characteristics f_R and σ. It is seen to be reasonably constant for these sets of parameters

$$0.162 \le \left(\frac{K}{R_H^2}\right)_\infty \le 0.195. \tag{6.36}$$

6.4 Other Phenomena

6.4.1 Nuclear Magnetic Resonance Diffusion Simulations in Two Phases

Time-dependent diffusion simulations, which can be measured by nuclear magnetic resonance (NMR), were numerically performed in consolidated reconstructed porous media saturated by two immobile fluids [23]. The two fluids are assumed to be oil and water denoted by the index $\alpha = $ o or w. The phase distributions were obtained by an immiscible lattice Boltzmann technique, which incorporates interfacial tension and wetting. Starting from an arbitrary configuration of phases, a pressure gradient was applied onto the medium and the two fluids flew until a stationary state was reached; then, the pressure gradient was canceled and the two phases went to an equilibrium state under the action of surface tension.

The apparent diffusion coefficient in each phase in the equilibrium state was determined by a random walk algorithm. A large number of particles N_p were released in the chosen phase at time zero. The initial distribution is uniform. During all the subsequent elementary time steps δt, the position of each particle i is updated by a displacement δ_D of a given modulus $\delta_D = |\delta_D|$, but of a random direction

$$x_i(t + \delta t) = x_i(t) + \delta_D, \tag{6.37}$$

where

$x_i(t)$ is the position at time t
$x_i(t + \delta t)$ is the position at time $t + \delta t$

The modulus δ_D of the random jump is constant and related to the molecular diffusivity D_m by

$$\delta_D^2 = 6 D_m \delta t. \tag{6.38}$$

Since the fluid phase through which the random walk simulations are performed, can be anisotropic, the apparent diffusion coefficient in phase α is defined as a 3×3 tensor

$$\mathbf{D}^\alpha(t) = \frac{1}{2t} \begin{pmatrix} \langle x^2(t) \rangle & \langle xy(t) \rangle & \langle xz(t) \rangle \\ \langle yx(t) \rangle & \langle y^2(t) \rangle & \langle yz(t) \rangle \\ \langle zx(t) \rangle & \langle zy(t) \rangle & \langle z^2(t) \rangle \end{pmatrix}, \tag{6.39}$$

where $\alpha = $ w, o; the brackets are defined in a particular manner; for instance,

$$\langle xy(t) \rangle = \frac{1}{N_p} \sum_{n=1}^{N_p} (x_n(t) - x_n(0))(y_n(t) - y_n(0)), \tag{6.40}$$

where $x_n(t)$, $y_n(t)$, and $z_n(t)$ are the coordinates of the diffusing particle n at time t. The total number of particles is N_p.

This tensor has known limits for long and short times. For long times, it is related to the macroscopic conductivity tensor $\mathbf{D}^\alpha$ in phase α by

$$\lim_{t\to\infty} \mathbf{D}^\alpha(t) = \frac{\overline{\mathbf{D}^\alpha}}{\varepsilon S_\alpha}, \quad \alpha = w, o. \tag{6.41}$$

It is convenient to display $D(t)$ as a function of the dimensionless time

$$T_{j\alpha} = \frac{4}{9\sqrt{\pi}} \frac{A^\alpha}{V_\alpha} \sqrt{D_m^\alpha t}, \tag{6.42}$$

where $\alpha = o, w$.

Finally, Valfouskaya et al. [21] showed that all the data relative to the apparent diffusion could be superposed on a single curve for very different porous media. In this purpose, recall the definitions

$$g_x^w(\tau_{jw}) = \frac{\left(D_x^w(\tau_{jw})/D_m^w\right) - \left(1/\alpha_x^w\right)}{1 - \left(1/\alpha_x^w\right)}, \quad \text{where } \tau_{jw} = \frac{T_{jw}}{1 - \left(1/\alpha_x^w\right)}, \tag{6.43}$$

where $\left(1/\alpha_x^w\right) = \lim_{t\to\infty}\left(D_x^\alpha(t)/D_m^\alpha\right) = \left(D_{xx}^\alpha/\varepsilon S_w D_m^\alpha\right)$ is the limit value of the apparent diffusion coefficient of phase α along the x-axis.

Finally, all the data obtained in single phase by Valfouskaya et al. [21] for water and oil in two phases are superposed in Figure 6.10. The single-phase

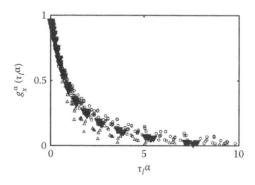

FIGURE 6.10
Superposition of the reduced representations $g_x^\alpha(\tau_{j\alpha})$. (○) water along the x direction for the samples $\varepsilon = 0.2, 0.3$, and 0.4 for $S_w \in [0.3, 1]$; (△) oil for the sample $\varepsilon = 0.3$ for $S_w \in [0.3, 0.8]$; (▼) average $g_i^\alpha(\tau_{j\alpha})$ ($i = x, y, z$) along the three directions for single phase for the samples of porosities $\varepsilon = 0.2, 0.3, 0.4$, and 0.5. (From Valfouskaya, A.M. and Adler, P.M., *Phys. Rev. E*, 72, 056317, 2005.)

data are seen to fall in the middle of the two-phase data. Such an average position is expected on a physical basis.

Therefore, the apparent diffusion coefficients are quite close, whatever the physical situation for values of g larger than 0.5 and (6.36a) of [21] still holds

$$g_x^\alpha(\tau_{|\alpha} = 0.75) = 0.5. \tag{6.44}$$

Such a relation can again be used to derive estimates of S^α/V_P^α and of K^α in the conditions discussed in this reference.

6.4.2 Mechanical Properties

The determination of the macroscopic mechanical properties of porous media was already addressed in several of our previous publications (see [1]).

6.4.2.1 Basic Equations

The basic equations are the equation of motion, the strain-displacement relation and the stress–strain relation

$$\nabla \cdot \boldsymbol{\sigma} = \rho \frac{\partial^2 \mathbf{u}}{\partial t}$$

$$\mathbf{e} = \frac{1}{2}\left[\nabla \mathbf{u} + (\nabla \mathbf{u})^t\right], \tag{6.45}$$

$$\boldsymbol{\sigma} = \mathbf{C} : \mathbf{e}$$

where
  ρ is the density
  t is the time
  $\boldsymbol{\sigma}$, $\mathbf{u}$, $\mathbf{e}$, and $\mathbf{C}$ are the stress, the displacement, the strain, and the stiffness tensor, respectively

In addition, the stress tensor $\boldsymbol{\sigma}$ on the solid interface Γ should verify

$$\boldsymbol{\sigma} \cdot \mathbf{n} = 0, \tag{6.46}$$

where $\mathbf{n}$ is the unit normal to Γ.

6.4.2.2 Macroscopic Mechanical Properties

The general form of the stress–strain relation for linear elastic materials is given by

$$\sigma_{ij} = C_{ijkl} e_{kl}, \qquad (6.47)$$

where

C_{ijkl} is the component of the fourth-order stiffness tensor
σ_{ij} is the stress tensor component
e_{kl} is the strain tensor component

Equation 6.47 can also be rewritten in the matrix form

$$
\begin{Bmatrix} \sigma_{xx} \\ \sigma_{yy} \\ \sigma_{zz} \\ \sigma_{yz} \\ \sigma_{zx} \\ \sigma_{xy} \end{Bmatrix}
=
\begin{bmatrix}
C_{11} & C_{12} & C_{13} & C_{14} & C_{15} & C_{16} \\
& C_{22} & C_{23} & C_{24} & C_{25} & C_{26} \\
& & C_{33} & C_{34} & C_{35} & C_{36} \\
& & & C_{44} & C_{45} & C_{46} \\
& & & & C_{55} & C_{56} \\
\text{sym.} & & & & & C_{66}
\end{bmatrix}
\cdot
\begin{Bmatrix} e_{xx} \\ e_{yy} \\ e_{zz} \\ 2e_{yz} \\ 2e_{zx} \\ 2e_{xy} \end{Bmatrix},
\qquad (6.48)
$$

for an anisotropic material whose stiffness matrix is denoted by **C**. For an orthotropic material, the stiffness matrix **C** reduces to

$$
\mathbf{C} =
\begin{bmatrix}
C_{11} & C_{12} & C_{13} & 0 & 0 & 0 \\
& C_{22} & C_{23} & 0 & 0 & 0 \\
& & C_{33} & 0 & 0 & 0 \\
& & & C_{44} & 0 & 0 \\
& & & & C_{55} & 0 \\
\text{sym.} & & & & & C_{66}
\end{bmatrix},
\qquad (6.49)
$$

with nine independent components. An isotropic medium is described by only two independent coefficients, namely, the Lamé constants λ and μ. Equation 6.49 reduces to

$$
\begin{aligned}
C_{12} &= C_{13} = C_{23} = \lambda \\
C_{44} &= C_{55} = C_{66} = \mu \\
C_{11} &= C_{22} = C_{33} = \lambda + 2\mu.
\end{aligned}
\qquad (6.50)
$$

The components of **C** can be determined by uniaxial compressions and simple shears. The remaining components of **C** in Equation 6.48 can be similarly determined by imposing $\langle e_{xx} \rangle$, $\langle e_{yy} \rangle$, and $\langle e_{xy} \rangle$ to be 1.

Homogenization is employed to derive the macroscopic properties of porous media by solving successively local field equations of various orders. These macroscopic properties are denoted by brackets; for instance, the two macroscopic Lamé coefficients are denoted by $\langle\lambda\rangle$ and $\langle\mu\rangle$. The celerities of the shear and compression waves denoted by $c_\perp$ and $c_\parallel$ are related to $\langle\lambda\rangle$ and $\langle\mu\rangle$ by the classical relations

$$c_\parallel = \sqrt{\frac{\langle\lambda + 2\mu\rangle}{\langle\rho\rangle}}, \quad c_\perp = \sqrt{\frac{\langle\mu\rangle}{\langle\rho\rangle}}. \tag{6.51}$$

6.4.2.3 Application

Applications are made to the consolidated packing of quartz spheres with a calcite filling, which was introduced in Section 6.2.5. The samples are generated by superposing a packing of polydisperse spheres and a reconstructed porous medium.

The calcite filling is a reconstructed porous medium with a porosity varying from 0% to 100% and with a correlation length $l_c = 50$ (in arbitrary units). The final porosity ranges from 0% to 40%. Figure 6.7b through d shows the illustrative samples with 30%, 70%, and 100% of calcite filling in the intergranular space, respectively. Hereafter, the symbol S_c denotes the calcite volume fraction in the pores.

In the numerical calculations, five realizations are generated for each case and the macroscopic properties of each sample are determined and averaged over all samples. The five samples for each final porosity are generated independently, i.e., different packings of quartz spheres and different random numbers for generating the calcite filling are employed for each realization. However, the samples with the various S_c in the same realization are not totally independent, since they are derived from the same initial grain packing and the filling distribution is generated from the same random field thresholded at different levels.

Figure 6.11 shows the effective elastic coefficients in the x-direction as functions of S_c. These coefficients are normalized by the ones of quartz (denoted by the subscript q). As expected, they all increase with S_c.

Figure 6.12 plots the wave celerities of the shear and compression waves as functions of S_c. Again they are normalized by the corresponding velocities in quartz. Both $c_\perp$ and $c_\parallel$ increase with S_c.

6.4.3 Wave Propagation through Saturated Porous Media

6.4.3.1 Basic Equation

The porous matrix is an elastic porous material filled by a fluid. The material is assumed to be statistically homogeneous and therefore it can be regarded as spatially periodic, i.e., it is made of identical unit cells Ω of size L. The

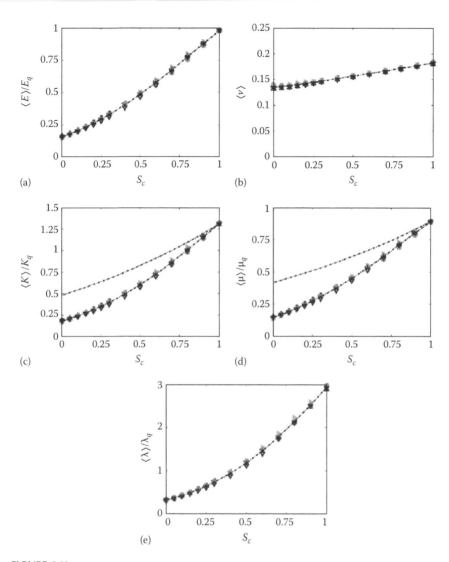

FIGURE 6.11
The effective Young modulus $\langle E\rangle/E_q$ (a), the effective Poisson ratio $\langle v\rangle$ (b), the effective bulk modulus $\langle K\rangle/K_q$ (c), the effective shear modulus $\langle\mu\rangle/\mu_q$ (d), the effective Lamé coefficient and, (e), $\langle\lambda\rangle/\lambda_q$ in the x-direction as functions of S_c. Data are for Sample 1 ($\triangle$), Sample 2 ($\triangleleft$), Sample 3 ($\circ$), Sample 4 ($\triangleright$), Sample 5 ($\triangledown$), and the average values (—●—).

characteristic size of the pores is denoted by l. Ω is partitioned into a solid and a fluid phase denoted by Ω_s and Ω_f, respectively. Ω_f is filled by an incompressible Newtonian fluid.

Consider the propagation in this medium of a harmonic wave of pulsation ω and wavelength λ large with respect to L. Therefore, one can introduce the small parameter η

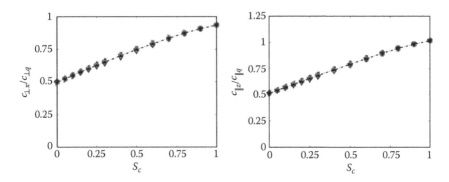

FIGURE 6.12
The shear wave celerity $c_{\perp x}/c_{\perp q}$ (a) and the compression wave celerity $c_{\parallel}/c_{\parallel q}$ (b) as functions of S_c. Data are for Sample 1 (△), Sample 2 (◁), Sample 3 (○), Sample 4 (▷), Sample 5 (▽), and the average values (—●—).

$$\eta = 2\pi l/L \ll 1. \tag{6.52}$$

Hereafter, all the quantities relative to the solid and fluid phases are denoted by the subscripts s and f. ρ, **u**, and σ denote the density, the displacement, and the stress tensor of the two materials, respectively. The displacements $\mathbf{u}_s$ and $\mathbf{u}_f$ are of the form

$$\mathbf{u} = \widehat{\mathbf{u}}e^{i\omega t}. \tag{6.53}$$

In the solid matrix Ω_s, wave propagation is governed on the microscopic level by the elastic equation

$$\nabla \cdot \boldsymbol{\sigma}_s = -\rho_s \omega^2 \mathbf{u}_s \quad \text{in } \Omega_s, \tag{6.54a}$$

where

$$\mathbf{e} = \frac{1}{2}\left[\nabla \mathbf{u} + (\nabla \mathbf{u})^t\right], \quad \boldsymbol{\sigma}_s = \mathbf{C}_{\{4\}} : \mathbf{e}, \tag{6.54b}$$

where $\mathbf{C}_{\{4\}}$ is the solid fourth-order elastic tensor. For isotropic materials, the expression of the stress tensor reduces to

$$\boldsymbol{\sigma}_s = \lambda_s{}^{tr}\mathbf{e}\mathbf{I} + 2\mu_s\mathbf{e}, \tag{6.54c}$$

where λ_s and μ_s are the Lamé coefficients.

The fluid velocity $\mathbf{V}$ is given by

$$\mathbf{V} = \frac{\partial \mathbf{u}_f}{\partial t} = i\omega \mathbf{u}_f. \tag{6.55}$$

Therefore, the linearized Navier–Stokes equations for the fluid motion in Ω_f can be written as

$$\nabla \cdot \boldsymbol{\sigma}_f = \rho_f \omega \, \mathbf{V}, \quad \nabla \cdot \mathbf{V} = 0, \tag{6.56}$$

together with

$$\boldsymbol{\sigma}_f = -P\mathbf{I} + 2\mu_f D(\mathbf{V}), \quad D(\mathbf{V}) = \frac{1}{2}\left[\nabla\mathbf{V} + (\nabla\mathbf{V})^t\right], \tag{6.57}$$

where
 P is the pressure
 μ_f is the dynamic viscosity

Continuity of displacements and normal stresses at the solid–fluid interface Γ implies

$$\mathbf{u}_s = \mathbf{u}_f, \quad (\boldsymbol{\sigma}_s - \boldsymbol{\sigma}_f)\cdot\mathbf{n} = 0 \quad \text{on } \Gamma, \tag{6.58}$$

where $\mathbf{n}$ is the unit normal to Γ.

The derivation of the macroscopic quantities and of the local equations at the various orders in η is a task too long and too delicate to be reported here. The interested reader may study [24].

6.4.3.2 Application

Various porous media were considered, and the influence of their microstructure on the wave celerities was analyzed.

For the slow wave, a unified description in terms of $\omega'_\Lambda = \omega\, \rho_f \Lambda^2/\mu^f$ is possible for all the media, as shown by Figure 6.13. In agreement with the predictions of Biot [25], Figure 6.13 shows that the slow wave celerity is a function of $\omega'^{1/2}_\Lambda$ and that the penetration depth h is a function of $\omega'^{-1/2}_\Lambda$.

The fast compressional and the shear wave celerities for all the media almost do not depend on ω'_Λ and only small variations appear when $\omega'_\Lambda > O(1)$. The fast compressional and the shear wave celerities for the model media with $\nu = 0.275$ can be roughly approximated by linear fits (Figure 6.14a)

$$c'_{||} = -0.7\varepsilon + 0.7, \quad c'_\perp = -0.35\varepsilon + 0.39. \tag{6.59}$$

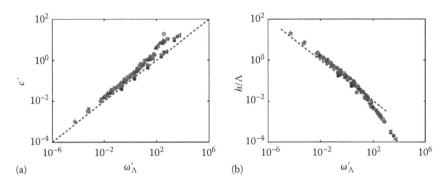

FIGURE 6.13
The slow wave celerity $c'_\parallel$ (a) and its penetration depth compared to the wave length h/Λ (b) as functions of the frequency ω'_Λ; $L/a = 128$. The dash lines in (a) and (b) correspond to $0.1\sqrt{\omega'}$ and to $0.1/\sqrt{\omega'}$, respectively. Data are for sandstone (*), the real carbonate (★), unimodal media HP (△), and HV (◁), bimodal media NPV (○), and PV (□); unimodal reconstructed porous media with $l_c = 6$ (▲), $l_c = 12$ (●), and $l_c = 24$ (■); and with $\varepsilon = 0.15$, $\varepsilon = 0.2$, and $\varepsilon = 0.3$.

For the real media, $c'_\perp$ follows well the fit (6.59) for both sandstone and carbonate (Figure 6.14a). $c_\parallel$ deviates more significantly from the fit (6.59).

The lengths h/Λ as functions of ω'_Λ for fast compressional and shear waves are almost the same for all media (Figure 6.14b).

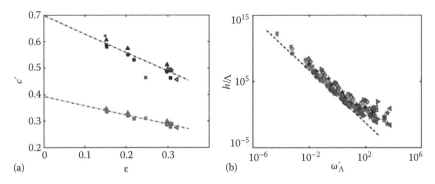

FIGURE 6.14
(a) The fast compressional (black) and the shear (grey) wave celerity c' as functions of the porosity ε for the real sandstone (×) and for the real carbonate (+), for unimodal media HP (△) and HV (◁), bimodal media NPV (○) and PV (□), and unimodal media with $l_c = 6$ (▲), $l_c = 12$ (●), and $l_c = 24$ (■). The dash-dotted lines correspond to the fits (6.59). (b) The penetration depth h/Λ for the compressional (empty symbols) and the shear (filled symbols) waves as functions of ω'_Λ for unimodal media HP (△) and HV (◁), bimodal media NPV (○) and PV (□), and unimodal reconstructed porous media with $l_c = 6$ (◇), $l_c = 12$ (▷), and $l_c = 24$ (★) and with porosities $\varepsilon = 0.15$, $\varepsilon = 0.2$, and $\varepsilon = 0.3$, the real sandstone (*) and the real carbonate (×). The dash line corresponds to $500/\omega'^2$.

6.5 Concluding Remarks

This short review shows the success of the approach, which was initiated almost 20 years ago by Adler [3]. The characterization and reconstruction of porous media have been considerably improved, thanks partly to microtomography. Many phenomena have been studied. However, a lot remains to be done in this field.

References

1. P.M. Adler and J.-F. Thovert, *Appl. Mech. Rev.*, 51, 537 (1998).
2. P. Spanne, J.-F. Thovert, C.J. Jacquin, W.B. Lindquist, K.W. Jones, and P.M. Adler, *Phys. Rev. Lett.*, 73, 2001 (1994).
3. P.M. Adler, *Porous Media: Geometry and Transports.* Stoneham, MA: Butterworth-Heinemann (1992).
4. P.M. Adler, C.G. Jacquin, and J.-F. Thovert, *Water Resour. Res.*, 28, 1571 (1992).
5. P.M. Adler, C.G. Jacquin, and J.A. Quiblier, *Int. J. Multiphase Flow*, 16, 691 (1990).
6. A. Moctezuma-Berthier, O. Vizika, and P.M. Adler, *Transp. Porous Media*, 49, 313 (2002).
7. N. Losic, J.-F. Thovert, and P.M. Adler, *J. Coll. Interf. Sci.*, 186, 420 (1997).
8. V.V. Mourzenko, J.-F. Thovert, and P.M. Adler, *Eur. Phys. J. B*, 19, 75 (2001).
9. J.-F. Thovert, F. Yousefian, P. Spanne, C.G. Jacquin, and P.M. Adler, *Phys. Rev. E*, 63, 061307 (2001).
10. D. Coelho, J.-F. Thovert, and P.M. Adler, *Phys. Rev. E*, 55, 1959 (1997).
11. V.V. Mourzenko, J.-F. Thovert, O. Vizika, and P.M. Adler, *Phys. Rev. E*, 77, 056307 (2008).
12. S. Békri, K. Su, F. Yousefian, P.M. Adler, J.-F. Thovert, J. Muller, K. Iden, A. Psyllos, A.K. Stubos, and M.A. Ioannidis, *J. Pet. Sci. Eng.*, 25, 107 (2000).
13. A.B. Yu and N. Standish, *Powder Technol.*, 55, 171 (1988).
14. A.B. Yu and N. Standish, *Powder Technol.*, 52, 233 (1987).
15. J.F. Thovert, F. Wary, and P.M. Adler, *J. Appl. Phys.*, 68, 3872 (1990).
16. D.L. Johnson, J.L. Koplik, and L. Schwartz, *Phys. Rev. Lett.*, 57, 2564 (1986).
17. R. Lemaître and P.M. Adler, *Transp. Porous Media*, 56, 325 (1990).
18. R. Peyret and T.D. Taylor, *Computational Methods for Fluid Flow*, Springer Series in Computational Physics. Berlin, Germany: Springer-Verlag (1985).
19. D.B. Pengra, S. Li, S.X. Li, and P.Z. Wong, Dynamics in small confining systems II, J.M. Drake, J. Klafter, K. Kopelman, and S.M. Troian, Eds, *Mat. Res. Soc. Symp. Proc.*, 366, 201–206 (1995).
20. G.E. Archie, *Trans. AIME*, 146, 54 (1942).

21. A.M. Valfouskaya, P.M. Adler, J.-F. Thovert, and M. Fleury, *J. Appl. Phys.*, 97, 1 (2005).
22. P.C. Carman, *Trans. Inst. Chem. Eng. (Lond.)*, 15, 150 (1937).
23. A.M. Valfouskaya and P.M. Adler, *Phys. Rev. E*, 72, 056317 (2005).
24. I. Malinouskaya, V.V. Mourzenko, J.-F. Thovert, and P.M. Adler, *Phys. Rev. E*, 77, 056307 (2008).
25. M.A. Biot, *J. Acoust. Soc. Am.*, 28, 168 (1956).

7

Characterization of Macroscopically Nonhomogeneous Porous Media through Transient Gas Sorption or Permeation Measurements

J.H. Petropoulos and K.G. Papadokostaki

CONTENTS

7.1 Introduction

In porous adsorbent/catalyst pellets or separation membranes (barriers), nonhomogeneity may be introduced either deliberately (laminates, "asymmetric" membranes) or spuriously (e.g., "skin layers" on powder compacts or extruded pellets), leading in each case to corresponding spatial dependence of the relevant gas sorption and diffusion coefficients, S and D_T, which are defined here as follows (Petropoulos and Roussis 1967a,b; Petropoulos 1985):

$$S = C/a \tag{7.1}$$

where C is the concentration of sorbed gas and a is the appropriate activity, which (for dilute gases of interest here) may be taken as effectively equal to the gas-phase concentration C_g at equilibrium with C, that is, $a \cong C_g = p/RT$ (p=gas pressure) and

$$\frac{\partial C}{\partial t} = \frac{\partial}{\partial X}\left(\frac{D_T C}{RT}\frac{\partial \mu}{\partial X}\right) = \frac{\partial}{\partial X}\left(D_T S \frac{\partial a}{\partial X}\right) = \frac{\partial}{\partial X}\left(P \frac{\partial a}{\partial X}\right) \qquad (7.2a, b, c)$$

Equation 7.2a is the unidimensional diffusion equation formulated on the basis of the chemical potential gradient $\partial \mu / \partial X$ of sorbed gas as driving force (e.g., Petropoulos and Roussis 1967a,b; Petropoulos 1985), which reduces to Equation 7.2b and 7.2c by virtue of the relation $d\mu = RT\,d\ln a$; X represents distance in the direction of flow; t is the time; $P = D_T S$ is the permeability coefficient. The diffusion coefficient D_T defined in this way is most commonly qualified as "thermodynamic," to distinguish it from the diffusion coefficient D, defined in terms of Fick's equation

$$\frac{\partial C}{\partial t} = \frac{\partial}{\partial X}\left(D \frac{\partial C}{\partial X}\right)$$

which is commonly encountered in the literature. However, for macroscopically homogeneous barrier-gas systems characterized by concentration-independent sorption and diffusion coefficients, $D \equiv D_T$ (e.g., Petropoulos and Roussis 1967a,b; Petropoulos 1985).

Variation of S and D_T of a porous barrier as a function of X, which leads to non-Fickian diffusion requiring application of Equation 7.2, is of particular interest and may be detected and characterized (at least semi-quantitatively) by suitable analysis of transient sorption or permeation data (e.g., Petropoulos and Roussis 1967b; Petropoulos 1985). Then, with the aid of an adequate physical structure–property model, it is possible to use various salient features of the resulting $S(X)$ and $D_T(X)$ functions, as diagnostics of the nature and degree of the underlying (axial) macroscopic structural inhomogeneity of the said barrier (e.g., Petropoulos and Roussis 1968).

Here, we present a scheme of "comprehensive permeation time-lag analysis" suitable for this purpose, and then show how analogous confirmatory or complementary information can be extracted from transient sorption and permeation kinetics, in the case of concentration-independent S and D_T parameters.

We also refer briefly to the diagnostic significance of the phenomenon of "directional" steady-state permeability in systems with concentration-dependent S and D_T.

7.2 Steady-State Permeability Properties

7.2.1 Concentration-Independent Systems

In the absence of any prior knowledge of the dependence of S and D_T on X (as is usually the case in practice), the sorption and steady-state permeation properties of the aforesaid porous barrier would be described by effective solubility (S_e = const.) and permeability (P_e = const.) coefficients, deduced from measurements of overall equilibrium concentration of sorbed gas (C_e) as a function of the corresponding activity a (identified here with the external gas concentration C_g; see above) and of steady-state permeation flux $q_s = dQ_s/dt$ (see Equation 7.5 below noting that Q denotes quantity of gas), respectively, which turn out to be the arithmetic mean of $S(X)$ and the harmonic mean of $P(X)$, respectively (Petropoulos and Roussis 1967b; Petropoulos 1985):

$$S_e = \frac{1}{l} \int_0^l S(X) dX = C_e / C_g \tag{7.3}$$

$$P_e = \frac{l}{\int_0^l dX/P(X)} = S_e D_e \tag{7.4}$$

where
    l is the thickness of the barrier
    D_e (= const.) is the corresponding effective diffusion coefficient

7.2.2 Concentration-Dependent Systems

The situation is more complex if S and/or D_T are also functions of C (or a). Equation 7.3 is still valid; however, unless $P(a, X)$ is separable into a product of pure functions of C (or a) and of X, P_e cannot, in general, be expressed analytically. Under these conditions, the value of P_e will, in general, also tend to differ according to whether flow is in the $+X$ or $-X$ direction. Theoretical investigation of the kind of $P(a, X)$ function required for maximization of this interesting "flow asymmetry" effect in artificial membranes (for potential use as a practical diagnostic tool) indicates (Petropoulos 1973, 1974) that "flow asymmetry factors" $f_a = \vec{P}_e / \overleftarrow{P}_e$ can, in principle, attain values well in excess of those suggested by early theoretical studies (e.g., Peterlin and Olf 1972) but still well below those achievable in biological systems. It was further shown that experimental directional permeability data from the literature (Rogers et al. 1957; Rogers 1965) on asymmetrically grafted polymeric membranes could be interpreted semiquantitatively by means of suitable model calculations. Thus, for example, it was found (Petropoulos 1974) that the value of $f_a = 3.4$ obtained for water vapor permeating through a laminate of Nylon 6-ethyl cellulose (Rogers et al. 1957) was reasonable and close to the maximum achievable with those polymeric materials. On the other hand, a value as high as $f_a \approx 6$ (or even more), measured at higher upstream water vapor boundary pressures in a polyethylene membrane with a gradation in vinyl alcohol graft content (Rogers 1965), largely exceeded the expected limit and could be accounted for only by invoking (Petropoulos 1974) the generation of microcracks in the membrane, as a result of the strong differential swelling stresses set up, when the high water vapor pressure was on the side rich in the graft component (a mechanism mimicking the action of mechanosensitive channels in biological cell membranes, which play a crucial role in the regulation of the water content of the cell; see, e.g., Jeon and Voth 2008; Vasquez et al. 2008). As an example of prospective technological applications (see Price 1996), we cite the interesting problem arising in the conservation of culturally valuable monuments, which are subject to stone decay induced by moisture imbibed in the pores of the building material. Here the standard approach of laying a protective external-surface coating to keep out atmospheric moisture, has the crucial disadvantage of becoming a trap preventing the egress (when conditions of low atmospheric humidity prevail) of moisture, which may have previously found its way into the pores (either after prolonged exposure to a humid atmosphere or by seepage up the walls of the monument from the ground). It is obvious that, in principle, a double surface coating of reasonably high flow asymmetry, analogous to the laminate studied by Rogers et al. (1957) (see above), would provide an elegant practical solution to this problem.

7.3 Permeation Time Lags (Concentration-Independent Systems)

Use of the above effective coefficients S_e and P_e or $D_e = P_e/S_e$, given by Equations 7.3 and 7.4, respectively, does not lead to the correct description of transient diffusion for a system characterized by $S(X)$, $D_T(X)$ and hence designated as "non-Fickian." However, the behavior of the ideal Fickian system defined by $S = S_e$ and $P = P_e$ or $D = D_e$ constitutes a useful standard of reference. Given the appropriate theoretical background, one may then deduce information about $S(X)$, $D_T(X)$ from the nature and magnitude of the deviation of suitable observed kinetic parameters from the calculated Fickian values.

For this purpose, a systematic approach of *general* applicability to systems consisting of a penetrant permeating through a solid barrier has been developed, in the form of a scheme of "comprehensive permeation time-lag analysis" (Petropoulos and Roussis 1967a,b; Petropoulos 1985), based on suitable elaboration of the standard experimental permeation setup (as illustrated in Figures 7.1a and 7.2a for the case of gaseous permeants under consideration here). In particular, the upstream (at $X=0$) and downstream (at $X=l$) surfaces of the barrier are maintained at effectively constant penetrant activities a_0 and a_l, respectively, by allowing only very small changes of the corresponding gas pressures p_0, p_l in the respective adjoining gas reservoirs. One measures the amount of penetrant, $Q(0, t)$, entering the barrier

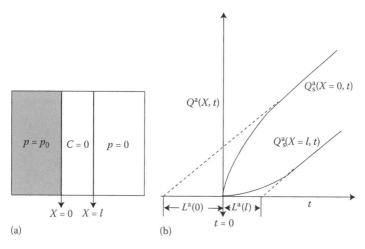

FIGURE 7.1
Schematic illustration of (a) the start-up ($t=0$) boundary conditions applicable to absorptive permeation experiments with $p_l=0$; (b) measured permeation curves (Q) and their linear steady-state asymptotes (Q_s) back-extrapolated to yield the corresponding time lags (L).

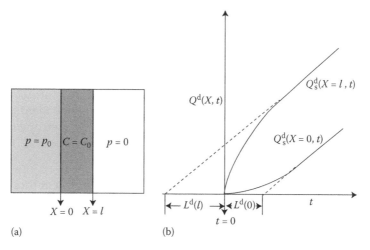

FIGURE 7.2
As Figure 7.1, for desorptive permeation experiments.

at $X=0$, as well as the amount, $Q(l, t)$, which leaves the barrier at $X=l$. The barrier is pre-equilibrated at uniform penetrant activity $a_1 = a_l$ ("absorptive permeation" denoted by superscript a; see Figure 7.1a) or $a_1 = a_0$ ("desorptive permeation," superscript d; see Figure 7.2a). These experimental conditions may be expressed in the form of boundary conditions for Equation 7.2, as follows:

$$a(X, t = 0) = a_1; \quad a(X = 0, t > 0) = a_0; \quad a(X = l, t > 0) = a_l$$

where $a_0, a_l = \text{const}$. Thus, it is possible to determine two "upper" permeation curves $Q^a(0, t)$, $Q^d(l, t)$ and two "lower" permeation curves $Q^a(l, t)$, $Q^d(0, t)$, as shown in Figures 7.1b and 7.2b. The corresponding steady-state linear asymptotes (denoted by Q_s) yield four time lags (intercepts on the t axis; cf. Figures 7.1b and 7.2b) $L^a(0) < 0$, $L^d(l) < 0$, $L^a(l) > 0$ (the only one measured in standard permeation experiments), and $L^d(0) > 0$, respectively. The same measurements may be repeated for flow in the $-X$ direction (denoted by Q^*, L^*).

The expressions required for the experimental determination of P_e, D_e, and S_e are

$$q_s = dQ_s(0, t)/dt = dQ_s(l, t)/dt = UP_e(a_0 - a_l)/l \tag{7.5}$$

$$L^a(l) - L^a(0) - L^d(l) + L^d(0) = l^2/D_e \tag{7.6}$$

$$S_e = P_e/D_e \tag{7.7}$$

where U is the cross-sectional area of the barrier and, in the particular experiments referred to below, $a_l = 0$.

It is noteworthy that the effective coefficients P_e, D_e, and S_e are all obtainable in this way, *without* recourse to equilibrium sorption measurements. This amounts to a kinetic method of measuring S_e, exactly analogous to that applicable to ideal systems (where measurement of $L^a(l)$ suffices for this purpose). This method is also applicable to concentration-dependent systems, as has been shown both theoretically and experimentally (Petropoulos and Roussis 1967a,b; Ash et al. 1979; Roussis et al. 1980; Amarantos et al. 1983).

7.3.1 Sorption and Permeation Apparatus for a Gaseous Penetrant

An apparatus, suitable for determination of the requisite time lags for subatmospheric gas–porous barrier systems, under conditions of $p_l = 0$, is illustrated schematically in Figure 7.3.

Measurement of $Q(l, t)$ and $Q(0, t)$ is effected by a constant volume-variable gas pressure technique, using suitable sensitive gas pressure gauges. The

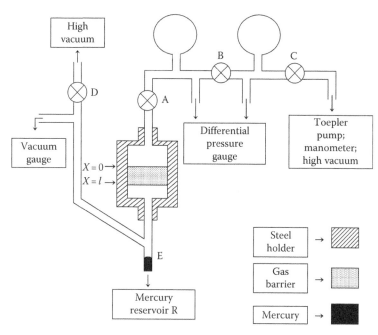

FIGURE 7.3
Salient features of pertinent gas transport apparatus (see the text for further details).

relevant upstream (V_0) and downstream (V_l) reservoir volumes are made large enough to ensure that the resulting boundary pressure changes, $\Delta p(0, t) = p(X=0, t=0) - p(X=0, t)$ and $\Delta p(l, t) = p(X=l, t) - p(X=l, t=0)$, are measurable with adequate accuracy without significantly disturbing the constancy of p_0, p_l. The latter requirement amounts to keeping $\Delta p(0, t)/p_0$ and $\Delta p(l, t)/p_0$ sufficiently small. Thus, the upstream reservoir consists of a large known volume V_2 (between stopcocks A and B in Figure 7.3) and a dead volume V_1 (between stopcock A and the upstream surface of the barrier at $X=0$), equipped with a differential pressure gauge (working against a constant reference pressure p_{ref} close to p_0, in volume V_3 between stopcocks B and C). For the sake of accuracy, V_1 should be as small as possible (e.g., by filling with glass beads) and its size accurately measured (e.g., by filling with mercury after completion of the gas transport work). The downstream reservoir is equipped with a vacuum gauge and a reservoir of ultrapure mercury R (cf. Figure 7.3), from which mercury may be raised to block the downstream surface of the barrier, under a hydrostatic head exceeding p_0 but still far too low to induce any penetration of mercury into the pores of the barrier. During the course of the permeation experiment, the mercury level is kept at the mark E, as shown in Figure 7.3. Thus, V_l is bounded by the downstream surface of the barrier ($X=l$), mark E, closed stopcock D (see Figure 7.3), and the vacuum gauge.

To initiate an absorptive permeation experiment, gas at a suitable pressure $p_{in} > p_0$ is introduced into V_2 and then, at $t=0$, expanded into V_1 (by opening stopcock A) to yield $p(X=0, t=0) = p_{in}V_2/(V_1 + V_2)$. For a desorptive permeation experiment, the downstream surface of the barrier is first blocked with mercury from reservoir R (see above), the barrier is equilibrated with gas in the upstream reservoir, with stopcock A open, at pressure $p(X=0, t=0)$, and the experiment is started by lowering the mercury level at $X=l$ to the mark E, as rapidly as possible.

It is easy to see that by keeping the downstream surface of the barrier blocked at all times, one may measure sorption equilibria as well as sorption kinetics. For absorption kinetics, one follows the same procedure as for absorptive permeation. For a desorption (with simultaneous flow reversal) experiment, after prior equilibration at $p = p_{eq}$, V_2 is first evacuated with stopcock A closed, and then stopcock A is opened, at $t=0$, after first closing B, yielding $p(X=0, t=0) = p_{eq} V_1/(V_1 + V_2)$.

For more detailed discussion on the design and preliminary testing of the above apparatus, see Tsimillis and Petropoulos (1977). For more general information on apparatus so far used for present purposes, see Roussis et al. (1980), Roussis and Petropoulos (1976, 1977), Ash et al. (1979), and Gavalas (2008).

7.3.2 Comprehensive Time-Lag Analysis Procedure

Time lag analysis can be simplified by using the standard time lag $L^a(l)$ normally measured, in conjunction with "absorption–desorption" and "upstream–downstream" time-lag differences (δL, ΔL, respectively), as well as the double difference $\delta\Delta L$. In particular, use is made of the following time-lag parameters (e.g., Petropoulos and Roussis 1967a,b; Petropoulos 1985):

$$L^a \equiv L^a(l); \quad \Delta L^a \equiv L^a - L^a(0); \quad \delta L \equiv L^a - L^d(l);$$

$$\delta\Delta L \equiv \Delta L^a - \Delta L^d \equiv \Delta L^a - L^d(l) + L^d(0)$$

One may also define L^{a*}, ΔL^{a*}, δL^*, and $\delta\Delta L^*$ obtained upon flow reversal, but note that $L^{a*} \equiv L^a$ and $\delta\Delta L \equiv \delta\Delta L^*$. The corresponding parameters calculated for a reference system obeying Fick's law with $S = S_e$ and $D_T = D_e$ are designated L_s^a, ΔL_s^a, δL_s, and $\delta\Delta L_s$, respectively, and are given by (Petropoulos and Roussis 1967a,b; Petropoulos 1985)

$$6L_s^a = 2\Delta L_s^a = 2\delta L_s = \delta\Delta L_s = \delta\Delta L = l^2/D_{T_e} = l^2/D_e \qquad (7.8)$$

The discrepancies (L_E^a, etc.) between the measured time lag parameters (L^a, etc.) and the calculated ideal values (L_s^a, etc.), namely

$$L_E^a = L^a - L_s^a; \quad \Delta L_E^a = \Delta L^a - \Delta L_s^a; \quad \delta L_E = \delta L - \delta L_s$$

are then analyzed, noting that $\delta\Delta L_E \equiv \delta\Delta L - \delta\Delta L_s \equiv 0$ (see Equation 7.8).

In this analysis, one must take into account possible contributions from causes other than dependence of S, D_T on X. In particular, S and/or D_T become functions of time (thus giving rise to "time-dependent non-Fickian diffusion") in the presence of a parallel rate process, which (1) occurs on a timescale comparable with that of the diffusion (i.e., transient sorption or permeation) process; and (2) can affect the values of S and/or D_T materially. Examples of such rate processes include penetrant-induced structural changes in the solid sorbent (such as are observed in glassy polymer or unconsolidated porous solid–vapor systems; see e.g., Park 1953; Barrer and Strachan 1955; Frisch 1959; Kishimoto 1964; Petropoulos and Roussis 1967a; Crank 1975a,b), chemical reaction between penetrant molecules and reactive groups of the solid (e.g., Crank 1975c; Siegel 1986, 1995; Siegel and Cussler 2004) or diffusion into blind pores of a porous solid. This last possibility is the only one that might be appropriate for the concentration-independent systems under consideration here. Indeed blind porosity is, not infrequently, assigned a significant role in both modeling transport in, and routine characterization of, porous media (e.g., Rajniak et al. 1999; Jena and Gupta 2005) and has also been invoked specifically for the interpretation of experimental time-lag discrepancies $L_E^a \neq 0$ (e.g., Barrer and Gabor 1959, 1960; Ash et al. 1968). This interpretation has, however, been strongly disputed on the grounds that requirements (1) and (2) are not (or are highly unlikely to be) fulfilled (Petropoulos and Roussis 1967a; Tsimillis and Petropoulos 1977; Galiatsatou et al. 2006a; see also below).

Analysis of the case of simultaneous dependence of S, D_T on X and t (with X and t dependences not affecting each other materially) shows that (Petropoulos and Roussis 1967b; Crank 1975d; Petropoulos 1985)

$$L_E^a = L^a - L_s^a = L_T^a + L_h^a; \quad \Delta L_E^a = \Delta L_T^a + \Delta L_h^a; \quad \delta L_E = \delta L_T + \delta L_h \tag{7.9}$$

where
 L_E^a is the observed discrepancy or "non-Fickian time lag increment"
 L_T^a, L_h^a are the corresponding increments due to time- and X-dependence, respectively

Study of the properties of the individual time-lag increments shows (Petropoulos and Roussis 1967a,b; Crank 1975d; Petropoulos 1985) that

$$\Delta L_T^a = 0 \tag{7.10a}$$

$$\Delta L_h^a = -\delta L_h = \delta L_h^* = -\Delta L_h^{a*} \tag{7.10b}$$

These properties enable one to separate time-dependence from X-dependence effects, for example through the relation that follows from Equations 7.9

through 7.10b (Petropoulos and Roussis 1967b; Tsimillis and Petropoulos 1977; Petropoulos 1985; Galiatsatou et al. 2006a)

$$\delta L_E = \delta L_h + \delta L_T = -\Delta L_h^a + \delta L_T = -\Delta L_E^a + \delta L_T \tag{7.11}$$

7.3.3 Experimental Application of the Comprehensive Time-Lag Analysis Approach

Development of the above theoretical background (Petropoulos and Roussis 1967b; Petropoulos 1985) was followed by experimental verification of time lag expressions for laminated media of the AB type (e.g., Choji et al. 1981). Laminated media (of the AB or ABA type) are particularly useful for this purpose, because the full $S(X)$, $D_T(X)$ functions can be determined precisely by measuring the (constant) sorption and diffusion coefficients in each component of the laminate separately.

Application of the full diagnostic method of time lag analysis described above has so far been restricted to simple gases permeating (in the Knudsen flow regime) through porous barriers constructed by compaction of powders in cylindrical dies (as illustrated in Figure 7.4a), which are expected to exhibit macroscopically nonhomogeneous structure along X (Roussis and Petropoulos 1976, 1977; Tsimillis and Petropoulos 1977; Savvakis and Petropoulos 1982; Galiatsatou et al. 2006a,b). The most detailed and accurate experimental data have been obtained for N_2 permeating through graphite or active carbon porous compacts.

A typical example of such data is given in Figure 7.5. As illustrated in Table 7.1, the experimental δL_E, ΔL_E^a resulting from this work obeys Equation 7.10b. Thus, $\delta L_T = 0$ according to Equation 7.11, indicating that the observed non-Fickian behavior is, at least predominantly, attributable to X-dependence of S, D_T. The absence of any evidence for time dependence is not surprising: on one hand, the work of Goodknight et al. (Goodknight et al. 1960; Goodknight and Fatt 1961), properly interpreted (Petropoulos and Roussis 1967a), has shown that slow diffusion into blind pores can affect macroscopic transient-state diffusion but not the time lag; on the other hand, the length of blind pores should normally be of microscopic dimensions; hence the rate of filling such pores would normally be expected to be very fast on the macroscopic diffusion timescale (Tsimillis and Petropoulos 1977; Galiatsatou et al. 2006a).

The salient properties of L_h^a and $\Delta L_h^a = -\delta L_h$ have been shown (Petropoulos and Roussis 1967b; Petropoulos 1985) to be governed by the compound transport function

$$\psi(X) \equiv S(X)^2 D_T(X)/S_e^2 D_e \equiv S(X)P(X)/S_e P_e$$

If $\psi(X)$ is symmetrical about the middle of the barrier, $\Delta L_h^a = 0$; but the presence of X-dependence is still detectable thanks to the fact that $L_h^a \neq 0$ under

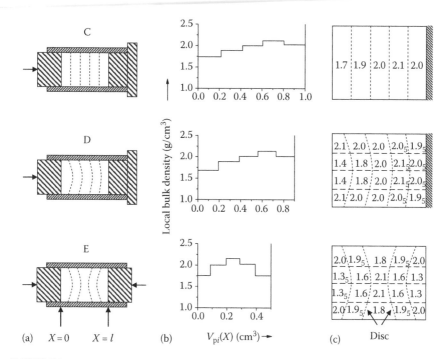

FIGURE 7.4

(a) Illustration of mode of construction of the porous barriers of Table 7.1 by unsymmetrical (barriers C, D) or symmetrical (barrier E) powder compaction in cylindrical dies (subsequently used as permeation cells). (b) Illustration of the resulting variation of the mean local densification, that is, of the reverse of $\varepsilon_L(X)$, from the upstream ($X=0$) to the downstream ($X=l$) side, using duplicates of barriers C, D, and E with intercalated pliable Pb discs. The discs, indicated by broken lines in Figure 7.4a, were inserted (1) after each compaction step in barriers constructed by multiple compaction (barrier of type C) or (2) at suitable positions along X in the powder before compaction in a single step (barriers of type D, E). In all cases, the weight of powder in the segments enclosed between successive disks δw_i (or between the barrier surfaces and the neighboring disks) is known (by preweighing the relevant portions of powder) and the corresponding volume occupied by each segment in the finished barrier δV_{pi} is measured by X-ray imaging of the discs. The resulting local bulk density of the ith segment ($\delta w_i/\delta V_{pi}$) is then plotted against the cumulative occupied volume $V_{pi} = \sum_{j=1}^{j=i} \delta V_{pj}$. (The replacement of the abscissa X by $V_{pi}(X)$ was necessitated by the bending of discs that accompanied single-step compaction). (c) Illustration of the presence of radial variation of the local degree of densification (and hence of ε_L) put in evidence by disc bending in single-step compaction (not visible in multistep compaction by the present method). Note that, in graphite barriers, gas transport is also influenced by the preferred orientation of the plate-like particles (see the text).

these conditions; furthermore a positive (negative) L_h^a shows that the turning point in the middle of the barrier is a maximum (minimum). As $\psi(X)$ deviates increasingly from symmetry, ΔL_h^a tends to increase in magnitude and is positive (negative), if the overall tendency of $\psi(X)$ is to assume higher (lower) values on the upstream side ($X \to 0$) of the barrier; at the same time, L_h^a, if

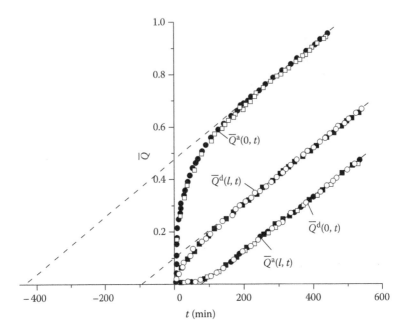

FIGURE 7.5
Experimental permeation curves for N_2 at 297 K in barrier D of Table 7.1 (points) and their corresponding steady-state asymptotes (lines) with Q given in dimensionless form $\bar{Q}$: $Q(0, t) = V_0 \Delta p(0, t)/Ulp_0$; $Q(l, t) = V_l \Delta p(l, t)/Ulp_0$. Absorptive permeation: (●) forward (+X) flow; (□) reverse (–X) flow. Desorptive permeation: (○) forward flow; (■) reverse flow. For clarity, only forward flow curves are labeled on the figure. Note conformity of the data to Equations 7.22 and 7.23. (From Savvakis, C. A study of diffusion and sorption of permanent gases through microporous graphite substrates in relation to their macroscopic structure. PhD thesis, Athens University, Greece, 1979.)

positive, changes progressively to negative; if negative, it remains so (see Petropoulos and Roussis 1967b; Galiatsatou et al. 2006a for further detail).

The advantage of studying porous barriers is that the spatial variation of S and D_T is relatively simply related to that of the local porosity ε_L (Petropoulos and Roussis 1968). In particular, S and P for N_2 were consistently found to be monotonic increasing functions of ε_L in all systems listed in Table 7.1 (Roussis and Petropoulos 1976, 1977; Tsimillis and Petropoulos 1977; Savvakis and Petropoulos 1982; Galiatsatou et al. 2006a,b). Hence, $\varepsilon_L(X)$ can be expected to display the same salient functional form as $\psi(X)$. When used in this manner, comprehensive time-lag analysis affords a nondestructive method of obtaining significant information concerning the macroscopic structural inhomogeneity of any diffusion barrier *in situ*. This claim is fully justified by the ensuing discussion of the data of Table 7.1.

Thus, the results $\Delta L_E^a \approx 0$, $L_E^a < 0$ for barriers B and E in Table 7.1, indicate that (1) $\psi(X)$ [and hence $\varepsilon_L(X)$] is nearly symmetrical about the middle of the barrier ($X = l/2$) (Petropoulos and Roussis 1967b, 1968), in accord with

TABLE 7.1

Permeation Results for N_2 at 297 K in Macroscopically Nonhomogeneous Porous Graphite (A–H) and "Carbolac" Carbon (K) Barriers

Barrier	A	B	C	D	E	G	H	K
Mode of construction	Nonsym	Sym	Nonsym	Nonsym	Sym	Nonsym	Nonsym	Nonsym
Compaction steps	6	5	5	1	1	6	5	13
Thickness (cm)	1.64	1.62	1.13	1.19_3	0.843	0.974	1.15	3.23
Porosity	0.15	0.13	0.13	0.15	0.15	0.13	0.24	0.50
$10^5 P_e$, (cm²/s)	7.80	8.81	2.64	2.49	4.85	3.05	3.7	117
S_e (a)	0.87	0.86	0.86	0.74	0.74	0.87	0.61	4.07
S_e (b)	0.91	0.86	0.86	0.72	0.75	0.84	0.61	3.96
L_E^a/L_s^a	−0.26	−0.38	−0.18	−0.31	−0.37	−0.18	−0.30	−0.11
$\Delta L_E^a/\Delta L_s^a$	0.30	−0.05	0.16	0.52	−0.03	0.26	0.47	0.18
$\delta L_E/\delta L_s$	−0.27	0.05	−0.16	−0.52	0.03	−0.26	−0.47	−0.19
L_E^{a*}/L_s^a	−0.26	−0.38	−0.17	−0.32	—	—	—	—
$\Delta L_E^{a*}/\Delta L_s^a$	−0.27	−0.05	−0.21	−0.51	—	—	—	—
$\delta L_E^*/\delta L_s$	0.31	0.05	0.20	0.51	—	—	—	—

Source: Data taken from Roussis, P.P. and Petropoulos, J.H., *J. Chem. Soc. Faraday II* 72, 737, 1976; Roussis, P.P. and Petropoulos, J.H., *J. Chem. Soc. Faraday II* 73, 1025, 1977; Tsimillis, K. and Petropoulos, J.H., *J. Phys. Chem.* 81, 2185, 1977; Savvakis, C. and Petropoulos, J.H., *J. Phys. Chem.* 86, 5128, 1982; Galiatsatou, P. et al., *Phys. Chem. Chem. Phys.* 8, 3741, 2006a; Galiatsatou, P. et al., *J. Memb. Sci.* 280, 634, 2006b.

Notes: (a) S, obtained from equilibrium sorption measurements; (b) S, obtained from permeation measurements by Equations 7.5 through 7.7.

the symmetrical mode of compaction of these barriers (produced by means of two symmetrically moving pistons, as illustrated for E in Figure 7.4a) (Roussis and Petropoulos 1976, 1977; Savvakis and Petropoulos 1982); and (2) the turning point is a minimum, as is confirmed by the corresponding maximum in the measured variation of the local bulk density along X (which is the reverse of $\varepsilon_L(X)$), displayed in the plot pertinent to E in Figure 7.4b (obtained by the X-ray imaging technique described therein).

Similarly, the results $\Delta L_E^a > 0$, $L_E^a < 0$ reported for the barriers (A, C, D, G, H, K) constructed by unsymmetrical compaction (keeping the piston at $X = l$ fixed, as illustrated in Figure 7.4a) indicate an overall tendency for $\psi(X)$ (and hence for $\varepsilon(X)$) to attain higher values on the upstream side of the barrier; a prediction that is again verified by the pertinent X-ray imaging plots of Figure 7.4b.

X-ray imaging also shows how the aforesaid variation in bulk density arises, given that, in all cases listed in Table 7.1, the barrier surface facing the mobile piston during unsymmetrical compaction, became the upstream surface ($X = 0$)

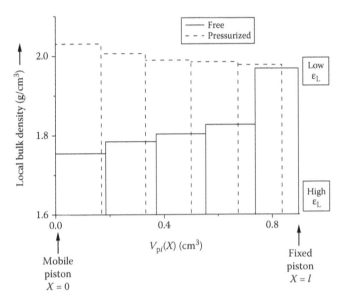

FIGURE 7.6
Typical example of variation of densification along X in an unsymmetrically compacted (alumina) barrier: (– – – –) in the pressurized state (local densification higher, hence local porosity ε_L lower, on the mobile piston side); (——) after the barrier has "sprung back" elastically, upon release of the compacting pressure (local densification lower, hence local porosity ε_L higher, on the mobile piston side).

in subsequent permeation runs. As illustrated in Figure 7.6 for an unsymmetrically compacted alumina barrier, the densification produced under compression is highest (i.e., ε_L is lowest) at $X=0$ and diminishes (i.e., ε_L rises) progressively up to $X=l$ (as might reasonably be expected on the grounds of increasing frictional compressive pressure losses); this pattern tends, in fact, to remain imprinted in compacts of ductile materials (e.g., in powder metallurgy; Goetzel 1949). However, compacts of hard materials, as is the case here, tend to recover or "spring back" elastically, upon decompression, and it is the extent to which such an expansion can occur (against frictional resistance), at various locations X, that determines the pattern of macroscopic structural inhomogeneity, which finally characterizes the finished barrier (see Savvakis and Petropoulos 1982 and Galiatsatou et al. 2006a for further details). This is also the reason why the turning point in the plot of local bulk density for E in Figure 7.4b is a maximum [(i.e., a minimum for $\varepsilon_L(X)$].

Furthermore, the possibility that the magnitude of the reduced parameter $\Delta L_h^a / \Delta L_s^a$ might serve as a useful indication of the overall degree of inhomogeneity of a barrier, for a *given* unsymmetrical functional form of $\psi(X)$, is suggested by the fact that the highest values of this parameter recorded in Table 7.1 pertain to barriers D and H, which were constructed under conditions known to intensify nonuniformity of densification, namely single-step compaction for D (Barrer and Gabor 1959, 1960; Savvakis and Petropoulos 1982) and lower overall densification (higher ε, due to lower applied compressive pressure) for H (Galiatsatou et al. 2006b), as compared with multistep compaction to high overall densification (low ε) for barriers A, C, G, H, K.

On the other hand, the technique of X-ray imaging with interposed X-ray-opaque discs used by Savvakis and Petropoulos (1982) (see legend of Figure 7.4b) reveals that the time-lag behavior of barriers constructed by one-step compaction is subject to an additional complication, namely the presence of a radial component of macroscopic inhomogeneity (also known in powder metallurgy; see, e.g., Goetzel 1949), resulting in two-dimensional macro-scopic variability of local densification (cf. plots for barriers of type D and E in Figure 7.4c). Nevertheless, there is a clear pattern of variation along X of the *mean* densification within each segment delimited by the broken lines in Figure 7.4c (cf. the resulting plots for barriers of type D and E in Figure 7.4b), which parallels the intersegmental variation of densification along X displayed by barriers of similar $\varepsilon_L(X)$ functional form prepared by multistep compaction (exemplified by the plots for barriers of type D and C in Figure 7.4b). This gross axial structural similarity is duly reflected in the observed qualitative similarity of time-lag behavior shown in Table 7.1.

Note that radial intrasegmental variability of local densification is also found in multistep compacts. It is not detectable by the above X-ray imaging technique, but may be estimated indirectly via suitable X-ray diffraction measurement (Savvakis and Petropoulos 1982) of the degree of orientation (alignment) of graphite particles at various locations in the barrier. (Due to the fact that the said particles are platelets, which pack most efficiently when parallel to one another, their degree of orientation is closely correlated with the extent of densification.) The results obtained (see Savvakis and Petropoulos 1982 for details) reveal a radial pattern of mild rise in densification from the center to the periphery of the barrier, which changes little with X, and should, therefore, not give rise to any appreciable time-lag discrepancies (in view of the fact that *pure* radial space dependence has *no* effect on ideal time-lag behavior; e.g., Petropoulos 1985). This is in marked contrast to the two-dimensional pattern exhibited by the corresponding single-step unsymmetrical barrier (D in Figure 7.4c). Here, the radial rise of densification from the center to the periphery of the barrier is very marked near the mobile piston ($X=0$), becomes gradually attenuated with increasing X, and is finally reversed near the fixed piston ($X=l$). For completeness, it is worth noting that the aforementioned X-ray diffraction results also yield the direction of the local orientation of the graphite platelets, which should affect the effective local value of D_T (via a microscopic "pore orientation factor" included therein; Savvakis et al. 1982; Galiatsatou et al. 2006b). The said particle orientation was found (Savvakis and Petropoulos 1982) to vary radially (in both single- and multistep compacts) from effectively normal to X (minimum local D_T), in the central region, to effectively parallel to X (maximum local

D_T) at the wall of the die. This radial variation of D_T is much the same at all X and should not, therefore, contribute significantly to deviation from the ideal time-lag behavior.

Accordingly, the observed non-Fickian time-lag behavior of multistep compacts should be representative primarily of intersegmental X-dependence of S and/or D_T (see also Petropoulos and Roussis 1969a in this respect), while that of single-step compacts would be expected to include an effect stemming from the X-dependence of the radial variability of these transport parameters (to an extent that cannot be estimated, at present, due to lack of appropriate theoretical background).

It follows that assessment of the potential diagnostic value of time-lag analysis for elucidation of such complex inhomogeneity effects requires further detailed study, complemented with parallel investigation of transient diffusion kinetics, which *are* sensitive to radial inhomogeneity. Furthermore, the above discussion illustrates the need to take due account also of the space variability of other relevant structural parameters, such as the orientation factor in graphite barriers considered above (see Galiatsatou et al. 2006b in particular) and the general pore structure factors κ_g, κ_s (cf. Barrer and Gabor 1959; Nicholson and Petropoulos 1982; Savvakis et al. 1982; Kanellopoulos et al. 1985; Petropoulos and Papadokostaki 2008), derived from appropriate sorption and permeability or relative permeability data. For examples of the fruitful combination of information derived from the said pore structure factors with time-lag analysis, see Petropoulos and Petrou (1992) and Petropoulos and Papadokostaki (2008).

For examples of other discussions of non-Fickian time lag analysis or uses thereof, see Crank (1975a), Ash et al. (1968, 1979), Ash (2006), Rutherford and Do (1997).

7.4 Transient Diffusion Kinetics (Concentration-Independent Systems)

An analogous method of analysis of transient state diffusion kinetics has been proposed (Tsimillis and Petropoulos 1977; Amarantos et al. 1983; Grzywna and Petropoulos 1983a,b; Petropoulos 1985), based on the consideration that, in any experiment, the kinetic behavior of the system represented by $S(X)$, $D_T(X)$ will generally deviate from that of the corresponding ideal system represented by $S = S_e$, $D = D_e$, in either of two ways: (1) ideal kinetics is obeyed, but with a different effective diffusion coefficient D_n, where $n = 1$, 2,... denotes a particular kinetic regime (D_n is usually deduced from a suitable linear kinetic plot) or (2) ideal kinetics is departed from, in which case one is reduced to comparison between the (nonlinear) experimental plot and the corresponding calculated ideal (linear) one.

The following effective diffusion coefficients D_n may be defined on the basis of standard sorption and permeation experiments (Grzywna and Petropoulos 1983a,b; Petropoulos 1985) (when absorption or desorption conditions need to be specified, superscripts a or d, respectively, are again used):

1. *Sorption*

   The barrier is initially equilibrated at $a = a_1$.

   a. In "unsymmetrical" sorption experiments (Tsimillis and Petropoulos 1977; Grzywna and Petropoulos 1983a), the barrier is exposed to penetrant at $X = 0$ and is blocked at $X = l$, namely

   $$a(X, t = 0) = a_1; \quad a(X = 0, t) = a_0; \quad \frac{\partial a(X = l, t)}{\partial X} = 0$$

   Denoting by Q_t, Q_∞ the amounts of penetrant sorbed (absorbed if $a_0 > a_1$ or desorbed if $a_0 < a_1$) at time t and at equilibrium, respectively, the simple expressions applicable at short and long times, respectively, are ($I_2 = $ const.)

   $$\frac{Q_t}{Q_\infty} = 2\left(\frac{D_1 t}{\pi l^2}\right)^{1/2} \tag{7.12}$$

   $$\ln\left(1 - \frac{Q_t}{Q_\infty}\right) = I_2 - \frac{\pi^2 D_2 t}{4 l^2} \tag{7.13}$$

   b. In "symmetrical" sorption experiments (Grzywna and Petropoulos 1983a), the barrier is exposed to penetrant at both $X = 0$ and $X = l$, namely

   $$a(X, t = 0) = a_1; \quad a(X = 0, t) = a_0; \quad a(X = l, t) = a_0$$

   with corresponding expressions for short and long times, respectively ($I_{2M} = $ const.)

   $$\frac{M_t}{M_\infty} = 4\left(\frac{D_{1M} t}{\pi l^2}\right)^{1/2} \tag{7.12a}$$

   $$\ln\left(1 - \frac{M_t}{M_\infty}\right) = I_{2M} - \frac{\pi^2 D_{2M} t}{l^2} \tag{7.13a}$$

2. *Permeation*

For the lower permeation curves and noting that $Q^a(l, t) = Q^d(0, t)$ [see Equation 7.23 below] we have ($I_5 = $const.)

$$\ln \frac{\sqrt{t}\ dQ^a(l, t)}{dt} = \ln\left(2USa_0\sqrt{\frac{D_4}{\pi}}\right) - \frac{l^2}{4D_3 t} \qquad (7.14)$$

$$\ln \frac{Q^a(l, t) - Q_s^a(l, t)}{Q_\infty} = I_5 - \frac{\pi^2 D_5 t}{l^2} \qquad (7.15)$$

at short and long times, respectively.

For the upper permeation curves $Q^a(0, t)$, $Q^d(l, t)$ we have, for example, for $Q^a(0, t)$,

$$\frac{Q^a(0, t)}{Q_\infty} = 2\left(\frac{D_1^a t}{\pi l^2}\right)^{1/2} \qquad (7.16)$$

$$\ln \frac{Q_s^a(0, t) - Q^a(0, t)}{Q_\infty} = I_7^a - \frac{\pi^2 D_7 t}{l^2} \qquad (7.17)$$

at short and long times, respectively, where $I_7^a = $const. Alternatively, use may be made of the net amount sorbed during permeation $\Delta Q_t = |Q(0, t) - Q(l, t)|$ and $\Delta Q_\infty = |Q_s(0, t) - Q_s(l, t)|$, in which case we have ($I_8 = $const.)

$$\frac{\Delta Q_t}{\Delta Q_\infty} = 4\left(\frac{D_6 t}{\pi l^2}\right)^{1/2} \qquad (7.18)$$

$$\ln\left(1 - \frac{\Delta Q_t}{\Delta Q_\infty}\right) = I_8 - \frac{\pi^2 D_8 t}{l^2} \qquad (7.19)$$

The reader is reminded that in Equations 7.12 through 7.19 the ideal value of D_n is in all cases D_e. The intercepts I_n are also useful kinetic parameters. Their ideal values are

$$I_5^0 = I_7^0 = \ln(2/\pi^2), \quad I_2^0 = I_{2M}^0 = I_8^0 = \ln(8/\pi^2) \qquad (7.20)$$

Considerable progress has been made in developing the theoretical background necessary for experimental application of the above method of transient kinetic analysis. An important step in this direction was the use of

Wentzel-Kramers-Brillouin (WKB) asymptotics to derive approximate analytical expressions for short- and long-time transient sorption and permeation in membranes characterized by concentration-independent continuous $S(X)$ and $D_T(X)$ functions. The earlier papers dealing with this subject (Frisch and Bdzil 1975; Frisch 1978a,b) have been reviewed in Frisch and Stern (1983). Grzywna and Petropoulos (1983a,b) provide the correct final asymptotic expressions applicable to all kinetic regimes of interest here; the usefulness of these expressions is evaluated, on the basis of complete numerical solutions of carefully selected examples, and the relation between transient transport behavior and the nature and degree of the inhomogeneity of sorption and diffusion properties is investigated therein in some detail.

Laminates with concentration-independent solubility and diffusion coefficients are of interest from the point of view of amenability to analytical solution of transient-state sorption or permeation (e.g., Choji et al. 1978; Choji and Karasawa 1980; Spencer and Barrie 1980; Le Poidevin 1982; Spencer et al. 1982; Berner and Cooper 1983). Such solutions are useful for the calculation of transient transport behavior, given the sorption and diffusion properties of each component. The inverse problem can, in principle, be handled, but the procedure may be impractical except in very simple cases (such as a laminate incorporating only one unknown component, e.g., a stagnant liquid layer, which cannot be studied in isolation). However, from the point of view of the fundamental understanding of transient transport behavior, as a function of the nature and degree of the inhomogeneity of the composite membrane, the said analytical solutions are not very informative because of their complicated nonexplicit character.

Among the main results of the theoretical studies reviewed above, several general relations between the quantities measured in sorption and permeation experiments (Roussis and Petropoulos 1976, 1977; Tsimillis and Petropoulos 1977; Frisch 1978b; Grzywna and Petropoulos 1983a,b) are worth noting, namely

$$Q_t^a = Q_t^{d*}; \quad Q_t^d = Q_t^{a*}; \quad M_t^a = M_t^d \tag{7.21}$$

$$Q^a(0, t) = Q^{d*}(l, t); \, Q^d(l, t) = Q^{a*}(0, t) \tag{7.22}$$

$$Q^a(l, t) = Q^{a*}(l, t) = Q^d(0, t) = Q^{d*}(0, t) \tag{7.23}$$

$$\Delta Q_t^a = \Delta Q_t^{d*}; \quad \Delta Q_t^d = \Delta Q_t^{a*} \tag{7.24}$$

$$M_t = \Delta Q_t^a + \Delta Q_t^d = \Delta Q_t^a + \Delta Q_t^{a*} \tag{7.25}$$

where M_t denotes the amount of penetrant sorbed in a symmetrical sorption experiment.

In all late-time kinetic regimes, notably those represented by Equations 7.13, 7.15, 7.17, and 7.19, ideal kinetics is obeyed (Frisch 1978b; Spencer and Barrie 1980; Spencer et al. 1982; Grzywna and Petropoulos 1983a,b); whereas this is not so in the short-time regimes presented by Equations 7.12, 7.16, and 7.18 (Choji et al. 1978; Frisch 1978b; Choji and Karasawa 1980; Grzywna and Petropoulos 1983a,b), all of which convey essentially the same information (Grzywna and Petropoulos 1983b). The short-time behavior of lower permeation curves represented by Equation 7.14 appears to occupy an intermediate position, in the sense that ideal kinetics appears to be followed only to a first approximation (Grzywna and Petropoulos 1983b). The relation between permeation and symmetrical sorption indicated by Equation 7.25 is also notable. The respective kinetics become very similar at long times (Frisch 1978b), as indicated by the relevant relations (Grzywna and Petropoulos 1983b)

$$D_{2M} = D_5 = D_7 = D_8 \tag{7.26a}$$

$$I_{2M} = \ln \left[(\Delta Q_\infty / Q_\infty) \exp I_8 + (\Delta Q_\infty^* / Q_\infty) \exp I_8^* \right] \tag{7.26b}$$

Transient kinetic behavior, like time lags, essentially reflects the properties of the function (Petropoulos and Roussis 1967b; Roussis and Petropoulos 1976, 1977; Grzywna and Petropoulos 1983a,b)

$$\psi_\omega [\omega(X)] = D_{T\omega} [\omega(X)] S_\omega [(\omega(X)]^2 / D_e S_e^2 \tag{7.27}$$

where

$$\omega = (l S_e)^{-1} \int_0^X S(z) dz \tag{7.28}$$

The detailed study of Grzywna and Petropoulos (1983a,b) indicates that the information about $\psi_\omega(\omega)$ conveyed by different kinetic regimes is in part similar and in part different. The similarity between late-time permeation and symmetrical sorption kinetics has already been pointed out above. Symmetrical sorption kinetics at both short and long times are found to reflect primarily the properties of $d^2\psi_\omega/d\omega^2$; whereas short time unsymmetrical sorption is mainly sensitive to $d\psi_\omega/d\omega$. For more detailed information the original papers should be consulted.

Some preliminary kinetic analyses of transient sorption and permeation data have been reported (Tsimillis and Petropoulos 1977; Amarantos et al. 1983; Galiatsatou et al. 2006b). Experimental data, pertaining to kinetic regimes expected to obey ideal kinetics, yield appropriate plots of reasonable linearity, as illustrated in Figure 7.7. Examples of effective diffusion coefficients for early-time permeation (D_3, D_4) and unsymmetrical late-time

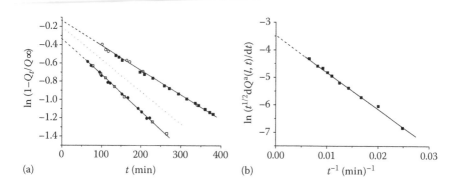

FIGURE 7.7

Experimental (a) long-time sorption, Q_t, and (b) short-time permeation, $Q^a(l,t)$, transient kinetic plots (points and corresponding lines extrapolated to $t=0$), according to Equations 7.13 and 7.14, respectively, for N_2 at 297 K in barrier C of Table 7.1. In Figure 7.7a: ($\bullet$) Q_t^a; ($\circ$) Q_t^d; ($\blacksquare$) Q_t^{a*}; ($\square$) Q_t^{d*} and the dotted line is the corresponding ideal long-time Q_t plot extrapolated to $t=0$. Effective diffusion coefficients ($\times 10^{-4} cm^2/s$): $D_e = 0.30$; $D_3 = 0.39$; $D_4 = 0.20$; $D_2^a = 0.34$, $D_2^d = D_2^{a*} = 0.23$. Intercepts: $I_2^a = -0.35$, $I_2^d = -0.15$ ($I_2^0 = -0.21$). (Data from Tsimillis, K. and Petropoulos, J.H. *J. Phys. Chem.* 81, 2185, 1977.)

sorption (D_2^a, D_2^d), determined from such plots, are shown in Table 7.2. For the multistep, unsymmetrical barriers C and H, differing only in overall degree of compaction and subject to predominantly axial inhomogeneity (see preceding section), the direction and fractional magnitude of the deviations of the observed transient D's from D_e, conform, as expected, to the

TABLE 7.2

Examples of Some Measured Transient-State Diffusion Coefficients of N_2 at 297 K in Some Porous Graphite Barriers of Table 7.1 in Comparison with the Corresponding Steady-State Effective Diffusion Coefficients D_e ($\times 10^{-4}$ cm²/s in All Cases)

Barrier	C	H	D	E
Compaction mode	Nonsym	Nonsym	Nonsym	Sym
		Multistep	Single-step	
Porosity	0.13	0.24	0.15	0.15
D_e	0.30	0.62	0.34	0.65
D_2^a	0.34	1.01	0.49	0.81
$D_2^{a*} = D_2^d$	0.23	—	0.20	—
D_3	0.39	0.94	0.63	1.70
D_4	0.20	0.35	0.13	0.11

Source: Data taken from Tsimillis, K. and Petropoulos, J.H., *J. Phys. Chem.* 81, 2185, 1977; Savvakis, C. A study of diffusion and sorption of permanent gases through microporous graphite substrates in relation to their macroscopic structure. PhD thesis, Athens University, Greece, 1979; Galiatsatou, P. et al., *J. Memb. Sci.* 280, 634, 2006b.

theoretical results of Grzywna and Petropoulos (1983a,b) corresponding to the general decreasing tendency of the compound transport function $\psi(X)$ inferred from time-lag analysis (and confirmed by X-ray imaging). For the single-step barriers D and E, the strong and complicated involvement of radial inhomogeneity revealed by X-ray imaging, coupled with the absence of appropriate theoretical background, precludes any meaningful interpretation. For the time being, one might take note of the occurrence of substantial (even at the qualitative level) deviation of the early-time permeation behavior of barrier E from that calculated for the corresponding case of pure X-dependence by Grzywna and Petropoulos (1983b).

Examples of analogous kinetic plots for a system with S and D_T exhibiting both spatial and concentration dependence have also been reported (Amarantos et al. 1983).

Transient sorption and permeation data may also be examined in the form of suitable moments. For example, in the case of sorption one may define the following moments (Frisch et al. 1979):

$$W_n = \int_0^{\infty} t^n Q_t dt \quad (n = 1, 2, \ldots) \tag{7.29}$$

These parameters are in many ways analogous to permeation time lags, but the relevant expressions for the case of $S(X)$, $D_T(X)$ are considerably more complicated (even in the case of the first moment, which is the simplest) (Frisch et al. 1979) than the corresponding time-lag formulae (Petropoulos and Roussis 1967b). Accordingly, moments represent a less efficient way of making use of the information contained in transient diffusion data than the methods discussed above. In spite of these limitations, further study of moments would be of interest.

The above discussion provides an illustration of the wealth of information that can be obtained from the transient transport behavior of concentration-independent, non-Fickian, solid barrier–penetrant systems. The methods discussed are capable of further development and refinement and are potentially applicable to a wide variety of experimental systems (see below).

7.5 General Conclusions and Future Research Prospects

Because most porous adsorbent/catalyst pellets or separation barriers used in practice are likely to be macroscopically inhomogeneous (cf. introductory section), and because their performance in various applications is inevitably (and often drastically) affected by this fact, it seems to us (cf. also Petropoulos and Petrou 1992) that characterization of porous solids, in terms of spatial distribution of porosity (and eventually of other salient structural parameters as well), should be no less important than present-day routine pore size distribution analysis, based on vapor absorption–desorption and analogous methods (e.g., Gregg and Sing 1982).

The work presented here demonstrates the (theoretical and practical) feasibility of extracting (hitherto largely unexploited) macroscopic structural information inherent in permeation time-lag and transient sorption or permeation kinetic data, which can be used for the above purpose, with the aid of appropriate (but not unduly elaborate) apparatus coupled with adequate theoretical background.

The most important emerging research need for the accomplishment of such a task appears to be the extension of the said theoretical background by the judicious and systematic study of the effect of macroscopic radial, and axial in-combination-with radial, structural inhomogeneity on time lags and transient diffusion kinetics. As pointed out above, the results of such studies are of general applicability, but, in the case of porous solid–gas systems of interest here, advantage can also be taken of the possibility of deriving additional useful information from permeability "structure factors" (see preceding section).

TABLE 7.3

Properties of "Non-Fickian Time-Lag Increments" and Other Relevant Parameters Pertaining to Axially Inhomogeneous Solid Substrate-Penetrant Systems Characterized by (1) $S(X)$, $D_T(X)$, (2) $S(a, X)$, $D_T(a, X)$ with Separable Dependence on a (or C) and X, and (3) $S(a, X)$, $D_T(a, X)$ with Nonseparable Dependence on a (or C) and X

$S = S(X),\ D_T = D_T(X)$		a, X Separable		a, X Nonseparable	
Nonsym	Sym	Nonsym	Sym	Nonsym	Sym
$L_h^a \neq 0$		$L_h^a \neq 0$		$L_h^a \neq 0$	
$\Delta L_h^a \neq 0$	$\Delta L_h^a = 0$	$\Delta L_h^a \neq 0$		$\Delta L_h^a \neq 0$	
$\delta L_h \neq 0$	$\delta L_h = 0$	$\delta L_h \neq 0$	$\delta L_h = 0$	$\delta L_h \neq 0$	
$\delta \Delta L_h = 0$		$\delta \Delta L_h = 0$		$\delta \Delta L_h = 0$	
$\delta L_h = -\Delta L_h^a$		$\delta L_h / \delta L_s = $ const.		$\delta L_h / \delta L_s \neq $ const.	
$L_h^a = L_h^{a*}$	N/A	$L_h^a \neq L_h^{a*}$	N/A	$L_h^a \neq L_h^{a*}$	N/A
$\Delta L_h^a = -\Delta L_h^{a*}$	N/A	$\Delta L_h^a \neq -\Delta L_h^{a*}$	N/A	$\Delta L_h^a \neq -\Delta L_h^{a*}$	N/A
$\delta L_h = -\delta L_h^*$	N/A	$\delta L_h = -\delta L_h^*$	N/A	$\delta L_h \neq -\delta L_h^*$	N/A
$q_s = q_s^*$	N/A	$q_s = q_s^*$	N/A	$q_s \neq q_s^*$	N/A
$\delta \Delta L = \delta \Delta L^*$	N/A	$\delta \Delta L = \delta \Delta L^*$	N/A	$\delta \Delta L \neq \delta \Delta L^*$	N/A

Source: Adapted from Petropoulos, J.H. and Roussis, P.P., *J. Chem. Phys.* 50, 3951, 1969b with typographical errors duly corrected.

The X-dependences may be unsymmetrical or symmetrical (at all $a_0 \leq a \leq a_1$) about $X = l/2$. Flow reversal effects are considered only in the former case because they are non-existent in the latter case. Note also that finite non-Fickian time-lag increments may assume zero values coincidentally.

Another important line of future research is suggested by the need to also deal with systems involving strongly adsorbed gases, wherein transport becomes concentration dependent. Comprehensive time-lag analysis schemes applicable to the cases of combined "separable" and "nonseparable" concentration- and X-dependences (see Section 7.2) have been worked out (Petropoulos and Roussis 1967b, 1969b) and are shown in Table 7.3, in comparison with the case of X-dependence for concentration-independent systems considered here. (The theoretical information on time-lag behavior of concentration dependent systems reported by Ash and Espenhahn 1999 and Ash et al. 2000, should also provide useful background in this respect).

Needless to say, a parallel study of transient diffusion kinetics is highly desirable, particularly in view of the fact that different transient D_n's are likely to be affected to different extents by concentration- and by X-dependence. A foretaste of what may be expected in this respect, in practice, is afforded by the example presented by Amarantos et al. (1983).

The study of other than slab geometries is also of interest.

Finally, the experimental demonstration of the applicability and practical utility of the theoretical results remains, of course, a prime research target in all cases.

References

Amarantos, S. G., K. Tsimillis, C. Savvakis, and J. H. Petropoulos. 1983. Kinetic analysis of transient permeation curves. *J. Memb. Sci.* 13:259–272.

Ash, R. 2006. A note on generalised time-lags. *J. Memb. Sci.* 270:196–200.

Ash, R. and S. E. Espenhahn. 1999. Transport through a slab membrane governed by a concentration-dependent diffusion coefficient. Part I. The four time-lags: Some general considerations *J. Memb. Sci.* 154:105–119.

Ash, R., R. W. Baker, and R. M. Barrer. 1968. Sorption and surface flow in graphitized carbon membranes. II. Time lag and blind pore character. *Proc. R. Soc. Lond. A* 304:407–425.

Ash, R., R. M. Barrer, H. T. Chio, and A. V. J. Edge. 1979. Measurements of adsorption for membranes in situ with the use of time-lags and steady-state flows. *Proc. R. Soc. Lond. A* 365:267–281.

Ash, R., S. E. Espenhahn, and D. E. G. Whiting. 2000. Transport through a slab membrane governed by a concentration-dependent diffusion coefficient. Part II. The four time-lags for some particular D(C). *J. Memb. Sci.* 166:281–301.

Barrer, R. M. and T. Gabor. 1959. A comparative structural study of cracking catalyst, porous glass, and carbon plugs by surface and volume flow of gases. *Proc. R. Soc. Lond. A* 251:353–368.

Barrer, R. M. and T. Gabor. 1960. Sorption and diffusion of simple paraffins in silica-alumina cracking catalyst. *Proc. R. Soc. Lond. A* 256:267–290.

Barrer, R. M. and E. Strachan. 1955. Sorption and surface diffusion in microporous carbon cylinders. *Proc. R. Soc. Lond. A* 231:52–74.

Berner, B. and E. R. Cooper. 1983. Asymptotic solution for nonsteady-state diffusion through oil-water multilaminates. *J. Memb. Sci.* 14:139–145.

Choji, N., M. Karasawa, H. Nagasawa, and I. Matsuura. 1978. Diffusion of dye in the composite media of skin/core type. *Sen-i Gakkaishi* 34:T274–281.

Choji, N. and M. Karasawa. 1980. Studies on the diffusion of dye in composite media. III. Penetrant diffusion in repeated laminates. *Sen-i Gakkaishi* 36:T451–458.

Choji, N., I. Matsuura, and M. Karasawa. 1981. Permeation of a penetrant through laminated composites. *Sen-i Gakkaishi* 37:T192–199.

Crank, J. 1975. *The Mathematics of Diffusion, 2nd edn.* Oxford, U.K.: Clarendon Press: (a) pp. 254–261; (b) p. 227; (c) pp. 347–350; (d) pp. 228–230.

Frisch, H. L. 1959. The time lag in diffusion. IV. *J. Phys. Chem.* 63:1249–1252.

Frisch, H. L. 1978a. Diffusion in inhomogeneous films and membranes. *J. Memb. Sci.* 3:149–161.

Frisch, H. L. 1978b. Permeation and sorption in the linear laminated medium. *J. Phys. Chem.* 82:1559–1563.

Frisch, H. L. and J. Bdzil. 1975. Sorption in inhomogeneous slabs. *J. Chem. Phys.* 62:4804–4808.

Frisch, H. L. and S. A. Stern. 1983. Diffusion of small molecules in polymers. *C.R.C. Crit. Rev. Solid State Mater. Sci.* 11:123–187.

Frisch, H. L., G. Forgacs, and S. T. Chui. 1979. Time moment analysis of sorption and permeation in linear laminated media. *J. Phys. Chem.* 83:2787–2792.

Galiatsatou, P., N. K. Kanellopoulos, and J. H. Petropoulos. 2006a. Comprehensive time-lag measurement as a diagnostic and analytical tool for non-Fickian transport studies: A salient porous barrier-gaseous permeant test case. *Phys. Chem. Chem. Phys.* 8:3741–3748.

Galiatsatou, P., N. K. Kanellopoulos, and J. H. Petropoulos. 2006b. Characterization of the transport properties of membranes of uncertain macroscopic structural homogeneity. *J. Memb. Sci.* 280:634–642.

Gavalas, G. R. 2008. Diffusion in microporous membranes: Measurements and modeling. *Ind. Eng. Chem. Res.* 47:5797–5811.

Goetzel, C. G. 1949. *Treatise on Powder Metallurgy: Vol. I – Technology of Metal Powders and Their Products.* New York: Interscience.

Goodknight, R. C. and I. Fatt. 1961. The diffusion time-lag in porous media with dead-end pore volume. *J. Phys. Chem.* 65:1709–1712.

Goodknight, R. C., W. A. Klikoff Jr., and I. Fatt. 1960. Non-steady-state fluid flow and diffusion in porous media containing dead-end pore volume. *J. Phys. Chem.* 64:1162–1168.

Gregg, S. J. and K. S. W. Sing. 1982. *Adsorption, Surface Area and Porosity, 2nd edn.* New York: Academic Press.

Grzywna, Z. J. and J. H. Petropoulos. 1983a. Transient diffusion kinetics in media exhibiting axial variation of diffusion properties. Part 1. – Sorption kinetics. *J. Chem. Soc. Faraday II* 79:571–584.

Grzywna, Z. J. and J. H. Petropoulos. 1983b. Transient diffusion kinetics in media exhibiting axial variation of diffusion properties. Part 2. – Permeation kinetics. *J. Chem. Soc. Faraday II* 79:585–597.

Jena, A. and K. Gupta. 2005. Pore structure characterization techniques. *Am. Ceram. Soc. Bull.* 84:28–30.

Jeon, J. and G. A. Voth. 2008. Gating of the mechanosensitive channel protein MscL: The interplay of membrane and protein. *Biophys. J.* 94:3497–3511.

Kanellopoulos, N. K., J. H. Petropoulos, and D. Nicholson. 1985. Effect of pore structure and macroscopic non-homogeneity on the relative gas permeability of porous solids. *J. Chem. Soc. Faraday Trans. I* 81:1183–1194.

Kishimoto, A. K. 1964. Diffusion and viscosity of polyvinyl acetate-diluent systems. *J. Polym. Sci. A* 2:1421–1439.

Le Poidevin, G. J. 1982. Non-steady-state transport of diffusants in ternary laminate slabs. *J. Appl. Polym. Sci.* 27:2901–2917.

Nicholson, D. and J. H. Petropoulos. 1982. Influence of macroscopic structure on the gas- and surface-phase flow of dilute gases in porous media. *J. Chem. Soc. Faraday Trans. I* 78:3587–3593.

Park, G. S. 1953. An experimental study of the influence of various factors on the time dependent nature of diffusion in polymers. *J. Polym. Sci.* 11:97–115.

Peterlin, A. and J. Olf. 1972. Steady flow through asymmetric membranes. *J. Macromol. Sci. Phys. B* 6:571–582.

Petropoulos, J. H. 1973. Flow reversal effects in a simple laminated membrane diffusion system. *J. Polym. Sci., Polym. Phys. Ed.* 11:1867–1872.

Petropoulos, J. H. 1974. "Directional" membrane permeability in polymer-vapor systems. *J. Polym. Sci., Polym. Phys. Ed.* 12:35–49.

Petropoulos, J. H. 1985. Membranes with non-homogeneous sorption and transport properties. *Adv. Polym. Sci.* 64:93–142.

Petropoulos, J. H. and K. G. Papadokostaki. 2008. Modeling of gas transport properties, and its use for structural characterization, of mesoporous solids. Paper resulting from lecture presented by J.H. Petropoulos at the First International Workshop "Nanoporous Materials in Energy and Environment," Chania, Greece, Oct. 12–15, 2008.

Petropoulos, J. H. and J. K. Petrou. 1992. Possibilities of structural characterization of porous solids by fluid flow methods. *Sep. Technol.* 2:162–175.

Petropoulos, J. H. and P. P. Roussis. 1967a. Study of "non-Fickian" diffusion anomalies through time lags. I. Some time-dependent anomalies. *J. Chem. Phys.* 47:1491–1496.

Petropoulos, J. H. and P. P. Roussis. 1967b. Study of "non-Fickian" diffusion anomalies through time lags. II. Simpler cases of distance- and time-distance-dependent anomalies. *J. Chem. Phys.* 47:1496–1500.

Petropoulos, J. H. and P. P. Roussis. 1968. Study of "non-Fickian" diffusion anomalies through time lags. III. Simple distance-dependent anomalies in microporous media. *J. Chem. Phys.* 48:4619–4624.

Petropoulos, J. H. and P. P. Roussis. 1969a. Study of "non-Fickian" diffusion anomalies through time lags. V. Simple distance-dependent anomalies in laminated media. *J. Chem. Phys.* 51:1332–1336.

Petropoulos, J. H. and P. P. Roussis. 1969b. Study of "non-Fickian" diffusion anomalies through time lags. IV. More complex cases of axial distance-dependent anomalies. *J. Chem. Phys.* 50:3951–3956.

Price, C. A. 1996. *Stone Conservation: An Overview of Current Research*. New York: Getty Conservation Institute, p. 22.

Rajniak, P., M. Soos, and R. T. Yang. 1999. Unified network model for adsorption-desorption in systems with hysteresis. *AIChE J.* 45:735–750.

Rogers, C. E. 1965. Transport through polymer membranes with a gradient of inhomogeneity. *J. Polym. Sci.: Part C* 10:93–102.

Rogers, C. E., V. Stannett, and M. Szwarc. 1957. Permeability valves – Permeability of gases and vapors through composite membranes. *Ind. Eng. Chem.* 49:1933–1936.

Roussis, P. P. and J. H. Petropoulos. 1976. Permeation time lag analysis of "anomalous" diffusion. Part 1. – Some considerations on experimental method. *J. Chem. Soc. Faraday II* 72:737–746.

Roussis, P. P. and J. H. Petropoulos. 1977. Permeation time-lag analysis of "anomalous" diffusion. Part 2. – Helium and nitrogen in graphite powder compacts. *J. Chem. Soc. Faraday II* 73:1025–1033.

Roussis, P. P., K. Tsimillis, C. Savvakis, S. Amarantos, N. Kanellopoulos, and J. H. Petropoulos. 1980. On the measurement of desorption-permeation time lags. *J. Phys. E* 13:403–405.

Rutherford, S. W. and D. D. Do. 1997. Review of time lag permeation technique as a method for characterization of porous media and membranes. *Adsorption* 3:283–312.

Savvakis, C. 1979. A study of diffusion and sorption of permanent gases through microporous graphite substrates in relation to their macroscopic structure. PhD thesis, Athens University, Greece.

Savvakis, C. and J. H. Petropoulos. 1982. Application of the method of time-lag analysis to the study of diffusion in solids of nonhomogeneous macroscopic structure. *J. Phys. Chem.* 86:5128–5133.

Savvakis, C., K. Tsimillis, and J. H. Petropoulos. 1982. Adsorption and diffusion of dilute gases in microporous graphite pellets in relation to their macroscopic structure. *J. Chem. Soc. Faraday Trans. I* 78:3121–3130.

Siegel, R. A. 1986. A Laplace transform technique for calculating diffusion time lags. *J. Memb. Sci.* 270:251–262.

Siegel, R. A. 1995. Comments on time lags and dynamic delays in diffusion/reaction systems. *J. Phys. Chem.* 99:17294–17296.

Siegel, R. A. and E. L. Cussler. 2004. Reactive barrier membranes: Some theoretical observations regarding the time lag and breakthrough curves. *J. Memb. Sci.* 229:33–41.

Spencer, H. G. and J. A. Barrie. 1980. Transient sorption by symmetrical multilaminate slabs in well-stirred semi-infinite and finite baths. *J. Appl. Polym. Sci.* 25:2807–2814.

Spencer, H. G., T. C. Chen, and J. A. Barrie. 1982. Transient diffusion through multilaminate slabs separating finite and semi-infinite baths. *J. Appl. Polym. Sci.* 27:3835–3840.

Tsimillis, K. and J. H. Petropoulos. 1977. Experimental study of a simple anomalous diffusion system by time-lag and transient-state kinetic analysis. *J. Phys. Chem.* 81:2185–2191.

Vásquez, V., M. Sotomayor, J. Cordero-Morales, K. Schulten, and E. Perozo. 2008. A structural mechanism for MscS gating in lipid bilayers. *Science* 29:1210–1214.

Part III

Fundamentals, Recent Advances and Improvements, Membrane, Catalytic and Novel Processes Involving Nanoporous Materials

8

Synthesis Processes of Nanoporous Solids

P. Cool, V. Meynen, and E.F. Vansant

CONTENTS

Since the first discovery of microporous zeolites, the field of porous materials has become a fast growing field, with rapid progress in a wide range of materials. Zeolites are among the best-known examples. However, due to the increasing interest in the processing of large bulky molecules, a lot of research is performed on mesoporous materials. This chapter describes the evolution of porous materials toward a new generation of materials with bimodal porosity. Furthermore, different synthesis routes are described for the formation of materials that combine mesopores with zeolite characteristics.

8.1 Introduction

Porous materials are known for more than a century. The presence of voids of controllable dimensions at the atomic, molecular, and nanometer scale makes them scientifically and technologically important. They are widely applied for various applications in industry [1–7], environmental applications [2,8–12], medicine [13–18], but also in other applications like sensors [19,20], packing materials for high-performance liquid chromatography (HPLC) [21–24], mirrorless lasers [25], food additives [26], etc. In each of these application areas, porous materials often serve different goals depending on their engineered design and synthesis. A large diversity in materials and their properties is developed over the years, studied and evaluated by an interdisciplinary community ranging from chemists and physicists to medical doctors and mathematicians. The International Union of Pure and Applied Chemistry (IUPAC) classified these porous materials into three categories based on their pore size: microporous materials (<2 nm), mesoporous materials ($2 < d < 50$ nm), and macroporous materials (>50 nm). Due to the interest in different application domains of porous materials with divergent properties, specifically those adjusted to their needs, a lot of research is put into the development and controlled modification of these materials. Also the mechanism of their formation is the subject of much research since it allows the control and the understanding of various synthesis parameters. This way, materials can be tailor-made based on the desired and postulated properties. In this chapter is provided a short overview of some of the best-known materials and their evolution into new materials with controlled properties. Moreover, an overview of the variety of synthesis routes for the formation of a new generation of bimodal materials that combine mesopores with zeolite characteristics is also given.

8.2 Evolution toward Materials with Larger Pores

Zeolites are without any doubt the best-known and most applied porous materials. A good deal of literature can be found on the synthesis, modifications, and various properties and applications of zeolites [1,27–34]. The big advantages of zeolites are their uniform pore sizes with molecular dimensions, good stability, selectivity, and activity due to their crystallinity and the presence of incorporated heteroelements in the structure [1,4,5]. Nevertheless, the microporous nature of zeolites limits their use in pharmaceutical, biological, and fine chemical applications. In these processes, large bulky molecules that exceed the size of the pores and cages of the zeolites are employed. Moreover, often liquid-phase systems are applied in these processes, which cause serious mass transfer limitations for microporous materials [4]. Due to this drawback, a lot of research on the synthesis of mesoporous materials was carried out during the last decades. The first attempts were based on the use of larger molecular templates that were used as void fillers in the zeolite synthesis. Unfortunately, this approach was only successful when using Al and P or Ga and P as framework elements while it failed for the synthesis of siliceous materials [5,35,36]. An alternative approach was found by the pillaring of clay minerals, the so-called pillared interlayered clays (PILC) materials. Although they were found as good candidates for heterogeneous catalysis, their pore size distribution is quite broad and is still largely in the microporous range [5]. In 1992, the publication of the M41S family of materials by a group of researchers from Mobil Oil formed the basis of a major breakthrough in the synthesis of mesoporous materials [37,38]. They described the use of supramolecular assemblies of organic molecules for the formation of the mesoporous materials instead of the molecular templates used for the production of zeolites. Since then, various mesoporous materials have gained growing interest and the research has focused on siliceous [4,5,7,39–42], non-siliceous [7], and organo-silica hybrid [43–47] mesoporous materials. Here, only siliceous materials will be discussed, although very interesting work has been done on non-siliceous and organo-silica hybrid materials and their field of application is growing rapidly.

Most of the synthesis approaches to form inorganic mesoporous materials in general are based on the use of organic template molecules that are used in different assembly processes around which the inorganic precursor can condense [7,40,42,48–50] (Figure 8.1). However, template-free synthesis mechanisms like the nanobuilding block mechanism and other approaches are also

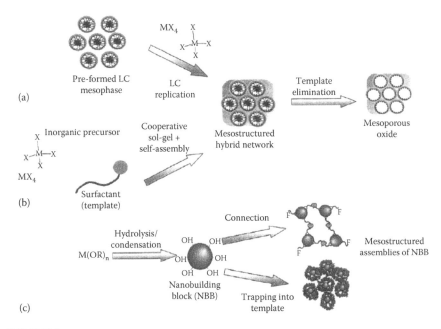

FIGURE 8.1
Schematic overview of the main synthetic approaches to mesostructured materials. (Adapted from Soler-Illia, G.J.A.A. et al., *Chem. Rev.*, 102, 4093, 2002.)

TABLE 8.1

Schematic Overview of the Main Synthesis Parameters to
Generate a Diversity of Mesoporous Materials

Surfactant	Mechanism	Interaction
MOS	LCT	Direct ionic
POS	Self-assembly	Direct non-ionic
Textual templates	Cooperative self-assembly	Indirect ionic
	Nanometric building blocks	Indirect non-ionic

Source: From Alexandridis, P. and Hatton, T.A., *Colloids Surf. A: Physicochem. Eng. Aspects*, 96, 1, 1995.

MOS, molecular-based organized systems; POS, polymeric-based organized systems; LCT, liquid crystal template.

studied [7] (Figure 8.1). A large diversity in synthesis approaches are known not only for the formation of different materials but also for similar materials (e.g., MCM-48) that can be made by different synthesis methods, each of them allowing other parameters to be altered and controlled. For this reason, knowledge of the synthesis methods and parameters that influence the final material will allow pore size engineering and control of the morphology and structure of the obtained material. Basically, the synthesis of mesoporous materials and its control can be limited to the altering of the combination of the chosen surfactant type, the specific synthesis mechanism, and the interaction of the silica source with the template molecules (if present) (see Table 8.1) [7]. For example, M41S is made by an S^+I^- direct interaction between an ionic, positively charged molecular organized systems disulfide (MOS) surfactant and negatively charged silica source in basic environment. Three types of mechanisms—liquid crystal template, self-assembly, and cooperative self-assembly—have been suggested for the synthesis of M41S materials based on the applied synthesis conditions [5,7]. Santa Barbara Acids (SBA) materials, on the other hand, have been made by use of polymeric organized systems (POS) surfactants that interact through an indirect reaction of the template with the positively charged silica source ($S^+X^-I^+$) in acid medium [41,62,70]. A neutral interaction between MOS surfactants and an inorganic source results in the formation of, for example, hexagonal mesoporous silica (HMS) materials (S^0I^0) [7]. Moreover, other parameters such as pH, additives (e.g., salts, swelling agents, cosolvents, and cosurfactants), the concentrations, the silica source, the solvent, temperature, etc. [48,49–52] will allow the fine tuning of the final material due to small changes in the characteristics of the surfactant, the mechanism or the interaction.

A detailed description of the different surfactants, mechanisms, and interactions has been reported in the review papers of Corma [5] and Soler-Illia et al. [7].

A general synthesis for the preparation of templated mesoporous materials can be described as the dissolving of the template molecules in the solvent (with attention for the pH, temperature, additives, cosolvents, etc.) followed by the addition of the silica source (tetraethylorthosilicate [TEOS], metasilicate, aerosil, etc.). After a stirring period at a certain temperature to allow hydrolysis and precondensation, the temperature is increased (sometimes combined with hydrothermal treatment, the addition of additives or changes in the pH) in order to direct the condensation process. In a next step, the products are recovered, washed, and dried. Finally, the template needs to be removed by calcination procedures or extraction methods. The latter is environmentally and economically the preferred procedure since it allows the recovery and recycling of the templates. However, extraction processes are often not 100% complete [53] and cannot be executed for all surfactants and materials. Moreover, in contrast to calcination procedures at high temperatures, they do not result in an extra condensation of the silica framework.

8.3 Combined Micro- and Mesoporous Materials

Combined micro- and mesoporous materials were claimed to have advantages compared to the exclusively micro- or mesoporous materials. They provide improved diffusion rates for transport in catalytic processes (faster reactions) [54], better hydrothermal stability [55], multifunctionality to process a large variety of feedstocks, capabilities of encapsulating waste in the micropores, controlled leaching rates for a constant and gradual release of active components, etc. [6,56]. Different porous materials that combine micro- and mesopores such as SBA-15, plugged hexagonal templated silica (PHTS), mesoporous aluminium silicates (MAS), mesoporous titanium silicates (MTS), UL-ZSM-5, zeotiles, etc. have been developed in the last few years, of which the latter four are mesoporous materials built up from zeolite precursors.

8.3.1 SBA-15 Materials

In 1998, a new family of highly ordered mesoporous silica materials were synthesized in an acid medium by the use of commercially available nonionic triblock copolymers ($EO_nPO_mEO_n$) with large polyethyleneoxide (PEO) $(EO)_n$ and polypropyleneoxide (PPO) $(PO)_m$ blocks [41,57]. Different materials with a diversity of periodic arrangements were prepared and denoted as SBA materials. A wide variety of SBA materials has been reported in literature, such as SBA-1 (cubic) [58,59], SBA-11 (cubic) [41,60], SBA-12 (3D hexagonal network) [41,60], SBA-14 (lamellar) [41], SBA-15 (2D hexagonal) (Figure 8.2) [41,57,60], and SBA-16 (cubic cage-structured) [41,60,61]. SBA-15 immediately attracted a lot of attention because of its desirable features and is now the most studied SBA structure.

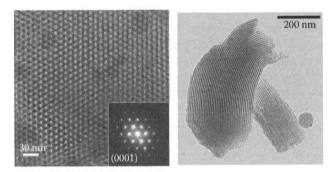

FIGURE 8.2
TEM images and diffraction pattern showing the hexagonal ordering of SBA-15 and the curved pores.

SBA-15 is a combined micro- and mesoporous material with hexagonally ordered tuneable uniform mesopores (4–14 nm) [62,63]. It consists of thick microporous silica pore walls (3–6 nm) responsible for the high hydrothermal stability of SBA-15 compared to other mesoporous materials with thin pore walls like Mobil composition of matter (MCM) and HMS [40,57,64]. Transmission electron microscopy (TEM) investigation of the SBA-15 materials revealed the curved nature of the pores [65,66] (Figure 8.2) although SBA-15 materials with short, straight channels have been reported by the addition of salts during the synthesis [67]. The shape and curvature of the pores was claimed to be of importance for the diffusion of molecules [65,68] through the structure and the ultimate adsorption capacity [68]. The micropores in the walls of the SBA-15 mesopores originate from the PEO blocks in the triblockcopolymer that are directed to the aqueous solution [63,69–72], whereas the PPO blocks are more hydrophobic and give rise to the internal structure of the mesopore [70–73]. A schematic representation of the structure directing assembly of the PEO and PPO blocks in SBA-15 can be seen in Figure 8.3.

By changing the lengths of the PEO blocks, different amounts of micropores and changes in the pore wall thickness could be obtained [62,72,73]. Moreover, the ratio of the number of PEO units to the number of PPO units are responsible for the mesophase (lamellar, hexagonal, cubic, etc.) of the structure [40,72,74]. On the other hand, altering the length of the PPO blocks will result in different pore diameters. Indeed, increased hydrophobic micelle cores formed by larger PPO blocks result in enlarged pore sizes [72]. Furthermore, synthesis parameters like temperature [40,63,72,75–77], pH [78], and the addition of additives such as cosurfactants, swelling agents, electrolytes, salts, etc. [40,52,79–80] will allow pore size engineering and tuning of the general properties and morphologies of SBA-15 to a large extent. A wide diversity of morphologies [52,79–83] have been reported for SBA-15 such as rods, fibers, gyroids, discoidlike, doughnutlike, spheres (micrometer

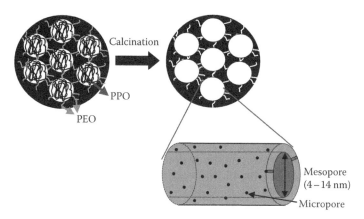

FIGURE 8.3
Schematic representation of SBA-15 before and after calcination.

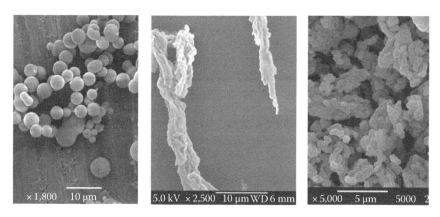

FIGURE 8.4
SEM images showing examples of a few different morphologies of SBA-15 (spheres, fibers, and rods).

and millimeters sized), ropelike, etc. (Figure 8.4). In addition, SBA-15 can be synthesized using low cost silicon sources [60,84–87] and fast synthesis procedures [76,88,89]. Due to these desirable features, SBA-15 has drawn a lot of attention.

Since the development of SBA-15, a lot of research has been performed on the development and modification of materials with a combined micro- and mesoporosity. SBA-15 has been modified with a wide diversity of transition metal oxides (V [90–93], Ti [90,94–97], Al [98–100], etc.) and organic functional groups [101–105] by post-synthesis and in situ processes [4]. This gives the active SBA-15 materials the possibility to be used in catalysis [4,90,92–106,107], controlled release [108,109], removal of heavy metals [110], photoluminescence [111,112], lithium batteries [113], immobilization of enzymes [68,114], etc. Moreover, one of the interesting applications of SBA-15 is their use as templates for the synthesis of (inverse) carbon replicas [6,82,115–118] and nanowires of various metals [119–126].

8.3.2 Plugged Hexagonal Templated Silica: An Analogue to SBA-15

By increasing the silica over surfactant ratio in the synthesis of SBA-15, PHTS is formed [6,56,65,127–129].

PHTS is analogous to a SBA-15 material as it has the same basic characteristics. It consists of hexagonally ordered mesopores with diameters that are similar to those of SBA-15. Moreover, it has thick pore walls (3–6 nm) perforated with micropores making PHTS a combined micro- and mesoporous material [130]. In addition, PHTS possess extramicroporous amorphous nanoparticles (plugs) in the uniform mesoporous channels. The pillaring effect of the nanoparticles gives PHTS a higher mechanical stability compared to the pure SBA-15 [54,56,127]. In addition, PHTS possesses a high hydrothermal stability, which improves when applying high synthesis temperatures and longer synthesis times [56,127,131]. Therefore, PHTS is put forward as a good candidate for industrial applications. Indeed, stability is one of the major factors that hinder prospective applications of mesoporous materials in the petrochemistry [131]. Furthermore, PHTS materials have a characteristic nitrogen sorption isotherm due to the presence of a tailorable amount of both open and narrowed sections induced by the dispersion of nanoparticles in the mesopores (see Figure 8.5 schematic representation) [6,56,127,128,132]. Nitrogen sorption isotherms at −196°C for both SBA-15 and

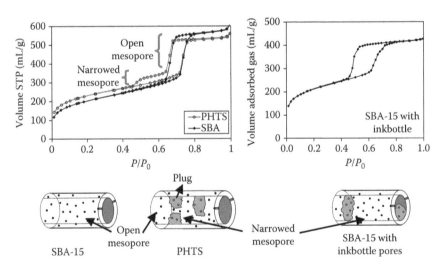

FIGURE 8.5
Nitrogen sorption isotherms at −196°C for SBA-15, PHTS, and SBA-15 with inkbottle pores. Schematic representation of the different pore structures.

PHTS are of type IV according to the IUPAC classification (Figure 8.5). The N_2 sorption isotherms of PHTS show in the adsorption branch a one-step capillary condensation, indicating the filling of the uniform mesopores. However, in contrast to SBA-15, PHTS exhibits a two-step desorption. The first desorption step is similar to desorption in pure SBA-15 and can be assigned to desorption of nitrogen from the open mesopores according to the normal Kelvin model. Desorption occurs at equilibrium conditions via a receding meniscus. The second desorption step can be attributed to the nanoparticles (plugs) within the mesopores, narrowing parts of the mesoporous channels, and creating inkbottle like sections. Therefore, nitrogen that is present between two nanoparticles can only be desorbed through the narrowed pore entrance. For these narrowed pores, desorption is delayed until the vapor pressure is reduced below the stability of the condensed nitrogen phase at the restriction resulting in a second desorption step at lower relative pressures compared to the open mesopores (first desorption step).

Interpretations of the size of the nanoparticles and therefore also the diameter of the pore at the constrictions, should be done with care. Indeed, the relative pressure at P/P_0 0.42 depends weakly on the pore size and pore geometry in nitrogen-sorption measurements due to the lower closure point of the hysteresis (P/P_0 0.42) for nitrogen [6,133–135]. When this lower closure limit is reached, capillary evaporation can no longer be delayed in the narrowed sections. Kruk et al. [6] proved that using argon adsorption–desorption isotherms (lower closure point limit at relative pressure ~0.3), all constrictions in the porous structure of PHTS, narrowing the mesopores, are likely to exhibit diameters above ~4–5 nm. By modification with hexyldimethylsilyl and decyldimethylsilyl, it was estimated that the largest diameter of the constrictions were larger than 2.4 nm but smaller than 3.4 nm [131]. Therefore, for materials with hysteresis loops closing at P/P_0 0.42, argon adsorption–desorption measurements or surface modification techniques should be carried out if information concerning the real size of the constrictions and pore entrances are necessary [6,91,131,134]. The ratio of open to narrowed pores can be tuned from 100% open pores to fully narrowed pores (inkbottle pores with nanoparticles at the pore mouth) by simply adjusting the synthesis parameters [6,127]. In addition, altering the synthesis conditions will allow controlling the size of the constrictions and the stability of the plugs [131]. It was found that the minimum time required to obtain good structural materials was 4 h [128]. Moreover, increasing synthesis temperatures result in increasing pore diameters and enlarged particle sizes. Changes in synthesis temperature result in similar phenomena observed in SBA-15 materials and PHTS. However, in contrast to SBA-15, the micropore volumes in PHTS increase when the synthesis temperature is raised, which was explained by the increase in the microporosity of the plugs [128]. By careful control of the stirring temperature and the amount of TEOS used for the synthesis of PHTS, different morphologies could be formed (Figure 8.6). At low temperatures and TEOS concentrations smooth rods are formed,

FIGURE 8.6
Different morphologies of PHTS synthesized by changes in the stirring temperatures. (From Van Bavel, E. et al., *J. Porous Mater.* 12, 65, 2005.)

whereas at high temperatures spherical morphologies will be obtained. The differences in morphologies were based on the cloud point of the surfactant and the balance between the rate of polymerization of the silica source and the rate of the mesostructure formation [129].

8.3.2.1 Synthesis Mechanism of PHTS Materials

Different mechanisms for the plug formation and the origin of the micropores in the plugs have been proposed. Van Der Voort et al. [56] suggested that the microporosity of the nanoparticles (plugs) are generated from the presence of impurities of lower molecular weights, appreciable amounts of diblock copolymers and even free propylene oxide in the commercial triblock copolymer mixtures [63,136–138]. Some of these components, especially the low molecular weight ones may not be involved in the actual templating of the mesopores. However, they still act as templates for the nanoparticles, inducing the microporosity of the nanoparticles [56]. The PEO chains however create the microporosity in the walls of the mesopores as previously suggested in literature and similar to SBA-15 [56,63]. Kruk et al. [6] proposed alternative mechanisms for the formation of the nanoparticles (plugs) inside the mesopores. The mechanism was based on the reported results that indicate that PHTS type materials are formed after post-synthesis modification of as-synthesized SBA-15 (with the template still present) with TEOS and subsequent calcination [139]. Therefore, Kruk et al. suggested that the formation of microporous nanoparticles in SBA-15 is a result of a secondary process of the reaction of excess TEOS with the preformed templated SBA-15 structure. The excessive amount of TEOS in the synthesis of PHTS that has no accessibility to the template will hydrolyze and/or condense more slowly since they are not in contact with the PEO blocks (the PEO blocks were reported to facilitate the hydrolysis and condensation of TEOS [140,141]). Subsequently, unhydrolyzed or partially hydrolyzed excess TEOS can dissolve in the copolymer micelles of the already formed silica structure.

Finally, in the internal micelle core, microporous silica nanoparticles (plugs) will be formed. Another possible and similar mechanism, proposed by the same authors, was that an excessive amount of TEOS would dissolve immediately in the PPO micelle core, wherein the hydrolysis and condensation of the species is retarded. This TEOS in the micelle core would then slowly undergo hydrolysis and condensation, leading to the formation of nanoparticles (plugs) [6]. The latter mechanism was supported by experimental results of Van Bavel et al. [128]. When the temperature was increased during the synthesis, an increase in the microporosity of the plugs was observed. Indeed, as the temperature increases, the PEO blocks dehydrates and become more hydrophobic. Therefore, these PEO blocks will turn toward the hydrophobic micelle core and into the plugs (nanoparticles) that are forming herein, rendering them more microporous [128]. Based on these proposed mechanisms, Van Der Voort et al. and Kruk et al. suggested that plugs can form at high TEOS/polymer ratios not in the case of SBA-15 only. They proposed that plug formation can be extended to all other synthesis of polymer or oligomer templated silica materials in which TEOS is used in an excessive amount [6,56]. This suggestion was confirmed by the formation of amorphous silica plugs in cubic $Fd3m$ mesoporous silica monoliths with highly controllable porous characteristics [142].

8.3.2.2 Modification of Plugged Materials

PHTS materials intrinsically consist of amorphous silica. However, this type of materials can be made catalytically active by post-synthesis modification methods. These post-synthesis methods can be divided in two main routes. On one hand, PHTS materials can be formed and calcined after which a layer of metal oxides is formed on their surface by use of existing modification methods such as the Molecular Designed Dispersion method (MDD) [91], grafting or impregnation [54]. The increased microporosity and the presence of the nanoparticles in the mesoporous channels were found to affect the dispersion and activity of the catalyst. Moreover, the lower deactivation of the modified PHTS materials in comparison to SBA-15 gave it superior performances in hydrodechlorination reactions [54]. On the other hand, PHTS materials can be made and activated simultaneously. In the latter one, SBA-15 materials are formed and calcined in which nanoparticles of metal oxides or zeolitic nanoparticles are deposited or formed. In this way, these nanoparticles act as plugs and introduce the active element at the same time. Different metal oxides nanoparticles have been introduced as plugs in order to form modified PHTS materials such as titania (anatase), zirconia, and other metal oxides [106,143–147]. The use of metal oxide nanoparticles as plugs in PHTS influences the porous properties of the PHTS materials to a great extent [146–148,154]. In relation to this, the performance in adsorption and catalysis will be altered [146,148]. In addition, zeolitic nanoparticles (vanadium silicalite-1) have also been introduced as plugs in SBA-15, which results in the formation of tuneable PHTS materials with different degrees of plugging (ratio of open to narrowed pores) and microporosity [148,149,154,155]. Moreover, in this way, zeolitic functions are introduced that could be beneficial for several applications. In a third method, an in situ formation of metal oxide nanoparticles (e.g., Ni [150], Nb [151]) in the mesoporous SBA-15 materials during the synthesis has resulted in the formation of PHTS materials with metal oxide nanoparticles as plugs and in the mesoporous walls [150,151]. By applying this method, a good control over the synthesis conditions is essential. Finally, Bao et al. [152] reported the formation of plugged periodic mesoporous organosilicas (PMOs) materials. By changing the SiO_2/surfactant ratio in the formation of ethylene-bridged SBA-15 PMOs, the porous characteristics of these materials could be controlled. This leads to the formation of plugged SBA-15 PMOs with a high microporous volume, thick pore walls and organic functionalities inside the walls and on the pore surface.

The combined micro- and mesoporosity, the high stability and the possibility to control the porous and morphological characteristics of PHTS materials, together with the possibility to create active PHTS type materials in different ways, could be very beneficial in several applications. In addition, PHTS can be made with low cost silicon sources [54]. Moreover, it was found that the nanoparticles (plugs) could influence the adsorption and diffusion of molecules within the mesoporous channels [65,91,148,153–155]. Therefore, PHTS is a promising candidate for processes involving catalysts [54], encapsulation media, controlled release, adsorption, etc.

8.3.3 Mesoporous Materials with Zeolitic Properties

Although mesoporous materials give accessibility to large molecules, make high mass transfers available and show slower deactivation, they still consist of amorphous silica. This makes them inferior to zeolites when selectivity, activity and stability are concerned [156–158]. Moreover, the low pH conditions for the synthesis of some stable mesoporous materials like SBA-15 make it difficult to incorporate heteroelements into the framework during their synthesis [159,160]. This puts some additional concerns in terms of their mass production and commercial applications. For this reason, novel routes are created for the formation of a new generation of combined micro- and mesoporous materials that exhibit the good qualities of zeolites as well as mesoporous materials. These materials have large pore sizes that warrant good accessibility for large molecules and high mass transfers similar to mesoporous materials. At the same time, they display zeolitic characteristics through the presence of zeolite building units in the framework in order to obtain high selectivities, activities and stabilities in combination with well incorporated heteroelements. A lot of different approaches for the synthesis of this new generation of bimodal porous materials have been described in literature. The most recent advances in the development of hierarchical zeolites are very well summarized in the

TABLE 8.2

Overview of the Different Existing Synthesis Routes for the Formation of
Mesoporous Materials with Zeolite Character

Method	Mesopore Formation		Zeolite Formation	References
Dealumination/ desilication	Steaming/acid leaching Alkaline treatment		Full-grown zeolite	[5,164–166] [167,198]
Recrystallization	Mesoporous amorphous silica structures	With template With carbon With carbonized template	TPAOH impregnation and temperature increase	[168,169] [168,170] [168]
Carbon based	Mesoporous carbon or inversed carbon replicas		Zeolite precursors	[159,171–178,198]
Deposition	Large pore silica structures		Zeolite precursors	[148,149,154, 155,157,179–181]
Templating approach	Meso-template and micro-template		Zeolite precursors	[156,159,166, 182–195]
	No meso-template, only micro-template TPAOH			[199–203]

review paper of J. Pérez-Ramirez [198]. An overview of the most important synthesis routes is given in Table 8.2.

The basic principles of these different methods (Table 8.2) can be described as follows.

8.3.3.1 Dealumination/Desilication

Dealumination of zeolites with low Al/Si ratios was one of the first techniques applied for the formation of mesopores (4–40 nm) [165] in zeolites in order to improve mass transfer, enhance accessibility and lower coke formation. Dealumination is a post-synthesis treatment based on the extraction of aluminum from the zeolite framework, thus causing a partial breakdown of the framework. The aluminum is extracted by acid leaching or by hydrothermal treatment (steaming), thus creating, mesopores in the zeolite crystals. A schematic representation of the method is given in Figure 8.7.

Depending on the method applied, small or high amounts of Al are leached from the structure. Removal of only small amounts of aluminum results in the preservation of most of the microporosity and activity of the material while the mesoporosity and accessibility remains low [5,164]. Severe aluminum leaching results in increased mesoporosity, accessibility, and conversion. However, the microporosity is destroyed to some extent leading to a decreased activity [5,164]. Other problems occur when regeneration of the catalysts is necessary at high temperatures. Regeneration at high temperature (e.g., in fluid catalytic cracking) of the dealuminated zeolites tends to alter the porosity of these catalysts in an uncontrolled way [5]. Moreover, it

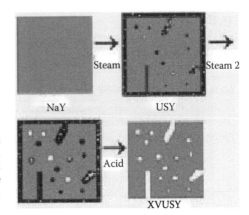

FIGURE 8.7
Model for the generation of mesopores in zeolite Y by steaming and acid leaching. The zeolite is gray, amorphous alumina is black, and the empty mesopores are white. (From Janssen, A.H. et al., *Angew. Chem. Int. Ed.*, 40, 1102, 2001.)

was found that the mesopores formed by dealumination of the zeolites are mostly cavities that are not interconnected to form a mesoporous network responsible for leading reactants to the internal and external of the zeolite [161,165]. Therefore, the diffusion of guest molecules in dealuminated zeolites is only enhanced to a small extent [161]. These disadvantages indicate the importance of the use of other methods that are based on the formation of zeolitic characteristics in mesoporous materials established from zeolite templates or zeolite precursors.

The desilication process is a reproducible methodology to obtain mesoporous ZSM-5 with preserved structural integrity [167]. A post-synthesis alkaline treatment responsible for the removal of silica from the structure will result in the formation of mesopores. A successive combination of posttreatments of desilication followed by dealumination enables a decoupled modification of the mesoporous and acidic properties [167]. In order to obtain the optimal presence of both micro- and mesoporosity, the Si/Al ratio should range between 25 and 50 after alkaline treatment. Surface areas up to $250 \, m^2/g$ with pores of around 10 nm can be obtained in this Si/Al range [167].

8.3.3.2 Recrystallization

The recrystallization of mesoporous materials with amorphous silica walls involves a templated solid-state secondary recrystallization into zeolites of mesoporous materials with corresponding elemental composition. Pure silica or silica alumina mesoporous materials are impregnated with tetra-propylammoniumhydroxide (TPAOH) or any other molecular template for the formation of zeolites. Subsequently, the impregnated solid materials are heated and/or subjected to hydrothermal treatments in an autoclave. For this method it is important that the walls of the amorphous mesoporous precursor materials are as thick as possible in order to obtain good quality materials [169]. These materials are denoted as semizeolitic materials

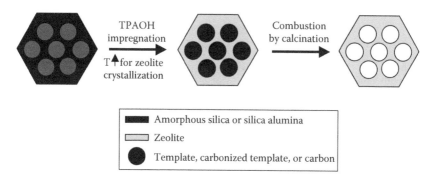

FIGURE 8.8
Schematic representation of the formation of mesoporous materials with zeolite character by recrystallization of amorphous mesoporous precursors. (From Wang, J. and Coppens, M.-O. *First International Workshop Inside Pores, European Network of Excellence*, March 19–23, 2006, La Grande Motte, France.)

(UL-zeolites) [162,169]. The degree of crystallinity can be altered by changes in the synthesis conditions [162]. The acid properties were estimated to be between that of the amorphous precursor and the reference zeolite structure [162]. In order to prevent collapse of the mesopores, the template [168,169] or any other filling agent like carbon [168,170] can be kept or synthesized in the mesopores of the amorphous mesoporous precursor material (Figure 8.8). The recrystallization process needs to be carefully controlled in order to prevent the formation of a separate zeolite phase [168].

8.3.3.3 Carbon-Based Methods

Generally, the carbon-based methods start with the formation of a mesoporous carbon matrix. Subsequently, the carbon matrices are impregnated with a zeolite precursor solution. After impregnation, the hybrid carbon-zeolitic nanoparticles phase is heated hydrothermally in order to initiate zeolite growth. Finally, the obtained material is calcined in order to remove the carbon template resulting in the formation of mesopores inside of the zeolite crystals. Different carbon sources have been applied as the template structure: carbon black aggregates [171,176], carbon nanotubes [176], carbon fibers [175,177,178], carbon aerogel [172], carbon molecular sieves [159,172] (denoted replicated [mesoporous materials RMM] after zeolitization), etc. Each of the applied carbon sources allow formation of different mesoporous materials with various zeolites and zeolite sizes in the mesoporous framework ranging from zeolite building units [159] to zeolite crystals of ~15 µm in size [177].

8.3.3.4 Deposition

This approach deals with the deposition of a clear solution containing zeolitic precursor particles in large pore mesoporous silica materials.

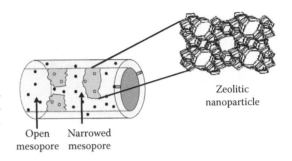

FIGURE 8.9
SBA-15 with vanadium silicalite-1 zeolite-like nanoparticles impregnated in the mesopores. A tuneable plugged material is formed. (From Meynen, V. et al., *Chem. Commun. 7*, 898, 2004.)

Open mesopore   Narrowed mesopore

Zeolitic nanoparticle

Trong-On et al. described the formation of a zeolite nanoparticles coating within siliceous mesostructured cellular foams (MCF materials) [157,179–181]. Another approach is the impregnation of zeolitic precursor solutions within SBA-15 materials resulting in the formation of tuneable PHTS type materials with zeolitic nanoparticles that form the plugs (Figure 8.9) [148,149,154,155]. This way, the micropore volume and other porous characteristics as well as the amount of heteroelements in the structure can be varied in a wide range.

8.3.3.5 Templating Approach

This method is based on the general methods for the formation of mesoporous materials [5,7]. The difference with the original mesoporous materials with walls of amorphous silica lies in the use of zeolitic precursor solutions as silica source. These zeolitic nanoparticles condense around the supramolecular assemblies instead of classically used organosilica or amorphous silica sources. A diversity in synthesis methods have been reported as summarized in Table 8.2, which differ in the utilized templates, zeolitic precursors (size of the zeolitic nanoparticles, type of zeolite, etc.), and synthesis conditions (hydrothermal or reflux, acid or basic, synthesis time, etc.). The formation of a micro- and mesoporous material is also possible in the absence of meso-templates, using only the micro-template (TPAOH) to form the zeolite nanoparticles solution. This type of structure has been reported for the first time by Stevens et al. [199] on the synthesis of acidified hydrothermal assemblies of silicalite-1 precursor nanoparticles. The porous characteristics of the so-called mesosil materials can be controlled in the pore size range of 5–15 nm and originates from the interparticle voids between the nanoparticles. The real framework mesoporosity present in ordered materials such as SBA and MCM is absent here. However, for most catalytic applications, the typical ordered mesopores are not necessary. The nanoparticles in the precursor solution will not grow into a pure crystalline silicalite-1 phase due to the acidification of the clear nanoparticle solution. Upon acidification, one propyl ligand of the TPAOH is lost; therefore, the further growth of the nanoparticles into a full-grown zeolite becomes disabled [200]. Other related types of hierarchically structured mesoporous zeolitic materials are the TUD-M (Technical University Delft) [201] and TUD-C [202,203] materials, where M stands for meso-structure and C stands for crystalline, emphasizing the structure that is being formed. Also here the concept is that a single additive (TPAOH) acts as a template for the formation of zeolite ZSM-5 micropores, and as a scaffold for the generation of the mesopores. The synthesis involves a two-step procedure involving gelation and crystallization. By applying different conditions in the two synthesis steps, TPAOH directs the formation of ordered zeolite micropores and disordered mesopores separately. Compared with TUD-M, TUD-C is more hydrothermally stable and more acid and the mesopore size is more easily tuned. The method to synthesize TUD-M can be applied to a wider range of zeolite types.

For all the reported materials, the nature of the mesoporous framework of these solid materials remains still largely unknown and basic understanding of the assembling process of the zeolitic precursors and surfactants as well as a straightforward understanding of the properties of the zeolitic particles itself has not yet been fully achieved [195].

8.4 Conclusion

The new evolution toward materials with bimodal porosity with zeolitic character built in by design makes the study of zeolite formation more important. In order to fully control the final materials that need to be made, knowledge of the growth mechanism and behavior of the zeolite nuclei in different environments will be of utmost importance. At the same time, the formation mechanism of the mesophase needs to be fully understood. The influence of changes in porous characteristics on the final properties in adsorption, catalysis, diffusion, etc. will allow the design of new, adjusted, or improved mesoporous materials that can be implemented in industrial processes of various kinds.

Acknowledgments

The authors are grateful to the EU (NoE Inside Pores) and the University of Antwerpen (GOA) for financial support. V. Meynen would like to thank the FWO-Flanders (Fund for Scientific Research-Flanders) for her research grant.

References

1. H. Van Bekkum, E.M. Flanigen, P.A. Jacobs, and J.C. Jansen. *Introduction to Zeolite Science and Practice*, 2nd edn. Elsevier, New York, 2001. *Stud. Surf. Sci. Catal.* 137 (2001) 747.
2. G.Q. Lu and X.S. Zhao. *Nanoporous Materials: Science and Engineering*. Series on Chemical Engineering, World Scientific, Vol. 4, 2004.
3. G. Ertl, H. Knözinger, and J. Weitkamp. *Handbook of Heterogeneous Catalysis*. VCH Weinheim, Germany, 1997.
4. A. Taguchi and F. Schüth. *Microporous Mesoporous Mater.* 77 (2005) 1.
5. A. Corma. *Chem. Rev.* 97 (1997) 2373.
6. M. Kruk, M. Jarnoniec, S.H. Joo, and R. Ryoo. *J. Phys. Chem. B* 107 (2003) 2205.
7. G.J.A.A. Soler-Illia, C. Sanchez, B. Lebeau, and J. Patarin. *Chem. Rev.* 102 (2002) 4093.
8. A.M. Liu, K. Hidajat, S. Kawi, and D.Y. Zhao. *Chem. Commun.* 13 (2000) 1145.
9. M.J. Hudson, J.P. Knowles, P.J.F. Harris, D.B. Jackson, M.J. Chinn, and J.L. Ward. *Microporous Mesoporous Mater.* 75 (2004) 121.
10. Z. Yan, S. Tao, J. Yin, and G. Li. *J. Mater. Chem.* 16 (2006) 2347.
11. T. Kang, Y. Park, K. Choi, J.S. Lee, and J. Yi. *J. Mater. Chem.* 14 (2004) 1043.
12. D. Pérez-Quintanilla, I. Del Hierro, M. Fajardo, and I. Sierra. *J. Hazard. Mater.* 134 (2006) 245.
13. C. Charnay, S. Bégu, C. Tourné-Péteilh, L. Nicole, D.A. Lerner, and J.M. Devoiselle. *Eur. J. Pharm. Biopharm.* 57 (2004) 533.
14. L.P. Mercuri, L.V. Carvalho, F.A. Lima, C. Quayle, M.C.A. Fantini, G.S. Tanaka, W.H. Cabrera et al. *Small* 2(2) (2006) 254.
15. F. Qu, G. Zhu, S. Huang, S. Li, J. Sun, D. Zhang, and S. Qiu. *Microporous Mesoporous Mater.* 92 (2006) 1–9.
16. J.C. Doadrio, E.M.B. Sousa, I. Izquierdo-Barba, A.L. Doadrio, J. Perez-Pariente, and M. Vallet-Regi. *J. Mater. Chem.* 16 (2006) 462.
17. A.L. Doadrio, E.M.B. Sousa, J.C. Doadrio, J. Pérez Pariente, I. Izquierdo-Barba, and M. Vallet-Regi. *J. Control. Release* 97 (2004) 125.
18. V. Wernert, O. Schäf, H. Ghobardar, and R. Denoyel. *Microporous Mesoporous Mater.* 83 (2005) 101.
19. H.S. Zhou, T. Yamada, K. Aisi, I. Honma, H. Uchida, and T. Katsube. *Stud. Surf. Sci. Catal.* 141 (2002) 623.
20. G. Wirnsberger and G.D. Stucky. *Chem. Mater.* 12 (2000) 2525.
21. J. Zhao, F. Gao, Y. Fu, W. Jin, P. Yang, and D. Zaho. *Chem. Commun.* 7 (2002) 752.
22. K.J. Nakanishi. *J. Porous Mater.* 4 (1997) 67.
23. M.U. Martines, E. Yeong, M. Persoin, A. Larbot, W.F. Voorhout, C.K.U. Kübel, P. Kooyman et al. *Chimie* 8 (2005) 627.
24. R. Tian, J. Sun, H. Zhang, M. Ye, C. Xie, J. Dong, J. Hu et al. *Electrophoresis* 27 (2006) 742.
25. P. Yang, G. Wirnsberger, H.C. Huang, S.R. Cordereo, M.D. McGehee, B. Scott, T. Deng et al. *Science* 287 (2000) 465.
26. D. Papaioannou, P.D. Katsoulos, N. Panousis, and H. Karatzias. *Microporous Mesoporous Mater.* 84 (2005) 161.
27. A.W. Burton, S.I. Zones, and S. Elomari. *Curr. Opin. Colloid Interface Sci.* 10 (2005) 211.

28. E.E. McLeary, J.C. Jansen, and F. Kapteijn. *Microporous Mesoporous Mater.* 90 (2006) 198.
29. J.D. Epping and B.F. Chmelka. *Curr. Opin. Colloid Interface Sci.* 11 (2006) 81.
30. D.W. Breck. *Zeolite Molecular Sieves, Structure, Chemistry and Use.* John Wiley & Sons, New York, 1974.
31. H. Rogson and K.P. Lillerud. *Verified Synthesis of Zeolitic Materials*, 2nd edn. Elsevier, Amsterdam,The Netherlands, 2001.
32. Ch. Baerlcoher, W.M. Meier, and D.O. Olson. *Atlas of Zeolite Framework Types*, 5th edn. Elsevier, London, 2001.
33. R.M. Barrer. *Hydrothermal Synthesis of Zeolites.* Academic Press, London, U.K., 1982.
34. L.V.C. Rees. Zeolites and related microporous materials. *Stud. Surf. Sci. Catal. A–C* 84 (1994) 1133.
35. Q. Huo, R. Xu, S. Li, Z. Ma, J.M. Thomas, R.H. Jones, and A.M. Chippindale. *J. Chem. Soc. Chem. Commun.* 12 (1992) 875.
36. M. Esterman, L.B. McCuster, Ch. Baerlocher, A. Merrouche, and H. Kessler. *Nature* 352 (1991) 320.
37. J.S. Beck, C.T.-W. Chu, I.E. Johnson, C.T. Kresge, M.E. Leonowics, W.J. Roth, and J.W. Vartuli. WO Patent 91/11390, 1991.
38. C.T. Kresge, M.E. Leonowics, Q.J. Roth, J.C. Vartuli, and J.S. Beck. *Nature* 359 (1992) 710.
39. P. Selvam, S.K. Bhatia, and C.G. Sonwane. *Ind. Eng. Chem. Res.* 40 (2001) 3237.
40. P. Cool, T. Linssen, K. Cassiers, and E.F. Vansant. *Recent Res. Dev. Mater. Sci.* 3 (2002) 871.
41. D. Zhao, Q. Huo, J. Feng, B.F. Chmelka, and G.D. Stucky. *J. Am. Chem. Soc.* 120 (1998) 6024.
42. T. Linssen, K. Cassiers, P. Cool, and E.F. Vansant. *Adv. Colloid. Interface Sci.* 103 (2003) 121.
43. A. Stein, B.J. Melde, and R.C. Schroden. *Adv. Mater.* 12(19) (2000) 1403.
44. F. Hoffmann, M. Cornelius, J. Morell, and M. Fröba. *Angew. Chem. Int. Ed.* 45 (2006) 3216.
45. A. Vinu, K.Z. Hossain, and K. Ariga. *J. Nanosci. Nanotechnol.* 5(3) (2005) 347.
46. M. Choi, F. Kleitz, D. Liu, H.Y. Lee, W.-S. Ahn, and R. Ryoo. *J. Am. Chem. Soc.* 127 (2005) 1924.
47. P. Krawiec, C. Weidenthaler, and S. Kaskel. *Chem. Mater.* 16 (2004) 2869.
48. G.D. Stucky, D. Zhao, P. Yang, W. Lukens, N. Melosh, and B.F. Chmelka. *Stud. Surf. Sci. Catal.* 117 (1998) 1. In: L. Bonneviot, F. Béland, C. Danumah, S. Giasson, S. Kaliaguine (eds). *Mesoporous Molecular Sieves.* Elsevier, Amsterdam, The Netherlands, 1998.
49. A.E.C. Palmqvist. *Curr. Opin. Colloid Interface Sci.* 8 (2003) 145.
50. J. Patarin, B. Lebeau, and R. Zana. *Curr. Opin. Colloid Interface Sci.* 7 (2002) 107.
51. J.M. Kim, Y.-J. Han, B.F. Chmelka, and G.D. Stucky. *Chem. Commun.* 24 (2000) 2437.
52. D. Zhao, J. Sun, Q. Li, and G.D. Stucky. *Chem. Mater.* 12 (2000) 275.
53. D. Huang, G.S. Lou, and Y.J. Wang. *Microporous Mesoporous Mater.* 84 (2005) 27.
54. J. Lee, Y. Park, P. Kim, H. Kim, and J. Yi. *J. Mater. Chem.* 14 (2004) 1050.
55. A. Karlsson, M. Stöcker, and R. Schmidt. *Microporous Mesoporous Mater.* 27 (1999) 181.
56. P. Van Der Voort, P.I. Ravikovitch, K.P. De Jong, M. Benjelloun, E. Van Bavel, A.H. Janssen, A.V. Neimark et al. *J. Phys. Chem. B* 106 (2002) 5873.

57. D. Zhao, J. Feng, Q. Huo, N. Melosh, G.H. Frederickson, B.F. Chmelka, and G.D. Stucky. *Science* 279 (1998) 548.
58. H.-M. Kao, J.-D. Wu, C.-C. Cheng, and A.S.T. Chiang. *Microporous Mesoporous Mater.* 88 (2006) 319.
59. M.J. Kim and R. Ryoo. *Chem. Mater.* 11 (1998) 487.
60. J.M. Kim and G.D. Stucky. *Chem. Commun.* 13 (2000) 1159.
61. P. Van Der Voort, M. Benjelloun, and E.F. Vansant. *J. Phys. Chem. B* 106 (2002) 9027.
62. Y. Bennadja, P. Beaunier, D. Margolese, and A. Davidson. *Microporous Mesoporous Mater.* 44–45 (2001) 147.
63. M. Kruk, M. Jaroniec, C.H. Ko, and R. Ryoo. *Chem. Mater.* 12 (2000) 1961.
64. K. Cassiers, T. Linssen, M. Mathieu, M. Benjelloun, K. Schrijnemakers, P. Van Der Voort, P. Cool et al. *Chem. Mater.* 14 (2002) 2317.
65. A.H. Janssen, P. Van Der Voort, A.J. Koster, and K.P. De Jong. *Chem. Commun.* 15 (2002) 1632.
66. P. Schmidt-Winkel, P. Yang, D.I. Largolese, B.F. Chmelka, and G.D. Stucky. *Adv. Mater.* 11(4) (1999) 303.
67. C. Yu, J. Fan, B. Tain, D. Zhao, and G.D. Stucky. *Adv. Mater.* 14 (2002) 1742.
68. J. Fan, J. Lei, L. Wang, C. Yu, B. Tu, and D. Zhao. *Chem. Commun.* 17 (2003) 2140.
69. R. Ryoo, C.H. Ko, M. Kruk, V. Antochshuk, and M.J. Jaroniec. *J. Phys. Chem. B* 104 (2000) 11465.
70. S. Ruthstein, V. Frydman, S. Kababya, M. Landau, and D. Goldfarb. *J. Phys. Chem. B* 107 (2003) 1739.
71. Y. Zheng, Y.-Y. Won, F.S. Bates, G.T. Davis, L.E. Scriven, and Y. Talmon. *J. Phys. Chem. B* 103 (1999) 10331.
72. P. Kipkemboi, A. Fodgen, V. Alfredsson, and K. Flodström. *Langmuir* 17 (2001) 5398.
73. M. Impéror-Clerc, P. Davidson, and A. Davidson. *J. Am. Chem. Soc.* 122 (2000) 11925.
74. J.M. Kim, Y. Sakamoto, Y.K. Hwang, Y.-U. Kwon, O. Terasaki, S.-E. Park, and G.D. Stucky. *J. Phys. Chem. B* 106 (2002) 2552.
75. P. Yang, D. Zhao, D. Margolese, B. Chmelka, and G.D. Stucky. *Nature* 396 (1998) 152.
76. P.F. Fulvio, S. Pikus, and M. Jaroniec. *J. Mater. Chem.* 15 (2005) 5049.
77. P. Feng, X. Bu, and D. Pine. *Langmuir* 16 (2000) 5304.
78. X. Cui, W.-C. Zhin, W.-J. Cho, and C.-S. Ha. *Mater. Lett.* 59(18) (2005) 2257.
79. H. Zhang, J. Sun, D. Ma, X. Bao, A. Klein-Hoffmann, G. Weinberg, D. Su et al. *J. Am. Chem. Soc.* 126 (2004) 7440.
80. W.-H. Zhang, L. Zhang, J. Xiu, Z. Shen, Y. Li, P. Ying, and C. Li. *Microporous Mesoporous Mater.* 89 (2006) 179.
81. L. Wang, T. Qi, Y. Zhang, and J. Chu. *Microporous Mesoporous Mater.* 91 (2006) 156.
82. C. Yu, J. Fan, B. Tian, D. Zhao, and G.D. Stucky. *Chem. Mater.* 14 (2002) 1742.
83. W. Stevens, K. Lebeau, M. Mertens, G. Van Tendeloo, P. Cool, and E. Vansant, *J. Phys. Chem. B* 110 (2006) 9183.
84. S.S. Kim, T.R. Pauly, and T.J. Pinnavaia. *Chem. Commun.* 17 (2000) 1661.
85. S.H. Joo, R. Ryoo, M. Kruk, and M. Jaroniec. *J. Phys. Chem. B* 106 (2002) 4640.
86. P.F. Fulvio, S. Pikus, and M. Jaroniec. *J. Colloid Interface Sci.* 287(2) (2005) 717.
87. M. Choi, W. Heo, F. Kleitz, and R. Ryoo. *Chem. Commun.* 12 (2003) 1340.
88. B.L. Newalkar, S. Komarneni, and H. Katsuki. *Chem. Commun.* 23 (2000) 2389.
89. B.L. Newalkar and S. Komarneni. *Chem. Commun.* 16 (2002) 1774.

90. Y. Segura, P. Cool, P. Kustrowski, L. Chmielarz, R. Dziembaj, and E.F. Vansant. *J. Phys. Chem. B* 109 (2005) 12071.
91. V. Meynen, Y. Segura, M. Mertens, P. Cool, and E.F. Vansant. *Microporous Mesoporous Mater.* 85(1–2) (2005) 119.
92. Y.-M. Liu, Y. Cao, N. Yi, W.-L. Feng, W.-L. Dai, S.-R. Yan, H.Y. He et al. *J. Catal.* 224 (2004) 417.
93. K. Zhu, Z. Ma, Y. Zou, W. Zhou, T. Chen, and H. He. *Chem. Commun.* 24 (2001) 2552.
94. M.S. Morey, S. O'Brien, S. Schwarz, and G.D. Stucky. *Chem. Mater.* 12 (2000) 898.
95. F. Chiker, J.Ph. Nogier, F. Launay, and J.L. Bonardet. *J. Appl. Catal. A: Gen.* 240(2) (2003) 309.
96. A. Tuel and L.G. Hubert-Pfalzgraf. *J. Catal.* 217 (2003) 343.
97. G. Galleja, R. van Grieken, R. Garcia, J.A. Melero, and J. Iglesias. *J. Mol. Catal. A: Chem.* 182–183 (2002) 215.
98. S. Sumiya, Y. Oumi, T. Uozumi, and T. Sano. *J. Mater. Chem.* 11 (2001) 1111.
99. Y. Yue, A. Gédéon, J.-L. Bonardet, N. Melosh, J.-B. D'Espinose, and J. Fraissard. *Chem. Commun.* 19 (1999) 1967.
100. Z. Luan, M. Hartmann, D. Zhao, W. Zhou, and L. Kevan. *Chem. Mater.* 11 (1999) 1621.
101. C.-M. Yang, Y.Q. Wang, B. Zibrowius, and F. Schüth. *Phys. Chem. Chem. Phys.* 6(9) (2004) 2461.
102. Y. Wang, B. Zibrowius, C.-M. Yang, B. Spliethoff, and F. Schüth. *Chem. Commun.* 1 (2004) 46.
103. D. Margolese, J.A. Melero, S.C. Christiansen, B.F. Chmelka, and G.D. Stucky. *Chem. Mater.* 12 (2000) 2448.
104. B.-G. Park, W. Guo, X. Cui, J. Park, and C.-S. Ha. *Microporous Mesoporous Mater.* 66 (2003) 229.
105. M.A. Markowitz, J. Klaehn, R.A. Hendel, S.B. Qadriq, S.L. Golledge, D.G. Castner, and B.P. Gaber. *J. Phys. Chem. B* 104 (2000) 10820.
106. M.V. Landau, L. Vradman, A. Wolfson, P.M. Rao, and M. Herskowitz. *C.R. Chim.* 8(3–4) (2005) 679.
107. M.S. Kumar, J. Pérez-Ramirez, M.N. Debbagh, B. Smarsly, U. Bentrup, and A. Brüker. *Appl. Catal. B: Environ.* 62 (2006) 244.
108. F. Balas, M. Mazano, P. Horcajada, and M. Vallet-Regi. *J. Am. Chem. Soc.* 128 (2006) 8116.
109. F. Qu, G. Zhu, S. Huang, S. Li, J. Sun, D. Zhang, and S. Qiu. *Microporous Mesoporous Mater.* 92 (2006) 1.
110. A.M. Liu, K. Hidajat, S. Kawi, and D.Y. Zhao. *Chem. Commun.* 13 (2000) 1145.
111. Q. Jiang, Z.Y. Wu, Y.M. Wang, Y. Cao, C.F. Zhou, and J.H. Zhu. *J. Mater. Chem.* 16 (2006) 1536.
112. Z.C. Liu, H.R. Chen, W.M. Huang, J.L. Gu, W.B. Bu, Z.L. Hu, and, J.L. Shi, *Microporous Mesoporous Mater.* 89 (2006) 270.
113. J. Xi, X. Qiu, X. Ma, M. Cui, J. Yang, X. Tang, W. Zhu, et al. *Solid State Ionics* 176(13–14) (2005) 1249.
114. M. Miyahara, A. Vinu, K.Z. Hossain, T. Nakanishi, and K. Ariga. *Thin Solid Films* 499 (2006) 13.
115. Z. Li, J. Zhang, Y. Li, Y. Guan, Z. Feng, and C. Li. *J. Mater. Chem.* 16 (2006) 1350.
116. H.J. Shin, R. Ryoo, M. Kruk, and M. Jaroniec. *Chem. Commun.* 4 (2001) 349.
117. T.-W. Kim, R. Ryoo, K.P. Gierszal, M. Jaroniec, L.A. Solovyov, Y. Sakamoto, and O. Terasaki. *J. Mater. Chem.* 15 (2005) 1560.

118. P. Dibandjo, F. Chassagneux, L. Bois, C. Sigala, and P. Miele. *J. Mater. Chem.* 15 (2005) 1917.

119. Z. Liu, O. Terasaki, T. Ohsuna, K. Hiraga, H.J. Shin, and R. Ryoo. *Chem. Phys. Chem.* 4 (2001) 229.

120. X. Ding, G. Briggs, W. Zhou, Q. Chen, and L.-M. Peng. *Nanotechnology* 17 (2006) 376.

121. L. Gai, Z. Chen, H. Jiang, Y. Tian, Q. Wang, and D. Cui. *J. Cryst. Growth* 291 (2006) 527.

122. F. Gao, Q. Lu, X. Liu, Y. Yan, and D. Zhao. *Nano Lett.* 1(12) (2001) 743.

123. Y.J. Han, J.M. Kim, and G.D. Stucky. *Chem. Mater.* 12 (2000) 2068.

124. S.H. Joo, S.J. Choi, I. Oh, J. Kwak, Z. Liu, O. Terasaki, and R. Ryoo. *Nature* 412 (2001) 169.

125. J. Gu, J. Shi, L. Xiong, H. Chen, L. Li, and M. Ruan. *Solid State Sci.* 6 (2004) 747.

126. M.H. Huang, A. Choudrey, and P. Yang. *Chem. Commun.* 12 (2000) 1063.

127. P. Van Der Voort, P.I. Ravikovitc, K.P. De Jong, A.V. Neimark, A.H. Janssen, M. Benjelloun, E. Van Bavel et al. *Chem. Commun.* 9 (2002) 1010.

128. E. Van Bavel, P. Cool, K. Aerts, and E.F. Vansant. *J. Phys. Chem. B* 108 (2004) 5263.

129. E. Van Bavel, P. Cool, K. Aerts, and E.F. Vansant. *J. Porous Mater.* 12 (2005) 65.

130. C. Herdes, M.A. Santos, S. Abelló, F. Medina, and L.F. Vega. *Appl. Surf. Sci.* 252(3) (2005) 538.

131. E.B. Celer, M. Kruk, Y. Zuzek, and M. Jaroniec. *J. Mater. Chem.* 16 (2006) 2824.

132. J. Sauer, S. Kaskel, M. Janicke, and F. Schüth. *Stud. Surf. Sci. Cata.* 135 (2001) 315. In: A. Galarneau, F. Di Renzo, F. Fajula, J. Vedrine, J. Vedrine (eds). *Zeolites and Mesoporus Materials at the Dawn of the 21st Century.* Elsevier, Amsterdam, The Netherlands, 2001.

133. P.I. Ravikovitch and A.V. Neimark. *Langmuir* 18 (2002) 1550.

134. J.C. Groen, L.A.A. Peffer, and J. Perez-Ramirez. *Microporous Mesoporous Mater.* 60 (2003) 1.

135. F. Roquerol, J. Roquerol, and K. Sing. *Adsorption in Powders and Porous Solids: Principles, Methodology and Applications.* Academic Press, London, U.K., 1999.

136. M. Almgren, W. Brown, and S. Hvidt. *Colloid Polym. Sci.* 273 (1995) 2.

137. P. Alexandridis and T.A. Hatton. *Colloids Surf. A* 96 (1995) 1.

138. G. Wanka, H. Hoffman, and W. Ulbricht. *Macromolecules* 27 (1994) 4145.

139. V. Antochshuk, M. Jaroniec, S.H. Joo, and R. Ryoo. *Stud. Surf. Sci. Catal.* 141 (2002) 607.

140. C. Boissiere, A. Larbot, C. Bourgaux, E. Prouzet, and C.A. Burton. *Chem. Mater.* 13 (2001) 3580.

141. S.A. Bagshaw, E. Prouzet, and T.J. Pinnavaia. *Science* 269 (1995) 1242.

142. S.A. El-Safty, F. Mizukami, and T. Hanaoka. *J. Mater. Chem.* 15 (2005) 2590.

143. J. Sauer, S. Kaskel, M. Janicke, and F. Schüth. In: A. Galarneau, F. Di Renzo, F. Fajula, J. Vedrine (eds). Zeolites and mesopolous materials at the dawn of the 21st century, Elsevier, Amsterdam, The Netherlands, 2001. *Stud. Surf. Sci. Catal.* 135 (2001) 315.

144. A.H. Janssen, C.-M. Yang, Y. Wang, F. Schüth, A.J. Koster, and K.P. De Jong. *J. Phys. Chem. B* 17 (2003) 10552.

145. I. Yuranov, L. Kiwi-Minsker, P. Buffat, and A. Renken. *Chem. Mater.* 16 (2004) 760.

146. A.M. Busuioc, V. Meynen, E. Beyers, M. Mertens, P. Cool, N. Bilba, and E.F. Vansant. *Appl. Catal. A: Gen.* 312 (2006) 153.

147. A.M. Busuioc, V. Meynen, E. Beyers, P. Cool, N. Bilba, and E.F. Vansant. *Catal. Commun.* 7 (2006) 729.

148. V. Meynen, A.M. Busuioc, E. Beyers, P. Cool, E.F. Vansant, N. Bilba, M. Mertens et al. In: R.W. Buckley (ed). Nanodesign of combined micro and mesoporous materials for specific applications in adsorption and catalysis, *Progress in Solid State Chemistry Research Trends*. Nova Science Publishers, New York, 2007, 63–89.

149. V. Meynen, E. Beyers, P. Cool, E.F. Vansant, M. Mertens, H. Weyten, O.I. Lebedev et al. *Chem. Commun.* 7 (2004) 898.

150. T. Kang, Y. Park, and J. Yi. *J. Mol. Catal. A: Chem.* 244 (2006) 151.

151. I. Nowak. *Colloid Surf. A.* 241 (2004) 103.

152. X. Bao, X.S. Zhao, X. Li, and J. Li. *Appl. Surf. Sci.* 237 (2004) 380.

153. E. Van Bavel, V. Meynen, P. Cool, K. Lebeau, and E.F. Vansant. *Langmuir* 21(6) (2005) 2447.

154. V. Meynen, P. Cool, E.F. Vansant, P. Kortunov, F. Grinberg, J. Kärger, O.I. Lebedev et al. *Microporous Mesoporous Mater.* 99 (2007) 14.

155. P. Kortunov, R. Valiullin, J. Kärger, V. Meynen, and E.F. Vansant. *Diffusion Fundam. (online journal)* 2 (2005) 95.

156. C.S. Carr, S. Kaskel, and D.F. Shantz. *Chem. Mater.* 16 (2004) 3139.

157. S. Trong-On, A. Ungureanu, and S. Kaliaguine. *Phys. Chem. Chem. Phys.* 5 (2003) 3534.

158. S.P. Naik, A.S.T. Chiang, R.W. Thompson, F.C. Huang, and H.-M. Kao. *Microporous Mesoporous Mater.* 60 (2003) 213.

159. A. Sakthivel, S.J. Huang, W.-H. Chen, Z.-H. Lan, K.-H. Chen, S.-W. Kim, R. Ryoo et al. *Chem. Mater.* 16 (2004) 3168.

160. Y. Han, F.-S. Xiao, S. Wu, Y. Sun, X. Meng, D. Li, S. Lin et al. *J. Phys. Chem. B* 105 (2001) 7963.

161. P. Kortunov, S. Vasenkov, J. Kärger, R. Valiullin, P. Gottschalk, M.F. Elia, M. Perez et al. *J. Am. Chem. Soc.* 127 (2005) 13055.

162. H. Vinh-Thang, Q. Huang, A. Ungureanu, M. Eic, D. Trong-On, and S. Kaliaguine. *Langmuir* 22 (2006) 4777.

163. P. Alexandridis and T.A. Hatton. *Colloids Surf. A: Physicochem. Eng. Aspects* 96 (1995) 1.

164. A. Corma. *Stud. Surf. Sci. Catal.* 49 (1998) 49.

165. A.H. Janssen, A.J. Koster, and K.P. de Jong. *Angew. Chem. Int. Ed.* 40(6) (2001) 1102.

166. I.I. Ivanova, A.S. Kuznetsov, V.V. Yuschenko, and E.E. Knyazeva. *Pure Appl. Chem.* 76(9) (2004) 1647.

167. J.C. Groen, J.A. Moulijn, and J. Pérez-Ramirez. *J. Mater. Chem.* 16 (2006) 2121.

168. J. Wang and M.-O. Coppens. *First International Workshop Inside Pores, European Network of Excellence*, March 19–23, 2006, La Grande Motte, France.

169. D. Trong-On and S. Kaliaguine. *Angew. Chem. Int. Ed.* 40(17) (2001) 3248.

170. S.I. Cho, S.D. Choi, J.-H. Kim, and G.-J. Kim. *Adv. Func. Mater.* 14(1) (2004) 49.

171. I. Schmidt, C. Madsen, and C.J.H. Jacobsen. *Inorg. Chem.* 39 (2000) 2279.

172. Z. Yang, Y. Xia, and R. Mokaya. *Adv. Mater.* 17 (2005) 2789.

173. Y. Tao, H. Kanoh, and K. Kaneko. *J. Am. Chem. Soc.* 125 (2003) 6044.

174. S.P.J. Smith, V.M. Linkov, R.D. Sanderson, L.F. Petrik, C.T. O'Connor, and K. Keiser. *Microporous Mesoporous Mater.* 4 (1995) 385.

175. I. Schmidt, A. Boisen, E. Gustavsson, K. Stahl, S. Pehrson, S. Dahl, A. Carlsson et al. *Chem. Mater.* 13 (2001) 4416.

176. C.J.H. Jacobsen, C. Madsen, J. Houzvicka, I. Schmidt, and A. Carlsson. *J. Am. Chem. Soc.* 122 (2000) 7116.
177. A.H. Janssen, I. Schmidt, C.J.H. Jacobsen, A.J. Koster, and K.P. de Jong. *Microporous Mesoporous Mater.* 65 (2003) 59.
178. J. Garcia-Martinez, D. Carloz-Amoros, A. Linares-Solano, and Y.S. Lin. *Microporous Mesoporous Mater.* 42 (2001) 255.
179. D. Trong On and S. Kaliaguine. *J. Am. Chem. Soc.* 125(3) (2003) 618.
180. D. Trong-On, A. Nossov, M.-A. Springuel-Huet, C. Schneider, J.L. Bretheron, C.A. Fyfe, and S. Kaliaguine. *J. Am. Chem. Soc.* 126 (2004) 14324.
181. D. Trong-On and S. Kaliaguine. *Angew. Chem. Int. Ed.* 41(6) (2002) 1036.
182. P.C. Shih, H.P. Lin, and C.Y. Mou. Nanotechnology in mesostructured materials. *Stud. Surf. Sci. Catal.* 146 (2003) 557.
183. R. Garcia, I. Diaz, C. Marquez-Alvarez, and J. Pérez-Pariente. *Chem. Mater.* 18 (2006) 2283.
184. P. Prokesova, S. Mintova, J. Cejka, and T. Bein. *Mater. Sci. Eng.* C 23 (2003) 1001.
185. P. Prokesova, S. Mintova, J. Cejka, and T. Bein. *Microporous Mesoporous Mater.* 64 (2003) 165.
186. Y. Liu and T.J. Pinnavia. *Chem. Mater.* 14(1) (2002) 3.
187. Y. Liu, W. Zhang, and T.J. Pinnavia. *Angew. Chem. Int. Ed.* 40(7) (2001) 1255.
188. Y. Liu, W. Zhang, and T.J. Pinnavia. *J. Am. Chem. Soc.* 122 (2000) 8791.
189. Z. Zhang, Y. Han, L. Zhu, R. Wang, Y. Yu, S. Qiu, D. Zhao et al. *Angew. Chem. Int. Ed.* 40(7) (2001) 1258.
190. F.-S. Xiao, Y.U. Han, Y. Yu, X. Meng, M. Yang, and S. Wu. *J. Am. Chem. Soc.* 124(6) (2002) 888.
191. C.E.A. Kirschhock, S.P.B. Kremer, J. Vermant, G. Van Tendeloo, P.A. Jacobs, and J.A. Martens. *Chem. Eur. J.* 11 (2005) 4306.
192. S.P.B. Kremer, C.E.A. Kirschhock, A. Aerts, K. Viallani, J.A. Martens, O.I. Lebedev, and G. Van Tendeloo. *Adv. Mater.* 15(20) (2003) 1705.
193. S.P.B. Kremer, C.E.A. Kirschhock, A. Aerts, C.A. Aerts, K.J. Houthoofd, P.J. Grobet, P.A. Jacobs et al. *Solid State Sci.* 7(7) (2005) 861.
194. R. Garcia, I. Diaz, C. Marquez-Alvarez, and J. Pérez-Pariente. *Chem. Mater.* 18(9) (2006) 2283.
195. J. Pérez-Pariente, I. Diaz, and J. Agundez. *C. R. Chemie.* 8(3–4) (2005) 569.
196. D.F. Li, D.S. Su, J.W. Song, X.Y. Guan, K. Hoffman, and F.S. Xiao, *J. Mater. Chem.* 15(47) (2005) 5063.
197. Y. Liu and T.J. Pinnavaia. *J. Mater. Chem.* 14(7) (2004) 1099.
198. J. Pérez-Ramirez, C.H. Christensen, K. Egeblad, C.H. Christensen, and J.C. Groen. *Chem. Soc. Rev.* 37 (2008) 2530.
199. W.J.J. Stevens, V. Meynen, E. Bruijn, O.I. Lebedev, G. Van Tendeloo, P. Cool, and E.F. Vansant. *Microporous Mesoporous Mater.* 110 (2008) 77.
200. M. Chiesa, V. Meynen, S. Van Doorslaer, P. Cool, and E.F. Vansant. *J. Am. Chem. Soc.* 128(27) (2006) 8955.
201. J. Wang, J.C. Groen, W. Yue, W. Zhou, and M.O. Coppens. *J. Mater. Chem.* 18 (2008) 468.
202. J. Wang, J.C. Groen, W. Yue, W. Zhou, and M.O. Coppens. *Chem. Commun.* 14 (2007) 4653.
203. J. Wang, W. Yue, W. Zhou, and M.O. Coppens. *Microporous Mesoporous Mater.* 120 (2009) 19–28.

9

Sorption Processes

F. Rodríguez-Reinoso, A. Sepúlveda-Escribano, and J. Silvestre-Albero

CONTENTS

9.1 Introduction

When an atom or molecule hits a solid surface, several phenomena can appear. It may rebound with its starting energy, or it may be retained on the surface for a more or less long time. In this case, *adsorption* is produced, and the time during which the adsorbed species is retained on the surface is related to the strength of the interaction with the surface. The adsorbed atoms or molecules may stay fixed on a surface site, they may diffuse across the surface, suffer a chemical reaction, or be dissolved in the bulk of the solid, this latter process being known as *absorption*. This chapter will deal with adsorption processes and, more specifically, with those systems in which the interaction between the surface and the adsorbed species is relatively weak, that is, the so-called *physisorption* processes.

Physisorption of pure gases or gas mixtures on surfaces of solid porous materials are becoming increasingly important both in fundamental and applied research. This phenomenon is the basis for several important practical industrial processes, such as gas separation, gas storage, cleaning and purification of gas streams to remove pollutants or not desired components, drying processes, and also for the characterization of porous materials. Although the aim of the chapter is to review the use of adsorption in the characterization of solid porous materials, the application of adsorption to some practical processes will be discussed in the last part of the chapter. Before that, a short introduction to the fundamentals of the adsorption phenomenon will be presented, to continue with the experimental methods used to quantify the adsorption phenomena, and the mathematical models used to interpret the experimental results. Thus, although adsorption of gases and vapors is a classical and convenient approach to the general characterization of porosity in solids, the interpretation of the adsorption data is not always straightforward. It is well documented that the adsorbed molecules are strongly retained in narrow micropores (pores smaller than 2 nm) at low pressure due to the enhancement of the adsorption potential caused by the close proximity of the micropore walls. This causes distortion in the adsorption isotherm (the plot of the adsorbent amount as a function of the equilibrium pressure, at a constant temperature). On the other hand, activated diffusion effects may appear when the size of the adsorbed species is close to the pore size, mainly for low working temperatures. In this case, equilibrium is reached very slowly, and the probability of obtaining wrong nonequilibrium isotherms is very high. All these factors, together with some others, can introduce uncertainty and produce erroneous results in the characterization of solids adsorbents. Different approaches to limit or to remove these problems will be discussed in this chapter.

9.2 Fundamentals of Adsorption

When an atom or molecule approaches the surface of a solid, a balance between attractive and repulsive forces is established. The result of this balance, in adequate conditions of pressure and temperature, is the adsorption phenomenon.

From a macroscopic point of view, adsorption can be defined as the increase of the concentration of one or more components of a fluid (gas or liquid) at the interface between the fluid and a solid (i.e., at the surface of the solid). The solid is named *adsorbent* and the adsorbed substance is the *adsorbate*, but it is named *adsorptive* when it is still in the fluid phase.

The attractive forces that cause adsorption can show different degrees of intensity. Thus, they can move from an actual chemical interaction (very strong forces) among some surface sites and the adsorbate atom or molecule, to weak van der Waals type interactions between the surface (as a whole) and the adsorbate. In the first case we are dealing with chemical adsorption or *chemisorption*. In this process, true chemical bonds are generated between the adsorbate and specific sites at the surface, which modify their electronic state. In this way, when a molecule such as hydrogen (H_2) is chemisorbed on a platinum surface, the bond between both hydrogen atoms in the molecule is broken, and two new H–Pt bonds are formed. Chemisorption is a specific phenomenon, which only takes place between given adsorbent–adsorptive pairs, and can take place even at high temperatures. Chemisorption is the initial stage in any catalytic process.

When the interactions between the adsorbent and the adsorbate are relatively weak, similar in strength to those that cause gas condensation, it is called physical adsorption or *physisorption*. In this case, no chemical bonds are formed, although if the adsorptive exhibits polarity, more intense interactions can be produced with surfaces containing polar sites. Thus, a polar molecule such as water is more strongly adsorbed on a polar surface, (as that of a metal oxide (SiO_2)) than on a nonpolar surface such as that of an activated carbon. A similar effect can be observed when comparing adsorption of argon, a spherical nonpolar molecule, with adsorption of nitrogen, a molecule with quadruple moment. The interaction of this latter with polar surfaces is much stronger. In fact, the adsorbent–adsorbate interactions, measured through the *adsorption energy*, can be constituted by several contributions: nonspecific contributions (the dispersion van der Waals type interaction and the repulsive interactions) and specific interactions (polarization energy, polar interactions, and quadrupole interactions). However, although the interactions can be more or less intense, physisorption is a nonspecific phenomenon and it can be observed under adequate pressure and temperature conditions between any adsorbent–adsorbate pair.

Adsorption is an exothermic process. As it is spontaneous, ΔG is negative and, in this way, ($\Delta H - T\Delta S$) must also be negative. Thus, $T\Delta S$ has to be larger than ΔH. The entropy change, ΔS, is also negative (entropy diminishes), as adsorption produces an ordered system, with less degree of freedom. As a result of this, the enthalpy change ΔH must also be negative and thus, the process has to be exothermic. For chemisorption, values for ΔH are in the order of those for chemical reactions, between -40 and -400 kJ/mol. For physical adsorption values around -20 kJ/mol can be found; larger values can, however, be found for systems involving polar molecules on polar or ionic surfaces.

There are some other differences between chemisorption and physisorption processes. In chemisorption, just one monolayer of adsorption species can be formed, as the adsorbent–adsorbate interaction is much higher than the adsorbate–adsorptive interaction. When the surface is completely covered, no more available sites remain for adsorption. However, a multilayer can be formed in physisorption, as the adsorbent–adsorptive interaction (adsorption enthalpy) is similar to the adsorptive–adsorbate or adsorptive–adsorptive interaction (condensation enthalpy). Thus, when the surface is completely covered, more adsorptive species can be adsorbed by interacting with the already adsorbed species. On the other hand, physisorption is normally faster than chemisorption, as this latter is a chemical reaction, and may have a larger or smaller activation energy, which can make it a slow process. As no activation energy affects physisorption this process is faster, unless diffusional problems can affect the rate at which the adsorptive can reach the surface. This is the case, for example, of adsorption at low temperatures on microporous adsorbents, when the adsorptive species has a molecular size similar to that of the pore entrance.

Adsorption is a dynamic process. There are always some atoms or molecules being adsorbed and, at the same time, some atoms or molecules being desorbed from the surface. When the adsorption equilibrium is reached, the rate at which species are adsorbed and desorbed, respectively, is the same. The amount finally adsorbed at the equilibrium depends on a number of factors: equilibrium pressure, working temperature, and adsorbent–adsorbate interactions.

The most commonly used expression to quantify the extent of the adsorption phenomenon is the *adsorption isotherm*, which is defined as the relationship, at constant temperature, between the amount adsorbed, n_{ads}, and the equilibrium pressure, p:

$$n_{ads} = f(p)_T \tag{9.1}$$

If the working temperature is lower than the critical temperature of the adsorptive, the adsorption isotherm can be expressed as

$$n_{ads} = f\left(\frac{p}{p_0}\right)_T \tag{9.2}$$

where
  p_0 is the saturation pressure of the adsorptive at the working temperature T
  the ratio p/p_0 is the *relative pressure*

The International Union of Pure and Applied Chemistry (IUPAC) classifies the adsorption isotherms in six types [1], which are shown in Figure 9.1.

Type I isotherm is the so-called *Langmuir isotherm*. It is concave toward the relative pressure axis, the adsorbed amount steeply increasing with the

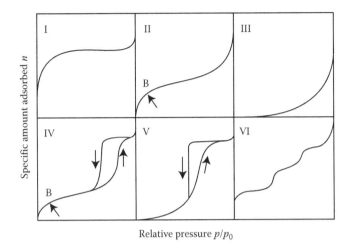

FIGURE 9.1
IUPAC classification of gas physisorption isotherms. (Adapted from Sing, K.S.W. et al., *Pure Appl. Chem.*, 57, 603, 1985.)

relative pressure for low p/p_0 values, and then reaching a plateau; that is, the amount adsorbed by unit mass of adsorbent asymptotically reaches a limit value as the relative pressure increases up to unity. This type of isotherm is typical for systems with strong adsorbent–adsorbate interactions. This is the case for adsorbents containing micropores, that is, pores not wider than 2 nm. In these systems, micropores become filled with the adsorbate in the liquid state as the similarity between the pore size and the adsorbent size does not permit the coverage of the surface. Thus, the amount adsorbed quickly increases with increasing pressure at low relative pressures, due to the high adsorbent–adsorbate interaction, and then it stops (plateau in the isotherm) when the micropores are completely filled.

Type II isotherm is at first concave toward the relative pressure axis, the amount adsorbed increases nearly linearly with the relative pressure and, finally, quickly increases at relative pressures close to 1. In the systems producing this type of isotherm, the surface is first covered by a monolayer of adsorbate at low relative pressures. When the relative pressure increases, an adsorbed multilayer is formed, the thickness of which increases progressively and then, at a relative pressure close to 1 (the working pressure is then the saturation pressure of the adsorptive at the adsorption temperature), the adsorbent condenses on the surface. Type II isotherms are characteristics of adsorption on nonporous or macroporous solids (with pores larger than 50 nm), in which the formation of a multilayer with infinite thickness at relative pressures close to unity is not restricted. If the adsorption temperature is equal or lower than the boiling temperature of the adsorptive, the process is reversible over the whole relative pressure range. That is, the adsorption isotherm fits with the desorption isotherm.

Type III isotherms are obtained for systems with a very weak adsorbate–adsorptive interaction. The relative pressure has to be increased up to a certain value to obtain some degree of adsorption. This type of isotherm is obtained, as an example, for nitrogen adsorption on ice and for water adsorption on graphite.

Type IV isotherm, with a low pressures region similar to that of Type II isotherm, shows a hysteresis loop: the desorption isotherm does not fit with the adsorption isotherm in a wide range of relative pressures. This hysteresis loop is associated to the filling and emptying of mesopores by capillary condensation (pores with a size comprised between 2 and 50 nm). These isotherms are very common, although the shape of the hysteresis loop changes for one system to another and also depends on the working temperature.

Type V isotherms are similar to Type III isotherms, not only reflecting very low adsorbent–adsorptive interactions, but they also contain hysteresis loop, related to the presence of mesopores.

Finally, Type VI isotherms, or stepped isotherms, are not very common. They are associated with layer by layer adsorption on uniform surfaces, although they are also observed when a change of the packing density of the adsorbate is produced by increasing the relative pressure.

9.3 Adsorption as a Tool for Characterization of Porous Solids

9.3.1 Determination of Adsorption Isotherms

Considering the tutorial nature of this book, this section just describes in a simple way the experimental systems more commonly used in the determination of the adsorption isotherm from which the information about the porosity of a given solid can be assessed. The interpretation of the adsorption isotherm is not straightforward, and the routine work of determining the surface area and the pore size distribution is not easy, but the adsorption isotherm reflects the porous structure of the solid and the information obtained from it is essential for the application of the solid as adsorbent or catalyst.

Since an adsorption isotherm plots the amount adsorbed per unit weight versus the relative pressure of the gas at the temperature of measurement, the experimental determination is conditioned by the parameters involved. There are two general approaches to the determination of the amount of gas adsorbed: manometric (sometimes called volumetric, by historical reasons) and gravimetric.

In the case of manometric measurements, and assuming the equipment uses the more common discontinuous dosing of gas, a given dose of the gas, at a given pressure and temperature, is introduced into a thermostated chamber containing the solid, and the pressure is measured. The amount adsorbed is calculated from the difference in pressure before and after expansion to the sample holder. To do this properly, it is critical to know the exact volume occupied by the adsorbent in order to be able to determine the remaining dead space of the adsorption system. If the adsorption isotherm is carried out at liquid nitrogen temperature (–196°C), a large part of the adsorption chamber is immersed into the liquid nitrogen, but the rest will be at a higher temperature. The temperature of the tubing connecting the sample bulb to the dosing bulb has to be controlled and taken into account for the determination of the amount adsorbed. This is relatively simple if the system is enclosed in a temperature-controlled container. However, the most important volume is the dead space, but its determination is not a simple process. Most manometric equipments use the expansion of a known amount of nonadsorbing gas into the sample bulb containing the sample, which is kept at the same temperature to be used during the adsorption experiment. Care must be taken to learn about the volume of the tubing, which is placed between the top of the liquid nitrogen (–196°C) and the valve closing the adsorption bulb (around room temperature). The problem is considerably reduced for experiments carried out at around room temperature.

This way of measuring the dead space is appropriate for many adsorbents when using helium as nonadsorbing gas; however, it has been shown in the last few years that this is not always the case, because helium may be adsorbed on many microporous adsorbents (specially porous carbons) at –196°C. In this sense, care must be taken when measuring high-precision nitrogen adsorption isotherms on porous carbons with narrow constrictions/micropores. Helium adsorption during the dead volume determination is clearly discerned by a downshift in the amount adsorbed below $p/p_0 \sim 10^{-4}$ in the subsequent nitrogen adsorption isotherm due to the already filled narrow micropores (the vacuum treatment is not enough to remove He from these narrow constrictions). Measuring the death volume after completion of the nitrogen isotherm avoids the aforementioned problem.

Once the dead space is known, the experimental procedure to obtain the adsorption isotherm is as follows:

1. Outgassing the sample at relatively high temperature; this is a function of the nature of the sample, but it is rather common to use between 200°C and 300°C for several hours to ensure that the pressure inside the sample bulb is less than around 10^{-5} mbar.

2. A dose of the gas is admitted into the dosing bulb and the pressure measured; this allows for the calculation of the amount of gas contained into the bulb.

3. The valve connecting the dosing and sample bulbs is open, and the gas is expanded into the adsorption system. The evolution of

pressure is followed closely until it does not change after a given period of time. This is then considered the equilibrium pressure for that specific point of the adsorption isotherm.

The decrease in pressure will permit to calculate the amount of gas adsorbed on the sample. The process is repeated to obtain the full adsorption isotherm until the saturation pressure ($p/p_0 \approx 1.0$) is reached. The number of experimental points of the isotherm is determined by the volume of the dose introduced in each point. Of course, care should be taken to avoid condensation in any part of the adsorption system outside the sample cell.

Some experimental aspects for the determination of the adsorption isotherm (mass of sample, equilibrium time for each experimental point, etc.) will be described in detail when analyzing the problems found when working with some specific adsorbents. The control of temperature, both in the system cabinet and in the bath surrounding the sample bulb, is critical to ensure precision data. The same applies to the measurement of the gas pressure, and good quality pressure transducers should be used.

The second approach to determine the adsorption isotherm is the gravimetric one. In this case, the sample is suspended from either a classic fused silica spring (McBain balance) or an electronic microbalance. In this method, the volume of the adsorption system is not important, and the experimental points are given by the increase of weight of the sample for each equilibrium gas pressure remaining after each gas dose has been put into contact with the sample. Although this approach seems to be much simpler than the manometric one, this is not so because of the need for a buoyancy correction. Additionally, the usually large distance between the sample container hanging from the balance and the walls of the tube of the balance makes it relatively difficult to have the sample at the same temperature than the surrounding thermostat (this is especially important at low temperature, e.g., –196°C).

Since an adsorption isotherm is a plot of the amount adsorbed (volume or weight per unit mass) as a function of the equilibrium relative pressure for each point, it is very important to be able to know the saturation pressure. For this purpose, an empty bulb is introduced into the thermostat, as near as possible to the sample bulb and connected to the dosing system; thus, in the case of nitrogen, at –196°C condensation takes place inside this bulb and it permits to determine the exact temperature of desorption.

9.3.2 Analysis of Adsorption Isotherms

Since the adsorption isotherm is a consequence of the interaction of the solid surface with the gas, the shape of the isotherm may provide (to a trained eye) a rather complete information about the porosity of the adsorbent. As a typical example, Figure 9.2 includes three adsorption isotherms corresponding to three different activated carbons [2].

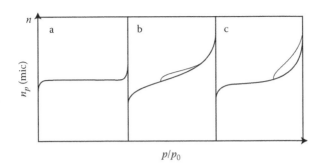

FIGURE 9.2
Nitrogen adsorption–desorption isotherms at 77 K in three different activated carbons. (Adapted from Rouquerol, F. et al., *Adsorption by Powders and Porous Solids*, Academic Press, London, U.K., 1999.)

As indicated in Section 9.1, the adsorption potential at the micropores is enhanced and, consequently, the presence of micropores is indicated by a steep increase of uptake at very low relative pressure. As the microporosity is widened, the increase in uptake is less steep. The presence of a range of micropores will result in a Type I adsorption isotherm, with an initial very steep uptake (filling of narrow micropores) and a knee resulting from the filling of wider micropores. If no further porosity is present the knee will lead to a plateau, parallel to the pressure axis, since no further adsorption is possible (Figure 9.2a). When some mesoporosity is present in addition to the microporosity, there is an increase in uptake at medium-to-high relative pressure, with or without hysteresis loop; there is no plateau now, and the knee of the isotherm provides information of the wide micropore size distribution of the sample (Figure 9.2b). In the case of Figure 9.2c, there is, in addition, a clear hysteresis loop, and its shape provides information about the shape of the mesopores.

It should be remembered here that the adsorption isotherm corresponding to a nonporous material is Type II, showing the formation of monolayer and multilayer on the surface (Figure 9.1). The presence of mesoporosity is shown by the capillary condensation in the mesopores, the adsorption on macropores taking place only at very high relative pressures, above $p/p_0 \approx 0.99$.

9.3.2.1 Determination of Surface Area

The adsorption data can be used for the determination of the surface area of any solid provided it is possible to determine the so-called monolayer volume and the specific solid–gas interactions are not strong; for this reason, the gases used for the determination of the surface area are mainly noble gases and nitrogen. Once the volume corresponding to the monolayer of adsorbed gas (adsorbate) is known, the surface area can be calculated from

$$S = V_m \cdot N \cdot A_m \qquad (9.3)$$

where
   N is the Avogadro constant
   A_m is the area occupied by each molecule in the monolayer

The critical parameter is A_m; Emmett and Brunauer [3] proposed in 1937 that it can be calculated from the density of the liquid adsorbate. In this way, the value for nitrogen at −196°C was estimated to be 0.162 nm², a value that is generally accepted nowadays. Although argon can also be used as a gas for the determination of the surface area, due to the fact that it is constituted by monoatomic spherical species, its use is not so popular because the temperature of the isotherm should be that of liquid argon, which is not as accessible as liquid nitrogen. An alternative would be to carry out the adsorption at −196°C, but this is below the triple point of argon (−184°C).

9.3.2.1.1 BET Method

Brunauer, Emmett, and Teller (BET) extended the simple Langmuir mechanism (monolayer condensation–evaporation) to the second and the rest of the absorbed layers. Following Langmuir postulates, they accepted that the surface of the solid is energetically uniform and they supposed the heat of adsorption of the layers above the first one to be identical to the latent heat of condensation of the gas. In this way, the first layer behaves as a series of active centers on which the second layer is formed and above this the third, and so on. Finally, they accepted that at relative pressure near saturation the vapor condenses as a liquid with infinite number of adsorbed layers. The linear form of the BET equation is

$$V = \frac{V_m \cdot C \cdot p}{(p_0 - p) \cdot [1 + (C-1) \cdot p/p_0]} \qquad (9.4)$$

which for plotting is written as

$$\frac{p}{V \cdot (p_0 - p)} = \frac{1}{V_m \cdot C} + \frac{C-1}{V_m \cdot C} \cdot \frac{p}{p_0} \qquad (9.5)$$

where
   p is the actual pressure
   p_0 is the saturation pressure
   V is the volume of gas adsorbed at pressure p
   V_m is the volume of gas adsorbed in the monolayer
   C is a parameter defined by

$$C = A \cdot \exp\frac{E_1 - E_L}{RT} \tag{9.6}$$

where
 A is a constant
 E_1 is the heat of adsorption on the first layer
 E_L is the heat of condensation of the gas

When $p/V(p_0-p)$ is plotted versus p/p_0, a straight line is obtained from which the values for V_m and C are easily calculated. It is a common practice to select for the plot the data in the relative pressure range 0.05–0.30, but there are many cases in which this range is not appropriate; as shown later, the range for most microporous materials must be limited to $p/p_0 \approx 0.05$–0.12. What is really important is to ensure that the experimental points fit the straight line and that its intercept is positive. Of course, the constant C must always have a positive value.

A general comment on the validity of the BET model. This method has been the subject of much criticism because of the assumption of energetically homogeneous and identical adsorption sites, and for not taking into account the lateral interactions among the adsorbed molecules. Furthermore, when the model is applied to microporous adsorbents on which the pore filling occurs at low relative pressures, below the monolayer, due to the strong adsorption potential caused by the proximity of the pore walls, the calculated monolayer volume corresponds to more than one adsorbed layer. If this value is converted into the BET surface area the actual figure is larger than the real value corresponding to the walls of the micropores, because the pore can adsorb several layers of adsorbate before the apparent monolayer is formed. It is to be noted that on cases such as this one the linearity of the BET plot is considerably reduced in the upper range, clear deviations being noticeable after $p/p_0 = 0.1$. For all these reasons it was suggested long time ago that the term "apparent" or "equivalent" surface area should be used when giving the BET surface area values; the range of linearity of the plot should also be provided to help better interpret the results. On the other hand, although the BET model does not depict the real situation, it is the method more commonly used to analyze the adsorption isotherm. This is so because being relatively simple it can be considered a good measure of the adsorption capacity of a given adsorbent, this being the main reason for the widespread use in industrial circles.

9.3.2.1.2 t-Method
The *t*-method, as it is the case for the α_S method, is based on the concept of standard isotherm for the assessment of the surface area as an alternative to the BET model. The main assumption here is that the monolayer is sensitive to the structure of the solid surface but the multilayer is not as dependent on

the structure. It is possible then to consider that the thickness of the adsorbed multilayer is a function of the pressure and that the density of the multilayer is that of the bulk liquid at the temperature of adsorption. For the case of nitrogen at −196°C, for which the nitrogen molecules are compacted, each one occupying an area of 0.162 nm², the value for *t* is

$$t = 0.354 \cdot \frac{V_a}{V_m}. \tag{9.7}$$

and the meaning is that an adsorbed layer of nitrogen has an average thickness of 0.354 nm (the kinetic diameter of the nitrogen molecule is 0.364 nm, a relatively similar value). This is the original equation proposed by Lippens and de Boer [4]. This means that when V_a is equal to V_m, when there is one layer of adsorbed nitrogen, the thickness of that layer is 0.354 nm. The dependence on the BET model is clear in the sense that the value of V_m is needed to calculate the so-called reduced isotherm (V_a/V_m), but there is no need to use the cross-sectional area of nitrogen for the calculation of the surface area.

Equation 9.7 is used to plot the values of *t* as a function of the relative pressures at which V_a is obtained for a standard nonporous material. The V_a data of the experimental isotherm for the problem material is plotted against *t*-values at the corresponding relative pressures. For the most favorable case, a nonporous solid, the *t*-plot will be a straight line going through the origin, the slope being a measure of the surface area. For the particular case of nitrogen at −196°C, the surface area will be given as

$$S_t = 15.47 \cdot \frac{V_a}{t} \,(m^2/g) \tag{9.8}$$

where V_a is the amount adsorbed expressed in mL(STP)/g.

The *t*-plot not only provides the value of the surface area of the adsorbent. The shape of the *t*-plot can also provide additional information. If the solid does not exhibit capillary condensation or pore filling, the surface is completely accessible to the nitrogen and there is multilayer adsorption, the slope of the plot providing the specific surface area. If, however, there is capillary condensation into the mesopores, the adsorption isotherm shows an upward deviation, the *t*-plot deviates upward from the straight line. On the contrary, if the microporosity is filled at low relative pressures, the *t*-plot exhibits a downward deviation of the straight line. The IUPAC, in 1985 [1] recommended that when using the *t*-plot the standard isotherm had to be determined on a nonporous solid having the same chemical structure and nature than the test solid.

9.3.2.1.3 α_S-Plot

The main problem found with the *t*-plot model is the need to know the thickness of the adsorbed layer. However, for the purpose of studying the deviations in respect to a standard isotherm the knowledge of the thickness is not necessary since it is possible to compare the shapes of the isotherms for the test and standard materials. Sing [5] used this approach when developing the so-called α_S-plot method. In this case, V_m is replaced by the amount adsorbed at a selected relative pressure, $p/p_0 = 0.4$, and the reduced isotherm α_S is V_a/V_S, so that the plot of V_a/V_S versus relative pressure is built without using the BET model. The α_S plot is then constructed as the t-plot using the α_S used in place of the t-data. If the result is a straight line through the origin, it is deduced that the problem isotherm is of the same shape as the standard and the slope $b\alpha$ will be equal to V_S. If capillary condensation in mesopores or micropores filling is present then the α_S plots will deviate from linearity, as in the case of the t-plot.

The calculation of the surface area of the sample is simple if the specific area of the standard reference sample is known (for instance by BET):

$$A(\text{sample}) = \frac{b_\alpha(\text{sample})}{b_\alpha(\text{reference})} \cdot A(\text{reference})(m^2/g) \qquad (9.9)$$

The main criticism to this method may lay in the uncertainty about the right relative pressure chosen, because it could be slightly lower or higher. The idea behind the method was to select $p/p_0 = 0.4$ because at that relative pressure the micropores should be filled and capillary condensation not yet started.

In principle, both the t-plot and the α_S-plot can be used to determine the volume of micropores of a solid provided the standard reference isotherm for a solid with similar surface structure and chemical properties is known. Figure 9.3 includes schematic α_S-plots for different types of solids [6].

If the solid is of the same nature than the nonporous reference sample, then the plot is line 1. If mesoporosity is present then plot 2 results, with a clear upward deviation at high relative pressure. When microporosity is present in the test solid, line 3, the α_S has a downward deviation at low relative pressures; if this deviation extrapolates to the origin, then this slope (OA) will provide the total surface area of the sample, whereas the extrapolation of the straight portion AC will provide the volume of micropores. On the other hand, the slope of the straight portion AC will provide the external surface area of the solid.

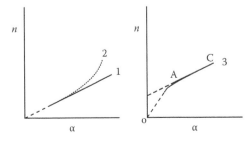

FIGURE 9.3
α_S Plots corresponding to different activated carbons. (Adapted from Rodríguez-Reinoso, F. et al., *Chemistry & Physics of Carbon*, Marcel Dekker, New York, 2001. With permission.)

9.3.2.2 Microporosity

As already indicated, the concept of surface area, as deduced from such methods as BET or t-/α_s-plot, does not have much physical meaning since the volume corresponding to the monolayer in microporous solids will include the layers of molecules adsorbed inside the micropores, as a consequence of the strong adsorption potential caused by the proximity of the pore walls. For this reason it was stated above that the term "apparent" or "equivalent" surface area should be used instead in these cases.

When a solid contains micropores (dimension smaller than 2.0 nm) the adsorption isotherm is Type I, with a very steep uptake at very low relative pressures, followed by an almost flat plateau. It is now generally admitted that the micropores are filled at very low relative pressure. This is the consequence of the overlapping of the adsorption potential caused by the proximity of the pore walls. It is to be noted here that Everett and Powl [7] calculated the enhanced adsorption energy as a function of the pore width/adsorbate dimension and the shape of the micropores (cylindrical or slit-shaped). When the ratio is small and the pore dimension is equal to only a few molecular dimensions, the enhanced adsorption potential leads to pore filling at very low relative pressure. On the other hand, they showed that the enhancement is larger for cylindrical than for slit-shaped micropores. Figure 9.2 included three typical examples of solids containing micropores. A sample constituted only by narrow micropores (very strong adsorption potential) will produce a Type I isotherm (case Figure 9.2a), with a horizontal plateau defining the volume of micropores since once the micropores are filled by primary micropore filling no further adsorption takes place due to lack of larger porosity. The isotherm in Figure 9.2b corresponds to a solid with a wide range of micropore size, this leading to a clear knee in the isotherm. The wider the knee the wider is the micropore size distribution and there is secondary micropore filling in the larger micropores. There is no clear cut between the primary micropore filling of narrow micropores and the so-called secondary micropore filling of wide micropores (between 1.0 and 2.0 nm dimension). In addition to microporosity, the solid of Figure 9.2b exhibits the presence of external surface and some mesoporosity, as shown by the presence of a small hysteresis loop. Figure 9.2c corresponds to the adsorption isotherm for a solid constituted by microporosity (rather uniform here because of the sharp knee) and mesoporosity and even some macropores (very steep upward deviation near saturation pressure).

Evaluation of the microporosity is simple when the adsorption isotherm is of exact Type I, with a large uptake at very low relative pressure, and a flat plateau up to near saturation pressure, since the volume of micropores is directly obtained from the uptake at the plateau converted into volume of liquid adsorbate (using the density of the gas in liquid state). The situation is more complex with most adsorbents since the microporosity is also accompanied by mesoporosity, macroporosity, and even extensive external surface. It has been shown in the previous section that the t- or α_s-plot can be used to determine the volume of micropores by extrapolation of the linear multilayer section.

9.3.2.2.1 Polanyi–Dubinin Equations

Dubinin, using the earlier potential theory of Polanyi, in which the adsorption isotherm data were expressed in the form of an invariant characteristic curve, advanced more than 50 years ago another method for the determination of the microporosity of a solid. Although several modifications were introduced to the original equation put forward by Dubinin and Radushkevich, this is the one more commonly used for the determination of the micropore volume. Since the details for the model can be found in any book related to adsorption, only the final form for plotting of the Dubinin-Radushkevich will be presented. This equation can be written as

$$\log V = \log V_0 - D \cdot \log^2 \frac{p_0}{p} \tag{9.10}$$

where

$$D = 2.303 \cdot \frac{K}{\beta^2} \cdot (RT)^2 = 0.43 \cdot B \cdot \frac{T^2}{\beta^2} \tag{9.11}$$

where

D is the so-called characteristic energy of the system
K is a constant depending on the porous structure
β is the "affinity coefficient," which Dubinin took as 1.0 for benzene, considered to be the reference adsorbate (in this way, it is possible to bring the so-called characteristic curves—temperature independent— of an adsorbent for different adsorptives into coincidence with that of benzene)

By introducing the appropriate β values and the adsorption temperature, the isotherms obtained for different adsorptives should be reduced to a single characteristic curve for the adsorbent. The characteristic curve is the plot of $\log V$ versus $(T/\beta)^2 \cdot \log^2(p_0/p)$ and a typical example from the literature is included in Figure 9.4 [8].

According to the DR equation, a plot of $\log V$ versus $\log^2 (p_0/p)$ will be a straight line with an intercept equal to $\log V_0$, from which the volume of micropores V_0 is deduced. At temperatures well below the critical point the density of the adsorbate can be taken as the normal density of the bulk liquid ($0.808\,g/cm^3$ for nitrogen at $-196°C$). The slope of the DR plot will give the value of D and since D is related to K and, consequently, to the microporous structure of the solid, it decreases as the micropore size decreases.

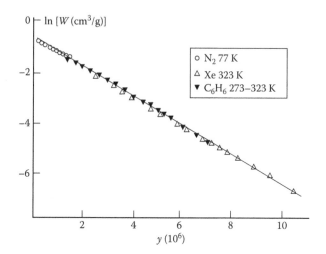

FIGURE 9.4
Adsorption of various gases on a carbosieve carbon: $y = (T/\beta)^2 \log^2 (p_0/p)$. (Adapted from Stoeckli, H.F. et al., *Characterization of Porous Solids*, Society of Chemical Industry, London, U.K., 1979, p. 31. With permission.)

The linearity for the DR plot covers a wide range of relative pressure for many experimental systems. Since the adsorption on micropores takes place at low relative pressures the straight line should be found in the range below 0.1–0.2. Although deviations from linearity have been described extensively in the literature, they could be reduced to mainly three, as shown in Figure 9.5.

Type a, with a clear upward deviation at higher relative pressures, corresponds to an adsorbent in which pores wider than micropores are filled by capillary condensation. Type b, with a downward deviation at low relative

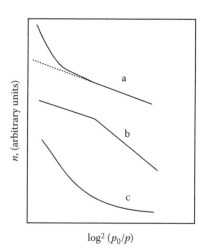

FIGURE 9.5
Schematic representation of a DR plot and some common types of deviation.

pressures, corresponds to an adsorbent with very narrow microporosity for which the access of the adsorptive is restricted at very low relative pressure in the normal time of measurement. Type c, with no apparent straight portion, corresponds to adsorbents for which the pores size distribution is so wide that no defined range of microporosity is detected. It is important to learn to decide which is the best straight line to be chosen for the determination of the micropore volume. This aspect will be further analyzed when describing in Section 2.3 the problems found when analyzing adsorption isotherms.

It is important to mention that in view of the many cases in which linearity was very limited, Dubinin and Astakov (DA) developed a more general form of characteristic curve using a Weibull rather than a Gaussian distribution of micropore size. For plotting purposes the equation is

$$\log V = \log V_0 - D' \cdot \log^n \left(\frac{p_0}{p} \right) \tag{9.12}$$

where D' is

$$D' = 2.303^{n-1} \cdot \left(\frac{RT}{E} \right)^n \tag{9.13}$$

and E is the characteristic free energy of adsorption. The DR equation is the especial case of the DA equation with $n = 2$. For microporous materials the n parameter is usually equal to 2, and it is near 3 for sieves with very narrow and uniform micropores.

At a later stage, Dubinin applied a two-term equation to describe a bimodal micropore size distribution [9]. The equation is a superposition of two microporous structures with their corresponding micropore volume and characteristic adsorption energy. Stoeckli et al. advanced in the application of the DR equation to heterogeneous porous structures [8,10,11] assuming a Gaussian distribution. They worked with a continuous distribution of pore dimension replacing the summation of the individual DR contributions by integration. Using a mathematical device similar to an error function to provide an exponential relationship between the micropores size distribution and the Gaussian half-width they then solved the integral transform.

Dubinin even proposed that there was an inverse relationship between the characteristic energy and the micropore width l:

$$l = \frac{K}{E} \tag{9.14}$$

where K is a proportionality empirical constant. The reader is addressed to the extensive work of Dubinin and Stoeckli to complete the information on these equations.

When the adsorbent is essentially microporous the DR or DA equations can provide the volume of micropores and indirect information about the pore size and the energetic of adsorption. When micropores are accompanied by larger pores such as mesopores, then the nonmicroporous contribution to the total porosity may be large. Consequently, there is a need for the separation of adsorption in micropores from the adsorption on nonmicroporous surface, in order to be able to calculate the so-called external (nonmicroporous) surface area and to get a better understanding of the porosity and surface area of the adsorbent. Besides the use of the t- and α_S-plots described in the previous section there are other methods to do this separation. Two of them will be very briefly described here.

Isotherm subtraction: When an adsorbent has a considerable contribution from micro-, meso-, and macroporosity, the DR plot exhibits deviations of Type c described in Figure 9.5. The isotherm subtraction method is based on the fact that at low relative pressures the adsorption is dominated by the large enhancement of the adsorption potential in micropores and the adsorption on nonmicroporous surfaces can be almost neglected. Therefore, the extrapolation of the low relative pressure linear region will give the microporosity contribution to the adsorption isotherm, which can then be subtracted from the total isotherm at higher relative pressures to obtain the microporous contribution

$$V_{ext} = V_{exp} - V_{DR} \tag{9.15}$$

where
  V_{ext} is the residual isotherm or external adsorption contribution
  V_{exp} is the total adsorption isotherm
  V_{DR} is the DR contribution

The resulting nonmicroporous isotherm may be now analyzed by the BET equation to yield the nonmicroporous surface area.

9.3.2.2.2 Pre-Adsorption

The most direct method to evaluate the microporosity of an adsorbent is the pre-adsorption technique, based on the assumption that the pre-adsorbate used is able to fill the micropores while leaving the rest of porosity free and accessible to adsorptive. A very common pre-adsorbate is n-nonane, because on out-gassing at room temperature it is desorbed from the nonmicroporous surface but it is retained in the microporosity. If the adsorption of nitrogen at $-196°C$ is measured before and after nonane pre-adsorption, the difference will correspond to the volume of micropores. Figure 9.6 shows the N_2 adsorption isotherms on a series of activated carbons with burn-off ranging from 8% to 80%, before and after n-nonane pre-adsorption [12]. In all cases, the amount of nitrogen adsorbed is much lower than in the original carbons. For carbons with up to 52% burn-off, the isotherms do not show a

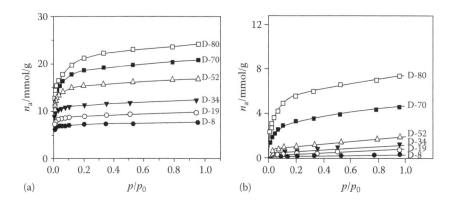

FIGURE 9.6

N_2 adsorption isotherms at 77 K on a series of activated carbons (a) before and (b) after *n*-nonane pre-adsorption. (Adapted from *J. Colloid Interface Sci.*, 106, Rodríguez-Reinoso, F., Martín-Martínez, J.M., Molina-Sabio, M., Torregrosa, R., and Garrido-Segovia, J., 315, Copyright 1985, with permission from Elsevier.)

well-defined knee and, additionally, exhibit a low amount adsorbed, thus reflecting the complete blockage of the microporosity by *n*-nonane. Above 52%, the burn-off samples exhibit a wide pore size distribution, the larger microporosity not being affected by *n*-nonane.

9.3.2.3 Mesoporosity

It was shown in Figure 9.1 that Type IV and V isotherms are initially coincident with nonporous Type II and III, but they level off at high relative pressure and they exhibit the presence of a hysteresis loop (adsorption and desorption loops are not coincident) that is usually associated with the filling and emptying of mesopores by capillary condensation. Although the presence of hysteresis loop is indicative of mesoporosity not always the mesoporosity will lead to hysteresis loop, as shown below.

The filling of mesopores takes place by capillary condensation and the Kelvin equation can be used to interpret the adsorption and desorption process. In fact, this equation is commonly applied in most methods devoted to provide the pore size distribution of a porous solid. Since capillary condensation starts at a relatively high relative pressure the pore walls are already covered with a physically adsorbed film of thickness t given by the relative pressure.

In the case of a cylindrical pore with a closed end, Figure 9.7 shows that capillary condensation occurs forming a hemispherical meniscus of radius r_c, which is smaller than the radius of the cylindrical pore, r_p ($r_p = r_c + t$). Since there is a given contact angle, θ, between the film and condensed adsorbate the real relation between r_c and r_p is given as

$$r_p = r_c \cdot \cos \theta + t \tag{9.16}$$

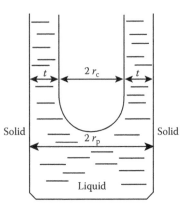

FIGURE 9.7
Relation between the Kelvin radius r_c and the pore radius r_p in a cylindrical mesopore. (Adapted from Rouquerol, F., Rouquerol, J., and Sing, K.S.W., *Adsorption by Powders and Porous Solids*, Academic Press, London, U.K., 1999, with permission from Elsevier.)

It is very difficult to measure the real contact angle θ, but for applications related to pore size distribution it is generally admitted that $\theta = 0$ and, consequently, $\cos \theta = 1$. The Kelvin equation is usually written as

$$\ln\left(\frac{p}{p_0}\right) = -\left[\frac{2 \cdot \gamma \cdot V_L}{RT}\right] \cdot \left[\frac{1}{r_m}\right] \tag{9.17}$$

where V_L is the molar volume of the liquid adsorbate. The equation may be simplified if $\gamma V_L / RT$ is written as K:

$$\left(\frac{p}{p_0}\right) = \exp\left[-\frac{2 \cdot K}{r_m}\right] \tag{9.18}$$

Capillary condensation starts at the bottom of the cylindrical pore forming a hemispherical meniscus in which r_m will be equal to r_c. The whole pore is filled when the relative pressure reaches the value

$$\left(\frac{p}{p_0}\right)_a = \exp\left[-\frac{2 \cdot K}{r_m}\right] = \exp\left(-\frac{K}{r_c}\right) \tag{9.19}$$

The desorption (evaporation) of the condensed liquid will start at the hemispherical meniscus near the open end of the cylinder and will then continue at the same relative pressure $(p/p_0)_a$ with the result that no hysteresis loop will be present.

If the cylinder is open at both ends condensation occurs on top of the adsorbed film of the inside wall of the cylinder, the result being that the meniscus is cylindrical. Condensation occurs at $(p/p_0)_a = \exp(-2K/r_c)$ since $r_m = 2r_c$ and the cylindrical pore becomes filled. However, evaporation from inside the pore can take place only from the hemispherical meniscus at the two ends of the tube and the pressure at which evaporation take place is

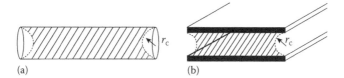

FIGURE 9.8
Schematic representation of the capillary condensation on (a) cylindrical and (b) slit-shaped pores.

$(p/p_0)_d = \exp(-2K/r_m) = \exp(-2K/r_c)$. Since condensation and evaporation occur at different relative pressures, a hysteresis loop will appear (see Figure 9.8a).

The situation may be more complex for other pore shapes. Let us consider the other important pore shape in adsorbents, slit-shaped, as it is the case for activated carbons (Figure 9.8b). In this case, the adsorbed film is situated in the two opposite walls of the pore and the film will increase in thickness until the two meet and then the liquid fills the pore. Evaporation will start at the hemi-cylindrical meniscus formed and since condensation and evaporation occur by different paths—multilayer condensation at the walls and evaporation from the hemi-cylinder—there will be a hysteresis loop.

Since capillary condensation in mesopores is almost always associated to a hysteresis loop it is very important to show the more frequent shapes that can be found experimentally, with the aim of showing the relationship between the loop shape and the shape of the mesopores. The four main types of hysteresis loop are selected in the IUPAC classification as H_1–H_4 (Figure 9.9).

Hysteresis loop type H_1 exhibits very vertical and parallel branches, thus indicating that capillary condensation takes place in mesopores of similar

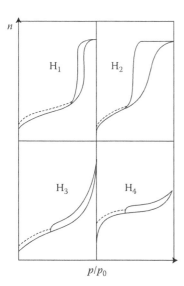

p/p_0

FIGURE 9.9
The IUPAC classification of hysteresis loops. (Adapted from Sing, K.S.W. et al., *Pure Appl. Chem.*, 57, 603, 1985.)

pore dimension, as it is the case of the new mesoporous silicas (e.g., MCM-41 and SBA-15). Because of the equal dimension of mesopores the desorption branch falls almost vertically since the pores are emptied at the same relative pressure.

Hysteresis loop type H_2 has a more progressive filling of mesopores (indicating that the larger size require a higher relative pressure to be filled) until all mesopores are filled (flat portion at a relative pressure near unity). However, the desorption branch is rather steep, almost vertical, thus indicating that evaporation takes place from mesopores of similar dimension. This could be the case for an adsorbent with ink-bottle mesopores, the entrance being smaller than the bulk of the pore.

Hysteresis loop type H_3 is usually attributed to solids constituted by platelets (as in some inorganic solids such as micas) or with slit-shaped mesopores. The only difference with type H_4 is the important contribution of microporosity in type H_4, this being very common in activated carbons. It is also very common to see the adsorption–desorption isotherm for activated carbons with a hysteresis loop between those of types H_3 and H_4; in many of these cases, the adsorption goes up very steeply at relative pressures near 1, where very probably a condensation of the adsorbate is taking place and the mesoporosity may not be as large as it seems to be. This is an aspect meriting more research in the near future.

A problem commonly found when using adsorption of gases and vapors for the characterization of solids is that the end of the adsorption isotherm, near saturation, is not always well defined. When, as noted above, the adsorption loop goes up almost vertically near saturation, condensation in the interparticle spaces or within platelets and slit-shaped particles is very possible and consequently the definition of the total pore volume adsorbed is not straightforward. For this reason, many authors take as total pore volume the amount adsorbed as liquid at a relative pressure near 1 (either 0.95 for some authors or 0.99 for other authors). There is no such problem when the adsorption branch of the isotherm exhibits a plateau very near saturation pressure, this being then the total volume of pores for the solid under test if the density of the adsorbate is taken to calculate the pore volume. This is the base for the so-called Gurvich rule since, if this total pore volume is determined by using different gases and vapors, the value will be independent of the adsorptive used. The rule is followed by many solids provided there is no restricted access to the porosity by any of the adsorptives used.

The Kelvin equation can be used to determine the pore size distribution from the adsorption or desorption isotherm. Although many methods have been proposed for this end, only a few of them remain to be used and included as part of the software of many commercial adsorption equipments. Among them the most popular is the Barrett, Joiner, and Hallenda (BJH), proposed in 1951 [13]. A typical example of a pore size distribution obtained with this method is given in Figure 9.10, where the maximum corresponding to different pore sizes is appreciated. It is also to be noted that more and

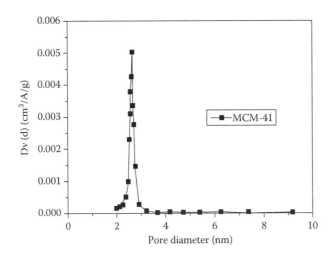

FIGURE 9.10
Pore size distribution obtained after application of the BJH method to the nitrogen desorption data at 77 K for mesoporous MCM-41 material.

more frequently some of the variations of density functional theory (DFT) are being used to calculate the pore size distribution when micropores are present in the test solid [14,15].

Given the fact that the previous paragraphs were devoted to the determination of the pore volume of a given solid from the adsorption isotherm, it may the appropriate to describe a simpler way to calculate the porosity of a given solid by measuring the density of the given solid with helium and mercury. The so-called real density is measured with helium since this gas will penetrate almost all pores of the solid given its small molecular size. Because of this, the inverse of this real density, $1/\rho_{He}$ will represent the volume of the solid (of course, plus any submolecular pores to which it cannot enter). On the other hand, the so-called apparent density is measured by mercury picnometry, which at atmospheric pressure cannot penetrate pores below several microns in diameter (the exact value is a function of the contact angle taken for mercury, but it is conventionally taken around 15 μm). Because of this, the reciprocal of the apparent mercury density $1/1/\rho_{Hg}$ is the volume of the material with all its pores. The consequence of these two reciprocal values is that the pore volume of the solid is given by $1/\rho_{Hg}-1/\rho_{He}$. This value is very similar to the value obtained from the Gurvich rule for many porous materials.

9.3.2.4 Macroporosity

The adsorption of nitrogen at −196°C cannot be used for the determination of pores in the range of macropores (above 50 nm dimension) because the relative pressure at which these macropores are filled is very close to unity.

For instance pores of 50 nm are filled at a relative pressure of around 0.96 but pores of 100 nm are filled at 0.98 and so on. It is almost impossible to distinguish all the macropore sizes in such a small range of relative pressure. It is for this reason that the development of mercury porosimetry took place. Since the contact angle of mercury is larger than 90°, external pressure has to be applied to force the mercury to enter the pores of the solid. The equation of Washburn [16] is used to relate the pore size to the intrusion pressure used, that is,

$$r_p = - \frac{2 \cdot \gamma \cdot \cos \theta}{\Delta P} \qquad (9.20)$$

where r_p is the pore radius. The actual commercial porosimeters are prepared to reach either 2000 or 4000 bars, this indicating that the minimum pore size that can be measured is either 7.5 or 3.75 nm, respectively. The maximum pore dimension measured at atmospheric pressure is 15 μm. Although this seems to be a relatively simple technique, the main problem is related to the toxicity of mercury and for this reason some other alternatives are now being searched for, although with no result till date.

9.3.2.5 Problems Commonly Found in Characterization by Adsorption

There are many experimental problems that can make the characterization of porous solids by adsorption a complex process. In what follows, a description of the main problems found by our research group along the last 30 years is briefly described.

9.3.2.5.1 Kinetics of Adsorption

Even being careful about the set of experimental parameters used in the adsorption of gases and vapors for the characterization of solid materials, it is very common to find manipulation errors that can lead to a set of not trustful data. The kinetics of adsorption of gas or vapor is very important. Although all commercial adsorption systems permit the control of the so-called time of equilibrium this is one of the main problems found. As a typical example, Figure 9.11a plots the adsorption isotherms for nitrogen at −196°C determined on a series of chars obtained from almond shells [17]. The preparation conditions were slightly changed in order to learn about the possible effect on the development of porosity by the subsequent activation with carbon dioxide.

The shape and the uptake for the chars are very different and unexpected and one reason could be that the equilibrium was not reached along the adsorption process. Figure 9.11b includes the isotherms for two of the previous chars exhibiting very different shape and uptake, obtained using a much larger time for equilibrium (to ensure equilibrium, a total of 6 weeks were taken to construct the isotherm). It can be observed that both isotherms are

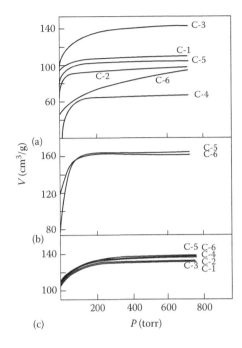

FIGURE 9.11

Adsorption of N_2 on carbonized almond shells: (a) apparent equilibrium (−196°C); (b) equilibrium at (−196°C); (c) at −183°C. (Adapted from *Carbon*, 20, Rodríguez-Reinoso, F., López-Gonzalez, J., and Berenguer, C., 513, Copyright 1982, with permission from Elsevier.)

very close together, this being a test for the lack of equilibrium in Figure 9.11a. Furthermore, in order to prove this, the adsorption of nitrogen was carried out at a higher adsorption temperature (−183°C) and not only the amount adsorbed was larger than in Figure 9.11b, but also all the isotherms were almost coincident (see Figure 9.11c). The fact that the uptake is higher at a higher temperature is a clear indication of the lack of equilibrium in the first set of isotherms. Adsorption is an exothermic process and an increase in temperature should mean a decrease in adsorption. If the adsorption uptake is larger it is because the higher temperature is facilitating the access of nitrogen to the interior of the micropores, which was restricted at −196°C because the size of the pore entrance is very similar to the molecular dimension. Consequently, increasing the adsorption equilibrium time has a similar effect to increasing slightly the adsorption temperature. This restricted diffusion of nitrogen to small micropores has been observed many times in the last few years.

The problem of equilibrium time and the restricted diffusion of nitrogen to narrow micropores (pores of around 0.4–0.5 nm dimension) led Rodríguez-Reinoso et al. [18,19] to propose in 1987 the use of adsorption of carbon dioxide at 0°C as a complementary technique to characterize porous carbons. Carbon dioxide is of similar kinetic dimension (0.33 nm) to nitrogen (0.36 nm) but, since the adsorption is at a much higher temperature, the restrictions should be minimum and the maximum relative pressure than can be reached at atmospheric pressure is 0.03, as compared to 1.0 for nitrogen at −196°C. Consequently, carbon dioxide at 0°C will be adsorbed only in very narrow micropores, the adsorption on larger pores and external surface area being almost negligible. This approach was generally accepted and most researchers working with porous carbons use it.

The surprise came to us when further testing the equilibrium conditions for carbon dioxide at 0°C, almost 200°C higher than the temperature of liquid nitrogen. The adsorption system considers that equilibrium is reached when several consecutive pressure readings agree to within 13.3 Pa, the time elapsed between measurements being set by the operator. In the case of carbon dioxide this time varied from 30 to 300 s. Figure 9.12 includes the carbon dioxide adsorption isotherms on two carbon molecular sieves, Takeda 3A and Takeda 5A [20].

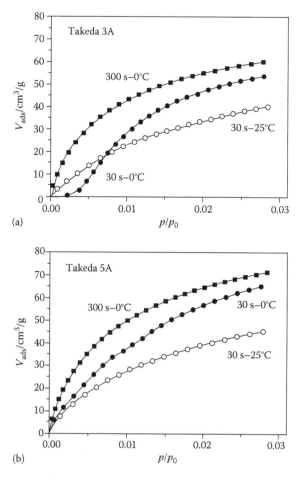

FIGURE 9.12
CO_2 adsorption isotherms at different equilibrium times (30 and 300 s) for carbon molecular sieves (a) Takeda 3A and (b) Takeda 5A at 0°C and 25°C. (Adapted from Rios, R.V.R.A. et al., *J. Phys. Chem. C*, 111, 3803, 2007. With permission.)

Carbon dioxide can be adsorbed on Takeda 3A at 0°C but this 0.3 nm molecular sieve does not adsorb nitrogen at –196°C under normal working conditions. However, at an equilibrium time of 30 s the adsorption is very weak and, as the time is increased, the shape and the uptake of the isotherm changes considerably. A further increase in the time above 300 s did not increase the uptake, thus confirming that a true equilibrium was being reached then. A further test of the restricted diffusion of carbon dioxide at 0°C was carried out by increasing the adsorption temperature to 25°C. The isotherm included in Figure 9.12a shows that the adsorption at 25°C and low relative pressures is higher than at 0°C, for the same reason given above for nitrogen. The diffusion is restricted at 0°C and the kinetics is increased with increasing temperature, facilitating the access to very narrow micropores. The restrictions are less evident for Takeda 5A (0.5 nm pore size), as clearly shown in Figure 9.12b.

These two particular cases can be extended to many others in which the solid material contains narrow micropores or when large molecules are adsorbed but they are kinetically restricted at the temperature of measurement. Consequently, this is an important aspect of adsorption that should be carefully checked when characterizing porous solids or when using them in adsorption processes, especially under dynamic conditions.

9.3.2.5.2 Adsorption Equipment

Another possible error when using adsorption equipment is related to the experimental setup to measure equilibrium pressure. A large proportion of adsorption systems are provided with a 1000 torr pressure transducer, this meaning that the precision for the pressure reading at low pressures is very limited. Figure 9.13 includes the adsorption isotherm for a silicalite sample obtained using such pressure transducer, together with the isotherm obtained for the same sample but using an experimental system that includes an additional 10 torr pressure transducer.

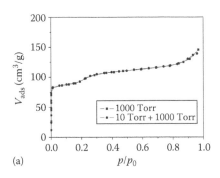

(a)

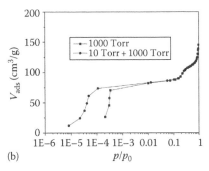

(b)

FIGURE 9.13
N_2 adsorption–desorption isotherms at –196°C on silicalite using conventional (1000 torr) and high precision (10 torr) pressure transducers.

Although both isotherms seems to be exactly coincident when plotted in a conventional way (Figure 9.13a), if the isotherms are plotted using a logarithm scale for relative pressure (Figure 9.13b), the differences are apparent. Consequently, if meaningful set of data are required the adsorption equipment should be equipped with either a 1 or 10 torr pressure transducer in addition to the conventional 1000 torr transducer.

9.3.2.5.3 *Type of Sample*

A third possible source of error is the own sample, especially when the particle size is large (aggregates, pellets, etc.). Figure 9.14 includes the adsorption isotherms for nitrogen at −196°C on a set of samples of Takeda 4A.

The reproducibility is relatively reasonable when using several pellets. However, the adsorption isotherm was repeated but using only one pellet in each adsorption experiment and the individual isotherms are shown in Figure 9.14. The large differences among the isotherms are apparent and they are due to the differences in the individual pellets, although the differences are compensated when several pellets are used for a single experiment. Although this may be an exceptional case it is possible to consider that something similar can be expected for samples with large particle size and that it could be expected not to be completely uniform due to the way in which the material is prepared (for instance, carbon molecular sieves).

9.3.2.5.4 *Recommendations*

With this background, it is convenient now to set an example of the use of gas adsorption for the characterization of materials. A well-defined sample

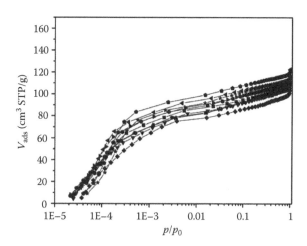

FIGURE 9.14

Nitrogen adsorption isotherms at −196°C on different individual pellets of carbon molecular sieve Takeda 4A.

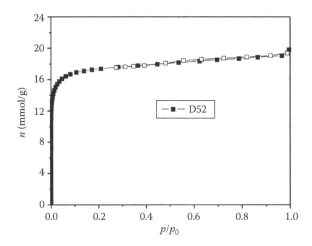

FIGURE 9.15

N$_2$ adsorption–desorption isotherm for activated carbon D52 at 77 K.

of activated carbon will be selected and the adsorption of nitrogen at −196°C and carbon dioxide at 0°C will be used for the characterization of the porosity. The results obtained when using the different approaches for the analysis of the porosity will be presented and the more convenient approach will be discussed for each case.

Figure 9.15 shows the N$_2$ adsorption isotherm at −196°C on an activated carbon prepared from olive stones with medium degree of activation (burn-off = 52 wt.%).

The isotherm exhibits the expected shape on an activated carbon, with an important uptake at very low relative pressures caused by the narrow microporosity. This steep branch is followed by an open knee that indicates the existence of wider micropores and an almost linear branch at higher relative pressures corresponding to adsorption in mesopores.

Figure 9.16 shows the BET plot for the activated carbon under test. It is clearly shown that if the classical BET range of relative pressures is used (0.05–0.30) the line is not straight, this meaning that one has to be careful when selecting the linear section of the plot.

In this particular case, the straight line is well defined if one considers the range 0.05–0.11 and the so-called monolayer volume is deduced from the slope. If the cross section of the nitrogen molecule is taken as 0.162 nm^2, the value for the surface area is immediate. It is also very important to check that the intercept is positive; otherwise the model will not apply to the given solid. This example for a typical activated carbon can be extended to any solid in which microporosity is present. It is then very important to check the linearity range and the positive intersection before providing any value for the apparent or equivalent surface area. This linearity range should be reported.

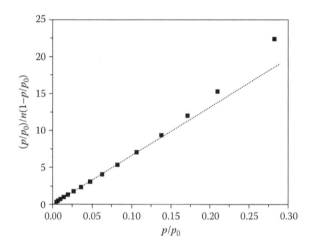

FIGURE 9.16
Typical BET plot obtained from the N_2 adsorption data at 77 K on activated carbon D52.

The next step is the calculation of the microporosity in the solid and three different approaches will be described: adsorption of N_2 at −196°C, adsorption of CO_2 at 0°C, and pre-adsorption of *n*-nonane.

Figure 9.17 shows the DR plot for the adsorption of nitrogen at −196°C using a different scale for the $\log^2 (p_0/p)$ axis in order to appreciate the main features. It is observed that a very low relative pressures—large values of $\log^2 (p_0/p)$—the experimental points deviate downward, this deviation being due to the lack of equilibrium at such a low region of relative pressure

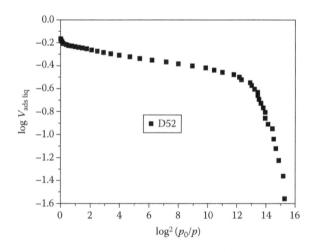

FIGURE 9.17
Typical Dubinin–Radushkevich plot obtained from the nitrogen adsorption data at 77 K on activated carbon D-52.

and/or to the remaining helium, used during the dead volume evaluation, and remaining in the narrow microporosity; these experimental problems were discussed in a previous section. When the region of the plot is reduced to the 0–10 range of $\log^2 (p_0/p)$, it is possible to draw a linear portion that can be extrapolated to the volume axis and deduce the volume of micropores expressed as volume of liquid adsorbate. However, it is clear that there is an upward deviation of the straight line at low values of $\log^2 (p_0/p)$, since larger pores are being filled. It is very common to find this type of deviation in carbons with a wide pore size distribution. On the contrary, if the carbon (or any other porous materials) contains only narrow micropores, the straight line will cover almost all the range of the DR plot. Sometimes, if the DR plot is examined more carefully it is possible to find an additional straight linear portion at lower values of $\log^2 (p_0/p)$, before capillary condensation produces the curved upward deviation (see Figure 9.18). If this shorter linear portion is extrapolated, the calculated volume will be greater than the one previously described, since it corresponds to the previously calculated volume plus that of larger micropores, which are being filled by a different secondary filling mechanism.

These deviations make the calculation of the micropore volume difficult, since one has to decide on the portion to be extrapolated for the determination. If an additional straight line can be defined below the value of $\log^2 (p_0/p) \approx 3$, its extrapolation will then provide the total volume of micropores of the solid. Rodríguez-Reinoso et al. [19] showed that this upper portion (below $\log^2 (p_0/p) \approx 3$) would lead to a micropore volume coincident with that obtained from the adsorption of hydrocarbons at 25°C; the DR plots for these vapors exhibited only one straight portion of the plot.

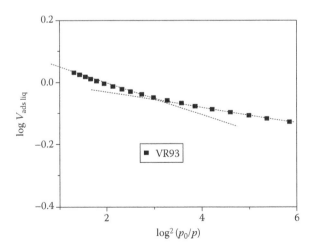

FIGURE 9.18
DR plot obtained from the N_2 adsorption data at 77 K for an activated carbon (VR-93) obtained from mesophase pitch.

When the data for the adsorption of CO_2 at 0°C are plotted according to the DR equation, a relatively linear plot is found for this and many other porous carbons, the extrapolation from which provides the volume of narrow micropores. The group of Rodríguez-Reinoso proposed in 1987–1989 the use of carbon dioxide at 0°C as the way to determine the volume of narrow micropores (up to two molecular dimensions, around 0.7 nm) [18,19]. It was shown that when the microporosity was narrow the adsorption of N_2 at –196°C and CO_2 at 0°C would lead to the same value of micropore volume. In such cases, the DR plots for the two gases were linear in most relative pressure range. As the micropore dimension is widened, the DR plot for CO_2 is still linear but that of N_2 starts to exhibit two straight lines. Nitrogen seems to be filling the wider micropores, although with a different mechanism as discussed in a previous section.

As a further proof of this idea, Figure 9.19 shows the characteristic curve for the same carbon, for both nitrogen and carbon dioxide as adsorptives at –196°C and 0°C, respectively.

It should be remembered here that the relative pressure range covered by both gases at their temperatures of adsorption are 10^{-5}–1.0 and 10^{-4}–3×10^{-2} for N_2 and CO_2, respectively. It is clearly shown that the characteristic curve leads to a common volume of micropores only if trustful low relative pressure data for nitrogen are available. At this point it must be stated that even when using high precision pressure transducer (10 torr) some experimental points at very low relative pressures must be removed due to the still existing kinetic limitations (see Figure 9.19b). This situation is even worse if the pressure transducer of the equipment is 1000 torr, since in this case the minimum relative pressure trustful is 5×10^{-3}; for the later case, the extrapolation leads to a value higher than for CO_2 data for the volume of micropores. The analysis of many microporous materials, especially activated carbons, has shown that when the microporosity is wide (in the case of carbons, when the burn-off is larger than 40 wt.%) the use of only data for relative pressures above 0.005—the range covered in may experimental systems using only the 1000 torr pressure transducer—would lead to values of V_0 larger than those for CO_2 at 0°C, corresponding to only narrow micropores.

Pre-adsorption of *n*-nonane can also be used to further test the volume of micropores in a solid material (see Figure 9.6). In this case, the sample was out-gassed in a conventional way and *n*-nonane adsorbed at –196°C and later out-gassed overnight at room temperature. The volume of micropores has been calculated as the difference (at a relative pressure of 0.6) between the parallel adsorption isotherms for N_2 at –196°C obtained before and after pre-adsorption. The resulting value is in very good agreement with the value obtained from the low relative pressure portion of the DR plot.

In order to have a more complete view of the characterization of the carbon D-52, Table 9.1 includes the volume of micropores as obtained by all methods described in this section; in order to facilitate the explanation, two additional

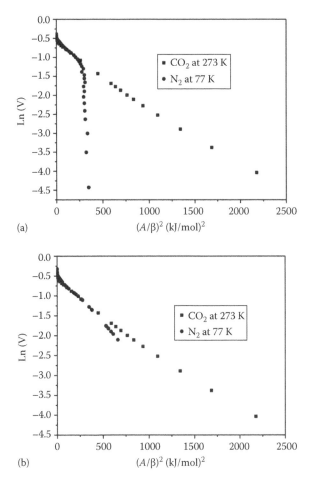

FIGURE 9.19

Characteristic curve obtained from N_2 adsorption data at 77 K and CO_2 adsorption data at 273 K on activated carbon D52 using (a) 1000 torr and (b) 10 torr pressure transducer for measuring the N_2 adsorption isotherm.

carbons have been added; carbon D-19, with low degree of activation, and carbon D-70, with a very high degree of activation.

Two values are given for the DR method applied to the nitrogen adsorption and the values for some DR plots corresponding to several hydrocarbons have also been included. When the adsorption data for N_2, hydrocarbons, and CO_2 are compared, one can conclude that for carbons with low burn-off the DR plots will give the same value of micropore volume, provided all of them are accessible to the microporosity. However, when the burn-off is high, hydrocarbons and N_2 measure the total volume of micropores if the higher straight portion (values of $\log^2 (p_0/p)$ below 3) of the DR plot is used to extrapolate to the volume axis, whereas CO_2 will only measure the narrow

TABLE 9.1

Micropore Volume (cm^3/g) Obtained using Different Methods for a Series of Activated Carbons

			Dubinin–Radushkevich					Langmuir	NP		α-Method
	N_2 77 K		Benzene	n-Butane	2,2-DMB	Iso-octane	CO_2				
Sample	Ip	hp	298 K	273 K	298 K	298 K	273 K	N_2 77 K	V_{NP}	V_n	N_2 77 K
D-19	0.31	—	0.25	0.28	0.25	0.21	0.30	0.32	0.31	0.24	0.31
D-52	0.50	0.55	0.49	0.49	0.50	0.50	0.41	0.54	0.51	0.41	0.53
D-70	0.57	0.67	0.69	0.66	0.66	0.68	0.48	0.68	0.52	0.47	0.64

Source: Rodríguez-Reinoso, F. et al., *Carbon*, 27, 23, 1989.

microporosity. All this is in the understanding that equilibrium is reached when measuring the corresponding adsorption isotherms.

If the data obtained from *n*-nonane pre-adsorption is included in the comparison, one can see that V_0 (CO_2) and V_{NP} are almost coincident for carbons D-19 and D-70 but V_0 (CO_2) is somewhat lower for D-52. The analysis of many series of activated carbons covering a wide range of activation has shown [21] that in a very large proportion of cases the two values agree rather well, thus validating the use of both approaches to determine the volume of micropores.

The values of micropore volume deduced from the application of the α_S method to the adsorption of N_2 at $-196°C$ have also being included in Table 9.1. An example of application to carbon D-52 can be found in Figure 9.20 using our own standard [22]; in this case the volume of micropore is obtained from back-extrapolation of the plot to the volume axis.

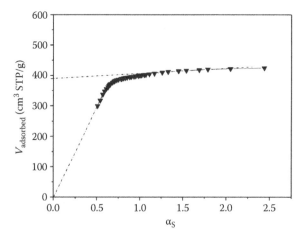

FIGURE 9.20
Characteristic α_s-plot for activated carbon D52.

In summarizing, it can be concluded that

1. If the microporosity is narrow and accessible to the adsorptive, all experimental methods described will yield almost the same volume of micropores.

2. If the microporosity is widely distributed, the α_S plot applied to the adsorption of N_2 at $-196°C$ will provide a volume of micropores almost in complete agreement with the value obtained if the DR plot is constructed with the adsorption of hydrocarbons data. If the DR plot for the adsorption of N_2 at $-196°C$ exhibits an additional straight portion at lower values of $\log^2 (p_0/p)$, its extrapolation will yield a volume of micropores in agreement with that from the α_S plot.

3. The adsorption of CO_2 and pre-adsorption of n-nonane are restricted to narrow microporosity.

4. The narrow microporosity can be evaluated by the adsorption of CO_2 at $0°C$ and the total microporosity by the adsorption of N_2 at $-196°C$. Pre-adsorption of n-nonane can be used as a further test.

9.4 Immersion Calorimetry

Although this chapter is devoted to adsorption, it is important to include a few basic concepts for an additional related technique that can be used in the characterization of porous solids, immersion calorimetry.

The enthalpy of immersion is defined as the enthalpy change, at constant temperature, that takes place when a solid is immersed into a wetting liquid in which it does not dissolve nor react. When the solid is out-gassed under vacuum previously to the calorimetric measurement, the enthalpy of immersion mainly depends on the following parameters:

1. Extent of the solid surface; for a given liquid–solid system the enthalpy of immersion increases with the surface area available.

2. Chemical nature of the liquid and of the solid surface; specific interactions between the solid surface and the liquid increase the enthalpy of immersion as in the case of, for instance, the wetting of polar surfaces with polar liquids.

3. Porous texture of the solid; when the liquid has a molecular size close to the dimensions of the pores, enhanced interactions can yield an increase in the enthalpy of immersion. Furthermore, molecules larger than some pores will not be able to access a given extent of the surface. Thus, the use of liquids with different molecular size permits the estimation of the experimental micropore size distribution

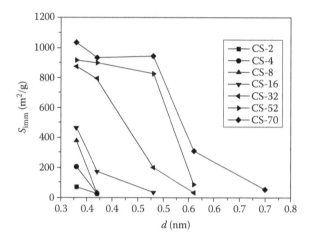

FIGURE 9.21
Surface area accessible to liquids with different molecular kinetic diameter, obtained by immersion calorimetry. (Adapted from *Colloid Surface A Phys. Eng. Aspect*, 187–188, Silvestre-Albero, J., Gómez-de-Salazar, C., Sepúlveda-Escribano, A., and Rodríguez-Reinoso, F., 151, Copyright 2001, with permission from Elsevier.)

of the solid. Of course, the molecular and pore shapes are also important. Benzene, a planar molecule of 0.37 nm width and 0.57 nm length, can enter a slit-shaped pore of say 0.4 nm pore dimension but cannot enter cylindrical pores of the same pore size.

A typical example of the immersion of several liquids (dichloromethane, 0.33 nm; benzene, 0.37 nm; cyclohexane, 0.48 nm; 2,2-dimethylbutane, 0.56 nm; α-pinene, 0.70 nm) into different carbon molecular sieves is shown in Figure 9.21 [23]. A nonporous carbon black, V3G, with a BET surface area of 62 m²/g was used as a reference to obtain the surface area available to the different liquids. For a given carbon the accessible surface area decreases as the molecular size of the liquid increases, but in all cases the pore size is kept below 0.7 nm.

An interesting comparison of adsorption and immersion is provided in Table 9.2, corresponding to carbons of the series D from which the carbon D-52 used in the previous section for the application of the different characterization methods was taken. The BET apparent surface areas are included for the sake of comparison. As it can be seen, except for carbons D-8 and D-19, for which the surface area available to the largest molecules (2,2-dymethylbutane and *iso*-octane) is lower, very similar values are obtained with both techniques for activated carbon with larger degree of activation not exhibiting molecular sieving toward the largest molecules of the liquids.

Carbon D-8 has mainly micropores that are narrower than 0.56 nm, the minimum dimension of 2,2-dimethylbutane, and, consequently, a high proportion of micropores are able to accommodate only one molecule of benzene or nitrogen (kinetic dimension, 0.37 and 0.36 nm, respectively). In this

TABLE 9.2

Specific Surface Area for a Series of Activated Carbons
Estimated from Immersion Calorimetry Measurements into
Different Liquids

	BET	Immersion Calorimetry		
Sample	N_2 (77 K)	Benzene	2,2-DMB	Iso-Octane
D-8	647	754	117	66
D-19	797	917	452	463
D-34	989	1114	958	928
D-52	1271	1402	1192	1243
D-70	1426	1426	1357	1460

BET surface area obtained from the nitrogen adsorption data is
included for sake of comparison.

case, immersion calorimetry provides the surface area actually accessible
to the molecules, whereas the nitrogen BET surface area is underestimated
by the fundamentals of the method themselves. The surface areas derived
from nitrogen isotherms and immersion into 2,2-dimethylbutane and *iso*-
octane are similar for carbons D-34, D-52, and D-70. This is so because the
micropores can now accommodate at least two molecules of nitrogen or at
least one molecule of the organic liquids. The presence of a residual narrow
microporosity could be the origin of the higher surface area obtained with
benzene.

9.5 Current Challenges in Adsorption Processes

9.5.1 Air Separation

The main components of air, nitrogen and oxygen, are the second and
third most produced chemicals since they are used in a very large number of industrial processes. Both gases have been traditionally produced by
cryogenic distillation, which is highly efficient for large volume production
(about 70% of the market). However, more than 20% of the two gases is produced by adsorption. Nitrogen and oxygen adsorbs to the same extent in
most adsorbents except on zeolites because the heat of adsorption for nitrogen is higher than for oxygen as a consequence of the quadrupole moment of
the former. When pressure swing adsorption (PSA) was invented in the early
1960s [24], the possibility of separating nitrogen and oxygen from air at room
temperature became possible. The development of vacuum swing adsorption (VSA) and the invention of the low silica zeolite LiLSX [25] took place
in the 1980s and since then, the production of both gases by the advancing
PSA/VSA cycles is continuously increasing.

In the case of obtaining oxygen from air, the apparently simplest separation since nitrogen is more strongly adsorbed in zeolites, the search was for a zeolite on which the interaction of nitrogen with the surface was greatly enhanced with respect to oxygen. The low silica X zeolite with Si/Al = 1 on which the exchanged small lithium ions are accessible to the nitrogen molecule was a very appropriate adsorbent for VSA cycles to obtain oxygen from air at room temperature. It was later found [26,27] that the silver cation Ag^+ exhibited a strong but reversible interaction with nitrogen and it was proved that just 1–3 wt.% of silver added to LiLSX can considerably improve the selectivity for oxygen separation by using PSA/VSA cycles.

Since the proportion of nitrogen in air is more than three times larger than oxygen, it should be easier to separate nitrogen from air by using an adsorbent selective to oxygen. Although this separation is carried out by industry using PSA, the volume is not as large as for oxygen and the purity of the resulting gas (around 99.5%) is lower than expected. In this case, the kinetic separation is based on the fact that oxygen is able to diffuse faster into the micropores than nitrogen because its kinetic dimension, 0.46 nm, is slightly smaller than that for nitrogen, 0.64 nm. The first adsorbent developed for the PSA separation of nitrogen from air was a carbon molecular sieve produced by Bergbau Forschung Co. in Germany, on which the diffusion of oxygen into the microporosity was much faster than the diffusion of nitrogen molecules (under equilibrium conditions, the adsorption isotherms for both gases were almost identical). Current industrial production of nitrogen by PSA uses carbon molecular sieves and in all cases the search is for a carbon porosity in which the diffusivity of oxygen is at least 30 times larger than for nitrogen. For some small units in which the interest is to produce enriched nitrogen it is possible to use a zeolite, NaA (4A), using the same principle, the faster diffusivity of oxygen.

The main challenge in this area is the development of advanced structured adsorbent materials for use and further development of the design in rapid-cycle PSA; the main application will be air separation to produce pure nitrogen and oxygen.

9.5.2 Hydrogen Purification

The actual industrial production of hydrogen is based on the steam reforming of natural gas, naphtha, or refinery gases and if hydrogen is the only product sought, the resulting gas is processed through a water–gas shift reaction to increase the proportion of hydrogen. Hydrogen purification is also necessary in the recovery of hydrogen from refinery and petrochemical plants, where it is accompanied by N_2, CO, CO_2, C_{1+}, etc. Purification is carried out using standard PSA technology and since the mixture of gases passing through it is constituted by several gases, it is common to use beds constituted by different layers of adsorbent. The more strongly adsorbed gas is retained first and the others are separated in the other adsorbent(s).

The main challenge in this area is the development of high-capacity CO_2- and CO-selective adsorbents that can operate in the presence of hydrogen and steam at elevated temperatures (working capacities in the range of 3–4 mol/ kg are of particular interest), along with the development of new PSA or temperature swing adsorption (TSA) cycle designs (possibly a PSA/TSA hybrid cycle design), at either ambient or elevated temperatures, which take advantage of these new adsorbents for the purification of hydrogen. Also, improved hydrogen separations with sorption enhanced reaction processes (SERPs), using a thermal swing regeneration and new materials—novel approaches— such as incorporating a high-temperature reversible metal hydride as a H_2 selective adsorbent in a SERP to drive the equilibrium, should be considered.

9.5.3 Other Gas Separations

Although there are many industrial processes requiring gas separation, the purification of natural gas (main component, methane) will be used here as a typical example. The composition of natural gas is a function of the region where it is being extracted. The main components of natural gas besides methane are carbon dioxide, nitrogen, and heavier hydrocarbons and the two main separation processes are for methane/carbon dioxide and methane/nitrogen. In both cases, the two gases reduce the heating value specifications of natural gas and their proportion has to be reduced to a minimum value. In the case of nitrogen, the problem is also present in the oil recovery processes since nitrogen is injected into the reservoir and, consequently, the petroleum gases become enriched in nitrogen.

The methane/carbon dioxide separation is much easier since there are a number of carbon molecular sieves that can be used for this purpose. In this particular case, the adsorption of carbon dioxide is much stronger at room temperature than the adsorption of methane. As a typical example, Figure 9.22 shows the kinetics of adsorption at 25°C on a carbon molecular sieve prepared from coconut shells [28].

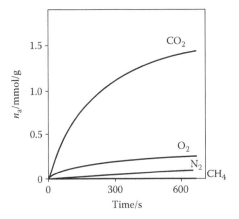

FIGURE 9.22
Adsorption kinetics of CO_2, CH_4, O_2, and N_2 at 25°C on carbon molecular sieve PS-HNO$_3$ (400). (Adapted from *Carbon*, 38, Gómez-de-Salazar, C., Sepúlveda-Escribano, A., and Rodríguez-Reinoso, F., 1879, Copyright 2001, with permission from Elsevier.)

The adsorption of methane is nil after 600 s, whereas the adsorption of carbon dioxide is large. Since the practical/industrial separation is carried out using a typical PSA system (the main reason for this is that the feed is already at high pressure), the separation factor can be defined as the ratio of gases adsorbed at, say, 120 s. It is clearly seen that in this particular case the separation factor is practically infinite. The reason for the use of carbon molecular sieves instead of a zeolite is that the adsorption of carbon dioxide is relatively strong in these materials and the desorption is, consequently, inhibited.

However, the nitrogen/methane separation is much more difficult because no appropriate adsorbent has been developed till date. Both molecules are nonpolar, although nitrogen exhibits a quadrupole moment and the equilibrium selectivity is slightly favored for methane. It is believed that the separation could be of kinetic type because the kinetic dimensions of both molecules are slightly different, 0.38 nm for methane versus 0.364 nm for nitrogen. Although carbon molecular sieves have been successfully used for some kinetic separations, no successful carbon molecular sieves have been prepared for this specific nitrogen/methane separation [29]. Much research is still needed to find the right adsorbent and the best PSA system to reach the removal of the nitrogen needed for the natural gas in the pipeline.

9.5.4 SO_2 and NO_x Removal

Adsorption has been playing an important role in environmental control of many industrial pollutants, a very wide range of adsorbents being used for this purpose, from activated carbon (the most popular) to polymeric resins, zeolites, etc.

Sulfur oxides (SO_x) are among the most toxic gases emitted into the atmosphere during the combustion of fossil fuels in industry (especially electric power stations) and, consequently, the restrictions on the emissions are becoming stricter. In a practical process, SO_2 is removed mainly by solid–gas reactions using calcite as adsorbent or in gas–liquid reactions using basic solutions [30]. However, these solutions are not appropriate because they require large investments and generate large amounts of by-products. SO_2 removal by adsorption provides a good alternative and several adsorbents have been investigated, including metal oxides, zeolites, and porous carbons. The latter have been widely investigated not only due to their large porosity and ability to remove a large amount of species in effluents but also because they can be used as support for catalysts. The main problem here is the reactivity of carbon in the presence of oxygen from air at medium temperatures. Metals can be used to reduce the reaction temperature and to minimize carbon loss by gasification, the more effective being Cu, V, and Fe. It has been recently found [31] that the catalytic adsorption of SO_2 on activated carbon provides an excellent alternative for the control of low concentrations of this pollutant. The surface complexes formed upon the adsorption on the carbon surface showed that the only species present was SO_3. The addition of copper to the carbon and the presence of oxygen assisted the transfer of oxygen to the carbon matrix, favoring the adsorption of SO_2 and increasing the break-through time, the adsorption capacity, and the formation of sulfur–oxygen complexes of higher thermal stability.

Nitrogen oxides (NO_x) are also very toxic gases emitted during the combustion of fossil fuels in both fixed and mobile plants. In the case of NO_x, the most used commercial techniques for removal are the selective catalytic reduction (SCR) by NH_3 and the selective noncatalytic reduction (SNCR). However, these processes have problems associated; for the former, the catalytic deactivation caused by catalyst poisoning by accompanying SO_2 and the high cost of maintenance, and for the later, the main problem is the high temperature required for the process [32–34]. It has been recently described that since both gases are simultaneous in flue gas, instead of attempting the individual removal of both from the effluent, their simultaneous removal could be advantageous [35]. Several technologies can be found in industry for this purpose. One is the SCR of NO_x with NH_3 in the presence of oxygen, followed by the adsorption or catalytic oxidation of SO_2 with supported metal oxide catalysts. However, this is an expensive process and there is deactivation of the catalyst at low temperatures. The Bergbau Forschung process uses the simultaneous removal of SO_2 and NO_x using a two-stage moving bed with activated coke [36]. Since the simultaneous removal of both species in a single process at room temperature is more desirable, it has been the object of much research. A recent report has shown that when adsorbing mixtures of SO_2 and NO on activated carbon, the presence of dispersed copper enhanced (as in the case of SO_2 catalytic adsorption described above) the chemical adsorption of both gases in such a way that SO_2 inhibited the adsorption of NO, whereas the adsorption of SO_2 was notably improved.

9.5.5 Gas Storage

The search for fuels to substitute for gasoline and diesel for vehicular use has been going on for many years. In the case of natural gas (constituted mainly by methane once it is purified from the presence of carbon dioxide, nitrogen, and hydrocarbons C_{2+}), the base of comparison is compressed natural gas (CNG), with an energy density of around 220 V/V, since liquefied natural gas is stored at −161°C at 1 atm pressure, and it is not recommended for passenger cars. The use of CNG implies the use of a very high pressure (around 20 MPa) and it is much more convenient to use adsorbed natural gas (ANG) provided this system is able to give an energy density similar to CNG but at a much lower pressure (it is actually accepted that a pressure of about 3.5 MPa would be adequate). The U.S. Department of Energy set the acceptable figure of 150 V/V at 3.5 MPa and room temperature. This requires an adsorbent with a very high adsorption capacity coupled with a high bulk density and the search for such adsorbent has produced a large amount of research in the last 30 years.

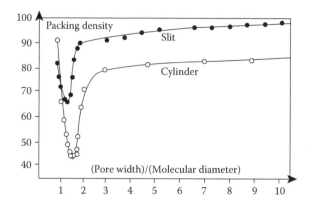

FIGURE 9.23
Packing of spherical molecules in narrow cylinders and slits. (Adapted from Rouquerol, F., Rouquerol, J., and Sing, K.S.W., *Adsorption by Powders and Porous Solids*, Academic Press, London, U.K., 1999, with permission from Elsevier.)

In order to prepare an adsorbent with high-adsorption capacity in a minimum volume one can think of zeolites and activated carbons, both with very large apparent surface areas. However, it is clear that the packing density of adsorbed molecules inside the solid is much larger in slit-shaped micropores than in cylindrical micropores of the same dimension (see Figure 9.23).

For this reason, activated carbon has been considered to be the best adsorbent possible since its micropores are slit-shaped. Furthermore, theoretical calculations have indicated that the optimum energy density for activated carbon would be reached when the carbon has a microporosity constituted by micropores in which two layers of adsorbed methane can fit; this means a pore dimension of around 0.8 nm. Although it is possible to prepare activated carbon with the needed porosity, the main problem when considering natural gas storage is the bulk volume, since the development of porosity occurring through the gasification of carbon reduces the solid carbon and decreases the bulk density. An interesting approach is to prepare monolithic activated carbon in which most of the porosity is microporosity (although some mesoporosity and macroporosity is needed to ensure a fast kinetics for the adsorption and desorption of methane) and the rest is solid carbon (skeleton carbon). Results by the group of Rodríguez-Reinoso [37–40] have shown that by conforming the carbon along the impregnation step for chemical activation with phosphoric acid it is possible to prepare self-binding monoliths of relatively high bulk density. In this way, it is possible to have monoliths with energy density close to the target provided by the U.S. Department of Energy.

In the case of hydrogen storage, the situation is very different because there are many uncertainties about some published results and the real storage target set by the U.S. Department of Energy, which is 6.5 wt.% for on-board storage in vehicles at room temperature has not been really reached yet. Although much research is being carried out with metal hydrides, there has been no major breakthrough in the last few years and, additionally, storage involves an absorption–desorption process and will not be described in this section.

9.5.6 Desulfurization of Fuels

The protection of the environment worldwide includes the use of cleaner transportation fuels; the new regulations in most advanced countries are asking for a drastic reduction in the sulfur content in gasoline and diesel. For instance, in Europe, United States, and Japan the average sulfur content in fuels is going to be reduced to around 10 ppm. The traditional hydrodesulfurization (HDS) process used in refineries is rather efficient to remove most sulfur-containing compounds except thiophene and derived compounds. The more recent deep desulfurization, being much more expensive, is more effective in sulfur removal but there still refractory compounds that cannot be removed. Because of this there has been a worldwide research for processes to treat the effluent from fluid catalytic cracking (FCC) by selective adsorption, since this process will work at room temperature and pressure. However, one has to take into account that the success of this process depends on the possibility of preparing an adsorbent that has to be (a) very selective to the adsorption of sulfur-containing molecules, (b) able to adsorb a large amount of these compounds, and (c) should be completely regenerated using very simple and inexpensive procedures. These three conditions are hard to be found in a commercial adsorbent and, consequently, research has been directed to the development of the appropriate adsorbents to satisfy the above requirements.

The main condition for a high selectivity of the solid surface toward the sulfur-containing species is a relatively high enthalpy of adsorption which, on the other hand, has to be overcome by the regeneration treatment. This principle was used in the development of the Irvad process [41,42] in which the adsorbent Alcoa Selexsorb was used. The main problem is that hydrogen has to be used for the regeneration of the spent adsorbent. A different approach was used by Yang et al. [43,44] in their search for π-complexation adsorbent, on which a weak bonding can be formed with the adsorbate. Chemical complexation bonds are usually stronger than van der Waals interactions, but weak enough to be reversible. Consequently, their selectivity is much higher. Most π-complexation adsorbents are constituted by Ag^+ and Cu^+ cations highly dispersed on high surface area of porous solids such as zeolites, silica, γ-alumina, etc., and the energy of adsorption of species such as thiophene may be of the order of 60–70 kJ/mol. This is stronger than van der Waals interactions, but not high enough to make the adsorption irreversible. Hernández-Maldonado and Yang [45] reported very good selectivity for the removal of thiophene using zeolites doped with silver and copper cations. The adsorbents could be regenerated by air oxidation followed by auto-reduction in an inert atmosphere.

Carbon has also been used as an adsorbent for the selective removal of sulfur-containing species found in gasoline and diesel [46–51]. In this case, some simple species such as thiols can be removed by the undoped carbon; however, for species such as thiophene or derived molecules metal doping proved to be necessary.

Acknowledgments

Financial support from MEC (Spain) (MAT2007-61734), Generalitat Valenciana (Spain) (PROMETEO/2009/002), and the Network of Excellence InsidePores (NMP3-CT2004–500895) is gratefully acknowledged.

References

1. Sing, K. S. W., Everett, D. H., Haul, R. A. W., Moscou, L., Pierotti, R. A., Rouquerol, J., and Siemieniewska, T. *Pure Appl. Chem.* 57 (1985), 603.
2. Rouquerol, F., Rouquerol, J., and Sing, K. S. W. *Adsorption by Powders and Porous Solids.* Academic Press: London, U.K., 1999.
3. Brunauer, S. and Emmett, P. H. *J. Am. Chem. Soc.* 59 (1937), 2682.
4. Lippens, B. C. and De Boer, J. H. *J. Catal.* 4 (1965), 319.
5. Gregg, S. J. and Sing, K. S. W. *Adsorption, Surface Area and Porosity.* Academic Press: New York, 1982.
6. Rodríguez-Reinoso, F. and Linares-Solano, A. Microporous structures of activated carbons as revealed by adsorption methods. In: *Chemistry & Physics of Carbon,* Radovic, L. R., Ed. Marcel Dekker: New York, 2001, Vol. 27.
7. Everett, D. H. and Powl, J. C. *J. Chem. Soc., Faraday Trans. I* 72 (1976), 619.
8. Stoeckli, H. F., Houriet, J. P., Perret, A., and Huber, U. In: *Characterization of Porous Solids.* S. J. Gregg, K. S. W. Sing and H. F. Stoeckli, eds., Society of Chemical Industry: London, U.K., 1979, p. 31.
9. Dubinin, M. M. *Progress in Surface and Membrane Science.* D. A. Cadenhead, Ed., Academic Press: New York, 1975, Vol. 9.
10. Huber, U., Stoeckli, F., and Houriet, J. P. *J. Colloid Interface Sci.* 67 (1978), 195.
11. Stoeckli, H. F. *J. Colloid Interface Sci.* 59 (1977), 184.
12. Rodríguez-Reinoso, F., Martín-Martínez, J. M., Molina-Sabio, M., Torregrosa, R., and Garrido-Segovia, J. *J. Colloid Interface Sci.* 106 (1985), 315.
13. Barrett, E. P., Joyner, L. G., and Halenda, P. P. *J. Am. Chem. Soc.* 73 (1951), 373.
14. Jagiello, J. and Thommes, M. *Carbon* 42 (2004), 1227.
15. Neimark, A. V., Lin, Y., Ravikovitch, P. I., and Thommes, M. *Carbon* 47 (2009), 1617.
16. Washburn, E. W. *Proc. Natl. Acad. Sci. U.S.A.* 7 (1921), 115.

17. Rodríguez-Reinoso, F., López-Gonzalez, J., and Berenguer, C. *Carbon* 20 (1982), 513.
18. Garrido, J., Linares-Solano, A., Martín-Martínez, J. M., Molina-Sabio, M., Rodríguez-Reinoso, F., and Torregrosa, R. *Langmuir* 3 (1987), 76.
19. Rodríguez-Reinoso, F., Garrido, J., Martín-Martínez, J. M., Molina-Sabio, M., and Torregrosa, R. *Carbon* 27 (1989), 23.
20. Rios, R. V. R. A., Silvestre-Albero, J., Sepúlveda-Escribano, A., and Rodríguez-Reinoso, F. *J. Phys. Chem. C* 111 (2007), 3803.
21. Marsh, H. and Rodriguez-Reinoso, F. *Activated Carbon*. Elsevier: London, U.K., 2006.
22. Rodríguez-Reinoso, F., Martín-Martínez, J. M., Prado-Burguete, C., and McEnaney, B. *J. Phys. Chem.* 91 (1987), 515.
23. Silvestre-Albero, J., Gómez-de-Salazar, C., Sepúlveda-Escribano, A., and Rodríguez-Reinoso, F. *Colloid Surf. A: Phys. Eng. Aspects* 187–188 (2001), 151.
24. Skarstrom, C. W. U.S. Patent 2,944,627 (1960).
25. Chao, C. C. U.S. Patent 4,859,217 (1989).
26. Chao, C. C., Sherman, J. D., Mullhaupt, J. T., and Bolinger, C. M. U.S. Patent 5,174,979 (1992).
27. Coe, C. G., Kirner, J. F., Pierantozzi, R., and White, T. R. U.S. Patent 5,152,813 (1992).
28. Gómez-de-Salazar, C., Sepúlveda-Escribano, A., and Rodríguez-Reinoso, F. *Carbon* 38 (2000) 1889.
29. Ackley, M. W. and Yang, R. T. *Aiche J.* 36 (1990), 1229.
30. Kaminski, J. *Appl. Energy* 75 (2003), 165.
31. López, D., Buitrago, R., Sepúlveda-Escribano, A., Rodríguez-Reinoso, F., and Mondragón, F. *J. Phys. Chem. C* 112 (2008), 15335.
32. Gómez-Garcia, M. A., Pitchon, V., and Kiennemann, A. *Environ. Int.* 31 (2005), 445.
33. Lee, Y. W., Kim, H. J., Park, J. W., Choi, B. U., Choi, D. K., and Park, J. W. *Carbon* 41 (2003), 1881.
34. Lee, Y. W., Choi, D. K., and Park, J. W. *Sep. Sci. Technol.* 37 (2002), 937.
35. López, D., Buitrago, R., Sepúlveda-Escribano, A., Rodríguez-Reinoso, F., and Mondragón, F. *J. Phys. Chem. C* 111 (2007), 1417.
36. Wilde, J. D. and Marin, G. B. *Catal. Today* 62 (2000), 319.
37. Prauchner, M. J. and Rodríguez-Reinoso, F. *Microp. Mesop. Mater.* 109 (2008), 581.
38. Rodríguez-Reinoso, F., Nakagawa, Y., Silvestre-Albero, J., Juárez-Galán, J. M., and Molina-Sabio, M. *Microp. Mesop. Mater.* 115 (2008), 603.
39. Almansa, C., Molina-Sabio, M., and Rodríguez-Reinoso, F. *Microp. Mesop. Mater.* 76 (2004), 185.
40. Molina-Sabio, M., Almansa, C., and Rodríguez-Reinoso, F. *Carbon* 41 (2003), 2113.
41. Irvine, R. L. U.S. Patent 5,730,860 (1998).
42. Ondrey, G. *Chem. Eng.* 106 (1999), 25.
43. Yang, R. T., Takahashi, A., and Yang, F. H. *Ind. Eng. Chem. Res.* 40 (2001), 6236.
44. Takahashi, A., Yang, F. H., and Yang, R. T. *Ind. Eng. Chem. Res.* 41 (2002), 2487.
45. Hernández-Maldonado, A. J. and Yang, R. T. *Ind. Eng. Chem. Res.* 42 (2003), 123.
46. Katoh, H., Kuniyoshi, I., Hirai, M., and Shoda, M. *Appl. Catal. B: Environ* 6 (1995), 255.
47. Dalai, A. K., Tollefson, E. L., Yang, A., and Sasaoka, E. *Ind. Eng. Chem. Res.* 36 (1997), 4726.

48. Bagreev, A., Bashkova, S., and Bandosz, T. J. *Langmuir* 18 (2002), 8553.
49. Bandosz, T. J. *Activated Carbon Surfaces in Environmental Remediation*. Elsevier: Amsterdam, The Netherlands, 2006, pp. 231–292.
50. Rios, R. V. R. A., Silvestre-Albero, J., Sepúlveda-Escribano, A., and Rodríguez-Reinoso, F. *Colloids and Surfaces A: Phys. Eng. Aspects* 300 (2007), 180.
51. Rios, R. V. R. A., Silvestre-Albero, J., Sepúlveda-Escribano, A., and Rodríguez-Reinoso, F. Liquid-phase adsorption/oxidation of sulfur-containing species by activated carbon. Mota, José Paulo, Lyubchik, and Svetlana eds., In: *Recent Advances in Adsorption Processes for Environmental Protection and Security*, Springer, Kiev, Ukraine, 2006, pp. 107–118.

10

Fundamental Approach to Supercritical Hydrogen Adsorptivity of Nanoporous Carbons

Shigenori Utsumi and Katsumi Kaneko

CONTENTS

10.1 Introduction

Carbon is a highly functional material that has a very kaleidoscopic structure and morphology, although it consists of a single element. Carbon atoms construct a three-dimensional solid structure in diamond and a two-dimensional graphene structure, leading to graphite. It is well known that diamond and graphite have extremely different properties, irrespective of the same carbon components. Of course, carbon has various morphologies other than graphite and diamond. At the same time, carbon materials are both old and new materials. Porous carbon is just one representative. The adsorption phenomenon of porous carbon has been used in human society since prehistoric times, in the bleaching and refinement of food and drink such as liquor, water, and sugar. Bones and burials in BC 178 have been preserved in an incredibly good state in China by the adsorption phenomena of charcoals.

Thus, humans had noticed the intensive power of porous carbons for centuries. Scientific research on the adsorption phenomenon of porous carbon designates Scheele's research in 1773 as the beginning, and it has currently become one of the main fields of carbon science. However, extensive and active studies on adsorption by porous carbons have been started relatively recently by Dubinin, Marsh, Stoeckli, Sing, Rodroguez-Reinozo, and others. Activated carbons in porous carbon play an active role in the fields of environmental technology and medicine as well as in refining, separation, and catalysis, taking advantage of these characteristics. Furthermore, over the last 20 years, superhigh surface area activated carbon and activated carbon fiber (ACF) have extended the application of porous carbon to new fields such as an electric double-layer capacitor. There is a stronger demand for better understanding of activated carbons from nanoscale levels as materials indispensable to constructing sustainable technologies, regardless of recent advances in structural characterization.

Fullerene and carbon nanotube (CNT) came to join this carbon material family in 1985[1] and 1991,[2] respectively, driving explosive developments in science and technology. As CNT has an intrinsic nanoscale pore structure and its bundled form has the characteristic nanopore nature, CNT can be a representative nanoporous adsorbent. Since Dillon et al. reported that a single-wall carbon nanotube (SWCNT) can adsorb 10 wt% hydrogen at room temperature, several experimental and simulation researches on the hydrogen storage of SWCNT have been conducted, accompanied by the establishment of popular fuel cell technology.[3-7] Unfortunately, their CNT sample contained CNT of only 0.1–0.2 wt%, and later studies using purer CNT samples could not yield the high adsorption amount.[8-15] However, the hydrogen storage fever on CNT has provided a structural understanding of gas adsorption on CNT-related carbons.

It is essentially important to understand the current status of adsorption science on nanoporous carbons, including CNT families, for wide applications to sustainable technologies such as energy and environmental technologies. In this chapter, we explain supercritical gas adsorption on nanoporous carbons using nanoscale structural concepts, which has a great potential for new applications of nanoporous carbons.

10.2 Fundamentals

10.2.1 Gas–Solid Interaction

The adsorption phenomenon is caused by the interaction between molecules and solids. There are four types of interactions between gas molecules and solids as summarized in Table 10.1: physical adsorption, chemisorption,

TABLE 10.1

Interaction Types between Molecules and Solids

	Structural Change on Interaction	
	Molecular Structure	Solid Structure
Physical adsorption	No change	No change
Chemisorption	Change	No change
Absorption	No change	Change
Occlusion	Change	Change

absorption, and occlusion, which has a narrow meaning of storage. Here, "storage" has a broad concept, embracing the above-mentioned four interactions. These interactions are distinguished by the structural changes in molecules and/or solids, which involves an atomic structural change.[16] No structural change in molecules and solids is caused by physical adsorption. On the other hand, only molecules change their molecular structure on chemisorption, since they form a new chemical bond with a solid surface. As the structural change of solids is restricted only to the surface, the whole structure of solids does not change. For example, SO_2 deprives oxygen ions from the surface of iron oxide and changes into SO_3^{2-} or SO_4^{2-} when SO_2 is chemisorbed by iron oxide. An oxygen defect (V_O^-) is formed in the iron oxide lattice, which can release a pseudo-free electron, as expressed by the following defect equations:

$$2O^{2-}(s) + SO_2 = O^- + V_O^- + SO_3^{2-}(s), \tag{10.1}$$

$$V_O^- = V_O^* + e^-. \tag{10.2}$$

where V_O^- and V_O^* are an oxygen vacancy having an excess electron and a neutral oxygen vacancy, respectively.[17] The generations of SO_3^{2-} and pseudo-free electrons are confirmed by the presence of the SO_3^{2-} vibration in the infrared spectrum and the electrical conductivity increase of the *n*-type iron oxide. The representative example of absorption is the interaction between water molecules and layered clay minerals. Water molecules are introduced into the interlayers interstices of the layered structure of montmorillonite, giving rise to a remarkable swelling of the montmorillonite crystals.[18] The adsorption isotherm at the pseudo-equilibrium state of the swelling of the montmorillonite crystals shows a step-wise uptake, since a predominant absorption of water vapor begins at relative pressure. Thus, only solids vary the structure on absorption of molecules. It is easy to understand occlusion if we consider hydrogen occlusion by palladium. A hydrogen molecule dissociates into hydrogen atoms, which penetrate into a palladium crystal lattice, providing a new crystal lattice according to an H/Pd ratio.[19] Since a solid-state metal hydride formation on the interaction of metal with hydrogen molecules basically belongs to this type, the stable solid form cannot be maintained by repeated occlusion operations. Thus, occlusion changes the structure of both molecules and solids. A new classification of the above-mentioned four interaction types is quite useful to avoid confusion in understanding an interaction of molecules with nanoporous carbons and to design better adsorbents for supercritical gas. However, there can be a medium interaction, depending on the interaction strength and a quasi-hybrid interaction having each feature. We mainly discuss physical adsorption here.

10.2.2 Physical Adsorption and Chemisorption

Generally, adsorption implies physical adsorption and chemisorption. The attractive interaction of physical adsorption originates from the dispersion interaction, which is a basis of the attractive interaction of the van der Waals equation, which will be described in detail later. Chemisorption stems from chemical bond formation because surface chemical compound formation typically occurs. Therefore, it is necessary to handle the chemisorptive interaction between gas molecules and solid surfaces in quantum mechanics. Such interactions between molecules and solid surfaces must be individually solved in principle by quantum mechanics. Chemisorption on insulating solid surfaces is difficult to handle with a general theory; therefore, the quantum mechanics approximation to the individual system of chemisorption sites and molecules is required. On the other hand, chemisorptions on metal and semiconductors can be generally described by the electronic band theory by the introduction of surface levels. The boundary layer theory is a representative chemisorption theory of semiconductors, making much of the electron transfer between semiconductor and chemisorbed molecules.[20–22] For example, when an electron-acceptor molecule such as O_2 is chemisorbed on an *n*-type semiconductor, O_2 is chemisorbed on the surface as O^- or O_2^- by the electron transfer from a semiconductor to an O_2 molecule. As a result, the conduction band near the surface bends upward due to the electronic repulsion from the negatively charged surface levels that electrons of adsorbed molecules occupy; the surface is positively charged through the electron transfer to the surface levels, leading to the surface boundary layer. Once the stable surface boundary layer is formed, chemisorption does not progress. This theory is useful in understanding the electrical conductivity change on chemisorption, in terms of the relationship between the number of chemisorbed molecules and the change in charge carriers. However, ab initio quantum chemical calculation has been actively carried out to understand the detailed nature of chemisorption sites in recent chemisorption researches for more a precise description of a surface reactive structure.

There is no clear distinction on whether or not the adsorption by the hydrogen bond is chemisorption. One of the established viewpoints of the hydrogen bond is to deal with it as a wave function considering the

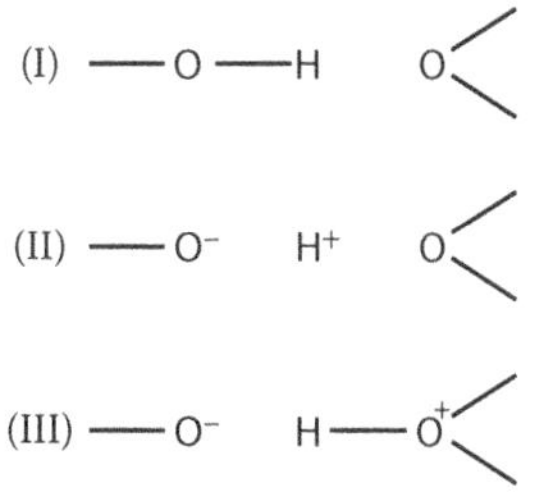

(I) ——O——H

(II) ——O⁻ H⁺

(III) ——O⁻ H——

O⟨

O⟨

O⟨

electrostatic arrangement, as shown in Figure 10.1. When wave functions depending on each arrangement I, II, and III are Ψ_I, Ψ_{II}, and Ψ_{III}, the wave function Ψ corresponding to the hydrogen bond is given by

FIGURE 10.1
Electron configuration of hydrogen bond.

$$\Psi = C_I\Psi_I + C_{II}\Psi_{II} + C_{III}\Psi_{III}, \qquad (10.3)$$

where C_i ($i=$ I, II, and III) is the constant. The general hydrogen bond is composed of 80%–90% of arrangement I, 10%–12% of arrangement II, and 2%–8% of arrangement III. Thus, it is necessary to consider the electrostatic interaction for the hydrogen bond. This is the reason why water exhibits unique properties; predominant intermolecular interactions except for a water molecule are not electrostatic interactions but dispersion interactions. Consequently, it is necessary to take the electrostatic interaction into account for water adsorption. As hydrogen bonding has been actively studied for elucidation of bio-functions, recent progresses in the hydrogen bond research may be referred to.[23,24]

The basis of the attractive interaction of physical adsorption is the dispersion force. We consider two neutral molecules, (a) and (b), each of which has one electron, numbered 1 and 2. The coordinates of these electrons 1 and 2 are assumed to be (x_1, y_1, z_1) and (x_2, y_2, z_2). When electrons 1 and 2 interact with each other on approaching different molecules, the perturbation term H' is expressed by

$$H' \approx \frac{e^2}{R^3}(x_1x_2 + y_1y_2 - 2z_1x_2), \qquad (10.4)$$

where R is the distance between both molecules. The primary perturbation energy of H' becomes zero, and the secondary perturbation is suggestive. When the wave functions of two molecules having electrons 1 and 2 are given by $\Psi_1(1)$ and $\Psi_2(2)$, respectively, the simplest expression of the wave function of the diatomic molecule system is

$$\Psi = \frac{1}{\sqrt{2}}\left\{ \Psi_1(1) + \Psi_2(2) \right\}. \qquad (10.5)$$

Next, the second order energy E_0'' can be obtained in the form of Equation 10.6.

$$E_0'' = -(\text{coefficient}) \times \frac{e^4}{R^6}\overline{r_1^2}\,\overline{r_2^2}, \qquad (10.6)$$

where

$$\overline{r_i^2} = \int \psi_i^*(i)\, r_i^2\, \psi_i(i)\, d\tau_i, \quad i = 1, 2. \tag{10.7}$$

If $\mu_i^2 = e^2\, \overline{r_i^2}, \mu\hat{i}$ is the mean value of a dipole moment when the electron is continuously moving in the neutral molecule. Thus, the dispersion interaction U_{dis} is expressed by

$$U_{\text{dis}} = -\frac{(\text{coefficient})\overline{\mu_1^2}\,\overline{\mu_2^2}}{R^6} = -\frac{C_6}{R^6}, \tag{10.8}$$

where C_6 is the constant. Because the dispersion force is in inverse proportion to the sixth power of the intermolecular distance R, the attractive interaction works effectively to the molecule at the distance of five times the molecular size. Here, R may be changed into r for a more general expression, since the electron coordinates of r_1 and r_2 are used no more in this chapter. In general, by approximating the repulsion term inversely proportional to the twelfth power of the distance between molecules, a pair of potential $U_{\text{ff}}(r)$ between molecules is expressed by

$$U_{\text{ff}}(r) = 4\varepsilon_{\text{ff}}\left[\left(\frac{\sigma_{\text{ff}}}{r}\right)^{12} - \left(\frac{\sigma_{\text{ff}}}{r}\right)^6\right], \tag{10.9}$$

which is known as the Lennard–Jones (LJ) potential. Here, σ_{ff} is a parameter of the molecular size at $U_{\text{ff}} = 0$, and ε_{ff} is the depth of the potential. The LJ potential is applicable to the interaction of neutral molecules and neutral solid surfaces, because it can describe the interaction between neutral molecules. Thus, the interaction $\psi_{\text{sf}}(r_i)$ between the neutral molecule f and i-th surface atom in terms of the mutual distance r is given by

$$\psi_{\text{sf}}(r_i) = 4\varepsilon_{\text{sf}}\left[\left(\frac{\sigma_{\text{sf}}}{r_i}\right)^{12} - \left(\frac{\sigma_{\text{sf}}}{r_i}\right)^6\right]. \tag{10.10}$$

Here, σ_{sf} and ε_{sf} have the physical meaning corresponding to Equation 10.9. The interaction between molecules and solid surfaces are obtained by adding the pair-interaction of Equation 10.10. By converting this summation to integration, the attractive interaction term can be expressed in terms of the vertical distance z between a molecule and the external layer of the solid surface:

$$\Phi_{sf}^{(z)} = A_{sf}\left(\frac{\sigma_{sf}}{z}\right)^{9} - B_{sf}\left(\frac{\sigma_{sf}}{r_i}\right)^{3}. \tag{10.11}$$

Here, A_{sf} and B_{sf} are the constants. The equilibrium value of z when this potential takes the minimum is 23% shorter than the equilibrium distance between a molecule and a surface atom. Consequently, a molecule is more strongly attracted to the solid surface than the single pair of one molecule and one surface atom. This interaction induces physical adsorption.

10.2.3 Diversity in Carbon Structure

It is a great benefit that carbon materials are light, since they are composed only of a carbon element that has the atomic weight of 12. Carbon materials are chemically stable, and their biocompatibility is also excellent, although there is a risk of combustion with oxygen at high temperature. Moreover, carbon materials produced from natural products can contribute to carbon dioxide (CO_2) sequestration in principle. Carbon materials have distinct diversity in their structure and property. The structural diversity in carbon derives from sp^3, sp^2, and sp valence states of the carbon atom. The representative carbon forms of sp^3, sp^2, and sp carbon are diamond, graphite, and carbyne, respectively, although the electronic structure of some actual carbon materials is different from the above-mentioned ideal hybrid orbital. A carbon atom in diamond makes an electron pair to form a stable covalent bond with four neighboring carbon atoms, having a very stable tetrahedron structure. Thus, a diamond is a typical insulator, since the stable valence band is perfectly filled with electrons and the band gap is too large, giving no carriers in the conduction band. The mutual orientation of tetrahedrons is random in diamond-like carbon. Carbyne is carbon that consists of sp orbital, which has two π-orbital and takes the structure in which single and triple bonds are alternately ranged or in which double bonds are ranged. This is produced in defluorination of polyvinylidene fluoride; for example, carbyne is unstable and therefore there is no sufficient knowledge on it. The sp^2 hybrid valence system, which is the most popular, forms the conjugated π electronic structure. The sp^2 carbon can be roughly divided into two groups: the graphite group based on the planar unit structure of graphene, and the fullerene and CNT group, which is called nanocarbon with a nanoscale-curved structure caused by the coexistence of non-hexagon-membered rings. Graphene is a conjugated system consisting of a two-dimensional network planar structure of carbon hexagons. Many carbon materials have a composition unit that consists of a two-dimensional carbon hexagon. Thus, the number density of carbon atoms per unit surface area is considerably large, and the dispersion interaction is strong enough to adsorb molecules on the surface in spite of the comparative light weight. The graphite intercalation compound has intercalants between graphene layers. The nanoscale spaces of the intercalation compound can adsorb hydrogen,[25] although the graphite intercalation structure is very unstable at ambient conditions. Various carbon blacks and carbon fibers are known, which are made up of a well-developed graphite structure or nanoscale graphite crystallites. Moreover, sp^3 carbons at the edge of the graphite crystallite are presumed to connect the graphite crystallites through a single C–C bond. In disordered carbon materials, such as glassy carbon, mesocarbon microbeads, and activated carbon, the interlayer spacing in the graphitic unit structures is larger than 0.335 nm, due to a distorted layer structure and the sp^3 bonding at the edge of the graphitic unit. The coexistence of sp^2 and sp^3 carbon atoms can introduce two- and three-dimensionalities, respectively, to produce complex atomic structures, as reported in porous carbon.[26] Moreover, pentagon and heptagon induce the positive and negative curvatures of the curved surface structures, respectively, in the π electron conjugated carbon sheets. Consequently, pentagon and heptagon of morphological defects can give localized bents to the plane carbon hexagonal network structure. These structural diversities can lead to new carbon allotropes one after another, stimulating carbon science and technology. For example, graphite nanoribbon[5] and cup-stack carbon nanofiber[27] have been expected to extend the application field of carbon materials. Even cokes and coals, as typical classical carbon materials, have attracted a great deal of attention with relevance to clean energy technologies. As mentioned above, the diversity of sp, sp^2, and sp^3 bonds in carbon atoms generates various carbon materials, extending the application of carbon to various technologies.

10.2.4 Nanoporous Carbon Materials

Pores are produced when defects occur in the atomic carbon structure and are coupled with each other. Carbon atoms in carbon precursors containing predominant sp^2 carbons can be gasified as carbon monoxide (CO) by the reaction with steam at high temperatures. This gasification reaction proceeds in the conjugated double-bonding system to produce slit-like pores. Thus, the pores in activated carbons are close to the slit shape, at least in nanoscale. The pore width w is defined by the effective distance between the envelope surfaces of parallel pore walls in case of a slit-shape pore and by the diameter in case of a cylindrical pore, as shown in Figure 10.2. The w is different from the physical width H that is used in the theoretical studies (see Equations 10.16 and 10.17). Pores are classified into three categories by w according to The International Union of Pure and Applied Chemistry (IUPAC): $w < 2$ nm are termed micropores; 2 nm $< w < 50$ nm in diameter are termed mesopores; $w > 50$ nm in diameter are termed macropores. Pores less than 0.7 nm ($w < 0.7$ nm) in diameter are idiomatically known as ultramicropores, which is not formally recommended by IUPAC. The mechanism of nitrogen adsorption at 77 K used for the characterization of porous materials depends on the

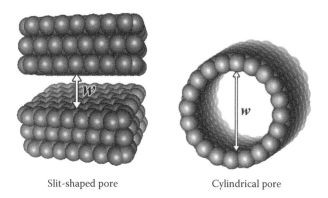

Slit-shaped pore   Cylindrical pore

FIGURE 10.2
Diameter w of cylindrical and slit-shaped pores.

type of pores. Adsorption on mesopores can be described by capillary condensation that is observed over 0.4 of the relative pressure. Adsorption on micropores begins from the extremely low relative pressure of less than 10^{-3}, which is called micropore filling. Nitrogen molecules cannot enter into ultramicropores at 77 K due to diffusion obstruction, so that fractional filling is often far from unity. However, ultramicropores work as strong adsorption sites for supercritical gas adsorption, because the diffusion obstruction is less marked at room temperature. The classification of pores by the IUPAC is inconvenient for the recent CNT family. A new concept is necessary for the CNT family in which pore width extends from micropores to mesopores. The boundary value 2 nm between micropore and mesopore was originally determined by the Kelvin relation, corresponding to the relative pressure 0.42 where the nitrogen condensation layer generated in pores at 77 K becomes unstable. It has been recently revealed that the Kelvin relation cannot be applicable to adsorption in pores of less than approximately 5 nm of the pore width in the case of ordered mesoporous silicas. The authors' term for a pore with a pore width of less than 5 nm is "nanopore."[28] The interaction of a molecule in the inside of a nanopore with pore walls restricts the random movement of molecules and easily forms a molecular assembly of high density. Thus, the nanopore is promising for gas separation and reaction fields, as well as for gas storage. Various carbon materials and their pore structures are listed in Table 10.2. The representative nanoporous carbon is an activated carbon. There are various activated carbons, due to their different source materials and activation methods. ACF with a diameter of approximately 10 μm has a sharper pore size distribution than activated carbon and is suitable for research on the effect of the pore width on gas adsorption characteristics. These activated carbons are composed of hexagonal network structures, approximately 1 nm in thickness and several nanometers in width, constructing pore walls. The carbon activated by potassium hydroxide (KOH) has a specific surface area over 2630 m²/g, known as superhigh surface area carbon.

TABLE 10.2

Various Carbon Materials and Their Porous Structures

Valence	Carbon Material	Pore Type
sp	Carbene	—
sp²	Graphite	—
	Carbon nanotube	Micropore, mesopore[a]
	Fullerene	Micropore[b]
	Graphite intercalation compound	Micropore[a]
	Carbon fiber	
	Carbon black	Micropore[a]
	Activated mesocarbon microbeans	Micropore
	Grassy carbon	
	Activated carbon (fiber)	Micropore, mesopore
	Carbon aerogel	Micropore[c], mesopore
	Diamond-like carbon	
sp³	Diamond	

[a] Small amounts; the amount depends on the preparation.
[b] Due to channeling of defects.
[c] Activation can add micropores.

The pore wall unit of this superhigh surface area carbon seems to be smaller than that of the above-mentioned activated carbon. Nanopores in activated carbons are approximately a slit shape reflecting the pore wall shape, at least in nanoscale. In some cases, it is better to regard it as a wedge shape. The pore width is generally smaller than 1.5 nm, and thus mesopores are intentionally added in order to enlarge the adsorption rate of molecules in micropores.[29] Very recent hybrid-reverse Monte Carlo simulation suggests a complex pore structure: the predominant structures in nanoscale are close to the slit shape. Recently reported carbide-derived carbon (CDC) prepared by removing metal from metal carbide has a slit-shaped pore smaller than 1 nm, which can be controlled with an accuracy of 0.1 nm order. Gogotsi et al.[30,31] found that the remnant carbon which is prepared by removing Si and Ti from Ti_3SiC_2 by reacting with chlorine gas as $SiCl_4$ and $TiCl_4$ has abundant slit-shaped pores, reflecting the layer structure of Ti_3SiC_2. The diameter of the slit-shaped pore in these carbons can be controlled very precisely by changing the chlorination temperature. In this case, the reaction under 873 K produces only micropores with pore size distribution as sharp as zeolite. Furthermore, ultramicropores less than 0.7 nm can be prepared by this method. Consequently, CDC is expected to be applied to supercritical gas storage for which pores less than 1 nm are necessary. Against these microporous carbons on the basis of the slit-shaped pores, porous carbons with a curvature structure of 1 nm scale have been prepared by the graphitization in the pores of zeolites. Carbon aerogel reported by Pekala et al. can be obtained by the carbonization of RF resin, which is made from the reaction of formaldehyde (F) and resorcinol (R) by adding sodium carbonate (Na_2CO_3).[32] RF resin originally has the mesoporosity of several decade nanometers. RF resin is the gel structure in which beads of several decade nanometers are ranged. The beads become carbons by the carbonization, and the gel-structure-derived porosity is maintained. Though carbon aerogel is a typical mesoporous carbon, micropores less than 1 nm can be added by the activation of the carbon beans units with CO_2.[33] The fabrication of nanoporous carbon using a template is also actively carried out. Microporous, mesoporous, and macroporous carbons have been designed using the template method. The template methods are roughly divided into two groups: introducing carbon precursors into pores of soluble materials such as zeolites and embedding inorganic compounds into carbon precursors.[34] In particular, the template method is the current route for preparation of ordered mesoporous carbons (MOCs). The mesoporous ordered carbon CMK-1 synthesized by Ryoo et al. is well known for using ordered mesoporous silica MCM-48 as a template.[35,36] CMK-3 and CMK-4 were prepared using templating of the hexagonal mesoporous silica SBA-15 and pipe-shaped nanoporous CMK-5 with a pore of 6 nm; these were also prepared with the templating method. However, the supercritical gas adsorption study on MOC is not so active, because their pores are too large to adsorb supercritical hydrogen. On the other hand, microporous-ordered carbon prepared by Kyotani et al. is expected to be a promising hydrogen adsorption material due to the narrow pores and high surface area.[37,38] They introduced polyacrylonitrile or polyfurfuryl alcohol into the pores of Y zeolite. Carbon with a surface area that exceeds $2000 \, m^2/g$ is obtained after carbonization and removal of the Y zeolite of the template by fluoric acid. Sharp peaks are observed in an x-ray diffraction pattern, reflecting the structure of Y zeolite. The zeolite-templated carbons (ZTC) probably consist of single carbon walls that have nanoscale curvature supported by the presence of the radial breathing mode (RBM) band of the Raman spectrum. This curvature structure in nanoscale should have a merit for hydrogen adsorption. Preparing ZTC in the smaller pores of zeolite has also been attempted. LTA zeolite with a pore of 0.7 nm is used as a template to prepare nanoporous carbon from methanol. Though it is not sufficiently clear yet, the hemispherical single-wall carbon structure with a curvature radius of 0.7 nm may offer the optimum adsorption sites for supercritical hydrogen.[39] The examples of the second group are not as many compared with the first group. Ozaki et al.[40] prepared mesoporous carbon of pore width 4 nm using blended polymers. Tao et al. succeeded in preparing carbons of hydrophilic micropores and hydrophobic mesopores using silica nanoparticles.[41]

10.2.5 Molecular Potential of Slit-Shaped Pores

Let us consider the interaction of a molecule with a slit-shaped pore. The potential energy of a molecule with a slit-shaped pore composed of two graphite sheets is given by the Steele's 10–4–3 potential $\Phi(z)$:[42]

$$\Phi(z) = 2\pi\varepsilon_{sf}\rho_C\sigma_{sf}\Delta\left[\frac{2}{5}\left(\frac{\sigma_{sf}}{z}\right)^{10} - \left(\frac{\sigma_{sf}}{z}\right)^4 - \frac{\sigma_{sf}^4}{3\Delta(z+0.61\Delta)^3}\right], \quad (10.12)$$

which is led on the basis of the pair LJ potential. Here, z is the vertical distance from the surface, Δ ($=0.335\,nm$) is the interlayer distance of graphite, and ρ_c ($=114\,nm^{-3}$) is the carbon atomic number density. Using LJ parameters for a gas molecule (ε_{ff}, σ_{ff}) and for a carbon atom (ε_{ss}, σ_{ss}), ε_{sf} and σ_{sf} are approximately given by

$$\varepsilon_{sf} = \sqrt{\varepsilon_{ff}\varepsilon_{ss}} \quad (10.13)$$

and

$$\sigma_{sf} = \frac{\sigma_{ff} + \sigma_{ss}}{2}. \quad (10.14)$$

If the slit width between the atoms of graphite surfaces is H, the interaction potential $\Phi_p(z)$ between the slit-shaped pore and a molecule located at the distance z from one graphite surface is given by

$$\Phi_p(z) = \Phi_{sf}(z) + \Phi_{sf}(H - z). \quad (10.15)$$

The slit width H is larger than the effective pore diameter w measured by adsorption measurements, that is,

$$w = H - 0.240 \text{ (nm)}, \quad \left(\frac{\varepsilon_{ff}}{k} = 101.5K; \sigma_{ff} = 0.3615\,nm\right), \quad (10.16)$$

for a nitrogen molecule and

$$w = H - 0.245\text{(nm)}, \quad \left(\frac{\varepsilon_{ff}}{k} = 34.2\,K; \sigma_{ff} = 0.296\,nm\right), \quad (10.17)$$

for a hydrogen molecule.[43] The pore width dependences of the interaction potential between a graphite slit pore and a one-center nitrogen molecule or a one-center hydrogen molecule are shown in Figure 10.3. The interaction potential becomes as deep as the decrease of the pore diameter, while the repulsion effect also appears when the pore diameter approaches the molecular size. This deep potential induces a predominant adsorption in micropores at very low pressure. However, the potential depth at $w=0.5\,nm$ for a

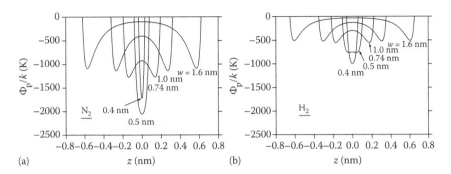

FIGURE 10.3
Molecular potential profiles of the slit-shaped pore of carbon materials with various diameters w for (a) nitrogen and (b) hydrogen.

hydrogen molecule is only $-760\,K$, while that of a nitrogen molecule reaches $-2054\,K$. This is why it is difficult to physically adsorb hydrogen abundantly in carbon pores at room temperature, although the graphitic structure has very high atomic density, enabling a stronger interaction with molecules.

10.2.6 Molecular Potential of the Carbon Nanotube Family

The carbons mentioned above fundamentally have an amorphous pore wall structure without atomic periodicity in the long range. On the contrary, CNTs are cylindrical and have an ordered wall structure. They are basically composed of graphene structures and have pentagons at the cap part.[2,44] A nanotube consisting of one graphene sheet rolling up is known as SWCNT, a nanotube consisting of two graphene sheets rolling up is known as double-wall CNT (DWCNT), and a nanotube consisting of multiple graphene sheets identically rolling up is known as a multi-wall CNT (MWCNT).[45] SWCNT is a special material referred to as "bi-surface nature material" because the whole carbon atoms are exposed to the internal and external surfaces of the SWCNT wall. Moreover, the internal and the external surfaces of SWCNT have different surface characteristics, due to the different nanoscale curvatures in positive and negative. We can see this curvature dependence of surface characteristics from the viewpoint of the interaction between a molecule and a SWCNT wall.[46] The internal (Φ_{int}) and the external (Φ_{ext}) potentials of SWNCT are, respectively, expressed as follows:[47]

$$\Phi_{\text{int}} = \pi^2 \rho_s \varepsilon_{sf} \sigma_{sf}^2 \left[\frac{63}{32} \frac{F(-4.5, -4.5, 1.0; \beta^2)}{\left[R^*(1-\beta^2) \right]^{10}} - 3 \frac{F(-1.5, -1.5, 1.0; \beta^2)}{\left[R^*(1-\beta^2) \right]^4} \right], \quad (10.18)$$

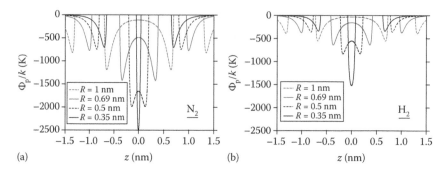

FIGURE 10.4
Molecular potential profiles of the cylindrical pore of the carbon nanotube family with various radii R for (a) nitrogen and (b) hydrogen.

$$\Phi_{ext} = \pi^2 \rho_s \varepsilon_{sf} \sigma_{sf}^2 \left[\frac{63}{32} \frac{F(-4.5, -4.5, 1.0; \delta^2)}{\left[R^*(1-\delta^2) \right]^{10}} \delta^{11} - 3 \frac{F(-1.5, -1.5, 1.0; \delta^2)}{\left[R^*(1-\delta^2) \right]^4} \delta^5 \right]. \quad (10.19)$$

Here, if r is the distance from the tube center and R the radius of SWCNT to the center of carbon atoms, we obtain $\beta = r/R$, $R^* = R/\sigma_{sf}$, $\delta = R/r$, and $\rho_s = 38.2\,\text{nm}^{-2}$. $F(a, b, c, d)$ is the hypergeometric function (a, b, c, and d are constants). Figure 10.4 shows the R dependence of the potential profiles for nitrogen and hydrogen molecules. This figure shows the interaction potential energy change from the zero of the SWCNT cylinder center toward the carbon wall; the potential profiles across the carbon wall are also shown. The potential is the deepest at the internal monolayer position except for the small pores whose width is comparable to the molecular size. The potential depth for nitrogen is much deeper than that of hydrogen. The potential minima for both nitrogen and hydrogen molecules become deeper with a decrease in R. This is because the overlapping effect of the molecular potential received from a cylinder pore wall becomes remarkable in a thin tube. Also, the external potential minimum corresponds to the monolayer adsorption site on the external surface. However, the potential energy difference at the internal and the external monolayer positions is almost a half value of the potential energy at the internal monolayer sites, indicating the importance of the nanoscale curvature. The potential minimum at $R = 0.35\,\text{nm}$ does not appear near the SWCNT surface but at tube center, indicating that a molecule is exposed to a large potential field and is stabilized at the center of SWCNT with $R = 0.35\,\text{nm}$. However, if the tube diameter is smaller than 0.7 nm, the effective tube diameter inside the tube becomes smaller than 0.4 nm, losing the effective space available for adsorption.

Ordinarily, SWCNT forms a bundled structure, which consists of tens of tubes by the dispersion force between tubes, as shown in Figure 10.5. The

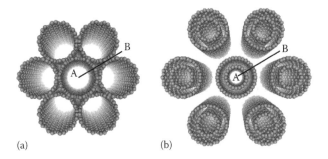

FIGURE 10.5
Bundled structures of SWCNT and DWCNT. (a) SWCNT bundle composed of (10×10) SWCNTs, the diameter of which corresponds to $R = 0.678$ nm. (b) DWCNT bundle composed of inner tubes with $R = 0.385$ nm and outer tubes with $R' = 0.715$ nm. The van der Waals gap g is 0.6 nm.

well-crystalline SWCNT bundle gives x-ray diffraction peaks stemming from the ordered bundled structure having a hexagonal symmetry. The molecular adsorption site of this bundle differs from that of one SWCNT. Figure 10.6 shows the potential profile for a hydrogen molecule in the (10, 10) SWCNT hexagonal bundle with an intertube distance of 0.34 nm.[48] As shown in Figure 10.6, three kinds of deep molecular potential fields are formed: internal tube, interstitial channel, and groove site. Therefore, the adsorption process should depend on these adsorption sites. The interstitial channel gives the deepest potential minimum, due to the overlapping effect of the molecular potential from three tubes. However, it is difficult to effectively

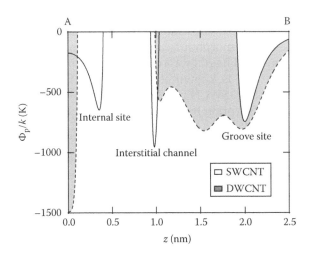

FIGURE 10.6
Molecular potential fields of SWCNT and DWCNT bundles for hydrogen. The potential fields on the lines connecting A and B in Figure 10.5a and b are shown.

utilize the interstitial channel as a molecular adsorption field at present, because the capacity of this interstitial channel of which the effective diameter can accept only small molecules is small. If the intertube distance can be controlled by chemical modification, the capacity of the interstitial channel becomes large enough to adsorb supercritical gases. A recent study showed that intercalation of SWCNT bundles with C_{60} remarkably increases hydrogen adsorptivity.[49] The groove sites that are surrounded by two tubes also have a considerably deep potential.

The opening of the nanotube is carried out using the reactivity of pentagons at the cap part. SWCNT can be prepared by arc discharge, laser ablation, and chemical vapor deposition (CVD) methods. The arc discharge and the laser ablation methods, which need very high temperature, can provide well crystalline SWCNT samples. Intensity ratio $I(G)/I(D)$ of the G-band and D-band in Raman spectrum is utilized as a degree of the crystallinity. The $I(G)/I(D)$ ratio of SWCNT prepared by the arc discharge and the laser ablation methods becomes approximately 100; that is, D-band is very weak and a single graphene wall is clearly observed by high-resolution transmission electron microscopy (HR-TEM). However, since the reaction temperature of the CVD method is below 1300 K, this method is suitable for large-scale synthesis, but the $I(G)/I(D)$ ratio is small. Therefore, the crystallinity of SWCNT prepared by the CVD method must be improved by high-temperature treatment. HiPCO-SWCNT prepared from high-pressure CO using an iron catalyst used as a standard SWCNT sample is an example of the CVD method. The catalytic iron content of HiPCO-SWCNT is more than 30 wt%, so that oxidation treatment and acid washing for purification are necessary. However, excessive purification treatment induces surface functional groups and defects on SWCNT, and thus the $I(G)/I(D)$ ratio decreases. Recently, Hata et al.[50] developed a new CVD method in which nanoscaled catalysts are separately embedded at a silicon substrate, and a small amount of steam is introduced to suppress the amorphous carbon formation. The SWCNT sample prepared by this method has 0.1 wt% of the catalytic iron and an unbundled structure. This SWCNT is known as supergrowth SWCNT (SG-SWCNT). The tube diameter of SG-SWCNT is 2–3 nm, which is twice that of HiPCO–SWCNT. Nitrogen adsorption measurements at 77 K and the grand canonical Monte Carlo simulation (GCMC) on SG-SWCNT reveal the process of physical adsorption of nitrogen molecules on SWCNT.[46]

As mentioned above, adsorption starts at the monolayer adsorption sites on the internal wall where the interaction potential energy is the strongest in one SWCNT. Subsequently, molecules are filled in the internal residual space as well as partial monolayer adsorption, accompanied by multilayer adsorption on the external wall. Figure 10.7 shows the nitrogen adsorption isotherms and the snapshots of GCMC. Monolayer adsorption on the internal wall almost finishes near relative pressure $P/P_0 = 6 \times 10^{-3}$, giving the step of isotherm. Filling in the internal nanospace and monolayer adsorption on the external wall are completed at $P/P_0 = 2 \times 10^{-1}$ after the next step.

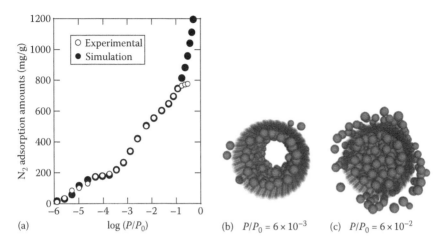

(a) log (P/P₀) (b) $P/P_0 = 6 \times 10^{-3}$ (c) $P/P_0 = 6 \times 10^{-2}$

FIGURE 10.7
(a) Nitrogen adsorption isotherms of unbundled SWCNT at 77 K (O: experimental, ●: simulation). (b, c) Snapshots of isotherms near steps.

Special attention should be paid to the difference in the radial distribution structures of the nitrogen monolayer adsorbed on the internal and the external walls, as shown in Figure 10.8. Figure 10.8 also shows the radial distribution function of the nitrogen monolayer adsorbed on a flat graphene for comparison. The ordered structure of the nitrogen monolayer on the internal wall reaches more than 1.5 nm, and thus long-range ordering like a solid structure is confirmed, due to the strong interaction potential. The radial distribution peaks of two other cases are limited below 1.3 nm. Thus, the ordering of the monolayer on the external wall is inferior to that on the graphene,

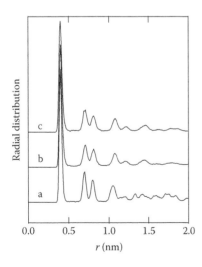

FIGURE 10.8
Radial distribution functions of adsorbed nitrogen molecules: (a) internal adsorption layer, (b) external adsorption layer, and (c) on graphene.

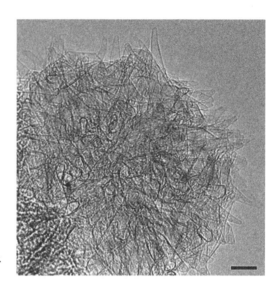

FIGURE 10.9
TEM image of an SWCNH assembly.
The bar length is 10 nm.

though the difference is only slight. Therefore, the sign of nanoscale curvatures of SWCNT has an explicit influence on the interaction with molecules.

A single-wall carbon nanohorn (SWCNH) has an assembly structure, as shown in an HR-TEM image (Figure 10.9).[51] Since the large-scale synthesis of SWCNH is possible without a catalyst, reliable gas adsorption research is available. Its tube is not long, being different from SWCNT. The cap part is a sharp structure. The graphene wall has many defects, showing a Raman spectrum greatly different from that of SWCNT;[52] the D-band of the Raman spectrum is almost comparable to the G-band. The electrical conduction property of SWCNT indicates an n-type response for gas adsorption due to the presence of many pentagons,[53] which is different from the p-type nature of SWCNT. This SWCNH has a partially oriented structure, with 0.7 nm of the quasi one-dimensional pores from the viewpoint of x-ray diffraction and GCMC-aided N_2 adsorption at 77 K.[54] Moreover, oxidation treatment can provide nanoscale holes on the SWCNH wall, known as nanowindows.[55,56]

The specific surface of SWCNH is controllable by the oxidation temperature.[57] Figure 10.10 shows the relationship between the specific surface area and the oxidation temperature. Though the specific surface of the untreated SWCNH is approximately $350 \, m^2/g$, that of SWCNH oxidized at 823 K (ox-SWCNH) dramatically increases to $1500 \, m^2/g$. Since this value of $1500 \, m^2/g$ is still small compared with the maximum geometric area of $2630 \, m^2/g$ of the infinite graphene structure, control of the assemble structure is desired. Moreover, the characteristic feature of SWCNH is in the utilization of nanowindows. A large capacity of nanospaces inside SWCNH particles, compared with SWCNT, is accessible for molecules by opening a nanowindow, showing the unique adsorptivity such as molecular sieving. The size control of the nanowindows clearly causes the molecular sieving effect and

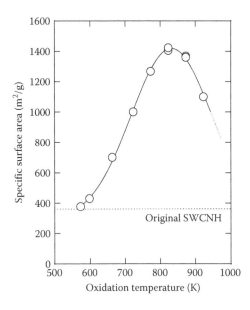

FIGURE 10.10
Oxidation temperature dependence of specific surface area of SWCNH.

is promising for efficient gas separation.[58] The internal nanospace of the sharp section has a stronger potential field than that of the tubular section. Furthermore, the methane adsorptivity of SWCNH of which the volume-filling rate is increased by compression shows a promising value of 160 vol.% at 35 atm and 303 K, being close to the DOE target value.[59] Thus, SWCNH is expected to have a great potential as a clean energy storage material.

Recently, many research studies on DWCNT have been carried out. The difference in the electronic states of the inner and the outer tubes give DWCNT unique characteristics that SWCNT does not have. DWCNT as well as SWCNT has a feature of "surface solid." However, the external surface of the outer tube and the internal surface of the inner tube of the concentric tube structure can be exposed to the gas phase. Therefore, there have been many successive efforts to prepare high-purity and uniform DWCNT. For example, Endo et al. have prepared DWCNT, in which the catalytic layer (Fe/MgO) content is less than 0.1 wt%, by the CVD method at 1148 K.[60] This DWCNT contains almost two kinds of concentric tube structures, that is, the diameters of the inner and the outer tubes are 0.72 and 1.48 nm and 0.91 and 1.61 nm, respectively, which is accurately determined by the RBM of Raman spectra.[61] Since this DWCNT forms a bundled structure with the triangular lattice regularity, three diffraction peaks assigned to the super lattice structure are observed in the low-angle region. The theoretical value of x-ray diffraction is consistent with the experimental value, when it is obtained under the assumption that the diameters of the inner and the outer tubes are 0.82 and 1.42 nm, and the van der Waals gap g is 0.6 nm.[62] This large value of g indicates that DWCNT is partially curved or has a wide distribution of tube diameter. The value $g = 0.6$ nm of DWCNT, twice that of SWCNT, leads the large capacity of the interstitial pores. Consequently, DWCNT shows excellent hydrogen adsorptivity, as mentioned below. Figure 10.5 shows the model diagram of the DWCNT bundled structure which takes a two-dimensional trigonal lattice (inner tube radius: 0.385 nm, outer tube radius: 0.715 nm, and van der Waals gap g: 0.6 nm). The molecular potential field of this DWCNT for hydrogen is shown in Figure 10.6. The internal pore of DWCNT exhibits a deep potential field: −1500 K for hydrogen. However, the internal pores, which are actually closed, do not contribute to measurable adsorption. Consequently, DWCNT has many sites that can adsorb hydrogen in the interstitial channels, due to a wide van der Waals gap; the hydrogen adsorption amount on DWCNT at 77 K is larger than that on HiPCO-SWCNT.

10.2.7 Gas and Supercritical Gas

We need to understand the difference between supercritical gas and vapor in order to elucidate supercritical gas adsorption on nanoporous carbons. We adopt the basic approach to show the characteristics of supercritical gases. The ideal gas is expressed by the ideal gas equation. When 1 mol of gas exhibits pressure P in the volume V_m at temperature T, the following equation is obtained:

$$PV_m = RT, \tag{10.20}$$

where R is the ideal gas constant. Of course, an ideal gas has no interaction between molecules. However, a real gas is condensed at low temperatures. The repulsive and the attractive interactions between gas molecules were corrected by means of the molecular exclusive volume and the compensation of gas pressure by van der Waals. Thus, the famous van der Waals equation can be lead to

$$\left(P + \frac{a_v}{V_m^2}\right)(V_m - b_v) = RT, \tag{10.21}$$

where
a_v is the constant related to the attractive interaction
b_v are the constants related to the repulsive interaction between molecules

Then, the compression factor z_g is used in order to show the difference in behavior of real gas from ideal gas.

$$z_g = \frac{PV_m}{RT} = 1 + \frac{1}{V_m}\left(b_v - \frac{a_v}{RT}\right) + \left(\frac{b_v}{V_m}\right)^2 + \cdots \tag{10.22}$$

Though z_g is 1 for ideal gas, the van der Waals equation is expanded to the term that contains constants a_v and b_v expressing interactions for real gas, as shown in Equation 10.22. This is known as the virial expansion, and the second term $b_v-(a_v/RT)$ is called a second virial coefficient that stems from the molecular interaction. In fact, the van der Waals equation is also insufficient for expressing real gas in detail. The virial equation, which consists of many terms, and other state equations with many parameters are used for a more complete description of real gases. It is possible to transform the van der Waals equation to the virial equation, which is useful for the expression of real gas considering molecular interaction. Then, is it possible to express the condensation process of real gas?

Under the condition that temperature is constant, the relationship between P and V exhibits $PV=$ constant at sufficiently high temperatures. However, a certain vapor pressure region appears at relatively low temperatures, as a result of the condensation of gas to liquid. The temperature at which the region of the constant vapor pressure of liquid becomes one point (critical point) can be seen in the PV relationship of real gas. Since this critical point is an inflection point of the PV relation, it is analytically obtained by using the following relations:

$$\left(\frac{\partial P}{\partial V}\right)_T = 0, \quad \left(\frac{\partial^2 P}{\partial V^2}\right)_T = 0 \tag{10.23}$$

Applying these relations to the van der Waals equation, the critical temperature T_C, pressure P_C, and volume V_C are obtained as follows:

$$T_C = \frac{8a}{27Rb}, \quad P_C = \frac{a}{27b^2}, \quad V_C = 3b \tag{10.24}$$

These equations do not exactly describe the critical constants of real gases, but they provide considerably reasonable values. The meaning of this critical point can be understood by Figure 10.11. The coexistence line of liquid and gas is shown by curve OC. As the temperature increases along curve OC and reaches point C, the gas and liquid states cannot be distinguished. This state at $P > P_C$ and $T > T_C$ is known as supercritical fluid. Molecules in supercritical fluid form clusters with short life, and the density of supercritical fluid fluctuates markedly in space and time. The state of $T > T_C$ is known as supercritical gas, which includes supercritical fluid. The saturated vapor pressure of supercritical gas cannot be defined, because supercritical gas cannot coexist with liquid. Furthermore, gas cannot become liquid at $T > T_C$, even applying high pressure, as shown in Figure 10.11. Physical adsorption is simply the phenomenon that gas becomes a liquid-like state on the solid surface by interaction with the solid surface. Thus, physical adsorption depends on vapor or

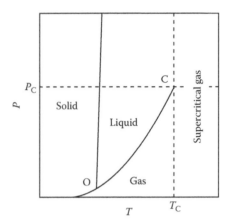

FIGURE 10.11
Phase diagram of substance.

supercritical gas. The saturated vapor pressure P_0 is defined for vapor. Vapor that can coexist with liquid transforms into supercritical gas above T_C. The physical adsorption depends sensitively on the temperature across T_C. It is easy for vapor to be physically adsorbed, due to greater interaction between molecules and solid surface than the condensation enthalpy, though it is difficult for supercritical gas. The physical adsorption of supercritical gas effectively occurs only in nanopores in which the molecular potential is great enough. Accordingly, the distinction of vapor and supercritical gas is key to understanding supercritical adsorption on nanopores, which is essentially important to develop better carbon adsorbents for hydrogen and methane as clean fuels.

10.3 Supercritical Gas Adsorption

Here, we focus on physical adsorption of supercritical hydrogen. Accordingly, hydrogen addition (chemisorption) to dissociated C=C double bonding by the mechanical treatment of graphite in a hydrogen atmosphere is not described. Hydrogen is a supercritical gas even at 77 K, which is the boiling point of nitrogen, since the critical point of hydrogen is 33 K. The adsorption experiment of hydrogen at 77 K is usually done from the standpoint of easy handling and the measurable adsorption amount of supercritical hydrogen. It is meaningful to know the hydrogen adsorptivity of the microporous activated carbon family.

Figure 10.12 shows the hydrogen adsorption isotherms of superhigh surface area carbon (specific surface area: 3090 m²/g) and pitch-based ACF (specific surface area: 1800 m²/g) at 77 K. These results show that each adsorption amount takes the maximum value exceeding 5 wt% below

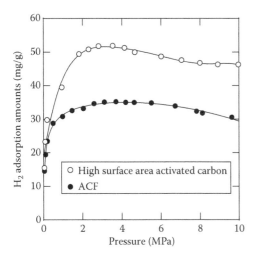

FIGURE 10.12
Hydrogen adsorption isotherms of high surface area activated carbon and ACF at 77 K.

4 MPa. This adsorption is reversible. The adsorption amounts on these activated carbons are largely due to their large micropore volume, although their molecular potential well with an average pore width of 1.1–1.3 nm is not so deep. Thus, hydrogen adsorption amounts in the supercritical state at 77 K and 0.10 MPa can exceed 5 wt% when the micropore volume is sufficiently large. Recently, almost 10 wt% of adsorption amounts has been reported.

Figure 10.13 shows the high-pressure hydrogen adsorption isotherms of the activated carbon family and the nanocarbon family at 303 K. The hydrogen adsorption amounts linearly increase with pressure for all isotherms. This indicates that there is no remarkable interaction between a hydrogen

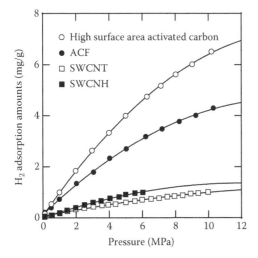

FIGURE 10.13
Hydrogen adsorption isotherms of various carbon materials at 303 K.

molecule and carbon materials at 303 K. The adsorption amount on high surface area activated carbon, which adsorbs 5 wt% of hydrogen at 77 K and 0.10 MPa is 0.7 wt% at 303 K and 10 MPa. A carbon material that has special-shaped pores adsorbs 0.9 wt% of hydrogen (not shown in the figure). On the other hand, the hydrogen adsorption amount of SWCNT is just 0.1 wt%, which is smaller than that of SWCNH. It is concluded that the interaction potential mentioned above is not sufficiently deep, compared with the thermal energy at 303 K. However, the measurement value of almost 1 wt% indicates that the hydrogen adsorptivity of carbon materials is hopeful, although the adsorption simulation by using the ultimate graphene slit model results in 0.7 wt% at 10 MPa. Hydrogen adsorptivity is not easily improved by the control of the nanopore structure because of insufficient hydrogen molecule–graphene interaction. Fundamental understanding by systematic research is required to design better carbon adsorbents for hydrogen at room temperature.

Figure 10.14 shows the nitrogen and hydrogen adsorption isotherms at 77 K of HiPCO-SWCNT.[62] The BET specific surface area determined by using the nitrogen adsorption isotherm is 660 m^2/g. The hydrogen adsorption amount at 77 K and 0.1 MPa is 6.9 mg/g. According to the geometric consideration by Eklund et al. one SWCNT with the diameter of 1.36 nm should give 2630 m^2/g.[63] The experimental value of the specific surface area of SWCNT is much smaller than the theoretical value, because the interstitial channel of a SWCNT bundle is not accessible by nitrogen molecules and the wall area of the internal space is not accurately estimated by nitrogen adsorption.[64] The hydrogen adsorption amount per the unit weight of SWCNT is not large, compared with activated carbons. This is because the micropore volume of SWCNT is smaller than that of activated carbons. Here, the monolayer

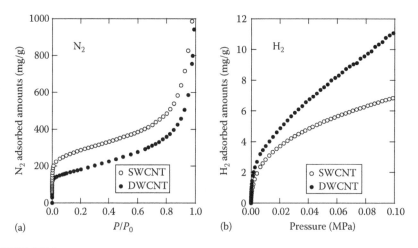

FIGURE 10.14
Nitrogen (a) and hydrogen (b) adsorption isotherms of SWCNT and DWCNT at 77 K.

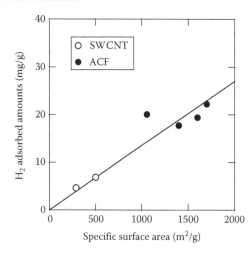

FIGURE 10.15
Relationship between surface area of SWCNT and ACF and supercritical hydrogen adsorption amounts at 77 K and 0.1 MPa.

adsorption sites generally have the deepest interaction potential well for hydrogen, and then the adsorption amount of supercritical hydrogen should be proportional to the surface area that is evaluated from the monolayer adsorption amount of nitrogen at 77 K. The relationship between the specific surface areas of SWCNT and ACF and the hydrogen adsorption amount at 77 K and 0.1 MPa is shown in Figure 10.15. Supercritical hydrogen adsorptivity can be sufficiently evaluated even by the adsorption amount at 77 K and 0.1 MPa. This clearly indicates that the hydrogen adsorption amounts should increase if the specific surface area of the SWCNT bundle becomes larger by increasing the intertube gap *g*. Thus, control of the intertube gap by chemical modification is a promising route for improvement of supercritical hydrogen adsorptivity. Figure 10.16 shows the hydrogen adsorption isotherms at 77 K

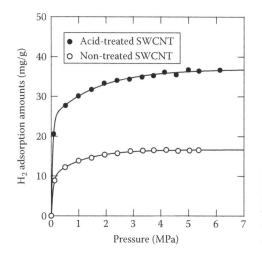

FIGURE 10.16
High-pressure hydrogen adsorption isotherms at 77 K of original HiPCO-SWCNT and HiPCO-SWCNT treated with an acid mix of nitric and sulfuric acid.

of original HiPCO-SWCNT and HiPCO-SWCNT treated by an acid mix of nitric and sulfuric acids to add micropores less than 1 nm, as much as four times the original HiPCO-SWCNT.[65] Correspondingly, the hydrogen adsorption amount multiplies by four, according to the change in the effective micropore volume of less than 1 nm. The density of the hydrogen adsorption layer for this case is 23 g/L at 0.1 MPa and 77 K. Supposing that the gradient of the adsorption isotherm per 0.1 MPa holds up to 5 MPa, the adsorbed density should be 150 g/L at 5 MPa and 77 K.[66] Accordingly, the increase of the interstitial pore volume is suitable for enhancement of the supercritical hydrogen adsorptivity of SWCNT. However, research on the reliable hydrogen adsorption on SWCNT of well-characterized nanostructures is insufficient, since the sufficient amount of SWCNT sample of high purity is not available, and the correct evaluation of supercritical hydrogen adsorption amount is not necessarily done.

As mentioned above, the interstitial site of an SWCNT bundle has the deepest potential for supercritical gas adsorption, although the volume of the interstitial nanospaces at the strongest sites is too small. Very recently, these authors have established a simple method of pillaring of SWCNTs with C_{60} for tuning the bundled structure to provide larger capacity of interstitial nanospaces, which is the co-sonication of C_{60} and SWCNT in toluene.[49] C_{60}-pillared SWCNTs have the expanded hexagonal and distorted tetragonal bundles revealed by HR-TEM (the models are shown in Figure 10.17), exhibiting twice the interstitial nanopore volume of the original SWCNT. Figure 10.18 shows the hydrogen adsorption isotherms of $C_{60}(x)$-pillared SWCNTs at 77 K, where x is the amount in gram of C_{60} doped to 1 g of SWCNTs. The H_2 adsorptivity of the C_{60}-pillared SWCNT bundles was markedly enhanced, providing almost twice the adsorption amount of SWCNTs in the low-pressure region, although SWCNTs used in this work have a small specific surface area (337 m^2 g^{-1}), because they are closed and their diameters are identical. Upward concave H_2 adsorption isotherms indicate the presence

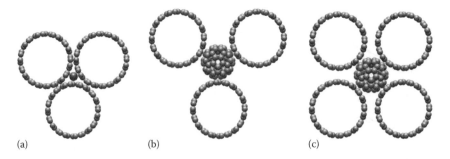

(a)   (b)   (c)

FIGURE 10.17
Models of bundled structures: (a) SWCNT bundle with a hexagonal arrangement, (b) C_{60}-pillared SWCNT bundle with an expanded hexagonal arrangement, and (c) C_{60}-pillared SWCNT bundle with a tetragonal array.

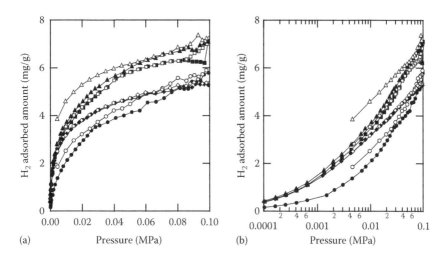

(a) Pressure (MPa) (b) Pressure (MPa)

FIGURE 10.18
H_2 adsorption isotherms of $C_{60}(x)$-pillared SWCNT (x; ●: 0, ■: 0.646, ▲: 1.68, and ◆: 3.58; solid symbols: adsorption, open symbols: desorption). The abscissas of a and b are expressed by the H_2 pressure and the logarithm of the H_2 pressure, respectively.

of strong adsorption sites, even for supercritical H_2. Moreover, the desorption branch is situated above the adsorption branch, indicating that there should be very narrow nanopores having strong interaction potential. As C_{60} molecules have a conjugated π-electron structure, a C_{60}-pillared SWCNT system can be regarded as a new nanocarbon, indicating a design route for SWCNT hydrogen storage.

Figure 10.19 shows the hydrogen adsorption isotherms at 77 K of oxidized SWCNH in which the internal nanospaces are accessible by molecules

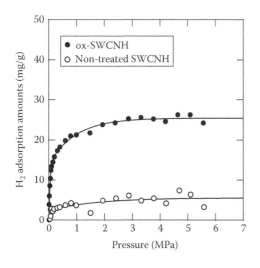

FIGURE 10.19
High-pressure hydrogen adsorption isotherms of non-treated SWCNH and oxidized SWCNH (ox-SWCNH) at 77 K.

through nanowindows.[67] The upward convex of the adsorption isotherm demonstrates that the interaction of SWCNH is enough strong even for the supercritical hydrogen at 77 K. The adsorption amount of the vertical axis is shown by the surface excess mass, because this adsorption isotherm is a high-pressure adsorption isotherm. We can estimate the density of the hydrogen adsorption layer from the surface excess mass and the pore volume determined from nitrogen adsorption at 77 K. The density of the adsorption layer in the internal spaces and the external pores, including interstitial pores, is almost equal to the bulk density (70 g/L) of liquid hydrogen at 20 K over 4 MPa.[67] Therefore, the concentration effect of supercritical hydrogen is remarkable at 77 K, which may be roughly equivalent to induce the bulk condensation of hydrogen at 20 K. The hydrogen adsorption amount of the oxidized SWCNH is 2.6 wt% at 5 MPa. The concentration effect of the supercritical hydrogen by SWCNH at 196 K is less marked than that at 77 K. The density of the hydrogen adsorption layer at 196 K and 5 MPa is 25 g/L, which is one-third of that at 77 K. The absolute adsorption amount is necessary for evaluation of storage amount of supercritical hydrogen, although the evaluation of the absolute adsorption amount has not been sufficiently established.[68,69]

Figure 10.14 also shows the nitrogen and hydrogen adsorption isotherms of DWCNT at 77 K.[62] Comparison of hydrogen adsorptivity of DWCNT with SWCNT is significant. The specific surface area of DWCNT is estimated to be only 320 m^2/g from the nitrogen adsorption isotherm. However, the supercritical hydrogen adsorption amount of DWCNT is almost twice that of SWCNT, as the intertube gap $g = 0.6$ nm of DWCNT is also twice that of SWCNT. The result of GCMC simulation is in good agreement with experimental results. Hydrogen molecules cannot be adsorbed at the interstitial pore of SWCNT, but can be adsorbed at the internal space. In contrast, hydrogen molecules are adsorbed at the interstitial pores of DWCNT, which has large interaction potential. It is noteworthy that even the hydrogen adsorptivity of DWCNT is superior to that of SWCNT in the adsorption amount per unit weight.

Recently, a carbon mesotube with an amorphous structure exhibits hydrogen adsorptivity exceeding 8 wt% at 77 K. The steady and systematic development of nanostructured carbons and research on the supercritical hydrogen adsorption are also necessary in the future. Table 10.3 summarizes the recent data on hydrogen adsorption on carbons. The latest reviews are referred to in order to understand the hydrogen adsorptivity of various carbon materials including carbon materials with many impurities.[74]

We have another related topic on hydrogen adsorption on nanocarbons. H_2 and D_2 must be regarded as quanta at lower temperatures.[75-77] The quantum nature depends on the mass of the molecule; therefore, nanocarbons show intensive molecular sieving effect for H_2 and D_2 below 77 K. The efficient D_2 separation can be a promising application route.

TABLE 10.3

Hydrogen Adsorptivity of Nanoporous Carbons

Sample	Surface Area (m²/g)	Adsorption Amounts		References
		At 77 K (wt%)	At Room Temperature (wt%)	
SWCNT	300	1.5 (6 MPa)	0.1	70
		0.6 (0.1 MPa)	—	
Open-SWCNT	1200	3.5 (10 MPa)	0.5 (10 MPa)	70
Acid-treated SWCNT	>1200	3.7 (6 MPa)	—	66
SWCNH	300	0.7 (6 MPa)	0.1 (6 MPa)	67
ox-SWCNH	1300	2.5 (6 MPa)	0.3 (6 MPa)	67
DWCNT	330	1.1 (0.1 MPa)	—	62
am-carbon mesotube[a]	1030	8.0 (10 MPa)	1.3 (10 MPa)	71
CDC[b]	1600	2.8 (0.1 MPa)	—	30,31
ZTC[c]	3200	5.3 (10 MPa)	—	72
MTC[d]	1800	3.7 (10 MPa)	—	73
KOH-AC[e]	3300	6.0 (10 MPa)	0.5 (10 MPa)	70
ACF	2000	6.0 (10 MPa)	0.4 (10 MPa)	70

[a] Amorphous carbon mesotube.
[b] Carbide derived carbon.
[c] Zeolite-templated carbon.
[d] Mesopore-templated carbon.
[e] KOH-activated carbon.

10.4 Conclusion

Physical adsorption of supercritical hydrogen on nanoporous carbons can be understood by use of the molecular interaction theory. Nanoporous carbons can store more than 5 wt% hydrogen at 77 K, an application that can be found in the transportation of hydrogen, although nanoporous carbons cannot store hydrogen of the DOE target level at an ambient temperature. Fundamental research studies on supercritical hydrogen on nanocarbons have led to new aspects in controlling nanopores and the interaction of molecules with nanoscale structures, although the research on hydrogen storage on nanocarbons could not attain the primary target of hydrogen storage at an ambient temperature. There are other important subjects on hydrogen adsorption on nanoporous carbons, such as quantum molecular sieving and *ortho-para* conversion. These subjects must be studied from their fundamental aspects in future in addition to the effect of embedding transition metal atoms in the carbon skeletons in order to enhance the hydrogen molecule–carbon surface interaction.

References

1. Kroto, H. W., Heath, J. R., O'Brien, S. C., Curl, R. F., and Smalley, R. E. 1985. C_{60}: Buckminsterfullerene. *Nature* 318: 162–163.
2. Iijima, S. 1991. Helical microtubules of graphitic carbon. *Nature* 354: 56–58.
3. Dillon, A. C., Jones, K. M., Bekkedahl, T. A., Kiang, C. H., Bethune, D. S., and Heben, M. J. 1997. Storage of hydrogen in single-walled carbon nanotubes. *Nature* 386: 377–379.
4. Chen, P., Wu, X., Lin, J., and Tan, K. L. 1999. High H_2 uptake by alkali-doped carbon nanotubes under ambient pressure and moderate temperatures. *Science* 285: 91–93.
5. Chambers, A., Park, C., Baker, R. T. K., and Rodriguez, N. M. 1998. Hydrogen storage in graphite nanofibers. *J. Phys. Chem. B* 102: 4253–4256.
6. Gupta, B. L. and Srivastava, O. N. 2001. Hydrogen adsorption in carbon nano-structures. *Int. J. Hydrogen Energy* 26: 831–835.
7. Chen, Y., Shaw, D. T., Bai, X. D. et al. 2001. Hydrogen storage in aligned carbon nanotubes. *Appl. Phys. Lett.* 78: 2128–2130.
8. Kajiura, H., Tsutsui, S., Kadono, K. et al. 2003. Hydrogen storage capacity of commercially available carbon materials at room temperature. *Appl. Phys. Lett.* 82: 1105–1107.
9. Shiraishi, M., Takenobu, T., and Ata, M. 2003. Gas–solid interactions in the hydrogen/single-walled carbon nanotube system. *Chem. Phys. Lett.* 367: 633–636.
10. Liu, C., Yang, Q. H., Tong, Y., Cong, H. T., and Cheng, H. M. 2002. Volumetric hydrogen storage in single-walled carbon nanotubes. *Appl. Phys. Lett.* 80: 2389–2391.
11. Ci, L., Zhu, H., Wei, B., Xu, C., and Wu, D. 2003. Annealing amorphous carbon nanotubes for their application in hydrogen storage. *Appl. Surf. Sci.* 205: 39–43.
12. Tarasov, B. P., Maehlen, J. P., Lototsky, M. V., Muradyan, V. E., and Yartys, V. A. 2003. Hydrogen sorption properties of arc generated single-wall carbon nanotubes. *J. Alloys Compd.* 356–357: 510–514.
13. Ritschel, M., Uhlemann, M., Gutfleisch, O. et al. 2002. Hydrogen storage in different carbon nanostructures. *Appl. Phys. Lett.* 80: 2985–2987.
14. Ma, R., Bando, Y., Sato, T. et al. 2002. Synthesis of boron nitride nanofibers and measurement of their hydrogen uptake capacity. *Appl. Phys. Lett.* 81: 5225–5227.
15. Dai, G.-P., Liu, C., Liu, M., Wang, M.-Z., and Cheng, H.-M. 2002. Electrochemical hydrogen storage behavior of ropes of aligned single-walled carbon nanotubes. *Nano Lett.* 2: 503–506.
16. Hanzawa, Y. and Kaneko, K. 2003. Gas adsorption. Chap. 20, 319, in *Carbon Alloys*, E. Yasuda, M. Inagaki, K. Kaneko, M. Endo, A. Oya, and Y. Tanabe, Eds. Elsevier, Amsterdam, The Netherlands.
17. Kaneko, K. and Matsumoto, A. 1989. The role of surface defects in the chemisorption of nitric oxide and sulfur dioxide on variable-sized crystalline.alpha.-iron hydroxide oxide. *J. Phys. Chem.* 93: 8090–8095.
18. Pinnavaia, T. J. 1983. Intercalated clay catalysts. *Science* 220: 365–371.

19. Bambakidis, G. and Bowman, R. C. 1985. *Proceedings of a NATO Advanced Study Institute on Hydrogen in Disordered and Amorphous Solids*, Plenum Press, New York.
20. Vokenstein, F. F. 1963. *The Electron Theory of Catalysis on Semiconductors*, Macmillan, New York.
21. Kiselev, V. F. and Krylov, O. V. 1989. Adsorption and catalysis on transition metals and their oxides, in Springer Series in Surface Science, Springer-Verlag, Berlin, Germany.
22. Kase, K., Yamaguchi, M., Suzuki, T., and Kaneko, K. 1995. Photoassisted chemisorption of NO on ZnO. *J. Phys. Chem.* 99: 13307–13309.
23. Marechal, Y. 2007. *The Hydrogen Bond and the Water Molecule: The Physics and Chemistry of Water, Aqueous and Bio-media*, Elsevier, Oxford, U.K.
24. Desiraju, G. R. and Steiner, T. 2001. *The Weak Hydrogen Bond: In Structural Chemistry and Biology*, Oxford Science Publication, Oxford, U.K.
25. Watanabe, K., Soma, M., Onishi, T., and Tamaru, K. 1971. Sorption of molecular hydrogen by potassium graphite. *Nat. Phys. Sci.* 233: 160–161.
26. Kaneko, K., Ishii, C., Kanoh, H. et al. 1998. Characterization of porous carbons with high resolution α_s-analysis and low temperature magnetic susceptibility. *Adv. Colloid Interface Sci.* 76–77: 295–320.
27. Yanagisawa, T., Endo, M., Lake, M. L., and Higaki, S. 2002. Electrode material for lithium secondary battery, and lithium secondary battery using the same. *Eur. Pat. Appl.* EP1262579.
28. Inoue, S., Hanzawa, Y., and Kaneko, K. 1998. Prediction of hysteresis disappearance in the adsorption isotherm of N_2 on regular mesoporous silica. *Langmuir* 14: 3079–3081.
29. Lei, S., Miyamoto, J., Kanoh, H., Nakahigashi, Y., and Kaneko, K. 2006. Enhancement of the methylene blue adsorption rate for ultramicroporous carbon fiber by addition of mesopores. *Carbon* 44: 1884–1890.
30. Gogotsi, Y., Nikitin, A., Ye, H. et al. 2003. Nanoporous carbide-derived carbon with tunable pore size. *Nat. Mater.* 2: 591–594.
31. Yushin, G., Dash, R., Jagiellom, J., Fisher, J. E., and Gogotsi, Y. 2006. Carbide-derived carbons: Effect of pore size on hydrogen uptake and heat of adsorption. *Adv. Funct. Mater.* 16: 2288–2293.
32. Pekala, R. W. 1989. Organic aerogels from the polycondensation of resorcinol with formaldehyde. *J. Mater. Sci.* 24: 3221–3227.
33. Hanzawa, Y., Kaneko, K., Pekala, R. W., and Dresselhaus, M. S. 1996. Activated carbon aerogels. *Langmuir* 12: 6167–6169.
34. Lee, J., Kim, J., and Hyeon, T. 2006. Recent progress in the synthesis of porous carbon materials. *Adv. Mater.* 18: 2073–2094.
35. Jun, S., Joo, S. H., Ryoo, R. et al. 2000. Synthesis of new, nanoporous carbon with hexagonally ordered mesostructure. *J. Am. Chem. Soc.* 122: 10712–10713.
36. Joo, S. H., Choi, S. J., Oh, I. et al. 2001. Ordered nanoporous arrays of carbon supporting high dispersions of platinum nanoparticles. *Nature* 412: 169–172.
37. Kyotani, T., Ma, Z., and Tomita, A. 2003. Template synthesis of novel porous carbons using various types of zeolites. *Carbon* 41: 1451–1459.
38. Nishihara, H., Yang, Q.-H., Hou, P.-X. et al. 2009. A possible buckybowl-like structure of zeolite templated carbon. *Carbon* 47: 1220–1230.

39. Song, L., Miyamoto, J., Ohba, T., Kanoh, H., and Kaneko, K. 2007. Novel nano-structures of porous carbon synthesized with zeolite LTA-template and methanol. *J. Phys. Chem. C* 111: 2459–2464.

40. Ozaki, J., Endo, N., Ohizumi, W. et al. 1997. Novel preparation method for the production of mesoporous carbon fiber from a polymer blend. *Carbon* 35: 1031–1033.

41. Tao, Y., Endo, M., and Kaneko, K. 2009. Hydrophilicity-controlled carbon aerogels with high mesoporosity. *J. Am. Chem. Soc.* 131: 904–905.

42. Steele, W. A. 1973. The physical interaction of gases with crystalline solids: I. Gas-solid energies and properties of isolated adsorbed atoms. *Surf. Sci.* 36: 317–352.

43. Kaneko, K., Cracknell, R. F., and Nicholson, D. 1994. Nitrogen adsorption in slit pores at ambient temperatures: Comparison of simulation and experiment. *Langmuir* 10: 4606–4609.

44. Iijima, S. and Ichihashi, T. 1993. Single-shell carbon nanotubes of 1-nm diameter. *Nature* 363: 603–605.

45. Endo, M. and Iijima, S., Eds. 2007. *Handbook of Nano Carbon*, NTS, Tokyo, *in Japanese*.

46. Ohba, T., Matsumura, T., Hata, K. et al. 2007. Nanoscale curvature effect on ordering of N_2 molecules adsorbed on single wall carbon nanotube. *J. Phys. Chem. C* 111: 15660–15663.

47. Tjatjopoulos, G. J., Feke, D. L. and Mann Jr., J. A. 1988. Molecule-micropore interaction potentials. *J. Phys. Chem.* 92: 4006–4007.

48. Tanaka, H., Murata, K., Miyawaki, J. et al. 2002. Comparative study on physical adsorption of vapor and supercritical H_2 and CH_4 on SWNH and ACF. *Mol. Cryst. Liq. Cryst.* 388: 429–435.

49. Arai, M., Utsumi, S., Kanamaru, M. et al. 2009. Enhanced hydrogen adsorptivity of single-wall carbon nanotube bundles by one-step C_{60}-pillaring method. *Nano Lett.* 9: 3694–3698.

50. Hata, K., Futaba, D. N., Mizuno, K. et al. 2004. Water-assisted highly efficient synthesis of impurity-free single-waited carbon nanotubes. *Science* 306: 1362–1364.

51. Iijima, S., Yudasaka, M., Yamada, R. et al. 1999. Nano-aggregates of single-walled graphitic carbon nano-horns. *Chem. Phys. Lett.* 309: 165–170.

52. Utsumi, S., Honda, H., Hattori, Y. et al. 2007. Direct evidence on C-C single bonding in single-wall carbon nanohorn aggregates. *J. Phys. Chem. C* 111: 5572–5575.

53. Urita, K., Seki, S., Utsumi, S. et al. 2006. Effects of gas adsorption on the electrical conductivity of single-wall carbon nanohorns. *Nano Lett.* 6: 1325–1328.

54. Murata, K., Kaneko, K., Steele, W. A. et al. 2001. Porosity evaluation of intrinsic intraparticle nanopores of single wall carbon nanohorn. *Nano Lett.* 1: 197–199.

55. Murata, K., Kaneko, K., Steele, W. A. et al. 2001. Molecular potential structures of heat-treated single-wall carbon nanohorn assemblies. *J. Phys. Chem. B* 105: 10210–10216.

56. Bekyarova, E., Kaneko, K., Kasuya, D. et al. 2002. Oxidation and porosity evaluation of budlike single-wall carbon nanohorn aggregates. *Langmuir* 18: 4138–4141.

57. Utsumi, S., Miyawaki, J., Tanaka, H. et al. 2005. Opening mechanism of internal nanoporosity of single-wall carbon nanohorn. *J. Phys. Chem. B* 109: 14319–14324.

58. Murata, K., Hirahara, K., Yudasaka, M. et al. 2002. Nanowindow-induced molecular sieving effect in a single-wall carbon nanohorn. *J. Phys. Chem. B* 106: 12668–12669.

59. Bekyarova, E., Murata, K., Yudasaka, M. et al. 2003. Single-wall nanostructured carbon for methane storage. *J. Phys. Chem. B* 107: 4681–4684.

60. Endo, M., Muramatsu, H., Hayashi, T. et al. 2005. Nanotechnology: 'Buckypaper' from coaxial nanotubes. *Nature* 433: 476.

61. Kataura, H., Kumazawa, Y., Maniwa, Y. et al. 1999. Optical properties of single-wall carbon nanotubes. *Synth. Met.* 103: 2555–2558.

62. Miyamoto, J., Hattori, Y., Noguchi, D. et al. 2006. Efficient H_2 adsorption by nanopores of high-purity double-walled carbon nanotubes. *J. Am. Chem. Soc.* 128: 12636–12637.

63. Williams, K. A. and Eklund, P. C. 2000. Monte Carlo simulations of H_2 physisorption in finite-diameter carbon nanotube ropes. *Chem. Phys. Lett.* 320: 352–358.

64. Ohba, T. and Kaneko, K. 2002. Internal surface area evaluation of carbon nanotube with GCMC simulation-assisted N_2 adsorption. *J. Phys. Chem. B* 106: 7171–7176.

65. Kim, D. Y., Yang, C.-M., Noguchi, H. et al. 2008. Enhancement of H_2 and CH_4 adsorptivites of single wall carbon nanotubes produced by mixed acid treatment. *Carbon* 46: 611–617.

66. Kim, D. Y., Yang, C.-M., Yamamoto, M. et al. 2007. Supercritical hydrogen adsorption of ultramicropore-enriched single-wall carbon nanotube sheet. *J. Phys. Chem. C* 111: 17448–17450.

67. Murata, K., Kaneko, K., Kanoh, H. et al. 2002. Adsorption mechanism of supercritical hydrogen in internal and interstitial nanospaces of single-wall carbon nanohorn assembly. *J. Phys. Chem. B* 106: 11132–11138.

68. Murata, K., El-Merraoui, M., and Kaneko, K. 2001. A new determination method of absolute adsorption isotherm of supercritical gases under high pressure with a special relevance to density-functional theory study. *J. Chem. Phys.* 114: 4196–4205.

69. Murata, K., Miyawaki, J., and Kaneko, K. 2002. A simple determination method of the absolute adsorbed amount for high pressure gas adsorption. *Carbon* 40: 425–428.

70. Xu, W.-C., Takahashi, K., Matsuo, Y. et al. 2007. Investigation of hydrogen storage capacity of various carbon materials. *Int. J. Hydrogen Energy* 32: 2504–2512.

71. Kawase, Y., Miyamoto, J., Noguchi, H. et al. 2006. High-pressure hydrogen adsorption on nanoporous carbons. Unpublished work.

72. Yang, Z., Xia, Y., and Mokaya, R. 2007. Enhanced hydrogen storage capacity of high surface area zeolite-like carbon materials. *J. Am. Chem. Soc.* 129: 1673–1679.

73. Yang, Z., Xia, Y., Sun, X., and Mokaya, R. 2006. Preparation and hydrogen storage properties of zeolite-templated carbon materials nanocast via chemical vapor deposition: Effect of the zeolite template and nitrogen doping. *J. Phys. Chem. B* 110: 18424–18431.

74. Ströbel, R., Garche, J., Moseley, P. T., Jörissen, L., and Wolf, G. 2006. Hydrogen storage by carbon materials. *J. Power Sources* 159: 781–801.
75. Wang, Q., Challa, S., Sholl, D., and Johnson, K. J. 1999. Quantum sieving in carbon nanotubes and zeolites. *Phys. Rev. Lett.* 82: 956–959.
76. Tanaka, H., Kanoh, H., Yudasaka, M. et al. 2005. Quantum effects on hydrogen isotope adsorption on single-wall carbon nanohorns. *J. Am. Chem. Soc.* 127: 7511–7516.
77. Hattori, Y., Tanaka, H., Okino, F. et al. 2006. Quantum sieving effect of modified activated carbon fibers on H_2 and D_2 adsorption at 20 K. *J. Phys. Chem. B* 110: 9764–9767.

11

Membrane Processes

A.F.P. Ferreira, M.C. Campo, A.M. Mendes, and F. Kapteijn

CONTENTS

11.1 Introduction

Millions of years ago, nature created biological membranes. The most important feature of a bio-membrane is that it is a **selectively permeable** layer. This means that the size, charge, and other physio chemical properties of the particles, molecules, atoms or substances attempting to permeate it will determine whether they succeed or not. This aptitude makes them the most efficient known separation "technology." Hence, to mimic such concept is the most logical path to follow when trying to integrate separation technologies. The available technologies present major drawbacks, separation processes like distillation tend to be energy demanding, and processes based on adsorption have the problematic feature of not operate in a continuous mode.

The synthetic membranes can be produced from organic compounds, and/or from inorganic materials. The bulk of the commercially employed membranes in the industry are made of polymers. Membrane separation processes have a very important role in the separation industry. However, they were not considered technically important until early 1980s. Membrane separation processes can be differentiated based on the separation mechanism itself and on the size difference of the separated particles. These processes include microfiltration (MF), ultrafiltration (UF), nanofiltration (NF), reverse osmosis (RO), electrolysis, dialysis, electrodialysis, gas separation, vapor permeation, pervaporation, membrane distillation, and membrane contactors [1–8]. Pervaporation is the only process that involves a phase transition (from liquid to vapor) over the membrane [8]. All processes are mainly pressure driven, except the (electro-) dialysis [8]. MF and UF are used in the food and beverage industry (e.g., beer MF, apple juice UF), biotechnological applications and the pharmaceutical industry (e.g., antibiotic production, protein purification), water purification and wastewater treatment, and others. NF and RO membranes are mainly used in water purification processes (e.g., water desalination). Dense membranes are used in proton and electron transport-related processes (e.g., fuel cells) and gas separation (e.g., H_2 or O_2 from air). Microporous membranes are utilized for gas separations (removal of CO_2 from natural gas, separating N_2 from air, organic vapor removal from air or nitrogen streams). Membranes can also be integrated within other separation processes, like in distillation to help in the separation of an azeotropic mixture, reducing the costs of distillation processes. In Figure 11.1 is presented a graphical resume of membrane types and their application versus their pore size.

11.1.1 Market

In a recent study by the Business Communication Company [2], the combined market, in the United States, for membranes is estimated to be worth 5 billion U.S. dollar (USD) and, rising at an average annual growth rate (AAGR) of 6.6%, is expected to reach 6.9 billion USD in 2009. This study, entitled Membrane

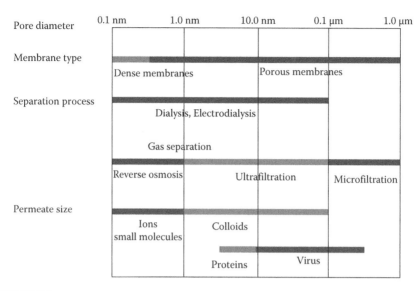

FIGURE 11.1
Membrane types, separation processes, and permeate size as function of membrane pore diameter.

Technology: A New Era, states that sales of membranes used in liquid and gas separations are valued at 2.2 billion USD and should expand at an AAGR of 9.9% through to 2009. The market for non-separating membranes, used in drug delivery, guided tissue regeneration, batteries, food packaging, and high-performance textiles, is currently worth about 2.8 billion USD and is expected to grow at an AAGR of 3.8% during the forecast period. In the same study, it is stated that the United States consumes about 40% of all membranes produced worldwide [2]. Water and wastewater treatment accounts for about half of the demand for membranes used in separating applications. RO and NF are experiencing double-digit growth rates, above all, an expression of the booming market in desalination, but also a reflection of increased application in process water treatment and an interest in reclaiming "used" water [2]. MF and UF are also in demand for treating potable and process water, but the most rapid growth in this sector is wastewater treatment with membrane bioreactors. Recording even stronger growth in the use of MF and UF technologies is the biotechnology and pharmaceuticals sector, which is benefiting from a surge in genomic and proteomic research, and the development of new biopharmaceuticals [2]. Steady growth of UF is also found in dialysis, an application with a large and increasing patient base. In other separation processes, membrane applications have yet to reach full market potential, but are the focus of very active commercialization efforts. Development of robust new membranes and membrane reactor processes are opening up a number of new gas application areas and promising new functions in catalytic membrane reactors. High growth rates are also expected for membranes that are used in fuel cell applications, although the market is still quite small. Because of their diversity, sales of membranes for non-separating applications are growing at vastly different rates. Although the markets are not the largest, the highest growth rates are found in food packaging applications and the manufacture of batteries [2].

In a previous study presented by the same company [3], the total predicted sales of inorganic membrane in 2003 was stated to be worth 228 million USD, of which 70% being ceramic membranes. In the same study was reported that in the large market of membranes (including polymeric membranes), the inorganic membrane market would represent 15% in 2003.

11.1.2 Commercial Application

Membranes are essential to a range of applications including potable water, process water, and wastewater treatment; tissue repair and other therapeutic procedures; pharmaceuticals production; food and beverage processing; and separation procedures needed for manufacturing chemicals, cars, electronics, fuels, and a range of other products. Primary drivers for membrane sales include consumer demand for higher-quality products, increased regulatory pressures, deteriorating natural resources, and the need for environmental and economic sustainability.

A membrane separation system is built from different parts and assembled into a module. Each step of the membrane module manufacturing (including the membrane itself) must satisfy to the application specifications. There are several points that must be considered for the success and acceptance of the membrane technology in the market on a commercial scale.

The main condition is the *low production cost* of the separation unit (the module itself and installation). When the membranes are being developed at laboratory scale, one must be concerned with the raw materials including high-quality supports, the scaling up to mass production, and its reproducibility.

The *reliability* is an important point to be concerned, to avoid process shutting down and its related costs. The separation unit in its whole must be reliable at the operating conditions. The difference between the thermal expansion coefficients (TECs) of the materials that compose the layers of a membrane can cause fractures/defects on the membrane when this is brought to the operation conditions. Other factor to be considered is the reliability of the sealing materials and procedure, especially for high-temperature applications. On Table 11.1 is resumed the operation temperature range for each type of membrane based on their porosity and composition.

The performance of a membrane separation unit must be constant in time. Thus, it is required to have *long-term stability*. For porous membranes, two main issues arise at this level. The first is fouling problems causing pore blocking. The second is the stability of the pore structure itself. When the membrane is being designed, thermal and chemical properties must be considered at the required operation conditions. Chemical reactivity can cause

TABLE 11.1

Operation Temperature Range for Different Type
of Membranes

Type	T-Range
Dense inorganic membranes	
Metallic (Pd)	300°C–600°C
Mixed oxides (ZrO_2, perovskites)	>500°C
Porous inorganic membranes	
Microporous oxides	<700°C
Carbons	<1000°C
Zeolites	<700°C
Polymers + hybrids	<200°C
Liquid phase	<100°C

fouling. Therefore, one must choose materials less prone to form carbon deposits, for example. The permeation/separation properties must be stable in time, not depending on factors that can change with continuous functioning of the unit, e.g., swelling in polymeric membranes. The mechanical stability of the membrane and separation layer is essential, especially under cyclic temperature and/or pressure operation.

A high *surface area-to-volume ratio* is required to minimize the investment cost for a certain production rate and the footprint of the equipment. This feature should be taken in consideration when the module architecture is designed. This approach will also lead to the prospect of the chemical plants miniaturization that would be significantly cheaper and safer than existing ones.

High selectivity and *high permeation* will determine the economical feasibility of a membrane separation process. These are controlled by the intrinsic (mass transport) properties of the membrane support and separation layer, and their limitations. However, the selectivity and permeation can also be affected by the process conditions, module architecture/design, separation layer thickness, surface area of the chosen supports. To improve the overall performance of a membrane unit, it is necessary to know the limiting mass transport phenomena.

11.2 Fundamentals: Basic Principles, Theoretical Background, and State of the Art

11.2.1 Membrane Types

According to International Union of Pure and Applied Chemistry (IUPAC) [1], a membrane is a structure, having lateral dimensions much greater than its thickness, through which mass transfer may occur under a variety of driving forces. Membranes can be classified according to their structure and composition into four main categories: inorganic dense membranes, inorganic porous membranes, polymers/hybrid membranes, and liquid phase membranes (see Table 11.1).

Dense inorganic membranes are referred to those membranes made of a polycrystalline ceramic or metal, which allows certain gas species to permeate through its crystal lattice [4]. Example of those membrane are metallic membranes (e.g., Pd membrane), and mixed oxides (e.g., ZrO_2, perovskites, dense silica, fluorite type ceramic membranes). **Porous inorganic membranes** include amorphous and crystalline ceramic membranes [4]. In this category, the following membrane types can be found: zeolite membranes, carbon membrane, and microporous oxides membranes. **Polymer membrane** according to the IUPAC Gold Book is a thin layer of polymeric material that acts as a barrier permitting mass transport of selected species. The polymer materials used on the membrane manufacturing include cellulose acetate, polysulfone, and nylon, among others. Cellulosic membranes, the least expensive membrane materials, account for the largest share of polymer membranes, with 58% of sales value in 2008 [5]. **Liquid phase membrane** is a liquid phase existing in either supported or unsupported form, which serves as a membrane barrier between two phases [1]. There are three basic types of liquid membranes, bulk liquid membrane, emulsion liquid membrane, and immobilized liquid membrane, also called a supported liquid membrane [6]. They are mainly used for metal ion recovery, transport of organic ions, gas separation, electron transport, and neutral molecules separation [6].

11.2.2 Membrane Configurations

The manifold assembly (module) containing the membrane or membranes to separate the streams of feed, permeate, and retentate is closely related to the membrane support geometry or to the membrane itself when not supported. The design of a membrane unit involves the modules, the membrane system (modules, pumps, piping, tanks, control/monitoring units, pretreatment, and cleaning facilities), and the operating mode (batch, continuous, or diafiltration) [7]. The module acts as membrane "housing," which simultaneously supports the membrane and provides effective fluid management [7]. Membranes can be produced in flat sheet or cylindrical shape and this configuration determines the module geometry. Since about 1955, and the creation of the first practical RO membrane, membrane devices have evolved into many configurations. Plate-and-frame and spiral-wound modules involve flat membranes, whereas tubular, capillary, and hollow fiber modules are based on cylindrical membrane configurations [7]. The most common configurations for a membrane module are [1]

- Hollow fiber/multi-tubular (polymer and inorganic membranes)
- Plate and frame (polymer and inorganic membranes)
- Spiral wound (polymer membranes)

The spiral-wound design (see Figures 11.2 and 11.3) is far and away the most common configuration used in RO and NF applications nowadays. This is largely due to the fact that it is the least expensive configuration, resulting from its relatively high packing density (membrane surface area per unit volume), as well as the fact that it is a very competitively priced design. Due to the close spacing of the membrane leaves, this design has the greatest susceptibility to fouling [9].

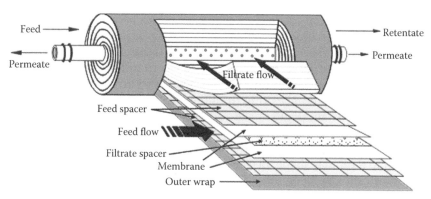

FIGURE 11.2
Cutaway drawing of a spiral wound membrane element.

FIGURE 11.3
A Degrémont membrane array similar to those for use in the new Perth Seawater Desalination Plant, Seawater Reverse Osmosis (SWRO), Kwinana, Australia. (From Water-technology.net, http://www.water-technology.net/projects/perth/perth1.html. With permission.)

The plate-and-frame membrane device has taken many shapes and sizes over the years and has been used primarily for the technologies of MF and UF. Plate-and-frame membrane devices are not commonly used for municipal water treatment but are more appropriate for certain waste or food processing applications where there is a high fouling tendency since they are basically quite fouling resistant compared to the spiral-wound configuration. However, the challenge of packing as much membrane area into as small a volume as possible, as well as problems associated with the sealing and gasketing to prevent leakage and cross-contamination, have limited the utilization of this kind of membrane packing configuration.

Capillary/hollow fiber membrane module consists of unsupported hollow fibers that have been extruded or spun from such polymers as polysulfone, polyethersulfone, polyacrylonitrile, polyvinylidene fluoride, or similar thermoplastic materials. Capillary membranes can be also produced from α-alumina or other inorganic materials. The fibers are constructed with inside diameters of several millimeters and wall thicknesses of fractions of millimeters. In most cases, the permeate flow is from the inside of the fiber to the outside (lumen feed), although the elements with the feed flow on the outside (and the permeate exiting from the lumen side) are considered to be more fouling resistant (Figures 11.4 and 11.5).

Tubular devices are differentiated from capillary fiber devices because they are constructed of a membrane-coated backing material, which is wound into a tubular shape. Because this construction is more pressure tolerant than capillary fibers, the diameters of tubular devices can be made larger, thereby improving their tolerance to suspended solids. Most of the

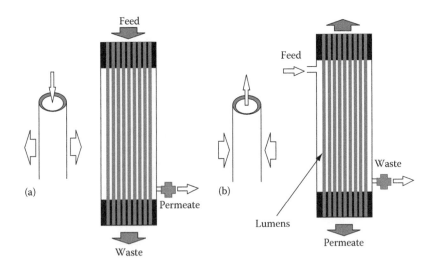

FIGURE 11.4
Schematic illustration shows hollow fiber membrane lumens and module configured in cross flow: (a) feed to inside of lumen and (b) feed to outside of lumen.

FIGURE 11.5
Six inch diameter KMS WINEFILTER™ hollow fiber ROMICON® cartridge used in a WF-3 crossflow MF system as obtained from Koch Membrane Systems. (Copyrighted by Koch Membrane Systems, Wilington, MA, http://www.kochmembrane.com/pdf/Wine%20 Brochure%206%2021%2006_FINALrev%201.pdf. With permission.)

ceramic devices today are in the tubular configuration. The inside diameters of tubular devices typically range from about 7 to 25 mm, and the flow is virtually always lumen feed.

From the perspective of cost and convenience, it is beneficial to pack as much membrane area into as small a volume as possible. This is known as "packing density." The higher the packing density, the greater the membrane area enclosed in a certain sized device, and (usually) the lower the cost of the membrane element (on the basis of permeate rate per membrane area). The downside of the high packing density membrane elements is their greater propensity for fouling.

11.2.3 Membrane Supports: Shapes and Materials

Depending on the final application, membranes can be classified in different types: biological or synthetic membranes, dense or porous, support

FIGURE 11.6
Different support materials: (a) stainless steel with α-alumina top layer, (b) tubular carbon membrane, and (c) tubular α-alumina membrane.

or self-supported, symmetric or asymmetric. Within the synthetic group, membranes can be divided according to their chemical composition into organic (polymers, hybrids, and liquids) and inorganic. Among inorganic membranes, there are metallic or mixed oxides (dense membranes) and microporous oxides, carbons, and zeolites (porous membranes).

In the asymmetric supported membranes, we can have different support materials and shapes. The most common shapes are disk, tubular, and multi-tubular/monolith (see Figure 11.7). The support available on the market porous stainless steel, α-alumina (sometimes with titania top layer), and carbon (see Figure 11.6).

Microstructure of porous support   Porous support with dense membrane

Membrane module fabricated by hydro

FIGURE 11.7
Monolith membrane close-ups and module.

11.2.4 Experimental Approach

The principle of separation of all membrane separation processes concerns the use of a membrane as a physical barrier, which selectively allows the passage of certain species, the permeants, presented in a mixture in detriment of others. The side where the mixture contacts with the membrane surface is called the feed side, whereas the side where the permeating species are collected is called the permeate side. The stream that does not permeate is designated by the retentate. To prevent ambiguous meanings, IUPAC terminology will be used in this chapter (Figure 11.8).

In the scope of this chapter, special attention will be given to microporous inorganic membranes (pores <2 nm) for gas separation as they are the areas of interest and expertise of the authors. The improvement of this technology is only possible if the material is exhaustively understood. For these reasons, the available experimental techniques usually used to characterize these membranes will be focused. Opposite to the dense materials where the material itself determines the performance, in porous membranes, both material and pore network are important, especially when molecular sieves are involved; species are only allowed to pass through pores if they can fit in, or be retained if they are larger than the pores. A review on transport models for different regimes of permeation will be also given.

First, membrane permeation definitions will be introduced. According to IUPAC [1,10], flux, permeance, and permeability can be defined by the following expressions:

Flux
$$N_i = \frac{\text{molar flow}}{\text{membrane area}} \quad \left[\text{mol} \cdot \text{s}^{-1} \cdot \text{m}^{-2} \right]$$

Permeance
$$\Pi_i = \frac{\text{flux}}{\Delta p_i} = \frac{N_i}{\Delta p_i} = \left[\text{mol} \cdot \text{s}^{-1} \cdot \text{m}^{-2} \cdot \text{Pa}^{-1} \right]$$

Permeability
$$P_i = \frac{\text{permeance}}{\text{membrane thickness}} = \frac{N_i}{\Delta p_i / \ell} \left[\text{mol} \cdot \text{s}^{-1} \cdot \text{m}^{-2} \cdot \text{Pa}^{-1} \cdot \text{m} \right]$$

The membrane permeance Π_i to a certain permeant i is defined by the flux of that same species N_i, divided by driving force here expressed in partial

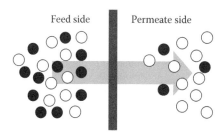

Feed side   Permeate side

FIGURE 11.8
Scheme of the permeation flow over a membrane.

pressure Δp_i. If the permeance is expressed per unit membrane thickness ℓ (selective layer), then the permeability (coefficient) P_i is obtained.

11.2.5 Membrane Characterization

Every material is defined by its properties. Morphology, topography, structure, chemical nature, chemical and thermal properties are all important characteristics that should be studied to improve materials, to suit a better performance. The combination of all these properties is what defines the success of a membrane.

Some of the existent techniques to characterize membrane-based materials are reported in characterization techniques chapter: scattering, spectroscopy, microscopy, calorimetry, and in situ and ex situ techniques are covered (Tables 11.2 and 11.3).

TABLE 11.2

Some of the Available Techniques for Membrane Morphology Characterization

Characterization Method	Principles and Description
SEM (Scanning electron microscopy)	Surface and cross section views
TEM (Transmission electron microscopy)	
FESEM (Field energy scanning electron microscopy)	Visualization of pore sizes within the resolution of the equipment
AFM (Atomic force microscopy)	
NMR (Nuclear magnetic resonance)	Determination of spin-lattice relaxations of condensed water inside the membrane pores
PGSE-NMR (pulse gradient spin echo NMR)	Determinations of diffusivity of the gas through pores
SANS (small-angle neutron scattering)	Assessment of water molecules in the pores by neutron scattering
XRD (x-ray diffraction)	Degree of cristallinity, interplanar distance, crystals sizes
ESR (Electron spin resonance)	Measurement of spin transitions of unpaired electrons

TABLE 11.3

Some of the Available Techniques for Pore Characterization

Characterization Method	Principles and Description
Mercury porosimetry	Mercury intrusion technique for the characterization of larger pores (mainly porous supports). Determination of macro-meso pore size distributions.
Permporometry [11–14]	Estimation of PSD under the assumption that noncondensable gas permeation is blocked by the capillary condensed vapor (0.5–30 nm). Only pores favorable for permeation are detected and no dead-end pores are accounted.
Gas adsorption	Determination of micropore size distribution based on the adsorption equilibrium data of probe gases. There is no distinction between active and dead end pores.

11.2.5.1 Scanning Electron Microscopy

Morphology (shape, size, texture, etc.) may be assessed by scanning electron microscopy, where the surface and cross-sectional roughness may be seen. This technique is very important in assessing defects such as pinholes in zeolite membranes or cracks in carbon membranes. When zeolite or carbon membranes are supported, fissures might also occur during preparation due to different TECs of the membrane and its support. It also allows the measurement of certain important dimensions, such as membrane thickness, diameters, etc. (Figure 11.9).

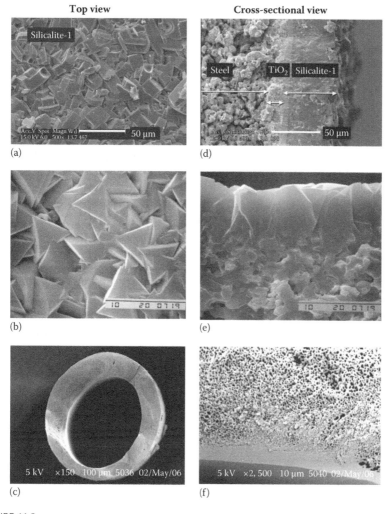

FIGURE 11.9
SEM pictures of (a), (d) Silicalite-1 membrane supported on stainless steel with titania top layer, (b), (e) DDR membrane supported on α-alumina, and (c), (f) carbon hollow fiber membrane.

11.2.5.2 Adsorption

Adsorption equilibrium isotherms and uptake curves are characteristic of each material. Using probe molecules different in size, polarity, and shape, very useful information may be collected to understand the porous structure of the membranes and to provide relevant data to further on modeling.

11.2.5.3 Permeation

Membranes are developed to preferentially permeate species. For this reason, permeation tests are probably the most powerful technique to characterize membranes. The performance or efficiency of a membrane process is dictated by the permeability and selectivity of the membrane. Permeation through membranes occurs by the establishment of a driving force; in gas permeation, the driving force is the difference in the partial pressure of the permeating species.

11.2.5.3.1 Single-Component Permeation

Single-component experiments of probing gases should be performed as a preliminary evaluation on the membranes' performance. The ideal selectivity for a given bicomponent separation is easily determined by the ratio between the single-component permeabilities of the more permeable and the less permeable species:

$$S_{ij} = \frac{\Pi_i}{\Pi_j} = \frac{P_i}{P_j}$$

To test a membrane for permeation, it is necessary to assemble it in a cell called the "permeate cell." Depending on the geometric configuration of the membrane, several approaches are used. In the case of flat disk-like membranes, the permeate cell is split in two by the membrane itself: one side of the membrane is exposed to the feed stream and called "feed side," and the other side where permeate is collected is called the "permeate side." As for cylindrical shapes, hollow fibers, or tubular membranes, the feed stream can circulate on the shell side of the membrane and the permeate collected from the bore side, or vice versa (Figure 11.10a and b).

11.2.5.3.2 Permeation Methods and Setups

There are two possible ways to perform permeation experiments [15]: (1) the continuous mode or (2) the batch mode.

In the first mode, the feed stream is at overpressure and the permeate side is at a lower pressure. Usually, fluxes through the membrane are high and can be measured directly with flowmeters with an adequate range according to the flowrate observed; a highly permeable species should require a flowmeter with a higher range due to its larger flux, while lower fluxes of less permeable species should be measured with flowmeters with an inferior

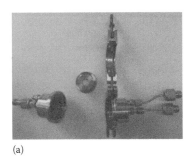
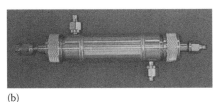

(a)   (b)

FIGURE 11.10
Test permeation cells: (a) for flat disk membranes and (b) for tubular membranes.

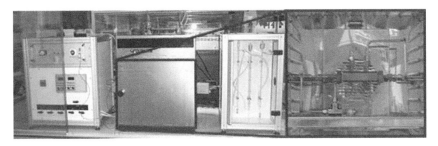

FIGURE 11.11
Membrane permeation experimental apparatus. Right picture shows the membrane module inside the thermostatted oven.

range. An example of an experimental apparatus is shown in Figure 11.11. An alternative configuration of this method may comprise the use of a sweep gas in the permeate side. A sweep gas has the role of decreasing the partial pressure of the permeating species (reduce its concentration) in the permeate side in order to keep the driving force constant.

The second method consists of keeping the feed compartment under pressure and the permeate side usually under vacuum. The volume of the permeate side has to be known, for what a previous calibration of the volume should be done. The flux through the membrane is then determined by the increment in pressure of the permeate species collected in this calibrated volume within a certain time interval. This time interval should be large enough to reduce the time measurement error; the pressure increment should be large enough to reduce the measurement error and at the same time small enough to assure the constancy of permeability (later on, it will be seen how the transport mechanism may be influenced by the pressure).

11.2.5.3.3 Multicomponent Permeation
Single-component permeabilities obtained provide a qualitative and a prior quantitative indication on the performance of the membranes. Nevertheless,

rather than single-component experiments, it is the multicomponent behavior that describes the real membrane performance or efficiency to separate (purify or concentrate) species.

For multicomponent measurements, the experimental setups must include the use of analysis equipment, such as a chromatograph, mass spectrometer, or specific detector, capable of recognizing and quantifying the species in permeate and retentate. The real selectivity of a membrane is then evaluated as follows:

$$S_{ij} = \frac{\left(y_i/y_j\right)_{permeate}}{\left(x_i/x_j\right)_{retentate}}$$

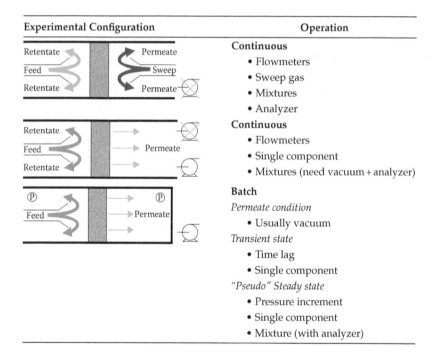

Experimental Configuration	Operation
Retentate / Permeate / Feed / Sweep / Retentate / Permeate	**Continuous** • Flowmeters • Sweep gas • Mixtures • Analyzer
Retentate / Feed / Permeate / Retentate	**Continuous** • Flowmeters • Single component • Mixtures (need vacuum + analyzer)
℗ / Feed / Permeate / ℗	**Batch** *Permeate condition* • Usually vacuum *Transient state* • Time lag • Single component *"Pseudo" Steady state* • Pressure increment • Single component • Mixture (with analyzer)

11.2.5.3.4 Support Effect and Operating Mode on the Concentration Gradients

The very thin zeolite membranes and carbon membranes suffer from a lack of mechanical stability. For this reason, these selective membranes must be strengthened by adding a porous layer, called the support. The thickness and porosity of the support differ from the selective layer and may actually interfere in the overall resistance to mass transfer over the membrane. The local concentration at the interface zeolite/support will then differ from the concentration at the permeate side. The first two cases illustrated below represent a Wicke–Kallenbach (WK) permeation mode: As the resistances of both selective and support layers are in series, the way the membrane is facing the feed stream will influence the final result. In the first case, the selective layer is in contact with the feed side where the concentration of the permeants is maximum. In the second case, the support layer is facing the feed side, yielding a decrease on the concentration of the permeating species at the boundary support/selective layer. As a result, the driving force through the selective layer will be lower, leading to a lower flux. The last case represents a batch operating mode where vacuum is applied to the permeate side. Therefore, the driving force is increased and, consequently, so the flux (Figure 11.12).

This effect of the orientation of an asymmetric membrane was demonstrated with methane/ethane permeation through a silicalite-1 membrane. For a 1:1 molar feed ratio, the WK-feed and WK-perm configurations were used with the 3 mm thick support membrane [16]. With the standard WK-feed operation mode, the selectivity was from 6 to 10, whereas with the WK-perm configuration, the selectivity was completely lost Figure 11.13. During steady-state permeation, concentration polarization in the support layer located at the feed side reduced the selectivity to 1. The methane concentration at the support/zeolite interface was much higher than in the feed

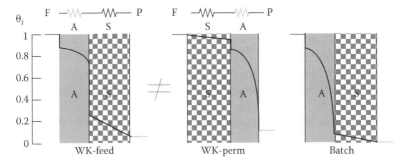

FIGURE 11.12
Graphical interpretation of the concentration profile within the membrane for the different permeation modes: A, selective layer; S, support.

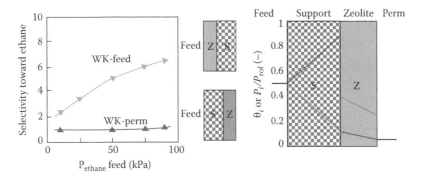

FIGURE 11.13
Graphical interpretation of the orientation effect of an asymmetric membrane.

due to the concentration polarization, thereby with the feed composition, and this counteracted the ethane selectivity of the selective layer. This clearly demonstrates why the zeolite layer should face the feed side rather than the support side, in order to avoid these polarization phenomena.

As a conclusion, in multicomponent separations, the orientation of the membrane to the feed side is crucial. The support resistance [15–17] does influence the permeance of the membrane and should then be included in the modeling.

11.2.6 Transport

As mentioned before, inorganic membranes for gas separations may comprise several different types such as, composite membranes, zeolite membranes, carbon membranes, metallic membranes, etc. [18] The structural properties of the membranes distinguish between porous or dense membranes. The main difference between them is mainly reflected on the comparable lower fluxes of dense membranes. However, within the porous categories, there are different ranges of pore sizes that influence the regimes of mass transport, together with other aspects as the physical properties of the species and the pressure. For membranes with large pores, fluxes are expectedly high but selectivities are low. As pore sizes decrease, selectivity is enhanced while fluxes become smaller.

A membrane often consists of a microporous selective layer over a porous support that provides the desired mechanical stability. Therefore, the mass transport through the membrane is the overall contribution of both selective and support layers. The most usual transport mechanisms through porous membranes for gas separations are the following described ones (Figure 11.14):

Viscous flow (>50 nm): Viscous flow characterizes the mass transfer through pores larger than the mean-free path of the molecules travelling across the membrane. With almost no selectivity, the permeation occurs due to the establishment of a pressure gradient between the two sides of the membrane. Usually, this type of flow occurs in the supports or large defects.

Molecular diffusion: Also called simply diffusion, happens when species move from a more concentrated side to a less concentrated one. Usually, happens in large pores or defects.

Knudsen diffusion (10–100 nm): This regime happens under low pressure conditions and when the pore size is smaller than the mean-free path; as a result, species start to collide more with the pore walls than with each other. The selectivity is not so high and is determined by the ratio of the molecular weights of the species.

Surface diffusion (0.3–1 nm): Surface diffusion takes place when species are adsorbed on the pore walls of the membrane; once in the adsorbed state, the adsorbates move through the pore network by jumping from site to site. The different rates of diffusion are based on the different rates of adsorption;

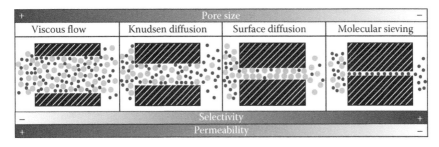

FIGURE 11.14
Scheme of transport mechanisms in pores.

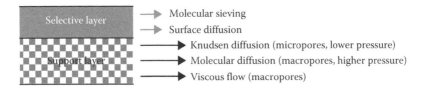

FIGURE 11.15
Schematic resume of the different transport mechanism within the membrane.

the strongly adsorbing species will preferentially cross the membrane at the expense of the species with lower affinity to the surface walls.

Molecular sieving (<1 nm): When pores have sizes very close to those of molecules or atoms diameters, the membrane acts as a sieve; species smaller than the pore size are allowed to pass through it, whereas species larger than the pores are retained. This regime is characterized by its high selectivity and low or moderate flux. Zeolite membranes as well as carbon molecular sieve membranes (CMSM) exhibit this type of mass transport regime.

These regimes are a result of the existent pore sizes of the membrane and on the operating conditions applied to the membrane. Usually, the selective layer of microporous membranes is characterized by a surface diffusion (zeolitic diffusion) or molecular sieving. On the other hand, supports normally have larger pores, where viscous flow or Knudsen diffusion takes place, depending on the applied conditions. Note that cracks or defects in the selective layer may also exhibit these transport modes (Figure 11.15).

11.2.6.1 Pressure Dependence of the Permeance

A prior information to reveal which transport occurs in a certain membrane may be easily determined by single-component experiments where permeance is studied as a function of pressure. The pressure dependence may qualitatively infer the mass transport in case: permeance of a membrane increases with pressure for viscous flow, is constant for Knudsen diffusion, and is inversely proportional to the pressure for molecular and surface diffusion, but more strongly in the last case (Figure 11.16).

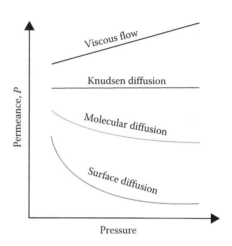

FIGURE 11.16
Pressure dependence of the permeance for
the different transport mechanisms.

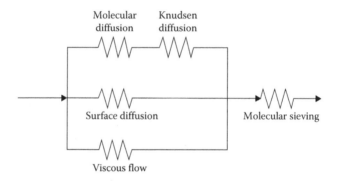

FIGURE 11.17
An electric analog for the different types of transport mechanisms in a membrane.

As mentioned before, to accurately predict the performance of a membrane, the support must be taken into account. The selective layer is associated to the support by an electric analog. However, the support itself may be formed by one or more layers, which may also be added in series. If the selective layer has defects, then a resistance in parallel must be included (Figure 11.17).

11.2.6.2 Support Characterization

The support resistance may play a significant role and should be quantified in order to develop a correct model to describe the membrane reality. This is the only way to accurately predict the transport phenomena through a membrane system. There are several experimental modes and methods to measure the permeance of a membrane. Depending on the conditions imposed by these methods, different approaches should be used to describe the fluxes:

11.2.6.2.1 Wicke–Kallenbach Method

Both feed and permeate compartments are kept at the same total pressure. To create a difference between the partial pressures, establishing a driving force for the permeating species, a sweep gas is introduced to the permeate side in order to keep its partial pressure low, removing the permeants. The sweep gas may act as a stagnant film in the support layer, over which the permeating species have to cross. For this reason, the transport regime over the support is usually described as molecular diffusion where the dusty gas model can be applied. For single-component permeation, the diffusion of the permeating species through the support is given by

$$N_{M_i} = -\frac{\varepsilon}{\tau} \frac{D_{iHe}}{\Re T} \nabla p_i$$

where D_{iHe} is the binary diffusivity of i in helium (e.g., diluent gas) and can be estimated by

$$D_{AB} = \frac{10^{-7} T^{1.75} \left(\dfrac{1}{M_A} + \dfrac{1}{M_B} \right)^{1/2}}{p_T \left(\left(\sum_A v_a \right)^{1/3} + \left(\sum_B v_a \right)^{1/3} \right)^2}$$

which is a relation provided by Füller et al. [19]. The porosity and tortuosity of the support are ε and τ, respectively.

11.2.6.2.2 Batch Method

In this method, the permeate side is submitted to vacuum and the driving force for permeation is also established by the difference in the partial pressures of the permeating species. However, a total pressure difference is also verified, and, for this reason, the transport through the support is often described by viscous flow

$$N_{V_i} = -\frac{\varepsilon}{\tau} \frac{B_0}{\eta} \frac{\bar{p}_i}{\Re T} \nabla p$$

Here, the permeability is defined as $B_0 = d_0^2/32$, d_0 is the pore diameter, $\bar{p}_i$ is the average pressure of permeant i, and η is the viscosity of the gas. If the pressure is low and the pores of the support are smaller, the transport may be defined by Knudsen diffusion rather than viscous flow, as follows:

$$N_{K_i} = -\frac{\varepsilon}{\tau} \frac{D_{K_i}}{\Re T} \nabla p_i$$

where the Knudsen diffusivity is defined by $D_{K_i} = (d_0/3)\sqrt{(8\Re T/\pi M_i)}$ and M_i is the molecular mass.

TABLE 11.4

Transport Mechanisms, Flux, and Permeance

Transport Mechanism	Flux, N_i	Permeance, $\Pi_i = N_i/\Delta p_i$
Viscous flow (support, large defects)	$N_{V_i} = \dfrac{\varepsilon}{\tau} \dfrac{B_0}{\eta_i} \dfrac{\bar{p}_i}{\Re T} \dfrac{\Delta p}{\ell}$ with $B_0 = \dfrac{d_0^2}{32}$	$\Pi_i = \dfrac{\varepsilon}{\tau} \dfrac{d_0^2}{32} \dfrac{1}{\ell} \dfrac{1}{\Re T} \dfrac{\bar{p}_i}{\eta_i}$
Molecular diffusion	$N_{M_i} = \dfrac{\varepsilon}{\tau} \dfrac{D_{AB}}{\Re T} \dfrac{\Delta p_i}{\ell}$ with $D_{AB} \propto \dfrac{T^{1.75}}{p_{tot}}$	$\Pi_i = \dfrac{\varepsilon}{\tau} \dfrac{D_{AB}}{\Re T} \dfrac{1}{\ell}$
Knudsen diffusion, (small pores, low p)	$N_{K_i} = \dfrac{\varepsilon}{\tau} \dfrac{D_{K_i}}{RT} \dfrac{\Delta p_i}{\ell}$ with $D_{K_i} = \dfrac{d_0}{3}\sqrt{\dfrac{8\Re T}{\pi M_i}}$	$\Pi_i = \dfrac{\varepsilon}{\tau} \dfrac{d_0}{3} \dfrac{1}{\ell}\sqrt{\dfrac{8}{\pi M_i \Re T}}$
Surface diffusion (small micropores)	$N_S = -D_S \rho \dfrac{\partial q_i}{\partial z}$ with $D_S = D_i(q_i)\Gamma_{ii}$	$\Pi_i = -\dfrac{\rho D_i \Gamma_{ii}}{\Delta p_i} \dfrac{\partial q_i}{\partial z}$

The transport through the support may also comprise the combination of both viscous and Knudsen flow; graphically, the dependence of the permeance with pressure will be a straight line. The intercept represents the Knudsen contribution and the slope of the line the viscous flow:

$$\Pi = \boxed{b} + \boxed{m\,p}$$

$$\downarrow \qquad\qquad \downarrow$$

Knudsen   Viscous

The quantitative measure of each transport regime is summarized in Table 11.4. More advanced descriptions of the transport through zeolite membranes can be found in [18,20]

11.2.7 State of the Art

11.2.7.1 Historical Perspective

The membrane concept dates back many years. Developments on membrane science started in the end of the eighteenth century with the first observations concerning osmosis (Nollet), electroosmosis (Reuss, Porret), and dialysis (Graham). The few primordial membranes were biological ones, such as bladders of pigs, cattle, or fish, and sausage casings made of animal gut. From this point on, other materials were discovered and research on membrane areas has started to grow and new processes with membranes were developed (see Table 11.5).

11.2.7.2 Emerging Membranes for Vapor/Gas Separation

Research on gas separation using membranes is a topic longing for more than a century when Thomas Graham started measuring permeation rates

TABLE 11.5

Historical Perspective of the Developments on Membrane Technology

Year	Developments on Membrane Technology	Process
1907	Preparation of nitrocellulose membranes (Bechhold)	Microfiltration
	Improvement of nitrocellulose membranes (Elford, Zsigmondy, Bachmann, Ferry)	Microfiltration
1930	Commercialization of microporous collodion membranes	Microfiltration
1950–	Cellulose acetate and others as filters for drinking water, during the World War II, sponsored by the U.S. Army. Millipore Corporation kept the research and was the first and one of largest U.S. producer of these filters.	Microfiltration
1960	Defect-free, highly permeable anisotropic membranes for water desalination (Loeb-Sourirajan)	Reverse osmosis
	Further developments and membrane commercialization;	Reverse osmosis Ultrafiltration
	Development of electrodialysis (U.S. Department of Interior, Office of Saline Water)	Microfiltration Electrodyalysis
1945	First developments on membranes for artificial kidneys	Artificial kidneys
1960–	Start up using membranes as substitutes of vital organs and drug delivery. Market explosion.	Artificial organs and drug delivery
1960–1980	Development of interfacial polymerization and multilayer composite membranes.	
1980	Development of membranes for hydrogen separation with industrial application (PRISM®);	Gas separation
	First commercial pervaporation units for alcohol dehydration	Pervaporation

of several gases; his work led to the first notions on solution–diffusion mechanism and thereafter to the Graham's Law of diffusion [21]. Developments on membrane science have been deeply taken, but until 30 years ago membrane technology was not industrially implemented as a separation process [22,23]. However, it has been considered an efficient low-cost alternative technology.

11.2.7.2.1 Polymer Membranes

The kickoff for the implementation of membrane technology in gas separation was done by polymer membranes. Despite the large number of polymers studied and used at laboratory scale, only a few polymers are industrially implemented; the most important and used ones are cellulose acetate, polysulfone, and polyimides [22]. Monsanto was the first company commercializing polymer membranes for large-scale gas separation in 1980, and is today a division of Permea Inc; their PRISM® membranes were developed for hydrogen separation [21,23]. Following, several other companies Grace, Separex, Cyanara [23,24], and others, launched dry cellulose acetate membranes to the market of membrane technology for CO_2 removal from natural gas. Almost by the same time, 1983, Ube Industries, Ltd developed polyimide hollow fibers to separate H_2 from hydrocarbons, and 2 years later, such membrane units were planted. The separation of nitrogen or oxygen from air using membranes was also rising in interest and the Asahi Glass Company started the production of membrane systems to concentrate oxygen up to 40%. A trade mark Spiragas® from Signal Company came out for ultrathin silicone film supported on a porous polysulfone shaped in a spiral-wound membrane module. The Dow Chemical Company also launched polyolefin made membranes by the trademark of GENERON® for producing a stream of 95% nitrogen from air [21,22]. The end of the 1980s was marked by the first commercial pervaporation system for alcohol dehydration due to the German company GFT [23].

11.2.7.2.2 Inorganic Membranes

Porous polymer membranes have high fluxes but low selectivities, while dense polymer membranes are extremely low permeable but with increased selectivities. However, polymer membranes in general suffer from fouling and present chemical and thermal instabilities [25]. Therefore, the use of inorganic membranes for gas separation should be a major step in this area; this last category of membranes has great chemical and thermal stabilities and combines high permeances and selectivities, when compared to polymer membranes. However, the industrial implementation of inorganic membranes is still very little; palladium dense membranes are one example for small-scale production of high-purity hydrogen, by Johnson Matthey [22]. Nevertheless, dense metallic membranes present relatively low permeance and high cost and the scientific community continued searching for less costly, highly permeable inorganic porous membranes. The available commercial porous membranes are ceramic ones, such as alumina, silica, titanium, porous glass, and sintered metals. These membranes present very high permeabilities but very limited selectivities. Despite the very promising results obtained in lab scale, microporous inorganic membranes, such as zeolite and carbon membranes, encounter their main obstacle in scale-up; they have reduced mechanical stability and are difficult to produce defect free and with reproducible performances. Furthermore, their large-scale production is very expensive when comparing to the manufacturing costs of polymer membranes [26]. Nevertheless, progress to overcome these barriers is being made as it is reflected by the research activity in this area (see Figure 11.18).

11.2.7.3 Carbon Membranes

Carbon membranes for gas separation have been successfully developed by Koresh and Soffer in the 1980s [27–30]. For a decade, several patents were filed by these authors [31–33] and by the end of the 1990s, the Israeli company, Carbon Membranes Inc., began commercializing CMSM. These authors have

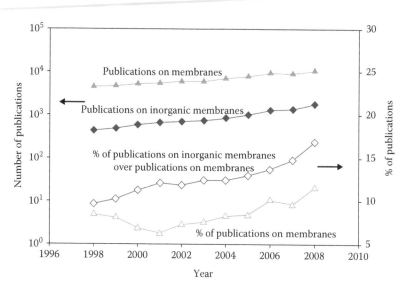

FIGURE 11.18
Publications containing the keyword "membranes" (triangles) and "inorganic membranes" (diamonds) limited to the area of chemical engineering from 1998 until 2008 (data obtained from SCOPUS database).

used cellulose and phenolic resin precursors to prepare CMSM with good permeabilities and selectivities [34]. These membranes present a very narrow pore size distribution, with pore sizes within the range of 0.3–0.5 nm, making possible the separation of molecules with similar diameters, such as nitrogen and oxygen. However, the company closed in 2001. Back to 1993, Rao and Sircar from the company Air Products have prepared nanoporous carbon membranes from the carbonization of poly(vinylidene chloride) supported on macroporous alumina to provide mechanical stability [35,36]. The pore sizes of the selective carbon layer were between the diameter of the more adsorbable species and about four times the diameter of the largest species. Strongly adsorbable species as hydrocarbons could be effectively separated from hydrogen due to selective adsorption and surface diffusion; selectivities of C_4H_{10} and C_3H_8 over H_2 of, respectively, 94.4 and 21.3, were obtained. Air Products Inc. has kept this research on SSF carbon membranes for more than a decade, trying to industrially implement them [35–38]; unfortunately, they ceased this attempt in 2001. In 1994, Jones and Koros studied a different approach by using polyimide hollow fibers as a precursor for CMSM to air separation [39,40]. The result was a membrane with a selectivity of oxygen over nitrogen of 14. Besides that, the carbon membranes also revealed good performances to other important separations such as, CO_2/N_2, CO_2/CH_4, and H_2/CH_4. However, these same authors verified that these carbonaceous materials suffer from a kind of deactivation due to adsorption of water vapor present in humidified streams leading to a decrease of performances

[41,42]. The same problem was also later reported by Menendez et al. [43] and Lagorsse et al. [44] for, respectively, phenolic resin- and cellulose-derived carbon membranes; they verified that the existence of oxygen-based functional groups on the membrane's surface could enhance the adsorption of water vapor blocking the pore network and reduce the performance of the membrane. This has been related to the preparation of the CMSMs at low temperatures; the higher the temperature, the more stable the carbon atoms of the pore network become and then less prone to oxygen chemisorption [45,46]. However, CMSM become more brittle when treated at higher temperatures. To overcome this disadvantage, the use of mixed matrix membranes, such as zeolite/carbon or γ-alumina/carbon membranes has been researched as a way to confer improved mechanical and selective properties to these membranes [47]. In 1998, Fuertes et al. [48] prepared asymmetric carbon membranes from poliamic acid. These authors also coated a macroporous support with a thin layer of a commercial polyimide (polyetherimide) and carbonized it [49]; the derived carbon membranes exhibited high ideal selectivity for permanent gas pairs (O_2/N_2, N_2/He, and CO_2/CH_4). After these reported studies, several works were performed concerning the preparation and characterization of carbon membranes for gas separation, as reviewed in literature [50–52]. Several other researchers also began carbonizing precursors loaded with metals such as silver [53], copper [54], platinum [55], etc., and alkali metals [56].

11.2.7.4 Zeolite Membranes

The development of zeolite membranes for gas separation also emerged in the beginning of the 1980s. Opposite to carbon membranes, zeolite membranes have a very crystalline and, hence, regular micropore structure. The framework of these microporous membranes may consist of different Al/Si ratios, conferring zeolite membranes their different properties of interaction with adsorbable species and different pore sizes; for this reason, zeolite membranes are very promising for several gas separations. This topic has been extensively studied by Caro and Noack, and a reading through their reviews is recommended [57–59].

The concept of mixed-matrix membranes was also developed by introducing zeolite crystals into polymers, metals, or silicas [60]. This was the first attempt to prepare zeolite-based membranes. Later, zeolite membranes were prepared by the deposition of a precursor gel on a porous support and subsequently autoclaved from 80°C up to 200°C for several hours. This is the direct *in situ* hydrothermal crystallization where the zeolite layer is formed on the top or inside the pores of the support. The main drawback of these zeolite membranes is their need to be supported. Other disadvantage is the waste of silicon and aluminum sources, and the impossibility to guarantee a homogenous synthesis of the membrane. Therefore, alternatives to this process have been studied in the subsequent years; self-supported zeolite films were prepared on Hg or Teflon by controlling pH, electrostatic forces, or even by laser ablation or sputtering [59,61]. During the last 10 years, many types of zeolite membranes were synthesized. There are about 150 different types of zeolites but about 15 have been claimed for zeolite membranes with different configurations from 1 to 3D pore networks. The interconnectivity between cages may comprise widths from 4 up to 13 Å, conferring to these membranes the ability to behave as molecular sieves, when smaller pores are present, to present surface diffusion, or for even larger pores, separation of gases based on their molecular weight (Knudsen flow). In literature, the most recently focused zeolite membranes are MFI and LTA [23,58]. Besides the extensive R&D around these zeolite membranes, MFI and LTA are also the most successful concerning scale-up for practical applications. LTA is already being commercialized by the Japanese companies, Mitsui Engineering & Shipbuilding Co. and the Nano-Research Institute Inc. (XNRI), a 100% subsidiary of Mitsui & Co., and most recently, by the European alliance between Smart (United Kingdom) and Inocermic (Germany) [61]. Due to its highly hydrophilic character, LTA is a very selective type of zeolite membrane and greatly recommended to separate water from organic solvents. The first industrial plants with LTA membranes have been implemented in the United Kingdom, India, Brazil, and Latvia for bio-ethanol dehydration. XNRI is also working on a different type of zeolite membrane that is chemically more stable than LTA: this is a high-flux FAU membrane of types X and Y on tubular alpha alumina support. Further developments concern shape-selective MFI that should be commercialized soon by NGK for xylenes or n/i-hydrocarbons separations; applications with MFI are already at pilot scale. Also, MFI membranes can be used to permeate selectively, ethanol from water/ethanol mixtures [14]. NGK has been working recently on the development of DDR-type membranes with narrow pore sizes (0.36–0.44 nm) able to separate CO_2 from CH_4 [58,62,63]. Although being strongly hydrophobic in character, this membrane exhibited an excellent performance in the removal of water from mixtures with ethanol, based on molecular sieving [64]. Further developments concerning small-pore zeolite membranes for H_2 separation and for chiral separations, paraffin/olefin and aromatic/aliphatic and CO_2-containing mixtures have continued and are still future challenges [57].

11.3 An Example of Industrial Application of Zeolite Membranes

In this section, we will briefly present some applications of zeolite membranes in the industry and developments at academic level of some promising cases. For more detailed information on applications, you should consult Chapter 11. We would like to call the reader's attention mainly to the application of MFI-type zeolite membranes on the separation of the C_4 olefin and paraffin isomers.

11.3.1 Alcohol Dehydration

The production of ethanol is expected to increase worldwide. The most common technology to produce ethanol is the hydration of ethylene in the gas phase at high temperatures and pressures, but increasingly bio-ethanol is produced. The production of ethanol from starch or sugar-based feedstocks is among man's earliest ventures into value-added processing. Bio-ethanol can also be produced from cellulose. It is more difficult to break down cellulose to convert it into usable sugars for ethanol production. Yet, making ethanol from cellulose dramatically expands the types and amount of available material for ethanol production. This includes many materials now regarded as wastes requiring disposal, as well as corn stover, rice straw and wood chips, or fast-growing trees and grasses. Consequently, this avoids ethical problems regarding rise of food prices due to the high demand of crops for the fuel industry. As a blend for gasoline, the water content on ethanol must be reduced to 2000 ppm, while, for the ethyl tertiary butyl ether production from *i*-butene and ethanol, the water content must be <500 ppm. The ethanol/water mixture resulting from a fermentation reactor reaches an azeotropic mixture with 93% ethanol when distilled. The conventional distillation processes for ethanol dewatering (azeotropic, extractive, or two-pressure distillation) are typically more energy intensive and require a complex process layout.

LTA with a pore size of 0.42 nm in the Na^+ form is the first zeolite membrane that has reached a commercial status in the dewatering of (bio) ethanol and *i*-propanol (IPA) by vapor permeation or pervaporation [65,66]. However, the successful application of LTA membranes in dehydration is rather based on the hydrophilic character of LTA resulting in a preferential adsorption of water from mixtures than on real size-exclusion molecular sieving; when tested for hydrogen separation from gas mixtures, the LTA membranes show only Knudsen separation [67]. The hydrophilic LTA zeolite layer is extremely selective in the separation of water from organic solutions, therefore, can be used for the production of water-free ethanol [58]. For the dewatering of the crude ethanol stream using membrane technology, Mitsui-BNRI (Bussan Nanotech Research Institute Inc.) played a pioneering role in the cost reduction of the membrane separation by integrating distillation and membrane separation in the so-called membrane separation and distillation (MDI process). By the MDI process, dehydrated ethanol of 0.4 wt% water is produced from fermented liquid containing 8 wt% ethanol starting from cellulose. The aim is the production of 1 L dehydrated ethanol with less than 1000 kcal (4200 kJ). LTA membranes were developed and produced by BNRI, which is a 100% subsidiary of Mitsui & Co. Ltd. Japan. These hydrophilic LTA membranes have been applied in industrial plants for dehydration [65]. The water flux measured in pervaporation operation for 90 wt% ethanol solution at

75°C is about 7 kg m^{-2} h^{-1} [58]. Ethanol scarcely leaks through the membrane, resulting in a separation factor α (water/ethanol) ≈ 10,000. Two different tubular ZeoSepA membranes are produced: large-size elements with 16 mm outer diameter and 1 m length for the dewatering of bio-ethanol and small-size elements of 12 mm outer diameter and 0.8 m length for the recovering of IPA. The supports are, in both cases, porous α-alumina tubes.

Mitsui-BNRI installed a large capacity at Daurala Sugar Works (Uttar Pradesh, India). The capacity of 30,000 L/day can be achieved with a LTA membrane area of 30 m^2 [58]. Each of the membrane is inserted into a sheath tube (i.d. 19 mm). The feed is evaporated by heating with steam. The feed is 93 vol% bio-ethanol, the product purity is 99.8 vol% ethanol for blending with automobile gasoline, the ethanol content in the permeate is <0.1 vol%. The operating pressure and temperature of the membrane are 600 kPa and 130°C, respectively. Extremely high ethanol fluxes are reported in vapor permeation: 11.9, 14.9, 17.6, and 22.4 kg m^{-2} h^{-1} at 100°C, 110°C, 120°C, and 130°C, respectively [68]. The plant is in permanent operation since January 2004 [58].

In Europe, Inocermic GmbH, which is a 100% subsidiary of the HITK, Hermsdorf, Germany, produces NaA membranes for dewatering processes, especially of bio-ethanol, by pervaporation and vapor permeation. The membrane layer is inside an α-Al$_2$O$_3$ four-channel support and is thus protected against mechanical damage [58]. Organic solutions can be dried to water levels down to 0.1% by pervaporation. For feed with 90 wt% ethanol, at 100°C–120°C fluxes of 7–12 kg H$_2$O m^{-2} h^{-1} with separation factors H$_2$O/ethanol > 1.000 are found [69]. The dewatering behavior of these semi-industrially produced NaA membranes was tested by pervaporation with bio-ethanol feeds from fermentation processes. The impurities of the bio-ethanol from grain fermentation or wine production lowered the specific permeate flux by 10%–15% compared with synthetic ethanol/water mixtures [70]. All bio-ethanol samples were dewatered until >99.5 wt% ethanol [69]. These LTA membranes are currently applied in a membrane unit in a demonstration plant in Latvia.

11.4 Future Challenges and Expected Breakthroughs, Identification of Research Needs

11.4.1 Breakthroughs on Recent Research

One of the major breakthroughs on membrane technology/research is the separation of CO$_2$ from various mixtures (e.g., CH$_4$, H$_2$) with DDR-type membranes. Recently, Tomita et al. [71] have successfully synthesized a DD3R membrane and demonstrated the capability of the DD3R membrane to separate an equimolar carbon dioxide/methane mixture with high separation factors (100–200). In 2007, van den Bergh et al. [63] reported excellent separation performance with DDR membranes for carbon dioxide/methane mixtures (selectivity 100–3000), a good selectivity for nitrogen/methane (20–45), carbon dioxide and nitrous oxide/air (20–400), and air/krypton (5–10).

Other recent breakthrough has been the use of DDR membranes for alcohol dehydration. Kuhn [72] presents an extensive work on this process. An all-silica deca dodecasil 3R (DD3R) membrane turns out to be well suited to separate water from organic solvents under pervaporation conditions, despite its hydrophobic character. Permeation of water, ethanol, and methanol through an all-silica DDR membrane was measured at different temperatures on his work. The DDR membrane shows high water fluxes and an excellent performance in dewatering ethanol and is also able to selectively remove water from methanol. This study revealed a major breakthrough, which is that small-pore all-silica zeolites are chemically stable membrane materials for selective water separation.

The continuous removal of ethanol from the fermentation broth (vapor phase) with MFI-type zeolite membranes was an interesting advance in separation technology, since the fermentation process stops at ethanol concentrations near 15%. Hydrophobic membranes such as the MFI type were developed to solve this problem. Typical result is a flux of about $1\,kg\,m^{-2}\,h^{-1}$ of 85 wt% ethanol from a feed with 8 wt% ethanol, which corresponds to a separation factor of 57 [73]. The relative low ethanol fluxes are due to the test conditions with real fermentation broths. By optimizing the support structure, reducing the membrane thickness, and increasing the Si/Al of the MFI membrane, the ethanol fluxes should be increased. By changing the top layer of the support from a 5 nm α-Al_2O_3 NF layer to a 250 nm α-Al_2O_3 MF layer, the ethanol-enriched flux (between 70 and 80 wt% ethanol) could be increased from 0.8 to $1.4\,kg\,m^{-2}\,h^{-1}$ [74]. Multichannel membranes have recently been presented with good results [14], opening up applications of this approach and which can be followed by the further dehydration by LTA membranes. This could result in an even more energy-efficient process.

An elegant example of molecular sieving has been recently given by Khajavi et al. for sodalite membranes. The windows of the sodalite cages consist of six oxygen anions forming openings of ~2.7 Å, and only water can permeate, as has been shown for water–alcohol mixtures [75–77] and salt solutions.

11.4.2 Research Needs in Membrane Technology: Future Challenges

Several topics are of critical importance for membrane technology to become a major player in the field of (gas) separation. The most important one is the development of inexpensive supports of high quality required for thin-film zeolite membranes. This point is extremely interconnected with another topic that needs attention, which is to develop a more continuous synthesis route, e.g., Aguado et al. [78], amenable for scale-up. To solve the problem of membrane defects by repairing them without flux losses, this is usually the case. Tackling this same problem resulted in the breakthrough of the polymeric membranes in the past.

From the separation point of view, the most interesting mixtures to undertake are applications for H_2 separation at high temperatures, alkane/alkene separation, aromatic isomers, linear from branched alkanes. Many of the classical separation challenges remain, being an open opportunity for membrane technology claim space in the separation processes field.

At the materials level, an interesting line of research would be to cast or *in-situ* crystallization of MOFs into membranes [79,80], with a large opportunity to vary pore sizes, interactions, etc. MOF materials present a novel approach in preparing porous materials, which exhibits a rational and even more flexible design of the network compared to the already known inorganic materials. Their flexibility even gives an outlook toward catalytic membrane processes, combining selective separation processes with selective conversions [81–84].

Acknowledgment

Dr. A. Ferreira and Prof. Dr. F. Kapteijn acknowledge the support of the European Network of Excellence INSIDEPOReS (NMP3-CT2004-500895).

References

1. W.J. Koros, Y.H. Ma, T. Shimidzu. Terminology for membranes and membrane processes. *Pure Appl. Chem.* 68 (1996), 1479–1489.
2. U.S. membrane market grows 6.6% per year. *Membr. Technol.* 10 (2005), 4.
3. A.J. Burggraaf, L. Cot (Eds.). *Fundamentals of Inorganic Membrane Science and Technology.* Elsevier Science, Amsterdam, the Netherlands, 1996.
4. Y.S. Lin. Microporous and dense inorganic membranes: Current status and prospective. *Sep. Purif. Technol.* 25 (2001), 39–55.
5. Membrane materials market set to grow. *Membr. Technol.* 8 (2004), 2.
6. T. Araki, H. Tsukube (Eds.). *Liquid Membranes: Chemical Applications.* CRC Press, Boca Raton, FL, 1990.
7. P. Vandezande, L.E.M. Gevers, I.F.J. Vankelecom. Solvent resistant nanofiltration: Separating on a molecular level. *Chem. Soc. Rev.* 37 (2008), 365–405.
8. M. Mulder (Ed.). *Basic Principles of Membrane Technology*, 2nd edn. Kluwer Academic Publishers, Dordrecht, 1996.
9. Membrane configurations. *Membr. Sep. Technol. News* (2003).
10. Table of units. *J. Membr. Sci.* 319 (2008), 3.
11. T. Tsuru, Y. Takata, H. Kondo, F. Hirano, T. Yoshioka, M. Asaeda. Characterization of sol-gel derived membranes and zeolite membranes by nanopermporometry. *Sep. Purif. Technol.* 32 (2003), 23–27.

12. M.L. Mottern, K. Shqau, F. Zalar, H. Verweij. Permeation porometry: Effect of probe diffusion in the condensate. *J. Membr. Sci.* 313 (2008), 2–8.

13. J. Caro, D. Albrecht, M. Noack. Why is it so extremely difficult to prepare shape-selective Al-rich zeolite membranes like LTA and FAU for gas separation? *Sep. Purif. Technol.* 66 (2009), 143–147.

14. J. Kuhn, S. Sutanto, J. Gascon, J. Gross, F. Kapteijn. Performance and stability of multi-channel MFI zeolite membranes detemplated by calcination and ozonication in ethanol/water pervaporation. *J. Membr. Sci.* 339 (2009), 261–274.

15. J.M. van de Graaf, F. Kapteijn, J.A. Moulijn. Methodological and operational aspects of permeation measurements on silicalite-1 membranes. *J. Membr. Sci.* 144 (1998), 87–104.

16. J.M. van de Graaf, E. van der Bijl, A. Stol, F. Kapteijn, J.A. Moulijn. Effect of operating conditions and membrane quality on the separation performance of composite silicalite-1 membranes. *Ind. Eng. Chem. Res.* 37 (1998), 4071–4083.

17. F.T. de Bruijn, L. Sun, Z. Olujic, P.J. Jansens, F. Kapteijn. Influence of the support layer on the flux limitation in pervaporation. *J. Membr. Sci.* 223 (2003), 141–156.

18. F. Kapteijn, W. Zhu, J.A. Moulijn, T.Q. Gardner. Zeolite membranes: Modeling and application. In: A. Cybulski, J.A. Moulijn (Eds.), *Structured Catalysts and Reactors.* CRC Taylor & Francis, Inc., Boca Raton, FL, 2006, pp. 700–746.

19. E.N. Fuller, P.D. Schettle, J.C. Giddings. A new method for prediction of binary gas-phase diffusion coefficients. *Ind. Eng. Chem.* 58 (1966), 19–35.

20. J.M. van de Graaf, F. Kapteijn, J.A. Moulijn. Modeling permeation of binary mixtures through zeolite membranes. *AIChE J.* 45 (1999), 497–511.

21. A.K. Fritzsche, J.E. Kurz. The separation of gases by membranes. In: M.C. Porter (Ed.), *Handbook of Industrial Membrane Technology.* Noyes Publications, Westwood, NJ, 1990, pp. 559–593.

22. S.P. Nunes, K.V. Peinemann. Gas separation with membranes. In: S.P. Nunes, K.V. Peinemann (Eds.), *Membrane Technology in the Chemical Industry.* Wiley, Weinheim, Germany, 2006, pp. 53–75.

23. R.W. Baker. *Membrane Technology and Applications,* 2nd edn. John Wiley & Sons Ltd, Chichester, CA, 2004.

24. R.W. Baker, E.L. Cussler, W. Eykamp, W.J. Koros, R.L. Riley, H. Strathmann. *Membrane Separation Systems: Recent Developments and Future Directions.* Noyes Data Corporation, NJ, 1991.

25. A. Basu, J. Akhtar, M.H. Rahman, M.R. Islam. A review of separation of gases using membrane systems. *Pet. Sci. Technol.* 22 (2004), 1343–1368.

26. A.M. Mendes, F.D. Magalhaes, C.A.V. Costa. New trends on membrane science. *Fluid Transp. Nanoporous Mater.* 219 (2006), 439–479.

27. J. Koresh, A. Soffer. Study of molecular-sieve carbons. 2. Estimation of cross-sectional diameters of non-spherical molecules. *J. Chem. Soc. Faraday Trans. 1* 76 (1980), 2472–2485.

28. J.E. Koresh, A. Soffer. Molecular-sieve carbon permselective membrane.1. Presentation of a new device for gas-mixture separation. *Sep. Sci. Technol.* 18 (1983), 723–734.

29. J. Koresh, A. Soffer. Study of molecular-sieve carbons. 1. Pore structure, gradual pore opening and mechanism of molecular-sieving. *J. Chem. Soc. Faraday Trans. 1* 76 (1980), 2457–2471.

30. J. Koresh, A. Soffer. Molecular-sieve carbons. 3. Adsorption-kinetics according to a surface-barrier model. *J. Chem. Soc. Faraday Trans. 1* 77 (1981), 3005–3018.

31. A. Sofer, J. Koresh, S. Saggy. Separation device. US004685940 (1987).

32. A. Soffer, J. Gilron, S. Saguee, R. Hed-Ofek, H. Cohen. Process for the production of hollow carbon fiber membranes. US005925591 (1999).

33. A. Soffer, M. Azariah, A. Amar, H. Cohen, D. Golub, S. Saguee, H. Tobias. Method for improving the selectivity of carbon membranes by carbon chemical vapor deposition. US005695818 (1997).

34. J. Koresh, A. Soffer. A molecular-sieve carbon membrane for continuous process gas separation. *Carbon* 22 (1984), 225–225.

35. M.B. Rao, S. Sircar. Nanoporous carbon membranes for separation of gas-mixtures by selective surface flow. *J. Membr. Sci.* 85 (1993), 253–264.

36. S. Sircar, M.B. Rao. Manporous carbon membranes for separation of gas-mixtures, Abstracts of Papers of the American Chemical Society, 207 (1994), 81-IEC.

37. S. Sircar, M.B. Rao, C.M.A. Thaeron. Selective surface flow membrane for gas separation. *Sep. Sci. Technol.* 34 (1999), 2081–2093.

38. M.B. Rao, S. Sircar. Performance and pore characterization of nanoporous carbon membranes for gas separation. *J. Membr. Sci.* 110 (1996), 109–118.

39. C.W. Jones, W.J. Koros. Carbon molecular-sieve gas separation membranes. 1. Preparation and characterization based on polyimide precursors. *Carbon* 32 (1994), 1419–1425.

40. C.W. Jones, W.J. Koros. Carbon molecular-sieve gas separation membranes. 2. Regeneration following organic-exposure. *Carbon* 32 (1994), 1427–1432.

41. C.W. Jones, W.J. Koros. Characterization of ultramicroporous carbon membranes with humidified feeds. *Ind. Eng. Chem. Res.* 34 (1995), 158–163.

42. C.W. Jones, W.J. Koros. Carbon composite membranes—A solution to adverse humidity effects. *Ind. Eng. Chem. Res.* 34 (1995), 164–167.

43. I. Menendez, A.B. Fuertes. Aging of carbon membranes under different environments. *Carbon* 39 (2001), 733–740.

44. S. Lagorsse, F.D. Magalhaes, A. Mendes. Aging study of carbon molecular sieve membranes. *J. Membr. Sci.* 310 (2008), 494–502.

45. H. Marsh, F. Rodríguez-Reinoso. *Activated Carbon*, 1st edn. Elsevier, Oxford, U.K., 2006.

46. M.C. Campo, S. Lagorsse, F.D. Magalhaes, A. Mendes. Comparative study between a CMS membrane and a CMS adsorbent: Part II—Water vapor adsorption and surface chemistry. *J. Membr. Sci.* 346 (2010), 26–36.

47. Y. Li, T.S. Chung. Exploratory development of dual-layer carbon-zeolite nanocomposite hollow fiber membranes with high performance for oxygen enrichment and natural gas separation. *Microporous Mesoporous Mater.* 113 (2008), 315–324.

48. A.B. Fuertes, T.A. Centeno. Preparation of supported asymmetric carbon molecular sieve membranes. *J. Membr. Sci.* 144 (1998), 105–111.

49. A.B. Fuertes, T.A. Centeno. Carbon molecular sieve membranes from polyetherimide. *Microporous Mesoporous Mater.* 26 (1998), 23–26.

50. A.F. Ismail, L.I.B. David. A review on the latest development of carbon membranes for gas separation. *J. Membr. Sci.* 193 (2001), 1–18.

51. A.F. Ismail, L.I.B. David. Future direction of R&D in carbon membranes for gas separation. *Membr. Technol.* (2003), 4–8.

52. S.M. Saufi, A.F. Ismail. Fabrication of carbon membranes for gas separation—A review. *Carbon* 42 (2004), 241–259.

53. J.N. Barsema, J. Balster, N.F.A. Van der Vegt, G.H. Koops, V. Jordan, M. Wessling. Ag functionalized Carbon Molecular Sieves membranes for separating O-2 and N-2. Membranes—Preparation, properties and applications 752 (2003), 207–212.
54. J.A. Lie, M.B. Hagg. Carbon membranes from cellulose and metal loaded cellulose. *Carbon* 43 (2005), 2600–2607.
55. R.D. Sanderson, E.R. Sadiku. Synthesis and morphology of platinum-coated hollow-fiber carbon membranes. *J. Appl. Polym. Sci.* 87 (2003), 1051–1058.
56. M. Yoshimune, I. Fujiwara, H. Suda, K. Haraya. Gas transport properties of carbon molecular sieve membranes derived from metal containing sulfonated poly(phenylene oxide). *Desalination* 193 (2006), 66–72.
57. J. Caro, M. Noack. Zeolite membranes—Recent developments and progress. *Microporous Mesoporous Mater.* 115 (2008), 215–233.
58. J. Caro, M. Noack, P. Kolsch. Zeolite membranes: From the laboratory scale to technical applications. *Adsorption* 11 (2005), 215–227.
59. J. Caro, M. Noack, P. Kolsch, R. Schafer. Zeolite membranes—State of their development and perspective. *Microporous Mesoporous Mater.* 38 (2000), 3–24.
60. S. Kulprathipanja, R.W. Neuzil, N.N. Li. Separation of fluids by means of mixed-matrix membranes. US4740219 (1988).
61. M. Noack, P. Kolsch, R. Schafer, P. Toussaint, J. Caro. Molecular sieve membranes for industrial application: Problems, progress, solutions. *Chem. Eng. Technol.* 25 (2002), 221–230. (Reprinted from *Chem. Eng. Technol.* 73 (2001), 958–967.)
62. J. Van Den Bergh, W.D. Zhu, F. Kapteijn, J.A. Moulijn, K. Yajima, K. Nakayama, T. Tomita, S. Yoshida. Separation of CO2 annul CH4 lay a DDR membrane. *Res. Chem. Intermed.* 34 (2008), 467–474.
63. J. van den Bergh, W. Zhu, J. Gascon, J.A. Moulijn, F. Kapteijn. Separation and permeation characteristics of a DD3R zeolite membrane. *J. Membr. Sci.* 316 (2008), 35–45.
64. J. Kuhn, K. Yajima, T. Tomita, J. Gross, F. Kapteijn. Dehydration performance of a hydrophobic DD3R zeolite membrane. *J. Membr. Sci.* 321 (2008), 344–349.
65. R. Qi, M.A. Henson. Optimization-based design of spiral-wound membrane systems for CO2/CH4 separations. *Sep. Purif. Technol.* 13 (1998), 209–225.
66. Y. Morigami, M. Kondo, J. Abe, H. Kita, K. Okamoto. The first large-scale pervaporation plant using tubular-type module with zeolite NaA membrane. *Sep. Purif. Technol.* 25 (2001), 251–260.
67. H. Kita. *Fifth International Conference on Inorganic Membranes, International Workshop on Zeolitic Membranes and Films*, Nagoya, Japan, June 1998, p. 43.
68. S. Inoue, T. Mizuno, J. Saito, S. Ikeda, M. Matsukata. *Proceedings of Ninth International Conference on Inorganic Membranes*, Lillehammer, Norway, 2006, p. 416.
69. H. Richter, I. Voigt, J.-T. Kühnert. *Proceedings of Ninth International Conference on Inorganic Membranes*, Lillehammer, Norway, 2006, p. 552.
70. H. Richter, I. Voigt, J.T. Kuhnert. Dewatering of ethanol by pervaporation and vapour permeation with industrial scale NaA-membranes. *Desalination* 199 (2006), 92–93.
71. T. Tomita, K. Nakayama, H. Sakai. Gas separation characteristics of DDR type zeolite membrane. *Microporous Mesoporous Mater.* 68 (2004), 71–75.
72. J. Kuhn. Zeolite membranes: Ozone detemplation, modeling, and performance characterization, Delft. PhD thesis, 2009.

73. M. Weyd, H. Richter, I. Voigt, C. Hamel, A. Seidel-Morgenstern. Transport and separation properties of asymmetrically structured zeolite membranes in pervaporation. *Desalination* 199 (2006), 308–309.

74. I. Voigt, H. Richter. Hermsdorf Institute for Technical Ceramics, HITK, personal communication.

75. S. Khajavi, F. Kapteijn, J.C. Jansen. Synthesis of thin defect-free hydroxy sodalite membranes: New candidate for activated water permeation. *J. Membr. Sci.* 299 (2007), 63–72.

76. S. Khajavi, J.C. Jansen, F. Kapteijn. Application of hydroxy sodalite films as novel water selective membranes. *J. Membr. Sci.* 326 (2009), 153–160.

77. S. Khajavi, J.C. Jansen, F. Kapteijn. Abs. Production of ultra pure water using H-SOD membranes. In: *Euromembrane 2009*, Montpelier, France, 2009.

78. S. Aguado, J. Gascon, J.C. Jansen, F. Kapteijn. Continuous synthesis of NaA zeolite membranes. *Microporous Mesoporous Mater.* 120 (2009), 170–176.

79. J. Gascon, S. Aguado, F. Kapteijn. Manufacture of dense coatings of Cu-3(BTC) (2) (HKUST-1) on alpha-alumina. *Microporous Mesoporous Mater.* 113 (2008), 132–138.

80. H.L. Guo, G.S. Zhu, I.J. Hewitt, S.L. Qiu. "Twin copper source" growth of metalorganic framework membrane: Cu-3(BTC)(2) with high permeability and selectivity for recycling H-2. *J. Am. Chem. Soc.* 131 (2009), 1646–1647.

81. J.M. van de Graaf, M. Zwiep, F. Kapteijn, J.A. Moulijn. Application of a zeolite membrane reactor in the metathesis of propene. *Chem. Eng. Sci.* 54 (1999), 1441–1445.

82. J. Gascon, U. Aktay, M.D. Hernandez-Alonso, G.P.M. van Klink, F. Kapteijn. Amino-based metal-organic frameworks as stable, highly active basic catalysts. *J. Catal.* 261 (2009), 75–87.

83. J. Gascon, M.D. Hernandez-Alonso, A.R. Almeida, G.P.M. van Klink, F. Kapteijn, G. Mul. Isoreticular MOFs as efficient photocatalysts with tunable band gap: An operando FTIR study of the photoinduced oxidation of propylene. *ChemSusChem.* 1 (2008), 981–983.

84. N. Nishiyama, K. Ichioka, D.H. Park, Y. Egashira, K. Ueyama, L. Gora, W.D. Zhu, F. Kapteijn, J.A. Moulijn. Reactant-selective hydrogenation over composite silicalite-1-coated Pt/TiO_2 particles. *Ind. Eng. Chem. Res.* 43 (2004), 1211–1215.

12

Diffusional Transport in Functional Materials: Zeolite, MOF, and Perovskite Gas Separation Membranes, Proton Exchange Membrane Fuel Cells, Dye-Sensitized Solar Cells

J. Caro

CONTENTS

12.1 Introduction

"Diffusion is the process by which matter is transported from one part of a system to another as a result of random molecular motions" [1]. In many cases, diffusion processes in porous materials limit the efficiency of catalytic or adsorption processes [2]. The function of numerous technical apparatus and processes can be diffusion controlled [3]. In four case studies, the diffusion of molecules, ions, and electrons in gas separation membranes, fuel cell membranes, and dye-sensitized solar cells (DSSCs) is discussed. In novel functional materials, often an overlap of transport due to a concentration and/or a potential gradient takes place. The transport parameters measured in material evaluation such as impedance spectroscopy can reflect a physical situation, which is different from that of the working device. A detailed fundamental knowledge of various factors is necessary to fully understand the nature of transport as a basis to optimize the corresponding functional materials.

12.2 Fundamentals

Irreversible thermodynamics suggests that isothermal mass transport of a component i is driven by the gradient of the electrochemical potential $\eta_i = \mu_i + z_i \cdot F \cdot \Phi$ with μ_i, z_i, F, and Φ denoting the chemical potential, the charge number, the Faraday constant, and the electric potential, respectively [4,5]. The resulting equation for the flux density of component i can therefore be described by

$$j_i = -L_i \cdot \nabla(\mu_i + z_i \cdot F \cdot \varphi) \tag{12.1}$$

where L_i are the Onsager transport coefficients (here neglecting the cross-correlations L_{ij}). The transport coefficients L_i are defined by

$$L_i = \frac{D_i \cdot c_i}{RT} \tag{12.2}$$

and for charged particles under the consideration of the Nernst–Einstein relation for the conductivity $\kappa_i = z_i^2 \cdot F^2 \cdot D_i \cdot c_i / RT$ by

$$L_i = \frac{\kappa_i}{F^2} \tag{12.3}$$

Assuming ideal behavior, i.e., $\mu_i = \mu_i^{\circledast} + RT \ln(c_i/c^{\circledast})$, we obtain from Equation 12.1 for the flux density of component i:

$$j_i = - \underbrace{L_i}_{D_i c_i/RT} \cdot \underbrace{\nabla \mu_i}_{RT(\nabla c_i/c_i)} + \underbrace{L_i \cdot z_i \cdot F}_{\kappa_i/F} \underbrace{\left(-\nabla \varphi\right)}_{E} = -D_i \cdot \nabla c_i + \frac{\kappa_i}{F} E \tag{12.4}$$

In the following section, we discuss the case studies for diffusional transport in zeolite and perovskite gas separation membranes, proton-conducting membranes for fuel cells, and DSSCs.

12.3 Selected Applications

12.3.1 Diffusion in Zeolite Gas Separation Membranes

Most zeolite membranes are supported; that is to say, a thin zeolite layer is crystallized on a porous support. This support can be a porous ceramic, sinter metal, or carbon. Sintered metal supports are relatively easy to mount to gas-tight modules, whereas the ceramic supports show similar thermal expansion coefficients (TECs) like the zeolite layer. This gives an asymmetric membrane that has a coarse porous support for the required mechanical strength and a thin 1–20 μm zeolite layer for sufficiently high fluxes. Depending on the membrane type, various intermediate layers are employed to establish the necessary surface properties (surface smoothness, sufficiently small pores, matching of the TECs between support and top layer) for successful coating of the membrane. For theory and praxis of preparation, characterization, testing, and application of zeolite membranes, see the recent review [6].

The most common supports for crystallization of zeolite layers are porous alumina and titania in planar and tubular geometry as plates, tubes, capillaries, multichannel tubes. It is the aim of the permeation tests to determine for the membranes under study the following crucial permeation parameters: single-component permeation experiments, the fluxes N_i, and the permselectivity (ideal selectivity) as a ratio of the single-component fluxes (Tables 12.1 and 12.2). For binary mixtures, the mixture separation factor α and the component permeate fluxes from a mixed feed are determined (Table 12.1). The permeation measurements of single- and mixed-component feeds require different experimental setups (Table 12.2). Whereas for single-component permeation studies, a pressure recording is sufficient, in mixture permeation, the change of the mixture composition of feed and permeate has to be analyzed (Table 12.2).

Recently, we published our results on the separation of the C_4 olefin [8] and alkane [9] isomers on silicalite-1 membranes as a function of the feed pressure up to 20 bar by shape-selective permeation. We have evaluated these silicalite-1 membranes also in the separation of the C_4 alkane isomers. From permeation experiments of a 50%/50% n- and i-butane mixture through a supported silicalite-1 membrane (measuring technique $p_1 > p_2$ without any sweep gas, undiluted feed, cf. Table 12.2), the flux densities j_i, the permselectivities PS, and the mixture separation factor α (see Table 12.1) were determined (Table 12.3). It was found unfortunately that with increasing transmembrane pressure Δp, both PS and α decreased (see Table 12.3). In the following section, possible reasons for this decrease are discussed.

For noncharged particles or in the absence of an electrical field, Equation 12.4 simplifies, and for the flux density j_i of a mixture component i through the membrane we get

$$j_i = -D_i \cdot \nabla c_i \qquad (12.5)$$

TABLE 12.1

IUPAC Definitions of Flux, Permeance, Permeability, Permselectivity, and Separation Factor

Flux, flux density j_i according to Equation 12.5	mol m^{-2} h^{-1} or m^3 (STP) m^{-2} h^{-1}
Permeance J_i = pressure-normalized flux	mol m^{-2} h^{-1} Pa^{-1} or m^3 (STP) m^{-2} h^{-1} bar^{-1}
Permeability[a,b] P_i = thickness-normalized permeance (permeance multiplied by the membrane thickness)	mol m m^{-2} h^{-1} Pa^{-1} or m^3 (STP) mm^{-2} h^{-1} bar^{-1}
Permselectivity[c] (ideal selectivity) PS(i,j)	Calculated as the ratio of the single-component fluxes PS $(i,j) = j_i/j_j$
Mixture separation factor $\alpha(i,j)$	Measured as $\alpha\ (i,j) = (y_i{:}y_j)/(x_i{:}x_j)$ with y and x as mole fractions i and j in the permeate (y) and feed (x)
Separation coefficient $S(i,j)$	Analyzing the retentate (downstream) and feed (upstream) concentrations $S\ (i,j) = z_i/z_j$(downstream)$/x_i/x_j$(upstream)

Source: Koros, W.J. et al., *Pure Appl. Chem.*, 68, 1479, 1996.

[a] The unit *Barrer* of gas permeability is the permeability represented by a flow rate of 10^{-10} cm^3 (STP) per second times 1 cm of membrane thickness, per square centimeter of area and centimeter of Hg difference in pressure, which is 1 *Barrer* = 10^{-10} cm^3 (STP) cm cm^{-2} s^{-1} cm Hg^{-1} or 10^{-10} cm^2 s^{-1} cm Hg^{-1} or in SI units 1 *Barrer* = 7.5×10^{-18} m^2 s^{-1} Pa^{-1}.

[b] Assuming a linear Henry-like absorption isotherm, for polymer membranes the permeability P_i (mol m m^{-2} h^{-1} Pa^{-1}) can be expressed as the product of solubility S (mol Pa^{-1} m^{-3}) and diffusivity D (m^2 h^{-1}). For pore membranes, however, this simple relation is not valid since due to the limited pore volume the amount adsorbed does not increase linearly with the pressure (Henry-like behavior only for permanent gases at relative high temperatures) and the adsorption isotherm is usually curved (Langmuir-like).

[c] Instead of the ratio of fluxes, the permselectivity PS can be calculated as well as the permeance or permeability ratio of the components i and j, respectively.

where D_i and c_i are the diffusion coefficient of component i in the mixture and its mixture adsorption equilibrium, respectively. From the analysis of the permeate composition, it follows that with increasing Δp, the i-butane flux density increases more steeply than the flux of n-butane (Table 12.3). As a consequence, the n- and i-butane ratio in the permeate becomes smaller with increasing Δp and as a consequence, PS and α decrease.

This experimental finding of decreasing C$_4$ separability with increasing pressure is difficult to understand on the basis of existing knowledge on mixture adsorption and mixture diffusion. For low/medium loading (low partial pressures, high temperatures), it can be expected according to Equation 12.5 that the separation effect is based either (a) on the different diffusivities D_i of the components to be separated or (b) on their different partial loadings c_i in the binary adsorbed mixture.

From near infrared studies, it is learned that for a 50%/50% n- and i-butane mixture at 100°C and 1 bar, the partial loadings in silicalite-1 are almost the same as the composition of the gas phase, namely, 0.50 and 0.54 mmol g^{-1}

TABLE 12.2

Measuring Principles for Membrane Permeation

Flux, permeance, permeability, permselectivity	Mixture separation factor α without pressure difference after Wicke and Kallenbach ($p_1 = p_2$)	Mixture separation factor α with pressure difference ($p_1 > p_2$)
Pressure increase in an evacuated volume is determined	Permeate is transported by sweep gas into a GC or MS	Feed pressure $p_1 > 1$ bar Permeate streams at $p_2 \approx 1$ bar without sweep gas into GC, MS
Flux, permeance, and permeability are calculated		
Permselectivity as the ratio of fluxes/permeances is calculated		

TABLE 12.3

Permeation Data for a Supported Silicalite-1 Membrane (20 μm Thick Layer) on an Asymmetric TiO$_2$ Support for an Undiluted 50%/50% n- and i-Butane Mixture at 130°C

Transmembrane Pressure Δp, Bar	Integral C$_4$ Flux Density, mol m^{-2} min^{-1}	Flux Density of n-Butane, mol m^{-2} min^{-1}	Flux Density of i-Butane, mol m^{-2} min^{-1}	Permselectivity PS	Mixture Separation Factor α
3	2.27	2.2	0.07	29.7	20.1
6	3.96	3.7	0.26	14.2	11.2
11	5.42	4.9	0.52	9.4	8.0
16	6.87	5.9	0.97	6.1	5.7
21	8.34	6.7	1.64	4.1	3.9

Source: Koros, W.J. et al., *Pure Appl. Chem.*, 68, 1479, 1996.

No sweep gas or vacuum was used to reduce the n-butane partial pressure on the permeate side. That is to say that the permeated n-butane was in adsorption equilibrium with the silicalite-1 membrane. Variable feed pressure but constant permeate pressure of 1 bar was applied. The permselectivity was calculated as the ratio of the n- and i-butane flux densities through the membrane from the 50%/50% C$_4$ alkane feed mixture. The mixture separation factor α was determined by gas chromatography as $\alpha(i,j) = (y_i{:}y_j)/(x_i{:}x_j)$ with y and x as mole fractions of the components i and j in the permeate (y) and feed (x).

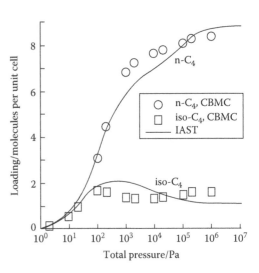

FIGURE 12.1
Loading of a 50%/50% mixture of n-butane and isobutane in silicalite. Comparison of the IAST prediction for total loading with CBMC (configurational-bias Monte Carlo) mixture simulation. (After Krishna, R. and Paschek, D., *Phys. Chem. Chem. Phys.*, 3, 453, 2001, by permission of The Royal Society of Chemistry.)

for i- and n-butane. However, at 2 bar, partial loadings of 0.52 mmol g^{-1} for i-butane and 0.71 mmol g^{-1} for n-butane were found [10]. One can see that there is some enrichment of n-butane in silicalite-1—compared with the gas phase composition—with increasing pressure. This experimental finding is supported by recent molecular modeling suggesting that under mixture conditions the linear hydrocarbon is increasingly adsorbed with increasing pressure, whereas the further adsorption of the branched component is stagnant [11]. Figure 12.1 shows the increase of the partial loadings of n- and i-butane with increasing total pressure of a 50%/50% gas mixture. Furthermore, it was found that the bulky i-butane is located in the channel intersections of the MFI structure where it is of maximum blocking efficiency on the mobile n-butane.

Pioneering self-diffusion studies of n- and i-butane mixtures in silicalite-1 by magic angle spinning (MAS)-pulsed-field gradient nuclear magnetic resonance (NMR) [13] are in complete accordance with the modeling work [11] and can explain the separability of the C_4 isomers showing that in the binary mixture, the diffusivity of n-butane is much higher than that of i-butane. It was found for the single gas diffusion of the C_4 isomers that the self-diffusion coefficient of n-butane is about three orders of magnitude larger than that of i-butane: At 90°C for a loading of four C_4 molecules per u.c. (1 u.c. = 96 SiO_2, which is 0.68 mmol g^{-1}) diffusion coefficients $D_{\text{n-butane}} = 6.4 \times 10^{-9}$ m^2 s^{-1} and $D_{\text{i-butane}} = 2.3 \times 10^{-12}$ m^2 s^{-1} were measured [13]. However, in the binary mixture, n-butane diffuses extremely slower and i-butane diffuses only slightly faster in comparison with the pure components. For the binary mixture (two n-butane in the presence of two i-butane molecules per u.c.), the self-diffusion coefficient of n-butane was found to be 4×10^{-11} m^2 s^{-1}, which is about two orders of magnitude lower than that of n-butane as single component [12]. Figure 12.2 shows the dramatic decrease of the n-butane diffusion coefficient

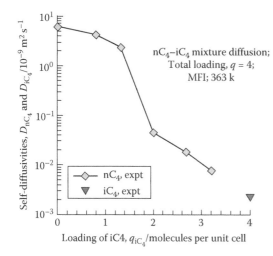

FIGURE 12.2

Decreasing self-diffusion coefficient of n-butane in the presence of increasing amounts of adsorbed i-butane. (From *Microporous Mesoporous Mater.*, 105, Fernandez, M., Kärger, J., Freude, D., Pampel, A., van Baten, J.M., and Krishna, R., 124, Copyright 2007, with permission from Elsevier.)

with increasing loading of i-butane. Snapshots from molecular simulations indicate that the i-butane molecules are preferentially located at the intersections between the straight and zig-zag channels of the MFI structure. These channel intersections are assumed to serve as traffic junctions. At an i-butane loading of two molecules per u.c., the molecular traffic along the straight channels is brought to a virtual standstill because of the obstructive influence of the slow-diffusing i-butane located at these junctions [13]. However, at high loading, n- and i-butane cannot diffuse independently. With increasing pore filling at elevated pressure, the mobile n-butane cannot move faster than the immobile i-butane. Thus, the similar low diffusivities rather than possible differences in the adsorbed mixture would lead to similar and low permeabilities of n- and i-butane, leading to a separability near 1.

This example demonstrates that reliable mixture adsorption and diffusion data are necessary for the molecular understanding of membrane separation. MAS-pulsed field gradient NMR technique can provide the mixture diffusion data [13]. Corresponding work to correlate permeation results with mixture adsorption and diffusion data has been done by Kapteijn et al. [14].

12.3.2 Diffusion in Perovskite Oxygen-Transporting Membranes

Oxygen transport through a mixed oxygen ion- and electron-conducting membrane can be limited by (a) the two surface exchange processes on the feed and permeate side of the membrane and (b) the bulk diffusion of the oxygen ions via vacancies in the oxygen framework [15]. Often a combination of the two processes is found to be rate determining [16]. Assuming the bulk diffusion of oxygen ions as a rate-limiting process, the oxygen permeation flux j_{O_2} through an oxygen-transporting membranes (OTMs) is described by the Wagner theory [17].

For a thick membrane, the rate of the overall oxygen permeation is usually determined by the lattice diffusion of oxygen or the transport of electronic charge carriers through the bulk oxide. The gradient of the chemical potential of the oxygen on the two sides of the membrane $\nabla\mu_{O_2}$ is the driving force for the oxygen transport. Assuming oxygen ion diffusion in the bulk as rate limiting (the electric conductivity is usually one or two orders of magnitude higher than the ionic one, i.e., $\sigma_i \ll \sigma_e$), the oxygen potential gradient depends at a given temperature and membrane thickness L only on the oxygen partial pressures on the feed (air) and permeate side of the membrane, p'_{O_2} and p''_{O_2} [18]:

$$\nabla\mu_{O_2} = \frac{\mu''_{O_2} - \mu'_{O_2}}{L} = \frac{RT}{L}\ln\frac{p''_{O_2}}{p'_{O_2}} \qquad (12.6)$$

With the ionic and electronic conductivities σ_i and σ_{el}, we obtain the oxygen flux j_{O_2} where F is the Faraday constant

$$j_{O_2} = -\frac{1}{4^2 F^2}\frac{\sigma_i \cdot \sigma_e}{\sigma_i + \sigma_e} \cdot \nabla\mu_{O_2} \qquad (12.7)$$

Integration of Equation 12.7 across the membrane thickness L using the relationship

$$\nabla\mu_{O_2} = \frac{\partial RT \ln p_{O_2}}{\partial x} \qquad (12.8)$$

yields the Wagner equation in the usual form

$$j_{O_2} = -\frac{RT}{4^2 F^2 L}\int_{\ln p'_{O_2}}^{\ln p''_{O_2}}\frac{\sigma_i \cdot \sigma_e}{\sigma_i + \sigma_e}\,d\ln p_{O_2} \qquad (12.9)$$

The investigation of the oxygen flux as a function of the membrane thickness is a simple but very efficient tool to identify the rate-limiting step [19,20]. For bulk diffusion-limited oxygen flux, a plot of j_{O_2} as a function of the reciprocal of the membrane thickness would give a straight line going through the origin. If the bulk diffusion of the oxygen ions is rate limiting, reducing the membrane thickness is a tool to increase the oxygen flux.

The rate-limiting step of oxygen flux through an OTM can be identified by studying j_{O_2} as a function of the oxygen partial pressure p''_{O_2} on the sweep gas side. This method is advantageous since only one membrane is studied and the experimental error is reduced in comparison with preparing a set of "identical" membranes, which differ only in their thickness. Different approaches are proposed. Following Shao et al. [21], an overall oxygen transport resistance $R_{overall}$ is defined as

$$R_{overall} = \frac{RT}{4^2F^2} \frac{1}{Sj_{O_2}} \left(\ln \frac{p'_{O_2}}{p''_{O_2}} \right) \tag{12.10}$$

If $R_{overall}$ decreases with decreasing oxygen partial pressure on the permeate side p''_{O_2} at fixed oxygen partial pressure p'_{O_2} on the feed side, oxygen bulk diffusion is mainly rate limiting. If $R_{overall}$ decreases with increasing oxygen partial pressure on the permeate or feed side, the surface kinetics on the low or high oxygen partial pressure side can be assumed to be rate limiting.

In another approach by Huang et al. [22], the oxygen flux j_{O_2} through an OTM is described according to the Wagner theory by

$$j_{O_2} = \frac{\sigma_i^0 RT}{4^2F^2nL} \left(p_1^n - p_2^n \right) \tag{12.11}$$

where
σ_i^0 is the oxygen ionic conductivity at $p_{O_2} = 10^5\,Pa$
L is the membrane thickness
p'_{O_2} and p''_{O_2} are oxygen partial pressures on the feed and permeate side, respectively
n is the fit parameter, which is derived from the steady-state oxygen permeation

From the value of n, the rate-limiting step of the oxygen permeation can be identified [16,22]. For negative values of n, the oxygen permeation is dominated by the bulk diffusion, while for $n \geq 0.5$, the exchange processes at the membrane surfaces can be assumed as rate limiting. In the case of the surface exchange as rate-limiting step, a catalytic coating of the membrane can accelerate the oxygen flux. In the surface exchange current model [23], j_{O_2} gives a linear relationship with $(p'_{O_2})^{0.5} - (p''_{O_2})^{0.5}$ if oxygen transport is controlled by the surface exchange reaction. In the case of bulk diffusion control, j_{O_2} is a linear function of $\ln(p'_{O_2} - p''_{O_2})$.

In the following section, two examples for novel applications of perovskite membranes, which are based on a high oxygen ion diffusivity, are discussed: (a) hydrogen production by thermal water self-dissociation and (b) catalytic N_2O abatement. In these two applications, a $Ba(Co_xFe_yZr_{1-x-y})O_{3-\delta}$ perovskite hollow fiber membrane obtained by a spinning process [ ] has been used.

In the case of the thermal water splitting according to the simple reaction $H_2O \leftrightarrows H_2 + \frac{1}{2}O_2$, one can produce hydrogen on the retentate side at an industrially interesting rate of more than $1\,m^3\,H_2$ (STP) $m^{-2}\,h^{-1}$ if the oxygen can be removed quickly from the reactor under equilibrium-controlled reaction condition [24]. However, at first sight, this concept seems not realistic since the equilibrium constant of the thermal water dissociation at 900°C is only

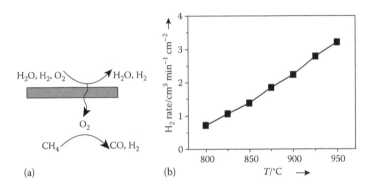

FIGURE 12.3
(a) Schema of the hydrogen production by thermal water splitting with simultaneous oxygen removal and (b) H_2 production rate on the retentate side of the membrane after water condensation. (Jiang, H., Wang, H.H., Werth, S., Schiestel, T., and Caro, J., Simultaneous production of hydrogen and synthesis gas by combining water splitting with partial oxidation of methane in a hollow-fiber membrane reactor, *Angew. Chem. Int. Ed.*, 2008, 47, 9341. Copyright Wiley-VCH Verlag GmbH & Co. KGaA. With permission.)

$K_p = 2 \times 10^{-8}$, which means that even at high temperatures the equilibrium is almost completely on the educts side. Nevertheless, the concept will work if (a) the water dissociation equilibrium establishes quickly and (b) the oxygen transport through the membrane is fast due to a high oxygen ion diffusivity in the perovskite. As Figure 12.3 shows, these prerequisites seem to be fulfilled. After condensation of nondissociated water on the retentate side, a hydrogen production rate $>1\,m^3\,H_2$ (STP) $m^{-2}\,h^{-1}$ is found. To increase the driving force for the oxygen transport through the membrane, on the permeate side of the hollow fiber the oxygen partial pressure has to be as low as possible. This can be achieved by an oxygen-consuming reaction like the partial oxidation of methane to synthesis gas according to $CH_4 + \frac{1}{2}O_2 \rightarrow CO + 2H_2$. From synthesis gas, it is relatively easy to obtain methanol or Fischer Tropsch products like diesel as transportable fuels.

Another innovative reaction in perovskite membrane reactors with high oxygen diffusivity is the decomposition of the nitrous gas N_2O into the elements according to $N_2O \rightarrow N_2 + \frac{1}{2}O_2$ [25]. The N_2O abatement is important since the green house effect of one molecule N_2O is equivalent to 300 molecules of CO_2. One could expect that this reaction could proceed easily since the change of standard free reaction enthalpy for this reaction $\Delta_R G^{\#} = -104.2\,kJ\,mol^{-1}$ is very negative. According to $\Delta_R G^{\#} = -RT \ln K$, under standard conditions (1 bar, 25°C), an equilibrium constant of $K = 1.8 \times 10^{18}$ is obtained, which means that the equilibrium should be completely on the nitrogen/oxygen side. However, conducting N_2O splitting on oxide surfaces such as on perovskite catalysts, one has to learn that this decomposition is product inhibited. The product molecule oxygen remains adsorbed and competes with N_2O for the catalytic sites on the perovskite surface. To get rid of this blocking oxygen species, we have developed the following concept in which perovskite fulfills a dual

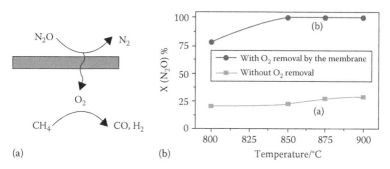

FIGURE 12.4
(a) Schema of the oxygen removal through a perovskite hollow fiber membrane during the N_2O decomposition and (b) influence of temperature on the conversion enhancement of the N_2O decomposition if the blocking product oxygen is removed. (Jiang, H., Wang, H.H., Liang, F., Werth, S., Schiestel, T., and Caro, J., Direct decomposition of nitrous oxide to nitrogen by in situ oxygen removal with a perovskite membrane, *Angew. Chem. Int. Ed.*, 2009, 48, 2983. Copyright Wiley-VCH Verlag GmbH & Co. KGaA. With permission.)

function: (a) it acts as a catalyst for the N_2O decomposition and (b) it removes in situ the blocking oxygen surface species. Figure 12.4 shows that the N_2O decomposition can be accelerated if the inhibiting oxygen is removed via a perovskite membrane with a high oxygen diffusivity [26]. For a high driving force for oxygen transport, the oxygen partial pressure was decreased by the oxygen-consuming synthesis gas formation.

12.3.3 Diffusion in Proton Exchange Membrane Fuel Cells

For the case of proton transport in proton exchange membrane (PRM) fuel cells (PEMFCs), there is a need to increase the proton conductivity through the membrane. A PEMFC transforms the chemical-free enthalpy liberated in the electrochemical reaction of H_2 and O_2 to H_2O into electrical energy. H_2 is delivered to the anode side of the membrane–electrode assembly (MEA). At the anode, H_2 is catalytically split into protons and electrons. The protons permeate by diffusion through the PEM to the cathode side. The electrons travel along an external load circuit to the cathode side of the MEA, thus creating the current output of the FC. Oxygen is delivered to the cathode where the formation of water with the protons permeated through the PEM and the electrons arriving via the external circuit take place.

There can be several diffusion-limiting steps in a PEMFC such as the gas supply to the electrodes or the gas transport through the porous electrocatalyst. However, in most cases, the proton diffusion through the PEM has been identified as rate limiting for a PEMFC. Therefore, a worldwide intense R&D for novel proton-conducting materials with a high proton diffusivity and chemical stability for $T > 100°C$ can be observed. A general overview about the challenges for new membranes with operation temperatures >100°C is found in Ref. [27]. Modified polymers like Nafion, polybenzimidazole, sulfonated polyetheretherketones, or polysiloxanes with imidazole or sulfonic acid groups are promising candidates. One strategy for improved PEM is to add nanoparticles with an intrinsic proton conductivity to a polymer. Examples are nanoparticles of TiO_2, SiO_2, zeolites, or mesoporous materials with sulfonic acid and/or imidazole groups on their outer and/or inner surface.

However, the direct detection of the proton conductivity is not possible since the PEM material blocks any electron transport as an electronic insulator. For the detection of the proton flux, the charge carrier proton must be transformed into measurable electric variables. Therefore, at the positive electrode, hydrogen atoms become oxidized according to $H \rightarrow H^+ + e^-$, and at the negative electrode reduction occurs according to $H^+ + e^- \rightarrow H$ [28]. To avoid polarization effects, the resistance of the PEM is determined under alternating current conditions (impedance spectroscopy).

As an example, Figure 12.5 shows the proton conductivity measured on a pressed disc of MCM-41* powder as a promising modifier for various polymers for PEM in comparison with Nafion as the standard polymer. The inner

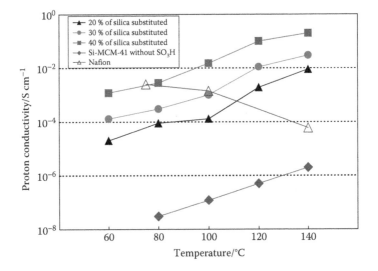

FIGURE 12.5
Proton conductivity of MCM-41 with surface SO_3H as a promising modifier for novel polymer-based PEMFC. (Redrawn after Gomes, D. et al., *J. Memb. Sci.*, 322, 406, 2008.) The impedance measurements were made on pressed discs of the MCM-41 powder at 100% RH for MCM-41 with different concentrations of SO_3H groups on its inner and outer surface: 20%, 30%, and 40% denote the percentage of the silica source $NaSiO_3$, which was replaced in the MCM-41 synthesis by (3-mercaptopropyl) trimethoxysilane acting as structure-directing agent (SDA). By microwave treatment of the as-synthesized MCM-41 with an HNO_3/H_2O_2 mixture, the SH groups of this SDA became oxidized into SO_3H groups. (With permission from Margolese, D., Melero, J.A., Christiansen, S.C., Chmelka, B.F., and Stucky, G.D., *Chem. Mater.*, 12, 2448, 2000. Copyright 2000 American Chemical Society.)

---

* Mobil Composition of Matter No. 41.

and external surface of the MCM-41 was modified by surface sulfonic acid groups (SO$_3$H groups anchored via a propyl chain to the surface). Ozin and coworkers [29] proposed for the first time SO$_3$H-modified MCM-41 as proton-conducting solid electrolyte. It can be seen in Figure 12.5 that SO$_3$H-modified MCM-41 shows for temperatures >100°C higher proton conductivities than Nafion. This high proton conductivity at elevated temperatures together with a higher hydrophilicity and an improved mechanical stability recommends SO$_3$H-modified MCM-41 as inorganic filler for novel hybrid membranes in combination with thermally stable polymers [30–32]. Consequently, an extremely high conductivity of 0.1 Scm^{-1} at the relative low temperature of 40°C was found for hybrid membranes of sulfonated MCM-41 in a Nafion matrix [32]. In polyoxadiazole membranes, the addition of SO$_3$H–MCM-41 particles could increase the ionic conductivity at 120°C and at only 15% relative humidity by one order of magnitude [33].

The question arises whether the proton conductivity determined by impedance spectroscopy as shown in Figure 12.5 is a real criterion to characterize the performance of a PEMFC. It is generally accepted that the current density of a PEMFC is limited by proton diffusion, assuming that all electrocatalytic processes and the gas supply are fast compared with the proton diffusion in the membrane. Therefore, the proton flux in a working PEMFC is under short circuit conditions mainly due to the proton concentration gradient as driving force. However, when the proton conductivity of a PEM material is characterized by impedance spectroscopy, different molecular processes contribute to the electric signal detected. First, the random walk diffusion of the protons under the concentration gradient results in a net diffusion flux of protons from and to the electrodes. Furthermore, in impedance spectroscopy, the alternating electrical field gives an additional field-driven proton flux. So Equation 12.4 describes the proton transport in impedance spectroscopy. By multiplying the proton diffusion flux density j_{H^+} by the Faraday constant F, we obtain the flux of electrical charge (charge/area × time), which is the current density i (current/area)

$$i = F \cdot j_{H^+} = -D \cdot F \frac{dc_{H^+}}{dz} + \kappa \cdot E \qquad (12.12)$$

This means that the conductivity determined from impedance measurements reflects a physical situation different from that we have in a working PEMFC. In impedance spectroscopy, we detect a superposition of concentration and field-driven transport and, in the working PEMFC, the current density is controlled by a concentration-driven proton diffusion. A quantitative estimate of the current density according to Equation 12.12—assuming a PEM thickness of 100 μm, a proton concentration of 2.3 mmol H$^+$g^{-1} as it is found in SO$_3$H-modified MCM-41, and an alternating voltage of 100 mV as used in impedance spectroscopy—shows that the field-driven transport under short circuit conditions is several orders of magnitude larger than the transport due to the random walk of protons assuming that the self-diffusion of water in these materials represents a lower limit of the proton self-diffusion (Grotthus mechanism). Nevertheless, impedance spectroscopy provides valuable insight into the proton mobility of a PEM and is an inevitable tool in material development.

12.3.4 Diffusion in Dye-Sensitized Solar Cells

In 1991, O'Regan and Grätzel [36] announced a breakthrough with 7.9% efficiency for a DSSC. The system consists of a nano-crystalline wide-bandgap semiconductor, an adsorbed dye with absorption in the visible region, and an appropriate electrolyte system. Electrons are injected from the photo-excited state of the dye into the conduction band of the wide-bandgap semiconductor like TiO_2 or ZnO. The resulting dye radical cation becomes reoxidized by the electron-transporting electrolyte [37]. Therefore, the electron lifetime τ_n must be longer than the diffusion-limited electron transit time τ_D to the back contact. By intensity-modulated photocurrent spectroscopy (IMPS), from the measurement of the phase shift and the amplitude of the photocurrent according to $\tau_D = 1/(2\pi f_{min,PS})$, the electron diffusion transit time τ_D can be determined. From measuring the photovoltage in intensity-modulated photovoltage spectroscopy (IMVS), from $\tau_n = 1/(2\pi f_{min,VS})$, the electron lifetime τ_n can be obtained with f_{min} as the frequencies where the imaginary component of the photovoltage or photocurrent, respectively, becomes minimum [38]. The relative values of τ_n and τ_D determine the collection efficiency of the injected photoelectrons. Only if τ_n is significantly larger than τ_D, most of the photoelectrons will reach the collector electrode by diffusion. On the contrary, if most of the charge carriers will recombine before they have reached the collector electrode, the efficiency will be low. Assuming that the photoelectrons are generated only at the outer part of the illuminated DSSC film of thickness d, by the Einstein–Smoluchowski relation, an effective electron diffusion coefficient D is defined by

$$D = \frac{d^2}{2 \cdot \tau_D} \tag{12.13}$$

The Grätzel cell operates with sintered or compressed nano-particulate TiO_2 layers as electron harvesting material. This several micrometers thick layer of semiconductor nanoparticles (TiO_2, ZnO, CuO) contains a high concentration of grain boundaries acting as recombination sites, thus reducing the lifetime τ_n. Therefore, the enhancement of the electron transport rate in the photoanode has attracted much attention for improving the efficiency of the DSSC [39]. It has been proposed that vertically grown single-crystalline nano-wire array morphology in the transparent conducting oxide (TCO) electrode would provide a direct conduction path for the photo-generated charge

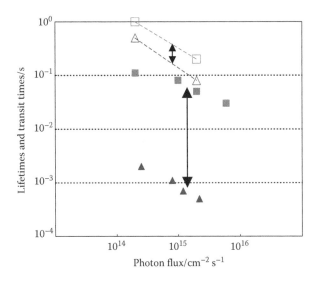

FIGURE 12.6
Electron diffusion transit times τ_D and lifetimes τ_n for an electrodeposited ZnO/eosinY film as shown in Figure 12.7b at different dc light intensities ($\tau_D = \blacktriangle$; $\tau_n = \blacksquare$). Transit times and lifetimes for a colloidal ZnO/eosinY film ($\tau_D = \triangle$; $\tau_n = \square$) are shown for comparison. The two arrows indicate the differences between τ_D and τ_n for both kinds of films. (Redrawn after permission from Oekermann, T., Yoshida, T., Minoura, H., Wijayantha, K.G.U., and Peters, L.M., *J. Phys. Chem. B*, 108, 8364, 2004. Copyright 2004 American Chemical Society.)

carriers, by avoiding the particle-to-particle hopping in nanoparticle TCO-based electrode. Novel concepts for anisotropic porous semiconductor films were looked for like the use of mesoporous TCOs [40], the electrochemical deposition of porous films [41,42], or the formation of ZnO nano-wire arrays [43,44] by using an aqueous chemical bath deposition method [45]. Figure 12.6 reflects the results of IMPS and IMVS measurements at electrodeposited as well as at nano-particular ZnO films.

Figure 12.6 gives a comparison of τ_n and τ_D for nano-particular colloidal ZnO and electrochemically deposited columnar ZnO films calculated from the f_{min} values of the IMVS and IMPS plots, as a function of the light intensity. For the ZnO nano-particular film, it was found that $\tau_n/\tau_D \approx 2$. This means that not all photoelectrons injected into the ZnO nano-particular film are collected at the back contact. On the other hand, a difference between τ_n and τ_D of nearly two orders of magnitude, as it is seen for the electrodeposited columnar ZnO, should lead to an electron collection efficiency of practically 100% since the electron lifetime τ_n is much longer than the diffusion transit time τ_D. Figure 12.7 shows schematically the two types of porous ZnO films, the nano-particular colloidal films, and the electrodeposited columnar ZnO films. Also in another study [46], a high ratio $\tau_n/\tau_D \approx 50$ was found for the electrodeposited columnar ZnO film, recommending the electrodeposition technology in comparison with coating techniques of nanoparticles.

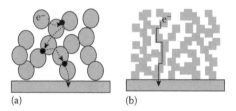

(a)   (b)

FIGURE 12.7

Simple qualitative model for the comparison of the electron transport in (a) colloidal ZnO and (b) electrodeposited ZnO as shown in Figure 12.7b. Especially at the positions where the nanoparticles touch each other, electron traps are found, which enlarge τ_D. (After permission from Oekermann, T., Yoshida, T., Minoura, H., Wijayantha, K.G.U., and Peters, L.M., *J. Phys. Chem. B*, 108, 8364, 2004. Copyright 2004 American Chemical Society.) By contrast, wire- or rod-like structures allow effective anisotropic electron transport.

The ratio τ_n/τ_D recommends not only the electrochemical deposition technique but also the absolutely higher diffusivities D due to a lower τ_D (see Equation 12.13). With the relative high electron diffusion coefficient $D \approx 10^{-5}\,\mathrm{cm^2\,s^{-1}}$ for electrodeposited ZnO films, the electron transport in the electrochemical ZnO film was found to be by the factor 4 higher than in the nano-particular layer with $D \approx 2.5 \times 10^{-6}\,\mathrm{cm^2\,s^{-1}}$ [47]. During the lifetime τ_n of the electron from the electron diffusion coefficient $D = 10^{-5}\,\mathrm{cm^2\,s^{-1}}$, a diffusion length of 5 μm can be estimated according to Equation 12.13, which is significantly longer than the film thickness with 2–3 μm [44]. The same finding is reported for TiO$_2$-based DSSCs [48].

Since the electron diffusion coefficient in an anatase TiO$_2$ nano-particular film is about $D = 10^{-8}$–$10^{-4}\,\mathrm{cm^2\,s^{-1}}$, which is several orders of magnitude lower than that in a single-crystal anatase TiO$_2$ with $D = 0.5\,\mathrm{cm^2\,s^{-1}}$ [49], the development of one-dimensional TiO$_2$-based DSSCs with nano-rods and nano-wires for improving the electron transport rates in the anodes is the objective of the current research [39,50–53]. Examples of novel nano-structured semiconductor depositions with anisotropic transport are given in Figure 12.8. Free-standing and ordered arrays of ZnO nano-wires with fine control over their size and structural quality could be synthesized [54–59]. These ZnO nano-structures are found to have low native defect density, and DSSCs have been fabricated based on these materials [60,61]. Slightly different is the situation when using ZnO as photocatalyst. A higher photocatalytic degradation found on ZnO nano-rods in comparison with ZnO nanoparticles is explained by a longer lifetime of the photo-excited electron-hole pairs. Intrinsic defects located in the ZnO nano-rods are proposed to act as electron capture centers that reduce the recombination rate of electrons and holes [62].

12.3.5 Diffusion in MOF Crystals of Type ZIF-8 to Forecast Gas Permeation

In polymer membrane science and practice, the rule of thumb holds that "permeability = diffusivity × solubility," with permeability in $(\mathrm{mol\,m^{-1}\,s^{-1}\,Pa^{-1}})$

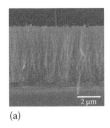 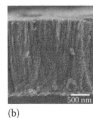 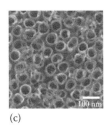 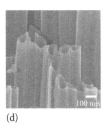

(a)   (b)   (c)   (d)

FIGURE 12.8

Examples for one-dimensional semiconductor films with anisotropic transport properties: (a) ZnO nano-wire array grown on seeded fluorine-doped tin oxide (FTO) using chemical bath deposition method. (From Perathoner, S. et al., *Catal. Today*, 122, 3, 2007.) (b) Compact ZnO/eosinY hybrid films obtained by electrochemical deposition. (From Gottschlich, G. and Oekermann, T., Leibniz University Hannover, unpublished.) (c) Ordered array of helical TiO_2 nano-coils obtained by anodization of Ti. (From Perathoner, S. et al. *Catal. Today*, 122, 3, 2007; Centi, G. et al., *Phys. Chem. Chem. Phys.*, 9, 4930, 2007.) (d) Vertically oriented TiO_2 nanotubes by anodization of metallic Ti. (From Macak, J.M. et al., *Curr. Opin. Solid State Mater. Sci.*, 11, 3, 2007; Macak, J.M. et al., *Phys. Stat. Sol. (RRL)* 1, 181, 2007; Albu, S.P. et al., *Nano Lett.* 5, 1286, 2007.)

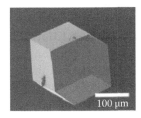

FIGURE 12.9

Large ZIF-8 crystal with rhombic dodecahedral morphology by conventional solvothermal synthesis. (From Bux, H. et al., *J. Am. Chem. Soc.* 131, 16000, 2009. With permission.)

as the product of the pressure-normalized flux density in (mol m^{-2} s^{-1}), called permeance in (mol m^{-2} s^{-1} Pa^{-1}), multiplied by the membrane thickness [7]. The diffusivity is expressed by the diffusion coefficient. Based on this concept, in a recent feature article [69], the separability of a CO_2/CH_4 mixture was forecasted, expressed by a permeation selectivity (= ratio of the CO_2 and CH_4 fluxes through the membrane from a binary feed mixture) by combining a diffusion selectivity (= ratio of the diffusion coefficient of component i in the presence of component j) with an adsorption selectivity (= ratio of the adsorbed amounts of i and j under mixture conditions). This concept will be evaluated in this chapter for the separation of CO_2/CH_4 on a newly developed ZIF-8 membrane. There are two independent ways to determine the diffusion selectivity and adsorption selectivity.

1. The mixture adsorption isotherms and the mixture diffusion coefficients of the binary mixture CO_2/CH_4 were determined on giant ZIF-8 single crystals (Figure 12.9) by infrared (IR) microscopy [70]. Giant ZIF-8 single crystals ($Zn(mim)_2$) (Figure 12.9) with SOD topology were obtained by the reaction of zinc chloride with 2-methyl-imidazole (Hmim) in a diffusion-controlled crystallization process [71]. On selected crystals as shown on Figure 12.9, the single component and mixture adsorption/desorption kinetics of CO_2 and CH_4 were studied time- and space-resolved by using IR microscopy [70].

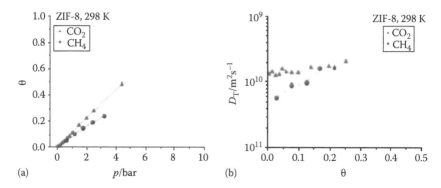

FIGURE 12.10
Relative amounts adsorbed (left) and transport diffusion coefficients (right) of CO_2 and CH_4 on the giant ZIF-8 single crystals shown in Figure 12.9 at room temperature, as determined by IR microscopy. (From Bux, H. et al., *Abstract, 16th International Zeolite Conference*, Sorrent, Italy, July 2010. With permission.)

The basis of these IR measurements is to monitor sorption uptake/desorption by measuring spotwise the absorption of IR radiation as a function of time and space. The concentration profiles of the individual components are determined by characteristic IR bands. By comparing the instationary concentration profile with solutions of second Fick law, diffusion coefficients for the component i in the presence of another component j are obtained. Further, after calibration, adsorption isotherm data of component i in the presence of j are obtained as shown in Figure 12.10. From the sorption/desorption studies on the giant ZIF-8 crystals by IR microscopy, it can be concluded that ZIF-8 shows both (a) an adsorption selectivity for CO_2 in a mixture with CH_4 (especially at high loadings) and (b) diffusion selectivity (especially at low loadings). As a result of this interplay of adsorption and diffusion selectivities, a clear permeation selectivity for CO_2 in a mixture with CH_4 can be expected.

2. Without any experiment, the mixture diffusion coefficients and mixture adsorption data can be calculated by means of molecular dynamic simulations using extensive data sets as described in [69]. Also molecular dynamics forecasts a CO_2 selectivity in a permeation experiment with CH_4 as second mixture component.

From both 1 and 2, an anti-Knudsen behavior is forecasted. That is to say, the ZIF-8 membrane is expected to show for the binary permeation CO_2/CH_4 a CO_2 selectivity. To test this forecast, such ZIF-8-type metal-organic framework (MOF) membrane was developed (Figure 12.11) and the single component and mixture permeation data were measured [71] (see Table 12.4). The ZIF-8 layer of about 30 μm thickness was prepared on an asymmetric titania support by microwave-assisted solvothermal synthesis [71]. The membrane

FIGURE 12.11
Cross section of the ZIF-8 membrane obtained by in situ crystallization using microwave heating. (From Bux, H. et al., *Abstract, 16th International Zeolite Conference*, Sorrent, Italy, July 2010. With permission.)

TABLE 12.4

Results of Single Gas and Binary Mixture Permeation on the ZIF-8-Type MOF Membrane Shown in Figure 12.10 at 298 K

	Permeance (10^{-8} mol m^{-2} s^{-1} Pa^{-1})	
Probe gas	Single component at 1 bar	1:1 binary mixture at 1 bar
CO_2	1.4	1.2
CH_4	0.4	0.4
Separation	3.5 (ideal separation factor)	3.0 (mixture separation factor)

Source: Koros, W.J. et al., *Pure Appl. Chem.*, 68, 1479, 1996.

According to IUPAC definitions, the ideal separation factors were determined as the ratio of the single-component fluxes measured by soap bubble counter and the mixture separation factor was determined by gas chromatography from analysis of feed and retentate.

exhibited a hydrogen separation factor of $\alpha = 12.2$ at 25°C, as determined from a 1:1 H_2/CH_4 mixture of 1 bar with a hydrogen permeance of 0.5 m^3(STP)m^{-2} h^{-1} bar^{-1}. Permeation was done following the Wicke–Kallenbach method with gas chromatographic gas detection. The mixture separation factor α is determined as the molar ratios of permeate and retentate [7].

Table 12.4 shows that the independent (a) measurement of adsorption and diffusion behavior by IR microscopy and (b) calculation of the adsorption and diffusion behavior through molecular dynamics are indeed a forecast of the permeation separation of a real membrane. The permeation experiments of Table 12.4 indicate clearly that the ZIF-8 membrane shows a CO_2 selectivity. The similarity of the single and mixture permeances as well as of the ideal selectivity and mixture separation factor can be understood if one assumes that at 1 bar the loadings are still relatively low and the mixture components can diffuse rather independently (diffusion-controlled separation).

It can be concluded, therefore, that for the permeation of CO_2/CH_4 through a new ZIF-8-type MOF membrane, it is shown that the permeation selectivity can be roughly approximated as the product of adsorption and diffusion selectivities. CO_2/CH_4 is separated in an anti-Knudsen manner, which would forecast a CH_4 selectivity with a mixture separation factor $\alpha = 1.6$. By contrast, the ZIF-8 membrane shows a CO_2 selectivity with $\alpha = 3.1$. This separation behavior can be expected due to both an adsorption and diffusion selectivity of CO_2 over CH_4 as determined by IR microscopy on selected giant ZIF-8 single crystals. Molecular dynamic simulations thus allowing to forecast the permeability of a membrane according to "permeability = solubility × diffusivity."

12.4 Future Challenges

Real breakthroughs for an industrial application of the functional materials discussed will take place if the following problems can be solved.

In the case of *zeolite membranes*, one should avoid the situation that the feed components to be separated can simultaneously enter the pores of the zeolite membrane. High fluxes and high selectivities can be obtained only in the case of real molecular sieving [6]. We must avoid the situation of an interplay of mixture diffusion and mixture adsorption, which takes place if the size of the zeolite pore is not in between the sizes of the components to be separated. For a given gas composition, temperature, and pressure, reasonable separation figures can be obtained. However, if the permeation conditions change, often a collapse of the separability can be observed since a slowly diffusing but strongly adsorbed component can hinder the diffusion of a more mobile component. Therefore, zeolite membranes with pore sizes should be developed which allow the simple separation of a mixture by size.

For a technical application of oxygen-transporting perovskite membranes, their long-time stability and different mechanical problems must be solved. Usually, there are earth alkaline ions like Ba^{2+} or Sr^{2+} on the A site of the ABO_3 perovskite structure. These ions can form with CO_2 and SO_2/O_2 carbonates and sulfates, thus destroying the perovskite crystal structure. On the B site, we usually have redox active ions like Fe^{2-4+} or Co^{2-4+} in different oxidation states with different ionic radii and, thus, different unit cell dimensions. On the air side of the membrane, these ions are in their highest oxidation state and a low concentration of oxygen lattice vacancies is found. On the membrane side with the low oxygen partial pressure, these ions are found in their lowest oxidation state and the concentration of oxygen vacancies is maximum. Therefore, perovskites show a non-regular TEC, whereas the TEC of the metal or ceramic supports remains constant. It is extremely difficult, therefore, to match the TEC of a thin perovskite membrane layer on a support.

For special applications like in submarines, *fuel cells* have become the state of the art. For thermodynamic reasons, from a long-term point of view, the solid oxide fuel cell (SOFC) is favored. However, the high-temperature operation requires a long-time stable gas-tight sealing of the components, which turns out to be difficult to realize. The first SOFC applications will be for stationary and not for mobile applications. So for the mobile applications such as cars, laptops, mobile phones, etc., the low-temperature PEMFC competes with charge storage systems like Li^+-based batteries. However, there exist so far no convincing concepts of how the latter ones will attain the necessary storage density. Therefore, PEM with improved proton mobility are believed to have a real chance in this competition with Li-based electricity storage.

Whereas single-crystal Si *solar cells* show under practical conditions a stable photoefficiency >20%, Grätzel-type DSSCs with nano-particular TiO_2 layers reach about 10% and DSSC with electrochemically deposited semiconductor layer about 5% photoefficiency. In this field, an extremely high R&D dynamics can be stated. Instead of dyes, also inorganic light-absorbing thin films such as $CuInSe_2$ deposited on suitable semiconductor supports like ZnO nano-wire arrays could do the job.

12.5 Conclusions

This chapter shows how a better design of devices containing porous materials requires a clear understanding of transport processes occurring in porous materials. The diffusion of molecules, ions, and electrons can be the decisive step in novel functional materials like gas separation membranes, fuel cells, or solar cells. For the molecular understanding of porous gas separation membranes, mixture adsorption and mixture diffusion coefficients are unavoidable. When evaluating the proton conductivity in proton-conducting membranes for fuel cells by impedance spectroscopy, under the testing conditions, the proton transport is field- and concentration driven, which is different from the proton transport in the working fuel cell. To increase the efficiency of DSSCs, the diffusion-controlled transport time of the photogenerated electrons to the electrodes must be shorter than their lifetime. This challenge can be solved by anisotropic semiconductor layers with one-dimensional electron diffusivity.

12.6 Questions on Basic Issues

Describe the flux density j_i according to the first Fickian law through a membrane of thickness L. The concentration difference of the component i between the two sides of the membrane is Δc_i and the diffusion coefficient of the component i in the membrane is D_i.

Describe the flux density j_i of oxygen ions through a perovskite membrane of thickness L with the oxygen partial pressures p'_{O_2} and p''_{O_2} on the two membrane sides assuming that the electronic conductivity σ_e is much larger than the ionic conductivity σ_i.

Discuss the efficiency of a DSSC for the case that the electron diffusion coefficient in the dye/semiconductor layer of thickness $L = 5\,\mu m$ is $D_e = 10^{-6}\,cm\,s^{-1}$. Assume that the photoelectrons are exclusively generated on the outer surface of the dye/semiconductor layer exposed to light and that they have to move to the opposite side of the layer that is attached to the electrode. The electrons have a medium lifetime of $\tau_n = 10^{-2}\,s$.

Acknowledgments

The author would like to thank M. Noack (Berlin), H. Richter and I. Voigt (Hermsdorf), and H. Voß (Ludwigshafen) for their cooperation in zeolite membranes; H.H. Wang (Kanton) in oxygen-transporting membranes; M. Wark and R. Marschall (Hannover) in proton exchange membrane fuel cells; and T. Oekermann (Hannover) in dye-sensitized solar cells. The author would also like to thank M. Martin (Aachen) for the stimulating discussion of the general diffusion theory. Thanks are also due to G. Centi (Messina), P. Schmucki (Erlangen), and A. Gottschlich (Hannover) for the SEM figures on 1D semiconductors. The author thanks DFG for its financial support within the European joint research project "Diffusion in Zeolites" (Ca 147/11) and the SPP 1181 NanoMat (Ca 147/13, Wa 1116/15). Finally, the author would like to thank the EU for financing the Network of Excellence "InsidePores."

References

1. J. Crank. *The Mathematics of Diffusion*. Oxford, Clarendon Press, 1956.
2. J. Kärger and D.M. Ruthven. *Diffusion in Zeolites and Other Microporous Solids*. Wiley & Sons, New York, 1992.

3. P. Heitjans and J. Kärger. *Diffusion in Condensed Matter: Methods, Materials, Models.* Springer, Berlin, Heidelberg, Germany, 2005; J. Kärger, P. Heitjans, and R. Haberlandt. *Diffusion in Condensed Matter.* Vieweg, Braunschweig/ Wiesbaden, 1998.

4. S.R. de Groot and P. Mazur. *Non-Equilibrium Thermodynamics.* North-Holland, Amsterdam, the Netherlands, 1962.

5. M. Martin. In: P. Heitjans and J. Kärger (eds). Diffusion in oxides, *Diffusion in Condensed Matter: Methods, Materials, Models.* Springer, Berlin, Heidelberg, 2005, p. 209.

6. J. Caro and M. Noack. *Microporous Mesoporous Mater.* 115 (2008) 215.

7. W.J. Koros, Y.H. Ma, and T. Shimidzu. Terminology for membranes and membrane processes, IUPAC recommendations from 1996. *Pure Appl. Chem.* 68 (1996) 1479.

8. H. Voß, A. Diefenbacher, G. Schuch, H. Richter, I. Voigt, M. Noack, and J. Caro. Butene isomers separation on titania supported MFI membranes at conditions relevant for practice, *J. Memb. Sci.* 329 (2009) 11.

9. J. Caro. Diffusion in porous functional materials: Zeolite gas separation membranes, proton ion exchange membrane fuel cells, dye sensitized solar cells, *Microporous Mesoporous. Mater.* 125 (2009) 79.

10. A.F.P. Ferreira. Diffusion and adsorption of linear and branched alkanes in zeolites. PhD thesis, TU Delft, 2007.

11. R. Krishna and J.M. van Baten. *Chem. Eng. J.* 140 (2008) 614.

12. R. Krishna and D. Paschek. *Phys. Chem. Chem. Phys.* 3 (2001) 453.

13. M. Fernandez, J. Kärger, D. Freude, A. Pampel, J.M. van Baten, and R. Krishna. *Microporous Mesoporous Mater.* 105 (2007) 124.

14. J.M. van de Graaf, E. van der Bijl, A. Stol, F. Kapteijn, and J.A. Moulijn. *Ind. Eng. Chem. Res.* 37 (1998) 4071.

15. H.J.M. Bouwmeester and A.J. Burggraaf. In: A.J. Burggraaf and L. Cot (eds). Dense ceramic membranes for oxygen seperation, *Fundamentals of Inorganic Membrane Science and Technology*, Vol. 4. of *Memb. Sci. Technol. Ser.* Elsevier, Amsterdam, the Netherlands, 1996.

16. H.H. Wang, T. Schiestel, C. Tablet, M. Schroeder, and J. Caro. *Solid State Ionics* 177 (2006) 2255.

17. H.J.M. Bouwmeester and A.J. Burggraaf. In: A.J. Burggraaf and L. Cot (eds). *Fundamentals of Inorganic Membrane Science and Technology.* Elsevier, Amsterdam, the Netherlands, 1996, pp. 435–528.

18. C.G. Guizard and A.C. Julbe. In: N.K. Kanellopoulos (ed.). *Recent Advances in Gas Separation by Microporous Ceramic Membranes.* Elsevier, Amsterdam, the Netherlands, 1996, pp. 435–471.

19. C.S. Chen, H. Kriudhof, H.J.M. Bouwmeester, H. Verweij, and A.J. Burggraaf. Thickness dependence of oxygen permeation through erbiastabilized bismuth oxide-silver composites, *Solid State Ionics* 99 (1997) 215.

20. X.F. Zhu, Y. Cong, and W.S. Yang. Oxygen permeability and structural stability of $BaCe_{0.15}Fe_{0.85}O_{3-\delta}$, *J. Memb. Sci.* 283 (2006) 38.

21. P.Y. Zeng, Z.H. Chen, W. Zhou, H.X. Gu, Z.P. Shao, and S.M. Liu. Re-evaluation of $Ba_{0.5}Sr_{0.5}Co_{0.8}Fe_{0.2}O_{3-\delta}$, *J. Memb. Sci.* 291(2007) 148.

22. K. Huang, M. Schroeder, and J.B. Goodenough. Oxygen permeation in co-containing perovskites: Surface exchange vs. bulk diffusion, *Electrochemical Society Proceedings*, Vol. 99-13 (1999) 95.

23. S. Kim, Y.L. Yang, A.J. Jacobson, and B. Abeles. Oxygen surface exchange in mixed ionic electronic conductor membranes, *Solid State Ionics* 121 (1999) 31.

24. H. Jiang, H.H. Wang, S. Werth, T. Schiestel, and J. Caro. Simultaneous production of hydrogen and synthesis gas by combining water splitting with partial oxidation of methane in a hollow-fiber membrane reactor, *Angew. Chem. Int. Ed.* 47 (2008) 9341.

25. H. Jiang, H.H. Wang, F. Liang, S. Werth, T. Schiestel, and J. Caro, Direct decomposition of nitrous oxide to nitrogen by in situ oxygen removal with a perovskite membrane, *Angew. Chem. Int. Ed.* 48 (2009) 2983.

26. T. Schiestel, M. Kilgus, S. Peter, K.J. Kaspary, H. Wang, and J. Caro. Hollow fibre perovskite membranes for oxygen separation, *J. Memb. Sci.* 258 (2005) 1.

27. J. Zhang, Z. Xie, J. Zhang, Y. Tang, C. Song, T. Navessin, Z. Shi et al. High temperature PEM fuel cells, *J. Power Sources* 160 (2006) 872.

28. R. Marschall. New solid proton conductors: Functionalized mesoporous SiO_2, materials for application in high temperature PEM fuel cell membranes. PhD thesis, Leibniz University Hannover, 2008.

29. J.D. Halla, M. Mamak, D.E. Williams, and G.A. Ozin. *Adv. Funct. Mater.* 13 (2003) 133.

30. R. Marschall, I. Bannat, J. Caro, and M. Wark. *Microporous Mesoporous Mater.* 99 (2007) 190.

31. M. Wilhelm, M. Jeske, R. Marschall, W.L. Cavalcanti, P. Tölle, C. Köhler, D. Koch et al. *J. Membr Sci.* 316 (2008) 164.

32. Y. Tominaga, I.-C. Hong, S. Asai, and M. Sumita, *J. Power Sources* 171 (2007) 530.

33. D. Gomes, R. Marschall, S.P. Nunes, and M. Wark. *J. Memb. Sci.* 322 (2008) 406.

34. R. Marschall, J. Rathousky, and M. Wark. Ordered functionalized silica materials with high proton conductivity, *Chem. Mater.* 19 (2007) 6401.

35. D. Margolese, J.A. Melero, S.C. Christiansen, B.F. Chmelka, and G.D. Stucky. *Chem. Mater.* 12 (2000) 2448.

36. B. O'Regan and M. Grätzel. *Nature* 353 (1991) 737.

37. Y. Shen, K. Nonomura, D. Schlettwein, C. Zhao, and G. Wittstock. *Chem. Eur. J.* 12 (2006) 5832.

38. T. Oekermann, T. Yoshida, H. Minoura, K.G.U. Wijayantha, and L.M. Peters. *J. Phys. Chem. B* 108 (2004) 8364.

39. J. Jiu, S. Isoda, F. Wang, and M. Adachi. *J. Phys. Chem. B* 110 (2006) 2087.

40. D. Fattakhova-Rohlfing, T. Brezesinski, J. Rathousky, A. Feldhoff, T. Oekermann, M. Wark, and B. Smarsly. *Adv. Mater.* 18 (2006) 2980.

41. T. Yoshida, K. Terada, D. Schlettwein, T. Oekermann, T. Sugiura, and H. Minoura. *Adv. Mater.* 12 (2000) 1214.

42. T. Oekermann, T. Yoshida, C. Boeckler, J. Caro, and H. Minoura. *J. Phys. Chem. B* 109 (2005) 12560.

43. M. Law, L.E. Greene, J.C. Johnson, R. Saykally, and P. Yang. *Nat. Mater.* 4 (2005) 455.

44. J.B. Baxter and E.S. Aydil. *Appl. Phys. Lett.* 86 (2004) 053114.

45. J.-J. Wu, G.-R. Chen, H.-H. Yang, C.-H. Ku, and J.-Y. Lai. *Appl. Phys. Lett.* 90 (2007) 213109.

46. K. Nonomura, D. Komatsu, T. Yoshida, H. Minora, and D. Schlettwein. *Phys. Chem. Chem. Phys.* 9 (2007) 1843.

47. K. Nonomura. Photochemical characterization of dye-modified ZnO hybrid thin films prepared by electrochemical deposition. PhD thesis, University Gießen, 2006.
48. S. Nakade, M. Matsuda, S. Kambe, Y. Saito, T. Kitamura, T. Sakada, Y. Wada et al. *J. Phys. Chem. B* 106 (2002) 10004.
49. L. Forro, O. Chauvet, D. Emin, L. Zuppiroli, H. Berger, and F. Lévy. *J. Appl. Phys.* 75 (1994) 633.
50. M. Adachi, Y. Murata, J. Takao, J. Jiu, M. Sakamoto, and F. Wang. *J. Am. Chem. Soc.* 126 (2004) 14943.
51. B. Tan and Y. Wu. *J. Phys. Chem. B* 110 (2006) 15932.
52. M. Durr, A. Schmid, M. Obermaier, S. Rosselli, A. Yasuda, and G. Nelles. *Nat. Mater.* 4 (2005) 607.
53. J.-J. Wu, G.-R. Chen, C.-C. Lu, W.-T. Wu, and J.-S. Chen. *Nanotechnology* 19 (2008) 105702.
54. Q. Li, K.W. Kwong, D. Ozkaya, and D.J.H. Cockayne. *Phys. Rev. Lett.* 92 (2004) 186102.
55. J. Wang, M.J. Zhou, S.K. Hark, Q. Li, D. Tang, M.W. Chu, and C.H. Chen. *Appl. Phys. Lett.* 89 (2006) 221917.
56. J. Wang, X.P. An, and Q. Li. *Appl. Phys. Lett.* 86 (2005) 201911.
57. J.-H. Yang, G.-M. Liu, J. Lu, Y.-F. Qiu, and S.-H. Yang. *Appl. Phys. Lett.* 90 (2007) 103109.
58. Y.-P. Fang, Q. Pang, X.-G. Wen, J.N. Wang, and S.-H. Yang. *Small* 2 (2006) 612.
59. X.-G. Wen, Y.-P. Fang, Q. Pang, C.-L. Yang, J.N. Wang, W.K. Ge, K.S. Wong et al. *J. Phys. Chem. B* 109 (2005) 15303.
60. S. Anandan, X.-G. Wen, and S.-H. Yang. *Mater. Chem. Phys.* 93 (2005) 35.
61. D.-F. Liu, W. Wu, Y.-F. Qiu, S.-H. Yang, S. Xiao, Q.-Q. Wang, L. Ding et al. *Langmuir* 24 (2008) 5052.
62. Y. Feng and Z.-Y. Yuan. *Nanocatalysis: Fundamentals and Applications*, Dalian, China, July 2008, Book of Abstracts, p. 226.
63. G. Gottschlich and T. Oekermann. Leibniz University Hannover, unpublished.
64. S. Perathoner, R. Passalacqua, G. Centi, D.S. Su, and G. Weinberg. Photoactive titania nanostructured thin films: Synthesis and characteristics of ordered helical nanocoil arrays, *Catal. Today* 122 (2007) 3.
65. G. Centi, R. Passalacqua, S. Perathoner, D.S. Su, G. Weinberg, and R. Schlögl. *Phys. Chem. Chem. Phys.* 9 (2007) 4930.
66. J.M. Macak, H. Tsuchiya, A. Ghicov, K. Yasuda, R. Hahn, S. Bauer, and P. Schmucki. *Curr. Opin. Solid State Mater. Sci.* 11 (2007) 3.
67. J.M. Macak, S.P. Albu, and P. Schmucki. *Phys. Stat. Sol. (RRL)* 1 (2007) 181.
68. S.P. Albu, A. Ghicov, J.M. Macak, R. Hahn, and P. Schmucki. *Nano Lett.* 5 (2007) 1286.
69. R. Krishna. *J. Phys. Chem. C* 113 (2009) 19773.
70. L. Heinke, C. Chmelik, P. Kortunov, D.M. Ruthven, D.B. Shah, S. Vasenkov, and J. Kärger. *Chem. Eng. Technol.* 30 (2007) 995.
71. H. Bux, F. Liang, Y. Li, J. Cravillon, M. Wiebcke, and J. Caro. *J. Am. Chem. Soc.* 131 (2009) 16000.
72. H. Bux, C. Chmelik, J. Kärger, S. Fritzsche, J. Cravillon, M. Wiebcke, R. Krishna et al. *Abstract, 16th International Zeolite Conference*, Sorrent, Italy, July 2010.

13

Zeolites and Mesoporous Aluminosilicates as Solid Acid Catalysts: Fundamentals and Future Challenges

Ana Primo, Avelino Corma, and Hermenegildo García

CONTENTS

13.1 Introduction

Heterogeneous catalysis is behind 85% of all industrial chemical processes.[1] Particularly, the high maturity and extraordinary efficiency of petrochemistry rely on the use of solid catalysts.[2] The paradigmatic examples of successful heterogeneous catalysts are zeolites and materials derived therefrom.[3] As widely recognized, the development of zeolites as large-scale catalysts

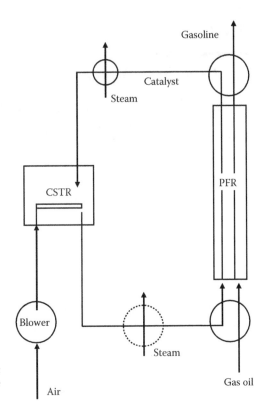

SCHEME 13.1
Simplified diagram of a fluid catalytic
cracking process that is based on the
use of acid zeolites as catalysts.

in petrochemistry has been one of the major achievements in the chemistry
of the twentieth century.[4] Scheme 13.1 shows the diagram of fluid catalytic
cracking that is currently the most important process based on acid zeolites
as catalysts.

Considering the large economical importance of the use of zeolites as solid
catalysts, any book dealing with nanoporous materials and their applications
should contain a chapter dealing with the use of zeolites as heterogeneous
catalysts. Taking into account the broad and wide range of examples describing the use of zeolites in catalysis, it would be impossible to summarize, even
at the minimal level of coverage, all the fields wherein zeolites are currently
being used as catalysts. The reader is referred to past and recent reviews for
a more in-depth coverage of this topic.[5–7]

In the present chapter, we limit to summarize briefly the use of zeolites
as solid acids, emphasizing the advantages of these materials to promote
acid-catalyzed reactions, the influence and role of the crystal structure on
the outcome of the reaction, the control and structure of the acid sites, and
representative examples of the use of zeolites as catalysts of organic reactions. Due to the fact that mesoporous-structured aluminosilicates constitute
a logical extension of the zeolite catalysis, we also include a brief description

of these types of materials. We finish this chapter with our view anticipating future breakthroughs in this area, which can be anticipated to cope with the present challenges and limitations of zeolites and mesoporous aluminosilicates as acid catalysts.

13.2 Green Chemistry and Acid Catalysts

There is currently a considerable concern on the impact of the human activity on the environment. Particularly, the chemical industry has always moved toward the improvement of the efficiency and toward the reduction of wastes. Since the 1990s, a new area of research has appeared based on the 12 principles enunciated by Anastas and Eghbali[8] according to which not every possible chemical process should be used but only those that are benign for the environment.[9] Scheme 13.2 lists the principles of green chemistry. These 12 principles include the minimization wastes, atom economy, and the use of reusable catalysts.[10,11]

Particularly, the most widely used chemical reactions that produce a large amount of waste are those catalyzed by acids in homogeneous liquid phase. The reason for this is that to separate products from the liquid acid, it is necessary to effect the neutralization of the catalyst during the reaction workup, and this step originates considerable stoichiometric amounts of salts typically dissolved in water. Both liquid Brönsted acids such as sulfuric acid and soluble Lewis acids such as $AlCl_3$ have a very high E-factor. This E-factor was introduced by Sheldon as a quantitative parameter to assess the "greenness" of one process, and it corresponds to the ratio of kilograms of wastes divided by kilogram of product.[12,13] For this reason, in acid-catalyzed reactions, particularly in fine chemistry, E-factors between 10 and 100 are not uncommon. Scheme 13.3 provides an average of E-factors according to the range of chemical process.

By contrast, for many zeolite-catalyzed acid reactions in petrochemistry, the E-factor is below 0.1. The reason for this low E-factor (high greenness) is that zeolites acting as solid acids can be easily separated from the products without the need of neutralization because they are in a separated phase and also they can be reused or the process can operate under continuous flow. Under these conditions, "the productivity" of zeolites in terms of kilogram of product per kilogram of catalyst can be in the range of 10^6 or even higher. Other advantages of zeolites as solid acids is the lack of equipment corrosion, the lack of toxicity, the wide availability, and the possibility to control the density and strength of the acid sites, reaching in some cases superacidic pKa. Scheme 13.4 lists the advantage of the use of zeolites and solid acids in comparison to liquid acids.

In view of all these considerations, the use of zeolites has been widely implemented in very large-scale industrial processes wherein the research

1. **Prevention**
    It is better to prevent waste than to treat or clean up waste after it has been created.
2. **Atom Economy**
    Synthetic methods should be designed to maximize the incorporation of all materials used in the process into the final product.
3. **Less Hazardous Chemical Syntheses**
    Wherever practicable, synthetic methods should be designed to use and generate substances that possess little or no toxicity to human health and the environment.
4. **Designing Safer Chemicals**
    Chemical products should be designed to effect their desired function while minimizing their toxicity.
5. **Safer Solvents and Auxiliaries**
    The use of auxiliary substances (e.g., solvents, separation agents, etc.) should be made unnecessary wherever possible and innocuous when used.
6. **Design for Energy Efficiency**
    Energy requirements of chemical processes should be recognized for their environmental and economic impacts and should be minimized. If possible, synthetic methods should be conducted at ambient temperature and pressure.
7. **Use of Renewable Feedstocks**
    A raw material or feedstock should be renewable rather than depleting whenever technically and economically practicable.
8. **Reduce Derivatives**
    Unnecessary derivatization (use of blocking groups, protection/deprotection, temporary modification of physical/chemical processes) should be minimized or avoided if possible, because such steps require additional reagents and can generate waste.
9. **Catalysis**
    Catalytic reagents (as selective as possible) are superior to stoichiometric reagents.
10. **Design for Degradation**
    Chemical products should be designed so that at the end of their function they break down into innocuous degradation products and do not persist in the environment.
11. **Real time analysis for Pollution Prevention**
    Analytical methodologies need to be further developed to allow for real-time, in-process monitoring and control prior to the formation of hazardous substances.
12. **Inherently Safer Chemistry for Accident Prevention**
    Substances and the form of a substance used in a chemical process should be chosen to minimize the potential for chemical accidents, including releases, explosions, and fires.

SCHEME 13.2
The 12 principles of green chemistry.

E-Factor	Industrial process
0.1	Petrochemical
1–10	Bulk chemicals, Polymers, Plastics
100	Fine chemicals
250	Pharmaceuticals, Electronic materials

SCHEME 13.3
E-factor corresponding to the different industrial processes.

and development effort has been very large. For this reason, they constitute the clearest examples of green catalysts. Upto 87 processes of the about 100 carried out in petrochemistry employs zeolites as catalysts.[14] In the next section, we describe the structure and nature of the acid sites in zeolites and structured mesoporous aluminosilicates.

Advantages of zeolites

Thermally and chemically robust
Chemical composition and structure controlled by synthesis and post-synthesis
Well characterizable solids (XRD, XRD, chemical analysis, spectroscopy, porosity and surface area)

Large variety of structures
Strongly acidic (superacidic?) and tunable distribution Green catalysts: No wastes, recoverable

---

SCHEME 13.4
The most important advantages of zeolites.

13.3 Structure of Zeolites

Zeolites are crystalline aluminosilicates wherein the primary building block is constituted by SiO_4^{4-} or AlO_4^{5-} tetrahedral sharing the corners.[5,15,16] These tetrahedra are arranged defining empty intracrystalline spaces forming long tubes (channels) or cavities (cages) that are interconnected and can be accessed from the exterior of the zeolite particle. This void intracrystalline space is denoted as micropores. Figure 13.1 presents the crystal structure of four common zeolites. For certain molecules having a molecular kinetic diameter smaller than the dimensions of the micropores, it should be possible to enter the interior of the channels and cages, reaching certain framework positions that can act as catalytic site.

Due to the similar ionic radius of silicon and aluminum (they are neighbors in the Periodic Table), the tetrahedra defined by four oxygen at the corners

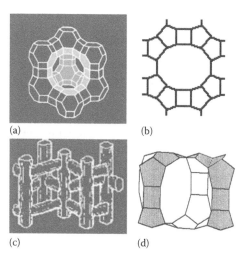

(a)   (b)

(c)   (d)

FIGURE 13.1
Crystal structure of (a) faujasite (zeolite X and Y): tridirectional, large pore (13 Å); (b) mordenite: unidirectional, large pore (7.4 Å); (c) pentasil (silicalite and ZSM-5): bidirectional, medium pore (5.4 × 5.6) Å; and (d) BEA (zeolite beta), tridirectional, large pore (12 Å).

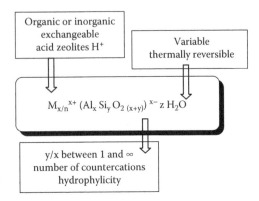

SCHEME 13.5
General elemental composition of the zeolites.

can accommodate equally in their interior one silicon or one aluminum ion. However, each time that one aluminum replaces isomorphically one silicon, a negative charge is generated in the framework due to the different valence between silicon and aluminum. This framework negative charge requires the presence of charge balancing cations that are occupying internal non-framework positions. Scheme 13.5 shows the general elemental composition of zeolites.

Although there are about 50 natural zeolites, many more (about 300) have been synthesized for the first time in the laboratory.[17] Indeed, the synthesis of classical and new zeolites represents an economical activity of large importance in the chemical industry. The nature of the charge-compensating cation depends on the gel composition used for the synthesis of these solids. Typically, they are alkali metal ions or quaternary ammonium salts. These quaternary ammonium salts are known to impart the geometry of the micropores by becoming occluded during the condensation of the silica species and, therefore, they are used as structure directing agents or "templates."[18,19]

13.4 Mesoporous Aluminosilicates

Zeolites are crystalline materials that can be prepared with different geometries and surface area. However, despite the large variety of crystal structure, there is a clear limitation in the maximum pore size that can be obtained for these materials, which is limited to around 0.7 nm. Therefore, there has been a continuous interest in developing zeolites with larger pore size, which would allow the incorporation of larger substrates. The IUPAC (International Union of Pure and Applied Chemistry) terms microporous materials those wherein the pore size is below 2 nm and mesoporous when the prose size ranges from 2 to 10 nm.

In this regard, the report by Mobil researchers of the synthesis of mesoporous materials (MCM series: Mobil Corporate Materials) and,

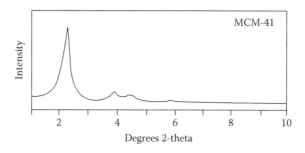

FIGURE 13.2
Example of MCM-41 XRD.

particularly, MCM-41 has constituted a breakthrough in material science.[20] These mesoporous silicates have amorphous walls but the internal void space is so regular that they exhibit characteristic X-ray diffraction (XRD) patterns at low 2θ angles as if they were crystalline solids. For this reason, these materials are denoted with the words "structured," "periodic," or "regular" mesoporous silicates to indicate that although the walls are amorphous, the material has a remarkable spatial order and arrangement. As an example, Figure 13.2 shows a XRD corresponding to MCM-41.

The periodicity and structuring of these mesoporous silicas can also be determined by transmission electron microscopy (TEM) wherein, due to the larger pore size, above a couple of nanometers, the openings of the pores can be seen. These TEM images also show the beautiful regularity of these mesoporous materials. These mesoporous materials can also be prepared with a certain amount of aluminum and the resulting structured mesoporous aluminosilicates exhibit the characteristic Brönsted and Lewis sites present in zeolites (see following sections). Considering the presence of acid sites and their large pore size, mesoporous aluminosilicates constitute a logical extension of zeolite catalysis because the nature of the acid sites is very similar; the composition of the materials is identical as that of zeolites but they overcome limitation in the pore dimensions.

In fact, the pore size of the mesoporous aluminosilicates can be controlled during the synthesis and can range from above 2 nm to more than 10 nm. The most widely used mesoporous aluminosilicate is MCM-41 whose structure is formed by parallel channels that are hexagonally arranged like in a honeycomb (Figure 13.3). In this case, a typical pore size is about 3.4 nm, but they can be obtained in larger dimensions by adding, during the synthesis of MCM-41 surfactants, with longer alkyl chains or cosurfactants.

The synthesis of MCM-41 requires the preparation of a gel containing silica and alumina under basic conditions in the presence of an alkyl trimethylammonium surfactant.[21,22] The most common one is cetyltrimethylammonium. The mechanism of formation of MCM-41 and related materials is well known (Scheme 13.6).[23,24] The key point is the aggregation of the surfactant forming micelles and rods in a liquid crystal state that

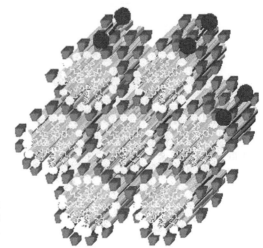

FIGURE 13.3
Model of the silica condensation around the surfactant crystal state leading to the formation of MCM-41 structure.

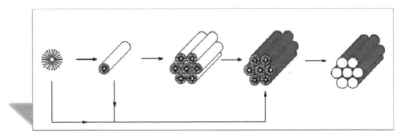

SCHEME 13.6
Accepted rationalization of the formation of MCM-41 aluminosilicates.

introduces anisotropy and structuring inside the liquid phase. Formation of the silicate occurs at the interphase between the water and the surfactant rods that act as "templates" leaving the shape to the silicate. The flexible liquid crystal surfactant produces in this way a rigid silicate skeleton that, after the synthesis and removal of the template, will form the mesopores. Scheme 13.6 illustrates the main events in the formation of a mesoporous aluminosilicates.

The main problem of mesoporous aluminosilicates, which arises from the thin walls and large porosity, is their poor structure stability, particularly when compared to crystalline zeolites. Thus, steaming of MCM-41 for a few minutes can cause the complete collapse of the mesopores.

The low hydrothermal stability of mesoporous aluminosilicates has been overcome particularly by the mesoporous materials described by Stucky (SBA-15), which is obtained using an amphiphilic triblock copolymer (Pluronics 123) as template.[25] Besides increasing stability, another feature of SBA-15 is that the pore sizes are generally larger than 10 nm.

13.5 Nature of Acid Sites in Zeolites and Mesoporous Aluminosilicates

Independently of the initial charge-compensating cation present in the as-synthesized solid and due to the electrostatic interaction with the framework, it is frequently possible to replace by ion exchange the original cation by another one. One particular case that is relevant to the present study is the exchange of the charge-balancing cations of the synthesis by a proton. In this way, Brönsted acid sites can be formed.[26] Scheme 13.7 shows three alternative strategies to obtain protonic zeolites.

The simplest procedure to introduce protons in a zeolite, and therefore the preferred method, is the thermal decomposition of quaternary ammonium salts used as templates in the synthesis of the solid (see Scheme 13.7). Since the zeolite crystal structure is thermally very robust, it is possible to effect the degradation of any organic component occluded in the zeolite. This thermal decomposition destroys the organic cation and, to compensate the framework negative charge, a proton has to be formed in the process that is reminiscent to the Hoffmann degradation of quaternary ammonium salts to form alkenes and tertiary amines.[27] One example of this way to obtain the acid form of a zeolite is the case of ZSM-5 in which the tetraethyl ammonium ion used as template in the synthesis is thermally removed, generating concomitantly the H^+ form of this zeolite. Thermal decomposition of the surfactant is also the general procedure to introduce Brönsted acidity in mesoporous aluminosilicates. For the synthesis of mesoporous silica, long-chain quaternary ammonium ion is used as templates. Removal of these surfactants by thermal decomposition renders the protic form of the mesoporous aluminosilicate.

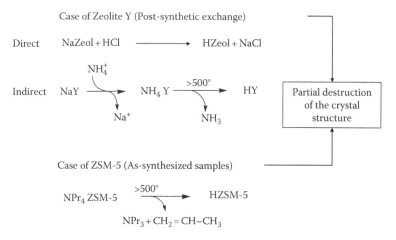

SCHEME 13.7
Different routes to obtain protonic zeolites.

The second way to generate an acid zeolite is by direct treatment of a previously synthesized zeolite containing any cation with concentrated liquid acid such as HCl or HNO_3 acids in aqueous phase. The main drawback of this approach is that although zeolites are also inert in the presence of most chemical reagents, they undergo certain structural damage (particularly, framework dealumination) when submitted to harsh acid conditions. Nevertheless, for many zeolites, the structural damage is relatively minor and, in several cases, for instance for beta zeolites, this direct treatment with concentrated aqueous acids is a convenient procedure to obtain the Brönsted form of the zeolite. In this regard, it should be commented that the only general way to completely dissolve the zeolite solid is using concentrated HF reinforced with some strong Brönsted acid. Conventional strong liquid acids such as nitric, phosphoric, or sulfuric acid only cause a partial damage that is reflected in the partial dealumination of the material and a partial decrease in the crystallinity as determined by the decrease of the number of counts of the material in powder XRD. In contrast to the chemical robustness of crystalline zeolites, mesoporous aluminosilicates are completely destroyed under much milder conditions and, therefore, they cannot be submitted to ion exchange in aqueous media with guarantee of stability.

Likewise, the most widely used acid zeolite, HY faujasite, cannot be prepared by any of the two previous methods. In this case, due to the lack of template in the synthesis and the large structural damage caused by liquid acids, it is necessary to use a third way to introduce protons in the material. This indirect procedure consists in effecting the intermediate ion exchange of the alkali metal ion by weakly acidic ammonium ions followed by the thermal decomposition of ammonium at about 500°C.

The structure of the Brönsted acid sites consists in bridging hydroxyl groups ≡Al–(OH)–Si≡.[28] Since aluminum ions are responsible for the framework negative charge, the positive counter cation, a proton, in the case of acid zeolites, is located as closely as possible to the negative center. By theoretical calculations, it has been estimated that the acid strength of the acid site depends on the "second coordination sphere" around the bridging hydroxyl groups.[29] Scheme 13.8 illustrates the structure of a Brönsted site with positive charge density at the hydrogen atom.

According to these calculations, the acid strength distribution of a given zeolite will depend on the Si/Al ratio. The higher the aluminum content, the higher is the density of acid sites (number of protons per gram), but the lower is the strength of the sites. On the contrary, the lower the aluminum content, the lower is the number of acid sites but the remaining sites exhibit stronger acidity. With respect to the strength of zeolite acid sites, there has been a certain debate in the 1980s about whether or not acid zeolites should be considered solid superacids.[30] In this context, it should be commented that, certainly

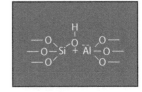

SCHEME 13.8
Representation of Brönsted acid sites.

not at room temperature, but upon heating at high temperatures, acid zeolites can behave as superacids able to protonate nonbasic carbon–carbon single bonds and effect efficiently the proteolytic cleavage of alkanes.[31,32]

13.6 Lewis Acid Sites

Besides Brönsted acid sites (protons of $\equiv$Al–(OH)–Si), zeolites can also have Lewis acid sites (Scheme 13.9). Initially, these Lewis acid sites were attributed to tricoordinated aluminum atoms.[33] However, the lack of supporting[27] Al NMR spectroscopy evidence for the existence of tricoordinated aluminum atoms and the observation of a considerably proportion of extraframework aluminum (EFAL) with octahedral coordination has led to propose this type of EFAL species as the existing Lewis acid sites in zeolites.[34]

Thus, while framework aluminum should be tetrahedrally coordinated,^{27}Al NMR reveals frequently the presence of octahedral aluminum atoms that cannot be accommodated in the framework and, therefore, has to correspond to EFAL. The same type of octahedral aluminum as Lewis acids sites is also present in mesoporous aluminosilicates. Furthermore, typically the population of EFAL species in a given zeolite sample increases as the sample is treated thermally or under steam. This increase of EFAL in many thermal treatments can be followed by MAS[27] Al NMR spectroscopy and is a consequence of the lower stability of framework aluminum compared to framework silicon. In this context, it is well known that the structural stability of the zeolites decreases as the aluminum content increases.[35,36] Indeed, as it has been commented earlier, faujasites having high aluminum content (Si/Al 2.4 for Y zeolite) are prone to collapse upon aqueous acid treatment, while in contrast, other zeolites having a higher Si/Al ratio such as beta zeolite are stable under identical conditions. Since high aluminum content introduces instability, the common way to increase the structural stability of faujasite Y is by dealumination and this modified faujasite having low aluminum content is typically denoted as ultrastable Y zeolite and is widely used in petrochemistry as a solid catalyst.

EFAL species behaving as Lewis acids can range from low-oligomeric aluminum species including monometal AlO$^+$ species and its dimers to light oligomers or can be heavily oligomerized aluminum oxo species. In fact, aging and

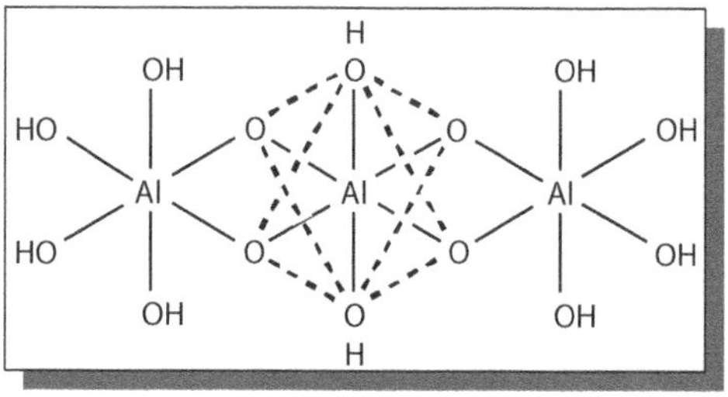

SCHEME 13.9
Representation of the octahedral aluminum non-framework.

continued use of an acid zeolite tend to transform initial low oligomerized EFAL species into higher EFAL oligomers. This tendency to grow due to condensation of EFAL species as well as the generation of different EFAL species by steaming and calcination of zeolites with high aluminum content can be followed by[27] Al NMR spectroscopy and is responsible for aging and variation of the acid properties of a zeolite upon use. It is possible to remove low condensed EFAL species from the intracrystalline space by washings with diluted acid aqueous solutions. In this way, and also by using highly dealuminated zeolites, it is possible to obtain purely Brönsted zeolites devoid of Lewis acid sites.

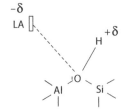

SCHEME 13.10
Cooperative effect of a Lewis acid site enhancing the acid strength of a Brönsted site by interacting with the lone electron pair of the oxygen atom and creating an extrapolarization ($+\delta$) on the proton.

However, the most frequent situation is to have samples of acid zeolites in which variable proportions of Brönsted and Lewis acids are simultaneously present. It has been observed that, as demonstrated by Olah with the magic acid in liquid phase,[37] the presence of Lewis acid sites increases notably the strength of the Brönsted acid sites by further polarization of the bridging OH group induced by the Lewis acid site. Scheme 13.10 illustrates the influence that the presence of Lewis sites exerts on the strength of the Brönsted acid sites.

13.7 Experimental Techniques to Measure Acidity in Solids

Compared to purely liquid acids, solid acids and, particularly, acid zeolites differ on the fact that solids may contain a wide distribution of sites with different strength and nature. As commented earlier, the positive charge of the proton in a bridging hydroxyl group depends on the first coordination sphere around the $\equiv$Al–(OH)–Si$\equiv$ site. Also, the Lewis acidity of EFAL species depends on the degree of condensation and oligomerization of the octahedral aluminum. Therefore, in order to determine the overall acid strength of a solid, it is important to develop experimental techniques to measure acidity (total number and sites and population distribution) and compare the acid strength of different solids. The two most common techniques to determine acidity in porous aluminosilicates are thermoprogrammed desorption (TPD) using ammonia as probe and the pyridine adsorption–desorption method monitored by IR spectroscopy.[38–40] TPD consists in determining the temperature profile of ammonia desorption. The higher the temperature needed to desorb ammonia, the stronger the acid sites. Figure 13.4 shows typical TPD profiles of zeolites.

The main limitation of TPD using ammonia is that this technique does not distinguish between Brönsted and Lewis acid sites. For this reason,

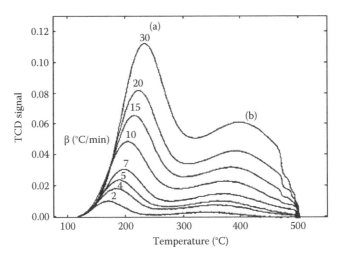

FIGURE 13.4
TPD of ammonia from ZSM-5. The labels (a) and (b) indicate the two desorption peaks corresponding to two different acid sites.

the preferred experimental technique to assess the acidity in porous aluminosilicates is the pyridine adsorption–desorption method. In this technique, the zeolite is firstly carefully dehydrated and then contacted with pyridine vapors. Then, the sample is outgassed at increasing temperatures while the population of pyridine interacting with Brönsted and Lewis acid sites is monitored by IR spectroscopy. Pyridine as a probe exhibits two specific peaks in IR spectroscopy appearing at 1550 and 1450 cm^{-1} corresponding to N-protonated pyridinium (Brönsted acid sites) or to the acid–pyridine adduct (Lewis acid sites), respectively. Figure 13.5 shows an example of the pyridine adsorption–desorption titration of an acid zeolite. The acid strength of the sites is determined by the relative decrease of the intensity of the pyridine peaks as the desorption temperature increases. Furthermore, knowing the amount of solid in the sample and the molar adsorptivity for the 1550 and 1450 cm^{-1} peaks, the method can be quantitative and the exact concentration of weak medium and strong sites determined.[41,42]

13.8 Influence of the Crystal Structure on the Acid Strength

We have previously commented that the chemical composition determines the acid strength of Brönsted and Lewis acid sites. However, besides the chemical composition, other factors can also play a role on the acid strength. One that is of particular importance is the crystal structure and dimension of the pores. In this way, acid sites in nonporous mixed $SiO_2 \cdot Al_2O_3$ oxide as

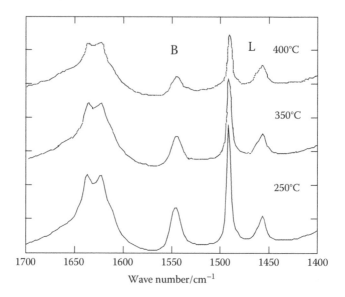

FIGURE 13.5

Aromatic region of the IR spectrum of pyridine adsorbed on HZSM-5. The letters B and L indicate the peak characteristic of pyridine interacting with Brönsted and Lewis acid sites, respectively. The spectra have been recorded at room temperature after adsorbing pyridine vapors at room temperature and then outgassing the wafer for 1 h under 10^{-5} torr at the increasing temperatures as indicated in the plot.

well as mesoporous aluminosilicates such as (Al)MCM-41 are always much weaker than the acid sites in zeolites of analogous composition. Furthermore, the strength of an acid site seems to increase as the dimension of the pore decreases. This effect has been termed as *"the confinement effect"* and has been interpreted considering that when the base is confined in a restricted space in which the frontier molecular orbitals, and particularly the HOMO, are limited by the zeolites walls similar models as those simply described as "the particle in a box" applied.[6,43] Then, the energy of the HOMO level of the base increases as a result of the confinement and orbital compression. This makes bases more easily protonated when the molecule is incorporated in a cavity where there is a tight fit between the molecule and the zeolite pore.[44] Figure 13.6 illustrates the concept of molecular confinement.

The general outcome of the confinement effect is that the apparent acid strength of a site is higher when the probe base fits tight in the cavity and, therefore, medium-pore-size zeolites and, particularly, H-ZSM-5 appear stronger acids than large-pore-size zeolites such as HY or mesoporous (Al) MCM-41 for the same composition. Considering the large pore dimensions of (Al)MCM-41 that are typically of 3.5 nm, only macromolecules will experience some degree of confinement. As consequence, the acid sites of (Al) MCM-41 are always weak or mild, while the acid sites of medium-pore-size zeolites appear always as much stronger.

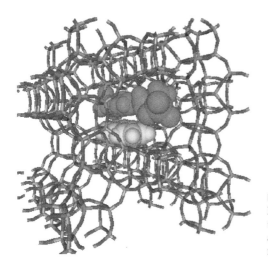

FIGURE 13.6
Molecular modeling showing a charge transfer complex between anthracene and methyl viologen confined inside a monodirectional large-pore zeolite.

13.9 Examples of Reactions Catalyzed by Solid Acids

In this section, we comment on three selected examples from our own work, which have been chosen to exemplify some of the most important issues in acid catalysis using solids. It is evident that acid catalysis is the main application of zeolites and that there is an enormous number of solid acid-catalyzed reactions impossible to cover in the limited space of this chapter. Therefore, the reader should consult books and reviews for a deeper coverage on this topic.[7,45–47] For our purpose here to give a brief introductory tutorial on the use of solid acids as catalysts, we have selected three reactions that will allow us to comment on some of the special features of porous aluminosilicates as solid acids such as influence of the nature and strength of the acid sites and catalyst deactivation by pore blocking, which are specific and distinctive of liquid acids.

13.10 Product Distribution of the α-Bromopropiophenone Cyclic Ethylene Acetal as a Test Reaction to Assess Brönsted and Lewis Acidity in Zeolites

The first property of solid acids is the concomitant presence of a broad distribution of Brönsted and Lewis acids.[6] The majority of the acid-catalyzed reactions can be promoted using both types of acid sites and the product distribution only exhibits minor changes in selectivity depending on the nature of the acid used.[48]

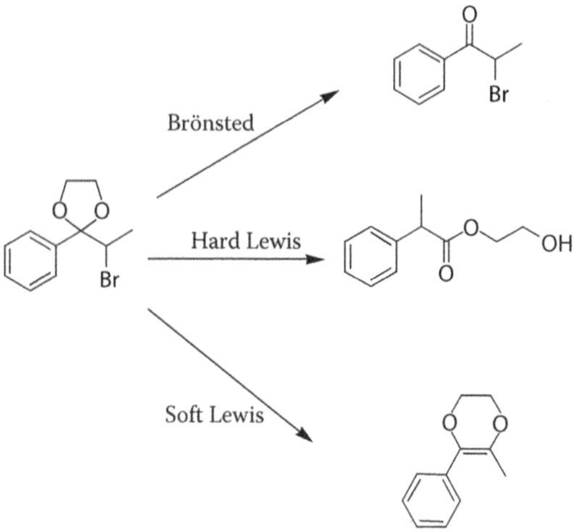

SCHEME 13.11
Specific products derived from α-bromopropiophenone ethylene acetal depending on the nature of the acid sites.

In this regard, we have published a probe molecule that is able to discriminate between Brönsted and Lewis solid and liquid acids just by monitoring the product distribution formed in the reaction.[49] The probe molecule is the cyclic ethylene acetal of α-bromopropiophenone. As indicated in Scheme 13.11, when Brönsted acid sites are present, hydrolysis of the acetal to yield the corresponding α-bromopropiophenone is the only product formed. The reason for this is because water, even in trace quantities, becomes activated preferentially by Brönsted acid sites attacking the ethylenedioxy group and producing acetal hydrolysis.

By contrast, under the same conditions, Lewis acid sites do not promote acetal hydrolysis but lead to two different rearrangement products depending on whether the phenyl ring or the ethylenedioxy is the migrating group (see Scheme 13.11).

Moreover, this test reaction can be used to discriminate between soft and hard Lewis acid sites just by determining the product distribution.[50] The origin of this remarkable product selectivity, depending on the softness/hardness of the sites, is the preferential interaction of the catalyst Lewis acid sites present in the zeolite with the bromine atom leaving group and the phenyl (hard Lewis acid) or the bromine and the oxygen atom of the acetal (soft Lewis acid groups), and the antiperiplanar conformation needed for the rearrangement.[51] Scheme 13.12 illustrates the different conformations prevailing depending on the softness/hardness of the acid sites and the different migrating group in each case as consequence of the preferential conformation determined by the Lewis acid sites.

Using the rearrangement of cyclic ethylene acetal of α-bromopropiophenone as a test reaction to monitor variations in softness/hardness as a function of the framework composition, it has been possible to demonstrate that the softness/hardness of the Lewis acid sites in zeolites can be tuned by

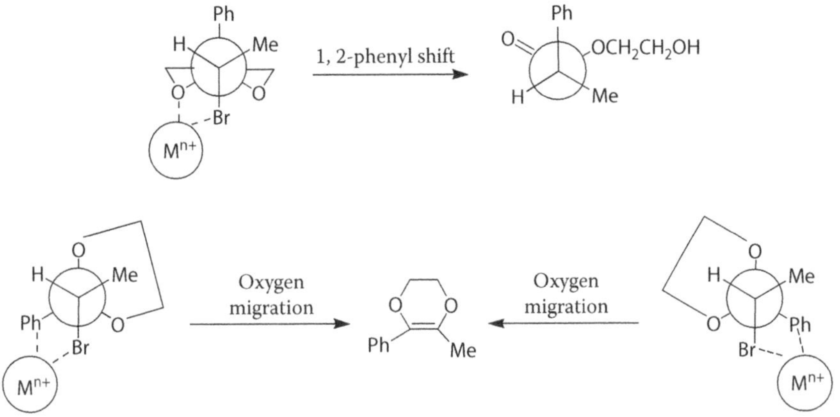

SCHEME 13.12
Preferential conformations of acetal from which phenyl shift or alkoxy migration occurs leading to different products.

controlling the silicon/aluminum framework ratio.[52] While in the simplest soft/hard Pearson theory for metal ions, the charge/ionic radius ratio can be used to assess the hardness/softness (larger ions with low charge behave as soft Lewis acids and small multicharged cations as hard Lewis sites),[53,54] the same concept in zeolites is less evident and high-level computational quantum mechanics calculations would be necessary to predict trends in softness/hardness for zeolites.[55] By contrast, by using the rearrangement of the ethylene acetal of α-bromopropiophenone in a consistent series of acid zeolites differing on the silicon-to-aluminum framework composition, it has been established that an increase in the aluminum content promotes the softness of the Lewis acid sites, the softest Lewis acid zeolite being that with the maximum possible aluminum content in the framework, i.e., faujasite X.[56] Table 13.1 presents some experimental data based on the product distribution of the propiophenone acetal rearrangement reflecting variations for different acid zeolites depending on the framework composition.[51]

13.11 Influence of the Acid Strength on the Catalytic Activity of Acid Zeolites

The Beckmann rearrangement of cyclohexanone oxime leading to ε-caprolactam is one of the most important industrial reactions carried out with liquid acids (Scheme 13.13). Typically, the industrial catalyst is oleum (concentrated sulfuric acid) that has to be neutralized during the reaction

TABLE 13.1

Results of the Reaction of Cyclic Acetal **1** in Chlorobenzene at 403 K for 20 h

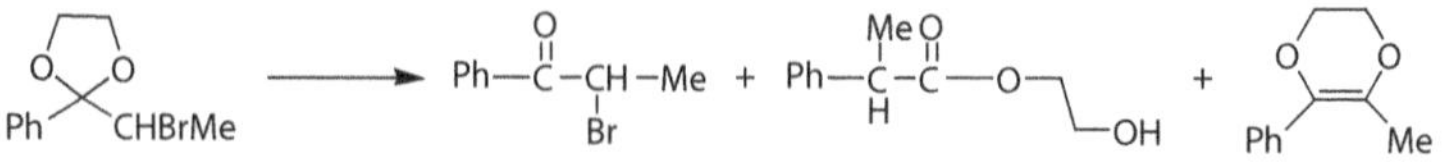

Catalyst	Transition Metal Content, mmol g^{-1}	Conversion, %	Selectivity			
			2	3	4	3/4
HY	—	100	91	0	0	—
ZnHY	0.78	98	66	8	2	4
ZnNaY	1.87	86	23	38	19	2
AgNaY	1.62	97.1	3	11	75	0.15
HgNaY	0.76	93.8	73	3	9	0.33

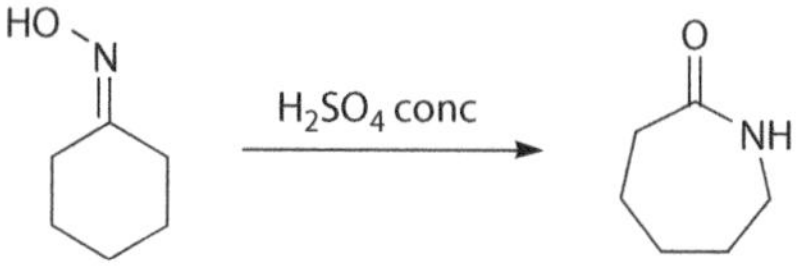

SCHEME 13.13
Beckmann rearrangement of cyclohexanone oxime to ε-caprolactam.

workup after the rearrangement in order to separate the cyclic lactam. ε-Caprolactam is the monomer of Nylon-6, an engineering plastic whose production has steadily increased since the last decades.[57] As the result of the sulfuric acid neutralization, stoichiometric amounts of the ammonium sulfate are formed. Although this ammonium salt is used as low-value nitrogen-containing fertilizer, it would be of interest to develop a solid acid as catalyst for this transformation because it would minimize the formation of wastes and will facilitate the workup.

While strong oleum acid in liquid phase at room temperature is used at the industrial level and it could be anticipated that strongly acidic zeolites would be needed for this process, many studies using zeolites as heterogeneous catalysts have worked in the gas phase using neutral silicalite as catalysts.[58,59] Silicalite is a pure silica MFI zeolite isomorphic with ZSM-5, which only contains ≡Si–OH silanol groups. These silanol groups (pKa 6.5) are very weak acid sites compared to the bridging ≡Al–(OH)–Si≡ groups. The problem of performing the reaction in the gas phase is the high boiling point of cyclohexanone oxime, which forces the process to be carried out at temperatures above 200°C and in the presence of solvents to dilute the cyclohexanone oxime concentration in the gas phase.[60–62]

The high-temperature requirement for the gas phase process catalyzed by zeolites and mesoporous aluminosilicates, as opposed to the room temperature Beckmann rearrangement using sulfuric acid, is responsible for the fact that acid sites with weak strength are sufficient and suitable to promote the process. This change in the acid strength requirements of the sites as a function of the reaction temperature illustrates that the weaker strength of the acid sites can be compensated by the temperature increase that can provide the activation to overcome the energy barrier of the process. Actually, if a ZSM-5 with strong acid sites is used as catalyst, the initial catalytic activity is very high but the catalyst becomes deactivated after a short time on stream. This result illustrates that in the case of continuous operation using microporous solids as catalysts, the initial reaction rate indicative of the intrinsic activity is one parameter, probably not the most important, to be considered among others. Under these conditions, catalyst stability becomes the most important parameter. For this reason, in order to minimize deactivation, mesoporous-structured silicas of MCM-41 type with much larger pore size can exhibit better performance.[63] Thus, it appears that the strongest acid sites are the most active initially but they tend to deactivate promptly, causing a short operation life of the material. On the other hand, sites of mild or weak acid strength could be intrinsically less active but they are less prone to become poisoned and deactivated. The lower intrinsic activity of mild acid sites can be compensated by increasing the reaction temperature.

In this context, working in the liquid phase with purely siliceous beta zeolites, it has been found that just silanol groups, typically considered as neutral sites in contrast to the bridging $\equiv$Si–(OH)–Al$\equiv$ groups can be sufficient to promote the Beckmann rearrangement cyclohexanone oxime. Furthermore, it has been found by studying pure silica beta samples with high and low crystallinity that the most active silanol groups are those in nests having hydrogen bonding interactions with other silanol neighbors. Figure 13.7

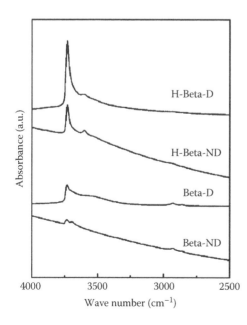

FIGURE 13.7

Different population (related to the intensity) and type (isolated or forming nests) of $\equiv$Si–OH groups (around 3700 cm^{-1}) depending on the nature of the beta zeolite: H-Beta-D—aluminated beta zeolite with defects; H-Beta-ND—aluminum-containing beta zeolite without structural defects; Beta-D—all-silica beta zeolite with defects; and Beta-ND—all-silica beta zeolite without structural defects.

shows the HO vibration peaks of a series of beta zeolites depending on the preparation procedure wherein remarkable differences can be observed. This different type and population of silanols are reflected on the catalytic performance. In fact, the Beckmann rearrangement of cyclohexanone oxime can also be performed using mesoporous (Al)MCM-41. In this case, the strength of the acid sites as determined by ammonia desorption is considerably much weaker than for analogous sites in zeolites. However, considering that no strong sites are required, the use of (Al)MCM-41 offers the advantage of larger pore dimensions, and therefore lower tendency to deactivation in MCM-41 compared to ZSM-5 or silicalite.

13.12 Friedel–Crafts Hydroxyalkylation of Arenes by Carbonylic Compounds

Friedel–Crafts reactions using aluminum chloride and other metal halides are the paradigmatic examples of acid-catalyzed reactions generating considerable amounts of wastes.[64] The reason for this is that, after the product is formed, the Lewis acid interacts strongly with the reaction product and destruction of this adduct is necessary to recover the products. Besides alkylation of aromatics with alkenes and alcohols and acylation of arenes by carboxylic acid derivatives, a third type of Friedel–Crafts reaction is the condensation of carbonilic compounds with arenes. Scheme 13.14 illustrates the reaction products and the corresponding intermediates for this type of Friedel–Crafts reaction.

This hydroxyalkylation of aromatics has been reported to be catalyzed by acid zeolites giving mixtures of triaryl and diaryl methanes.[65–68] One example of this is the reaction of paraformaldehyde with benzene, toluene, and other aromatic compounds.[69,70] For this reaction using a consistent set of acid zeolites, it has been determined that the process is not catalyzed by all the acid sites regardless their strength.[71]

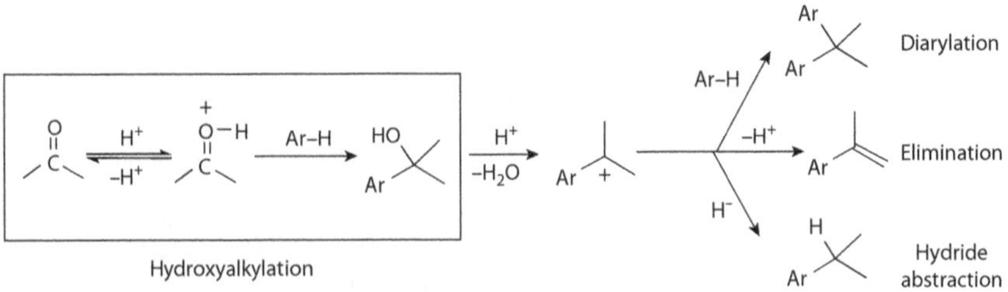

SCHEME 13.14
Mechanism of hydroxyalkylation and common consecutive processes.

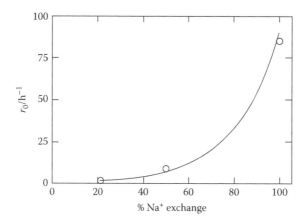

FIGURE 13.8
Exponential relationship between the initial reaction rate of diphenylmethane formation (r_0) at 80°C and the percentage of Na^+ to H^+ exchange of the NaHY catalyst.

Thus, for instance, if the initial reaction rate for the hydroxyalkylation of toluene with formaldehyde to give ditolyl methanes is plotted against the sodium-to-proton ion exchange percentage in HNaY zeolites, an exponential relationship is observed.[72] This indicates that when the number of acid sites introduced by replacing Na^+ by H^+ is doubled, the reaction rate is more than doubled and so on. Thus, the reaction rate is not proportional to the total population of acid sites in the zeolite but to the population of the strongest sites (see Figure 13.8).

Furthermore, as it has been commented in the previous sections, the silicon-to-aluminum ratio in the framework determines the strength of a given site, and therefore dealuminated zeolites have a higher population of stronger sites than the aluminated ones. Therefore, if the initial reaction rate for the hydroxyalkylation of toluene with formaldehyde is plotted against the total number of sites in zeolites with different framework-to-aluminum ratio, a straight line should be only observed in the case that all the sites have the same intrinsic activity. In other words, if only strong sites are active for the catalysis, the linear relationship between the initial reaction rate and the population of sites should be only observed when plotting the number of strong acid sites rather than the total (weak medium and strong) number of sites. The fact that experimentally the linear relationship between the initial reaction rate in the presence of zeolites with different silicon-to-aluminum ratio and the total number of sites is not observed (see Figure 13.9) demonstrates that not all the sites present are catalytically active.

As commented earlier in the case of the Beckmann rearrangement, one point of interest in acid catalysis working with zeolites is the long-term stability and the deactivation mechanism. In fact, this hydroxyalkylation of aromatics constitutes an example of a reaction that causes the strong

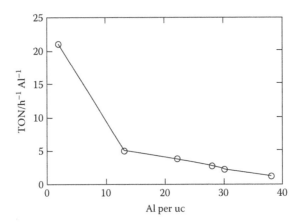

FIGURE 13.9
Turnover number, defined as initial rate/Al per uc) for the zeolite Y-catalyzed alkylation of benzene (50 mL) with formaldehyde (200 mg) at 80°C versus the number of Al per unit cell (uc) of the catalyst.

deactivation of the catalyst. In this case, it has been demonstrated that the main deactivation mechanism is the formation of large triaryl methylium cations replacing the acid sites and blocking the pores.[72] Scheme 13.15 describes the hydride transfer mechanism that produces the formation of these bulky carbocations.

In fact, the same reaction can be used in a different way to prepare materials wherein an organic cation is permanently immobilized ("incarcerated") inside the cavities of a large pore zeolite. When the topology of a zeolite contains large cavities and cages that are accessed through smaller windows, it is possible to envision a situation in which small precursors able to enter through the windows react into the cavity forming a bulky species that once formed cannot diffuse out of the cavity and becomes encapsulated inside the matrix. This type of reaction has been generically termed as *"ship in a bottle,"* alluding to those artistic bottles containing a large ship that cannot enter or exit through the bottle neck.[73] Scheme 13.16 shows the formation of *p*-substituted trityl ions encapsulated in zeolite Y by reaction of benzaldehyde with anisole.

In the case commented here of the Friedel–Crafts hydroxyalkylation of arenes by carbonyl compounds, an additional feature is that the species being formed is an organic cation that probably will not survive outside the zeolite framework due to its trapping by nucleophiles and water. In addition, as indicated in Scheme 13.16, the reaction forming trityl cations means, from the catalytic point of view, the deactivation of the catalyst since the trityl ion is really neutralizing the Brönsted acid site, i.e., replacing protons by the organic cation, and also blocking the zeolite pores. Thus, the same process, i.e., the reaction of electron-rich arenes with carbonyl compounds, can be viewed from two different perspectives, either to obtain products in

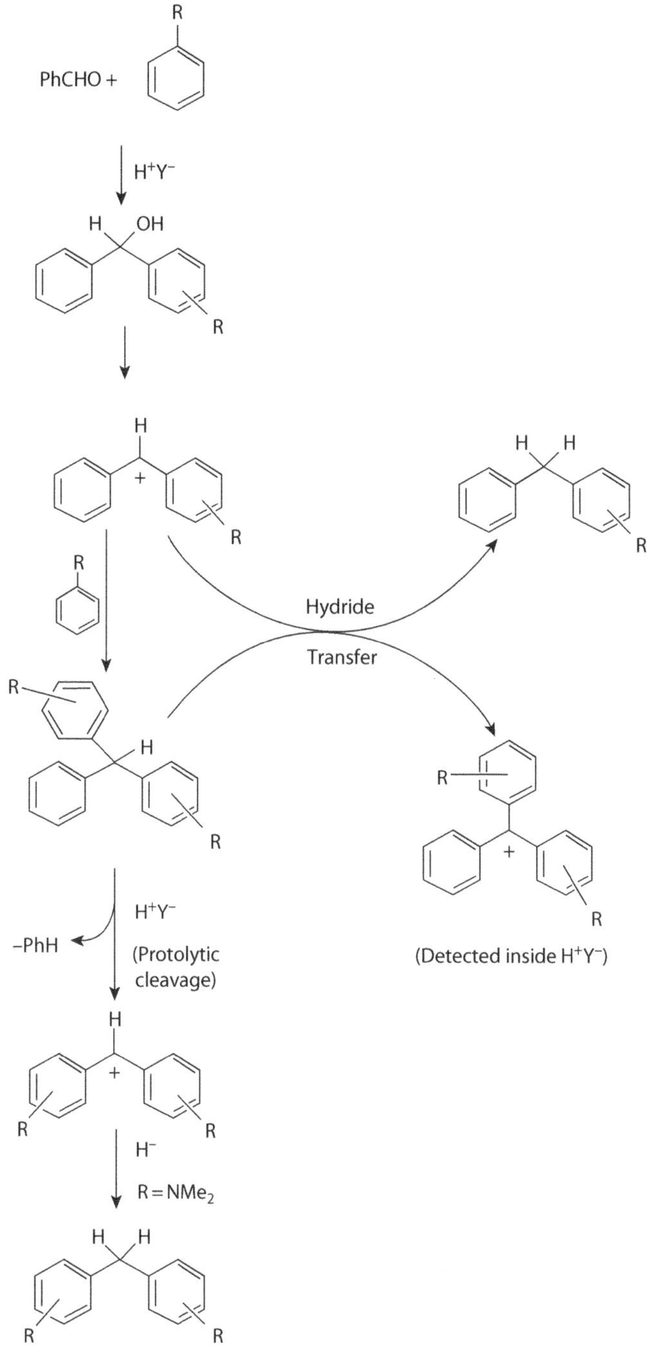

SCHEME 13.15
Generation of trityl cations through hydride transfer from triarylmethanes to diarylmethyl cation.

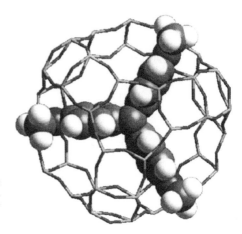

SCHEME 13.16
Visualization of the best Rietveld fit to the experimental HR-XRD of tris(4-methoxy-phenyl)methyl cation inside the zeolite Y supercage.

the liquid phase or to form a material containing an organic guest inside the cages.[74–76] In both points of view, the presence of protons in the zeolites plays a central role.

13.13 Conclusions and Future Trends

Acidity in zeolites is a field that has attracted the interest of heterogeneous catalysis since the middle of 1960s and has led to the development of vital industrial processes with a remarkable high efficiency. Due to the considerable amount of work that has been carried out in this area and the degree of understanding that has been achieved, it is unlikely that new breakthroughs could occur in this area. One of the past significant achievements has been the development of mesoporous aluminosilicates but in spite of the large porosity and surface area of these materials, their poor hydrothermal stability has precluded any application of this mesoporous aluminosilicates at the industrial level up to now. It is also very likely that these mesoporous aluminosilicates will not find any industrial use as catalysts.

However, particularly considering the more recent interest and development of green and sustainable chemistry, there is still a need of finding stable porous materials for fine chemistry and for biomass transformation. For this field, new stable zeolites with larger micropore size, as the one recently reported (Figure 13.10), are still necessary.

On the other hand, another type of zeolites that are missing is those that could be used as catalysts for enantioselective reactions. All the attempts based on the inclusion of chiral spectators inside zeolites have met with failure due to the low activity and enantiomeric excesses obtained.[77,78] In this regard, the report of the first enantiomerically pure zeolite[79] opens the door

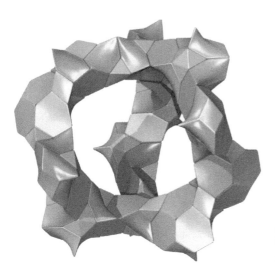

FIGURE 13.10
The large cavity defined by three 30-rings of ITQ-37.

to develop enantioselective acid catalysts using solid acids with high asymmetric induction.

In addition, other materials wherein systematically other heteroatoms are introduced in the framework to generate acid sites will be also developed and adapted to specific reactions and particularly to catalytic oxidations. One well-known example of this is the use of titanium-containing zeolites as Lewis acids to promote alkene epoxidation and benzene hydroxylation using H_2O_2 as oxidizing reagent.[80]

Therefore, new and exciting developments are expected to occur in the near future regarding one type of material that has already shown that it can be extremely efficient and selective in heterogeneous acid catalysis.

References

1. Kumar, R. Heterogeneous catalysts for highly selective and green industrial processes and chemical transformations. *Chemical Industry Digest* **2006**, *19*, 69–76.
2. Makhlin, V. A. Development and analysis of heterogeneous catalytic processes and reactors. *Theoretical Foundations of Chemical Engineering* **2009**, *43*, 245–259.
3. Corma, A., Martinez, A. Zeolites in refining and petrochemistry. *Studies in Surface Science and Catalysis* **2005**, *157*, 337–366.
4. Dantzenberg, F. M. New catalyst synthesis and multifunctional reactor concepts for emerging technologies in the process industry. *Catalysis Reviews: Science and Engineering* **2004**, *46*, 335–368.
5. Corma, A. From microporous to mesoporous molecular sieve materials and their use in catalysis. *Chemical Reviews (Washington, D.C.)* **1997**, *97*, 2373–2419.
6. Corma, A. Inorganic solid acids and their use in acid-catalyzed hydrocarbon reactions. *Chemical Reviews (Washington, D.C.)* **1995**, *95*, 559–614.

7. Corma, A. Zeolites as catalysts. Sheldon, R. A. and Van Bekkum, H., Eds., *Fine Chemicals Through Heterogeneous Catalysis*. 2001, pp. 80–91.
8. Anastas, P., Eghbali, N. Green chemistry: Principles and practice. *Chemical Society Reviews* **2009**, *39*, 301–312.
9. Anastas, P. The transformative innovations needed by green chemistry for sustainability. *Chemsuschem* **2009**, *2*, 391–392.
10. Anastas, P., Beach, E. S. Green chemistry: The emergence of a transformative framework. *Green Chemistry Letters and Reviews* **2007**, *1*, 9–24.
11. Anastas, P. Green chemistry design, innovation, solutions and a cohesive system. *Green Chemistry Letters and Reviews* **2007**, *1*, 3–4.
12. Sheldon, R. A. Catalysis: The key to waste minimization. *Journal of Chemical Technology & Biotechnology* **1997**, *68*, 381–388.
13. Sheldon, R. A. E-factors, green chemistry and catalysis: An odyssey. *Chemical Communications (Cambridge, England)* **2008**, *29*, 3352–3365.
14. Thomas, J. M. Uniform heterogeneous catalysts. The role of solid state chemistry in catalyst development. *Angewandte Chemie* **1988**, *100*, 1735–1753.
15. Breck, D. W. *Zeolite Molecular Sieves, Structure, Chemistry and Use*. John Wiley and Sons: New York, 1974.
16. van Bekkum, H., Flanigen, E. M., Jansen, J. C. *Introduction to Zeolite Science and Practice*. Elsevier: Amsterdam, the Netherlands, 1991.
17. Meier, W. M., Olson, D. H., Baerlocher, C. Atlas of zeolite structure types. *Zeolites* **1996**, *17*, 1–229.
18. Kuehl, G. H. ZSM-5 Zeolite, patent number EP 124364A219841107. In: *European Patent Application*, 1984.
19. Calvert, R. B., Rollmann, L. D. Synthesis of ZSM-5 and ZSM-11 utilizing a mixture of quaternary ammonium ions and amines, In: *European Patent Application*. EP 101183A1 19840222, 1984.
20. Beck, J. S., Vartuli, J. C., Roth, W. J., Leonowicz, M. E., Kresge, C. T., Schmitt, K. D., Chu, C. T. W. et al. A new family of mesoporous molecular sieves prepared with liquid crystal templates. *Journal of the American Chemical Society* **1992**, *114*, 10834–10843.
21. Huo, Q., Margolese, D. I., Stucky, G. D. Surfactant control of phases in the synthesis of mesoporous silica-based materials. *Chemistry of Materials* **1996**, *8*, 1147–1160.
22. Vartuli, J. C., Schmitt, K. D., Kresge, C. T., Roth, W. J., Leonowicz, M. E., McCullen, S. B., Hellring, S. D. et al. Effect of surfactant/silica molar ratios on the formation of mesoporous molecular sieves: Inorganic mimiary of surfactant liquid-crystal phases and mechanistic implications. *Chemistry of Materials* **1994**, *6*, 2317–2326.
23. Chen, C.-Y., Burkett, S. L., Li, H.-X., Davis, M. E. Studies on mesoporous materials II. Synthesis mechanism of MCM-41. *Microporous Materials* **1993**, *2*, 27–34.
24. Monnier, A., Schüth, F., Huo, Q., Kumar, D., Margolese, D., Maxwell, R. S., Stucky, G. D. et al. Cooperative formation of inorganic-organic interfaces in the synthesis of silicate mesostructures. *Science* **1993**, *261*, 1299–1303.
25. Zhao, D. Y., Feng, J. L., Huo, Q. S., Melosh, N., Fredrickson, G. H., Chmelka, B. F., Stucky, G. D. Triblock copolymer syntheses of mesoporous silica with periodic 50 to 300 angstrom pores. *Science* **1998**, *279*, 548–552.
26. Weibin, F., Wei, S., Yokoi, T., Inagaki, S., Li, J., Wang, J., Kondo, J. N., Tatsumi, T. Synthesis, characterization, and catalytic properties of H–Al–YNV-1 and H–Al–MWW with different Si/Al ratios. *Journal of Catalysis* **2009**, *266*, 268–278.

27. Xie, W., Gao, Z., Pan, W., Hunter, D., Singh, A., Vaia, R. Thermal degradation chemistry of alkyl quaternary ammonium montmorillonite. *Chemistry of Materials* **2001**, *13*, 2979–2990.
28. Ward, J. W. The nature of active sites on zeolites: I. The decationed Y zeolite. *Journal of Catalysis* **1967**, *9*, 225–236.
29. Soscún, H., Castellano, O., Hernandez, J., Hinchliffe, A. Acidity of the Brönsted acid sites of zeolites. *International Journal of Quantum Chemistry* **2001**, *82*, 143–150.
30. Farcasiu, D., Lukinskas, P. Similarities and differences in catalytic hydrocarbon conversions by liquid and solid acids. *Revue Roumaine de Chimie* **2000**, *44*, 1091–1099.
31. Planelles, J., Sanchez-Marin, J., Tomas, F., Corma, A. On the formation of methane and hydrogen during cracking of alkanes. *Journal of Molecular Catalysis* **1985**, *32*, 365–375.
32. Corma, A., Planelles, J., Sanchez-Marin, J., Tomas, F. The role of different types of acid site in the cracking of alkanes on zeolite catalysts. *Journal of Catalysis* **1985**, *93*, 30–37.
33. Patrylak, L. K. Chemisorption of Lewis bases on zeolites—A new interpretation of the results. *Adsorption Science & Technology* **1999**, *17*, 115–123.
34. Li, S., Zheng, A., Su, Y., Zhang, H., Chen, L., Yang, J., Ye, C. et al. Brönsted/Lewis acid synergy in dealuminated HY zeolite: A combined solid-state NMR and theoretical calculation study. *Journal of the American Chemical Society* **2007**, *129*, 11161–11171.
35. Fichtner-Schmittler, H., Lohse, U., Engelhardt, G., Patzelova, V. Unit cell constants of zeolites stabilized by dealumination. Determination of aluminium content from lattice parameters. *Crystal Research and Technology* **1984**, *19*, 1–3.
36. Lohse, U., Engelhardt, G., Alsdorf, E., Koelsch, P., Feist, M., Patzelova, V. Dependence of the adsorption capacity and thermal stability of Y zeolites upon silicon/aluminium ratio. *Adsorption Science & Technology* **1986**, *3*, 149–158.
37. Olah, G. A. Magic acid and superacid chemistry. *Magyar Kemikusok Lapja* **2002**, *57*, 6–9.
38. Miao, S., Liu, Z., Ma, H., Han, B., Du, J., Sun, Z., Miao, Z. Synthesis and characterization of mesoporous aluminosilicate molecular sieve from K-feldspar. *Microporous and Mesoporous Materials* **2005**, *83*, 277–282.
39. Zajac, J., Dutartre, R., Jones, D. J., Roziere, J. Determination of surface acidity of powdered porous materials based on ammonia chemisorption: Comparison of flow-microcalorimetry with batch volumetric method and temperature-programmed desorption. *Thermochimica Acta* **2001**, *379*, 123–130.
40. Meziani, M. J., Zajac, J., Jones, D. J., Partyka, S., Roziere, J., Auroux, A. Number and strength of surface acidic sites on porous aluminosilicates of the MCM-41 type inferred from a combined microcalorimetric and adsorption study. *Langmuir* **2000**, *16*, 2262–2268.
41. Caeiro, G., Lopes, J. M., Magnoux, P., Ayrault, P., Ramoa Ribeiro, F. A FT-IR study of deactivation phenomena during methylcyclohexane transformation on H-USY zeolites: Nitrogen poisoning, coke formation, and acidity-activity correlations. *Journal of Catalysis* **2007**, *249*, 234–243.
42. Pais de Silva, M. I., Lins da Silva, F., Tellez, S., Claudio, A. FT-infrared band analysis and temperature programmed desorption for the Y, L and ferrierite zeolites. *Spectrochimica Acta, Part A: Molecular and Biomolecular Spectroscopy* **2002**, *58A*, 3159–3166.

43. Gorte, R. J., White, D. Measuring sorption effects at zeolite acid sites: Pursuing ideas from W.O. Haag. *Microporous and Mesoporous Materials* **2000**, *35–36*, 447–455.

44. Derouane, E. G., Chang, C. D. Confinements effects in the adsorption of simple bases by zeolites. *Microporous and Mesoporous Materials* **2000**, *35–36*, 425–433.

45. Cejka, J., Perez-Pariente, J., Roth, W. J. *Zeolites: From model Materials to Industrial Catalysts*. Transworld Research Network, Trivandrum, India. 2008.

46. Jansen, J. C., Stoecker, M., Karge, H. G., Weitkamp, J. Advanced zeolite science and applications. *Studies in Surface Science and Catalysis* **1994**, *85*.

47. Corma, A., Martinez, A. Zeolites and zeotypes as catalysts. *Advanced Materials* **1995**, *7*, 137–144.

48. Corma, A., García, H. Organic reactions catalyzed over solid acids. *Catalysis Today* **1997**, *38*, 257–308.

49. Corma, A., García, H., Primo, A., Domenech, A. A test reaction to assess the presence of Brönsted and the softness/hardness of Lewis acid sites in palladium supported catalysts. *New Journal of Chemistry* **2004**, *28*, 361–365.

50. Algarra, F., Corma, A., Fornes, V., García, H., Martinez, A., Primo, J. Rearrangement of acetals of 2-bromopropiophenone as a test reaction to characterize the Lewis sites in large pore zeolites. *Studies in Surface Science and Catalysis* **1993**, *78*, 653–660.

51. Baldovi, M. V., Corma, A., Fornes, V., García, H., Martinez, A., Primo, J. Soft and hard acidity in ion-exchanged Y zeolites: Rearrangement of 2-bromopropiophenone ethylene acetal to 2-hydroxyethyl 2-phenylpropanoate. *Journal of Chemical Society, Chemical Communications* **1992**, *13*, 949–951.

52. Corma, A. Covalent interactions in zeolites: The influence of zeolite composition and structure on acid softness and hardness. *Studies in Surface Science and Catalysis* **1995**, *94*, 736–747.

53. Pearson, R. G. Hard and soft acids and bases, *Journal of the American Chemical Society* **1963**, *85*, 3533–3543.

54. Pearson, R. G. Acids and bases, *Science* **1966**, *151*, 172–177.

55. Corma, A., Sastre, G., Viruela, R., Zicovich-Wilson, C. Molecular orbital calculation of the soft-hard acidity of zeolites and its catalytic implications. *Journal of Catalysis* **1992**, *136*, 521–530.

56. Corma, A., Garcia, H., Leyva, A. Controlling the softness-hardness of Pd by strong metal-zeolite interaction: Cyclisation of diallylmalonate as a test reaction. *Journal of Catalysis* **2004**, *225*, 350–358.

57. Corma, A., García, H., Primo, J., Sastre, E. Beckmann rearrangement of cyclohexanone oxime on zeolites. *Zeolites* **1991**, *11*, 593–597.

58. Fois, G. A., Ricchiardi, G., Bordiga, S., Busco, C., Dalloro, L., Spano, G., Zecchina, A. The Beckmann rearrangement catalyzed by silicalite: A spectroscopic and computational study. *Studies in Surface Science and Catalysis* **2001**, *135*, 2477–2484.

59. Bordiga, S., Roggero, I., Ugliengo, P., Zecchina, A., Bolis, V., Artioli, G., Buzzoni, R. et al. Characterization of defective silicalites. *Dalton* **2000**, *21*, 3921–3929.

60. Conesa, T. D., Luque, R., Campelo, J. M., Luna, D., Marinas, J. M., Romero, A. A. Gas-phase Beckmann rearrangement of cyclododecanone oxime on Al,B-MCM-41 mesoporous materials. *Journal of Materials Science* **2009**, *44*, 6741–6746.

61. Forni, L., Tosi, C., Fornasari, G., Trifiro, F., Vaccari, A., Nagy, J. B. Vapour-phase Beckmann rearrangement of cyclohexanone-oxime over Al-MCM-41 type meso-structured catalysts. *Journal of Molecular Catalysis A: Chemical* **2004**, *221*, 97–103.

62. Ko, A. N., Hung, C. C., Chen, C. W., Ouyang, K. H. Mesoporous molecular sieve Al-MCM-41 as a novel catalyst for vapor-phase Beckmann rearrangement of cyclohexanone oxime. *Catalysis Letters* **2001**, *71*, 219–224.

63. Camblor, M. A., Corma, A., García, H., Semmer-Herledan, V., Valencia, S. Active sites for the liquid-phase Beckmann rearrangement of cyclohexanone, aceto-phenone and cyclododecanone oximes, catalyzed by beta zeolites. *Journal of Catalysis* **1998**, *177*, 267–272.

64. Chenier, P. J. *Survey of Industrial Chemistry.* VCH: New York, 1992.

65. Climent, M. J., Corma, A., García, H., Primo, J. Zeolites in organic reactions: Condensation of formaldehyde with benzene in the presence of HY zeolite. *Applied Catalysis* **1989**, *51*, 113.

66. Climent, M. J., Corma, A., García, H., Iborra, S., Primo, J. Mono- and tridirec-tional 12-membered ring zeolites as acid catalysts for carbonyl group reactions. *Studies in Surface Science and Catalysis* **1991**, *59*, 557.

67. Climent, M. J., Corma, A., García, H., Primo, J. Zeolites as catalysts in organic reactions: Condensation of aldehydes with benzene derivatives. *Journal of Catalysis* **1991**, *130*, 138.

68. Climent, M. J., Corma, A., García, H., Iborra, S., Primo, J. Acid zeolites as cata-lysts in organic reactions: Condensation of acetophenone with benzene deriva-tives. *Applied Catalysis* **1995**, *130*, 5.

69. March, J. *Advanced Organic Chemistry: Reactions, Mechanisms and Structures.* McGraw Hill: New York, 1993.

70. Algarra, F., Corma, A., García, H., Primo, J. Acid zeolites as catalysts in organic reactions. Highly selective condensation of 2-alkylfurans with carbonyl com-pounds. *Applied Catalysis, A: General* **1995**, *128*, 119.

71. Corma, A., García, H., Primo, J. *Journal of Chemical Research (S)* **1988**, 40.

72. Corma, A., García, H. A unified approach to zeolites as acid catalysts and as supramolecular hosts exemplified. *Journal of the Chemical Society Dalton Transactions* **2000**, 1381–1394.

73. Corma, A., García, H. Supramolecular host-guest systems in zeolites prepared by ship-in-a-bottle synthesis. *European Journal of Inorganic Chemistry* **2004**, *6*, 1143–1164.

74. Cano, M. L., Corma, A., Fornés, V., García, H., Miranda, M., Baerlocher, C., Lengauer, C. Triarylmethylium cations encapsulated within zeolite supercages. *Journal of the American Chemical Society* **1996**, *118*, 11006.

75. Scaiano, J. C., García, H. Intrazeolite photochemistry: Toward superamolecu-lar control of molecular photochemistry. *Accounts of Chemical Research* **1999**, *32*, 783.

76. Cano, M. L., Cozens, F. L., García, H., Martí, V., Scaiano, J. C. Intrazeolite Photochemistry. 13. Photophysical Properties of Bulky 2,4,6-triphenylpyry-lium and tritylium cations within large- and extra-large-pore zeolites. *Journal of Physical Chemistry* **1996**, *100*, 18152.

77. Davis, M. E. Ordered porous materials for emerging applications. Influence of the acid strength distribution of the zeolite catalyst on the tert-butylation of phenol. *Nature* **2002**, *417*, 813–821.

78. Corma, A. State of the art and future challenges of zeolites as catalysts. *Journal of Catalysis* **2003**, *216*, 298–312.

79. Sun, J., Bonneau, C., Cantin, A., Corma, A., Diaz-Cabanas, M., Moliner, M., Zhang, D. et al. The ITQ-37 mesoporous chiral zeolite. *Nature (London, United Kingdom)* **2009**, *458*, 1154–1157.

80. Corma, A., Diaz, U., Domine, M. E., Fornes, V. Ti-ferrierite and TiITQ-6: Synthesis and catalytic activity for the epoxidation of olefins with H_2O_2. *Chemical Communications (Cambridge, England)* **2000**, *2*, 137–138.

Part IV

Case Studies of Applications of Advanced Techniques in Involving Nanoporous Materials

14

Recent Developments in Gas-to-Liquid Conversion and Opportunities for Advanced Nanoporous Materials

Gabriele Centi* and Siglinda Perathoner

CONTENTS

14.1 Introduction

14.1.1 Backgrounds on Gas-to-Liquid Processes

Fast development of the world economy and increase of the international oil price have made the global energy and environmental problems increasingly serious. Gas-to-liquid (GTL) processes (Fischer–Tropsch [FT], methanol, and dimethyl ether [DME] syntheses) have become increasingly important and received much attention. Besides providing clean fuel, the

---

* Coordinator of the EU Network of Excellence IDECAT and of EU Large Collaborative Project NEXT-GTL.

products of GTL processes can be further processed to many other valuable chemical products.

The GTL process produces liquid fuels of excellent technical and environmental qualities, which often supersede those produced by conventional oil refinery. Key GTL products are diesel, liquefied petroleum gas (LPG), lube base stocks and waxes, and petrochemicals (naphtha and stream cracking). These products are typically of high quality with near-zero sulfur and high cetane for the diesel. Many other factors that have stimulated the increasing interest in alternative routes to produce liquid fuels for transport include [1–3]

- Reduction in cost of transport of natural gas (NG)
  - Monetization of stranded NG
  - Economic utilization of associated gas
- High current and projected demand for liquid transportation fuels
  - Higher costs tied in with crude markets and refining capacity issues
  - Need of clean fuels (sulfur free)
- Flaring reduction and environmental concerns

GTL fuels can be produced from NG, coal (coal-to-liquid [CTL]), and even biomass (biomass-to-liquid [BTL]). Although, strictly speaking, GTL refers only to NG, the concept is often generalized. It is common to include GTL processes in the production of liquid fuels from coal or biomass, because in all cases, there is a common key step in the formation of *synthesis gas* (CO/H_2 mixtures, whose relative ratio is depending on the raw material), and the processes after this step are similar. We will focus discussion here mainly on the conversion of NG to liquid fuels for the transport sector. Excellent recent reviews on CTL and BTL can be found in Refs. [4–10]. Economic aspects of these reactions, particularly BTL, were instead discussed in Refs. [11,12]. The acronym XTL instead indicates the conversion of all carbon sources (NG, coal, biomass, oil-share) to synthetic liquid fuels. GTL is often used as synonymous of the production of synthetic fuels via FT conversion, although more correctly it includes NG conversion to all liquid synthetic fuels. The entire process requires at least three stages. The first is the conversion of feedstock to synthesis gas (often indicated also as *syngas*). A second step is the FT synthesis, which converts the syngas to waxy hydrocarbons. The third step is product upgrading, which converts the waxy hydrocarbons into final products—naphtha and sulfur-free diesel. The resulting diesel can be used neat or blended with today's diesel fuel in existing diesel engines. The entire process is even more complex, because it includes also the preparation of the feed for syngas step, as summarized in Figure 14.1.

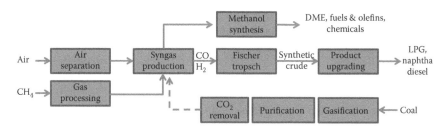

FIGURE 14.1
Main steps in the GTL process plant through indirect route.

Although GTL fuels can be produced by different feedstock, the use of coal or biomass requires more steps and has a higher impact on environment than starting from NG. For this reason, several companies have focused attention to NG to liquid, while starting from coal is mainly a primary interest of countries such as China that largely bases its energy politics on coal. Starting from biomass will be a possibility for future biorefineries based on a thermal route, but still, several concerns regarding the techno-economic feasibility exist.

The GTL technology is essentially established, although there are still improvements necessary. A large commercial scale GTL plant is coming into operation in Qatar, but a number of unresolved problems have still prevented full capacity operation. Some companies initially involved, such as Exxon, have stepped down, and others such as BP believe in opportunities to take conversion technologies even further to deliver a wider "gas-to-products" portfolio. Critical issues are the large size of reactors and investments necessary for FT and the cost of syngas and the related air separation step. GTL is a very capital-intensive process [13].

From syngas, it is possible to produce methanol that can be used in part as gasoline blend (with opportune additives) or transformed alternatively to DME to be used also as fuel component. The main advantage of methanol over FT products is that it is much easier converted to chemicals, either by transformation to olefins using small-pore zeolites (methanol-to-olefin [MTO] process), to other chemicals or to gasoline (methanol-to-gasoline [MTG] process). The route through methanol offers thus better possibilities for an integrated fuel and chemical production, an important concept for the future of refineries as well as of biorefineries [14].

In conclusion, interest of producing methanol from NG, particularly in remote areas (of the proven NG reserves, a large part cannot be exploited actually—see Section 14.1.3), is rising, for its potential to be a flexible intermediate for fuel and chemicals, thereby minimizing risks in a fluctuating market. Direct methane to methanol (MTM) is the grand challenge. However, large progress has been made recently as reviewed in various papers [15–17]. The number of patents is also rising, showing that several companies are looking seriously into this route.

There are other interesting possible routes for direct conversion of methane, in particular, the direct conversion of methane/NG to aromatics [18,19]. Under nonoxidative conditions, it is possible to form hydrogen as co-product, and this is a first element of interest. Second, high value-added products such as aromatics are generated, answering the increasing aromatics demand in the chemical industry. Third, the produced aromatics can be further converted to alkylaromatics [20] for the chemical industry and the gasoline pool by reaction with other NG components (ethane, propane) using catalysts in the presence of a H_2-permselective membrane.

14.1.2 Drivers and Issues

Still, many of the original factors that created large interest in GTL and associated processes during the past decade remain valid:

- There is still an increase in global energy demand. Energy demand is roughly associated one-third to transport, one-third to buildings, and one-third to industry. Energy saving based on new nanotech materials for more efficient lighting, better isolation, more energy-efficient processes, etc. can still significantly reduce energy consumption in the latter two areas, while the possibility to reduce the need of liquid fuels for the transport section will be much more limited, considering also the fast raising access to cars of population in many emerging economy countries. Therefore, the share of liquid transport fuels in the energy pool will increase in the future. GTL-based processes are considered a viable contribution to solve this issue, by making possible to convert NG into liquid fuels more suited for the transport sector, and having lower impact on the environment (sulfur-free, reduced particulate and emissions) with respect to those deriving from oil [21].

- Large volumes of stranded gas in several global areas having little access to the market (see also Section 14.1.3). About one-third of total NG reserves are located in remote areas from which transport by pipeline is impossible or not economic. New gas pipelines are under construction or programmed, but the investment costs are very high and there are sometimes additional geopolitical problems. Often, the on-site use of NG to produce electrical energy and then transport of electrical energy is impossible for the same motivations. Due to the density about three order of magnitude higher in a liquid with respect to a gas, transport of NG from these remote areas requires, thus, to convert it to a liquid (liquefied natural gas [LNG]). Methane liquefaction requires very low temperatures. An initial pretreatment is necessary to remove dust, acid gases, helium, water, heavy hydrocarbons (C_2–C_4 alkanes, indicated often as LPG and C_{5+} alkanes, which are liquid at room temperature) and sometimes also mercury.

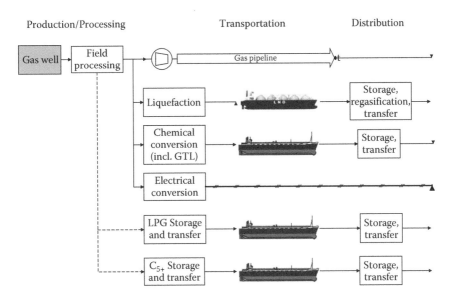

FIGURE 14.2
Natural gas transport mechanisms. (Adapted from Rahmim, I.I., Gas-to-liquid technologies: Recent advances, economics, prospects. *Presented at 26th IAEE Annual International Conference*, Prague, Czech Republic, June 2003.)

Methane (and minimal amounts C_2 and C_3 alkanes) is condensed into a liquid at close to atmospheric pressure (maximum transport pressure set around 25 kPa/3.6 psi) by cooling it to approximately −162°C. LNG accounted for 7%–8% of the world's NG demand, and will grow substantially in the future. On arrival, regasification is necessary. In addition to GTL and LNG, other options such as gas-to-chemicals facilities are either operational or in various stages of progress. Figure 14.2 summarizes these different options.

- Tightening quality specifications such as those for sulfur in fuel products, particularly diesel and gasoline, in many countries. These requirements are being met largely through improved hydrocracking, hydrotreating, and other refining processes, although FT-based synthetic products are excellent blend stocks due to their low sulfur and high cetane levels.

- The need to reduce gas flaring for economic, environmental, and legal reasons in many (e.g., Nigeria) countries. Because many of Nigeria's oil fields lack the infrastructure to produce and market associated NG, it is often flared. According to the National Oceanic and Atmospheric Administration (NOAA), Nigeria flared 593 billion cubic feet (Bcf) of NG in 2007, which, according to Nigerian National Petroleum Corporation (NNPC), cost the country US $1.46 billion in lost revenue. The NOAA estimation for gas flaring in Russia, Iran,

Iraq, Kazakhstan, Algeria, Libya, Angola, and Saudi Arabia in 2007 was 1763, 374, 246, 186, 185, 130, 125, and 119 Bcf, respectively. The total volume of gas flaring is thus very high.

- The increase in the market value of crude oil. An oil price that has crossed the psychological \$100/bbl mark multiple times. Although stabilization in the future for \$70–\$80/bbl (barrel, e.g., about 159 L) is expected, which is at the limit of economics for synthetic fuels, the uncertainty in the oil price has fostered R&D in this field.

- Worldwide resource security concerns, given the current global political climate. Only about 40% of the proved reserves of gas are in the Middle East and the United States has the largest reserves in the world of coal.

- Concern about the possibility of global climate change and the potential impact of CO_2 release. XTL technologies are often promoted as a contribution to solve this issue, but this is a controversial question. Jaramillo et al. [22] have used life cycle analysis (well-to-wheel) to determine the greenhouse gas (GHG) emissions of coal- and NG-based FT liquids, as well as to compare production costs. The results show that the use of coal- or NG-based FT liquids will likely lead to significant increases in GHG emissions compared to petroleum-based fuels. Figure 14.3 shows the high emissions scenario well-to-wheel GHG emission factors for gasoline. The error bars presented in these figures represent the uncertainty/variability in the upstream emission factors of coal, NG, and electricity reported

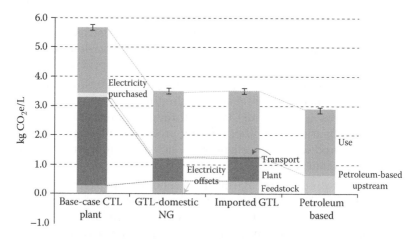

FIGURE 14.3

Comparison of well-to-wheel GHG emissions for gasoline produced from coal and NG (high-emissions scenario, e.g., based on the current U.S. fuel mix for electricity generation [50% coal, 20% NG, and 30% low-carbon sources] and not considering carbon sequestration for the FT plants). (Adapted from Jaramillo, P. et al., *Environ. Sci. Technol.*, 42 (20), 7559, 2008.)

by Jaramillo et al. [22], as well as variability in the emissions from liquid fuel transport. This figure shows the emissions for petroleum gasoline as a comparison to the GTL and CTL gasoline. A similar value was found for diesel. Gasoline and diesel produced from coal could emit about double the GHG emissions of petroleum-based gasoline and diesel. If domestic NG were used to produce gasoline, or if NG-based gasoline were imported from Qatar or Malaysia, an increase in emissions of 20%–25% would be seen. If LNG is used, an increase of around 50% in emission factors for both gasoline and diesel could be observed. Even in a best-case scenario, coal- or NG-based FT liquids have emissions only comparable to petroleum-based fuels.

In addition, the economic advantages of GTL fuels are not obvious [22]: there is a narrow range of petroleum and NG prices at which GTL fuels would be competitive with petroleum-based fuels. CTL fuels are generally cheaper than petroleum-based fuels. However, considering the uncertainty about the availability of economically viable coal resources, e.g., in the United States, the goal of increasing energy security, and at the same time, significantly reducing GHG emissions by promoting CTL and GTL technologies, appears questionable. GTL is thus preferable over CTL, while BTL appears still a longer-term option, due to a number of technological issues.

This discussion on the drivers and issues in the production of synthetic fuels by XTL technologies evidences, thus, that there are opportunities, but at the same time, issues and controversial aspects. In other words, this is an area in which it is still necessary to determine the preferable option, and the same considerations may be not valid for all countries.

14.1.3 Stranded Gas

World NG consumption has grown steadily since the 1970s, and today accounts for about a quarter share of the world energy demand. The International Energy Agency is expecting NG to become the second energy source, ahead of coal, due to its use in power generation in developed countries (particularly Europe). The estimated amount of NG reserves has been continuously increasing, moving from the 1995 estimation of 148 Tscm (trillion standard cubic meter) to over 182 Tscm in 2007. The total amount of stranded gas reserves has been estimated to be 66.5 Tscm, 36% of gas proven reserves (average of published data). Such an amount could, if converted to synthetic fuels, generate around 250 billion barrels of synthetic oil, e.g., a quantity equal to one-third of Middle East's proven oil reserves. For example, Alaska has large reserves of NG stranded in its Prudhoe Bay oil field, Canada in its Arctic Islands, Beaufort Sea, and Mackenzie Delta, Russia in Siberia and across the Bering Strait, Norway in the Nord Pole see area (Midgard, Smørbukk, and Smørbukk Sør). In all these cases, the exploitation of this NG is difficult.

All these factors play a role in the new strategic relevance of gas-to-market technologies to transport gas from the production area to the final market. In the last 20 years, thanks to technological evolution, the market for the NG has changed its nature from regional to international, e.g., there is an increasing globalization of the gas and derivatives markets, similarly to the oil market.

Different technologies are currently available to bring gas on the market covering long distance, as discussed in the previous section: high pressure/capacity pipelines, liquefaction and regasification of natural gas (LNG), electric power generation (gas-to-wire), and finally NG conversion into liquid hydrocarbons (GTL). Natural gas hydrate (NGH), e.g., the formation of a solid clathrate with water under pressure, is also an interesting option on which large attention has been given recently [23,24]. NGH can be considered modified ice structures enclosing methane and other hydrocarbons, but they can melt at temperatures well above normal ice. At 30 atmospheres pressure, methane hydrate begins to be stable at temperatures above 0°C and at 100 atmospheres, it is stable at 15°C. It is thus an alternative to transport methane in liquid form, although transporting NGH means to move large amounts of water. NGH represents a vast potential, though not presently commercial, source of NG, additional to proven NG (conventional) reserves.

The capacity of the reserve and the distance from use are the two main factors that roughly determine the choice of the preferred technology to transport NG. Figure 14.4 shows schematically the main ranges of these two parameters in which the different options for transporting NG are preferable. In between these regions, there is also an overlapping zone where it is difficult to indicate a preferable technology.

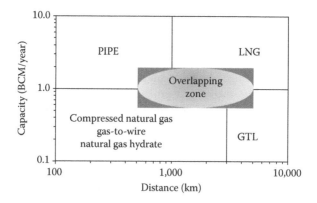

FIGURE 14.4
Capacity–distance diagram in the preferable technologies for transporting NG. (Adapted from Gudmundsson, J.S., Natural gas hydrate. *Presented at Gas Hydrates (Tekna) Conference*, Bergen (Norway), October 2008.)

14.2 Gas-to-Liquid Technologies

GTL is the process of NG conversion to transportable liquids, characterized by the intermediate step of synthesis gas production. Today, there are two main GTL technologies: the production of oxygenate liquid compounds (methanol and DME), and the FT synthesis (FTS) for the production of high-quality middle distillates (i.e., jet, kero, and diesel fuel), base-oil, or waxes. In the last decade, considerable push has been directly toward the GTL route of synthetic fuels via FT process. However, there is a renewed interest to produce methanol, for the possibility of a better integration between fuel and chemical production, an increasingly relevant factor in a market with fast-rising costs of oil. New catalysts have been developed to produce directly ethanol instead of methanol.

Fan et al. [25,26] discussed catalyst development for syngas conversion to C_2 oxygenates over Rh-based catalysts, and evidenced the effects of various additives and supports on the activity and selectivity for this reaction. The study was focused in particular on the role of nanocarbon-based materials. Carbon nanotubes (CNTs)-supported catalyst showed a high overall activity and yield of C_2 oxygenates (ethanol, acetic acid, and acetaldehyde) compared to the other carbon-supported catalysts. Rh–Mn bimetallic nanoparticles (1–2 nm) introduced into the CNT channels show a remarkable enhancement of the catalytic activity as compared to the catalyst with RhMn dispersed on the CNT exterior surfaces (Figure 14.5).

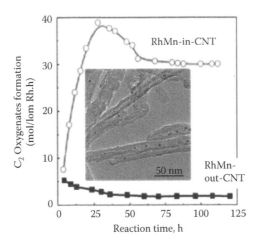

FIGURE 14.5
Activity of RhMn-in-CNT and RhMn-out-CNT in syngas conversion at 320°C and 30 bar. In the inset, TEM image of the CNT-confined bimetallic RhMn nanoparticles. (Adapted from Pan, X. et al., *Nat. Mater.*, 6, 507, 2007.)

RhMn in the CNT interior likely exists in a more reduced state than that on the exterior. When CO is adsorbed on Rh inside CNTs, the adjacent oxophilic Mn tends to attract the O of the adsorbed CO leading to tilted adsorption forming the Rh–C and Mn–O bonds. Such a tilted adsorption of CO facilitates the dissociation of CO, which improves the activity. In contrast, the electronic structure of exterior CNT surface approaches a planar graphite layer since the outer diameter is around 10–20 nm. Thus, the tendency of Mn to accept oxidic CO donor electrons could be reduced in comparison to that inside the electron-deficient interior of CNTs. Consequently, the CNT-confined catalyst could exhibit a higher activity in CO conversion than the catalyst dispersed on the CNT exterior surfaces.

The Bao's group [27] has made a more systematic analysis of the use of the concept of confinement of metal nanoparticles inside CNT to develop advanced catalysts. The localization of these nanoparticles inside the CNT channels does not only exert spatial restriction limiting the aggregation of metal particles, but also modifies the redox properties of metal and metal oxide particles. For example, the reduction of iron oxide particles is easier inside CNT channels with respect to those localized on the outer surface of CNTs, and the reduction temperature decreases within smaller nanotubes [28]. The improved reducibility was found to favor the formation of iron carbides under FTS conditions. Such CNT-inner Fe catalyst exhibited a higher activity in FTS in comparison to the iron particles dispersed on the exterior CNT surfaces [29]. In general, there are many interesting opportunities offered from the use of these nanocarbons to improve the performances of the catalysts in these novel sustainable energy processes [30].

Selectivity to ethanol obtained in the direct synthesis from syngas, however, is not optimal. An alternative route is via DME, which can be easily made from NG, coal, or biomass via syngas. Bifunctional catalysts based on conventional methanol synthesis catalysts (Cu–ZnO) and a zeolite material could be used [31]. Carbonylation provides then a convenient route to functionalize DME to ethanol. DME and CO can be converted to methyl acetate by halide-free carbonylation over zeolite, as shown originally from the Iglesia group [32]. Then, the formed methyl acetate can be hydrogenated to methanol and ethanol. DME carbonylation and methyl acetate hydrogenation steps could be combined in one single step. The formed methanol can be recycled as reactant of DME synthesis. Therefore, DME and syngas can form ethanol with water as (formal) unique byproduct. Figure 14.6 shows the conceptual scheme for a novel ethanol synthesis method via DME and syngas by combination of carbonylation and hydrogenation.

Different routes to convert methane into useful liquid hydrocarbons (fuels, for the transport sector) include (1) indirect methane conversion and (2) direct methane conversion. Figure 14.7 schematically shows the different options, and also the different possibilities to convert NG (methane) to fuels, hydrogen, or chemicals. At present, demonstrated technologies are based on indirect route that consists of multistep reaction processes.

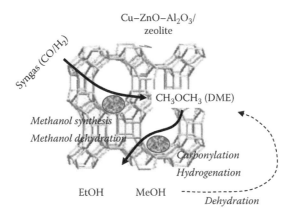

FIGURE 14.6
Conceptual scheme for a novel ethanol synthesis method via DME and syngas by combination of carbonylation and hydrogenation.

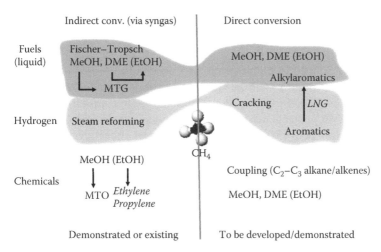

FIGURE 14.7
Scheme of the possible routes for converting methane to fuel (components for gasoline or diesel pool), and produce H_2 or chemicals. *Note:* MTG (methanol-to-gaoline), MTO (methanol-to-olefins), MeOH (methanol), EtOH (ethanol), DME (dimethyl ether), LNG (liquefied natural gas—C_2–C_4 alkanes).

These processes include five key steps: (a) air separation, (b) gas processing, (c) syngas production, (d) conversion of syngas to a syncrude (FTS) or oxygenate, and (e) upgrading syncrude by hydroprocessing to marketable products.

Currently, the only industrial feasible option for GTL passes through the first step of conversion of NG to syngas followed by conversion of syngas to either hydrocarbon via FT synthesis or oxygenated products (CH_3OH/DME, or higher alcohols) via well-established technologies. While FT projects are

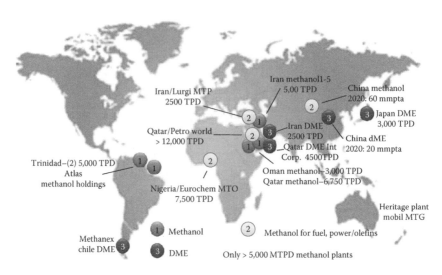

FIGURE 14.8
Methanol/DME projects: transition from chemicals to fuels.

concentrated mainly in Qatar (SASOL "Oryx" and Shell "Pearl," while others in the same region such as the ExxonMobil, ConocoPhillips, etc., were stopped or postponed for the risks and too large investment), there are many methanol/DME projects, which clearly evidence a transition from chemicals to fuels (Figure 14.8) [33]. Note that in terms of thermal/carbon efficiency, GTL today is about 60%–77%, while methanol/DME is about 70%–83% [33]. In terms of economic competitiveness of methanol and DME in fuel markets, with respect to FT products, they are comparable.

A typical FT plant flow scheme is shown in Figure 14.9a, while Figure 14.9b shows the cost breakdown for the main components [34]. The largest part of the cost (36%–48%) is associated to air fractioning/syngas production. To reduce this cost, very large plants are necessary for FT production, but this is a critical aspect, because (a) large investments and risks are necessary and (b) there are only few gas fields for which these large investments could be justified.

As a consequence, the valorization of small/medium size gas fields for which pipelines and LNG are not convenient (Figure 14.4), requires the development of (a) alternative syngas technologies suited for this size, and (b) direct routes for NG conversion. The industrial challenges are thus the following:

- To introduce a novel, less costly, and more energetically efficient scheme for syngas, based, e.g., on membranes (as discussed later) to develop GTL processes which are cost effective on small-medium size

- To develop new direct routes for GTL, which avoid the costly syngas step

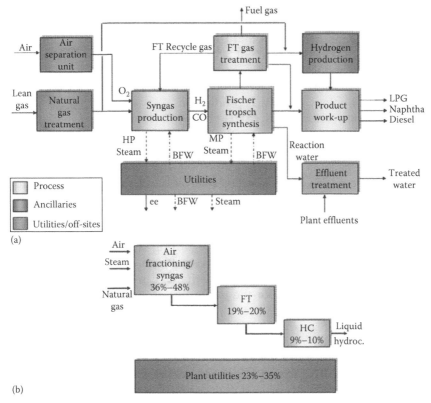

FIGURE 14.9
(a) FT plant flow scheme and (b) costs breakdown for FT process. (Adapted from Zennaro, R. et al., The Eni-IFP/Axens GTL technology: From R&D to a successful scale-up. *Presented at "Syngas Chemistry Symposium"* (*DGMK-SCI Conference*), Dresden, Germany, October 2006.)

- To synthesize liquid products that can be used either in fuel or chemical production, and that realize a better balance between diesel versus gasoline pools

14.2.1 Challenges in GTL Processes

The syngas production is a critical element either to go to FT products or to methanol. In both cases, it is the more costly stage of the production chain and more energy intensive. Therefore, this is the stage that requires further development, even if a well-established process. Recent developments in membrane technology to remove H_2 and/or CO_2, and in O_2 separation from air open novel perspectives to reduce costs in syngas production [35].

The largest part of the capital cost in GTL technologies is associated with the syngas plant (Figure 14.9b). A major breakthrough is required to substantially reduce the cost. The Argonne National Lab., in cooperation with Amoco, has pioneered the use of membrane technology in the production of syngas [36], and has shown that the membrane process could lower the cost of syngas production by about 30%. Recently, an alliance of Amoco, BP, Praxair, Statoil, and SASOL announced to develop this technology. The U.S. Department of Energy (DoE) announced an $84 million project to develop membrane technology for syngas production.

The direct conversion of CH_4 to CH_3OH/DME [15–17] is also another challenge and opens new routes to synthesize easily transportable liquid products suitable both for the fuel and the chemical market. Although, for a long time, this has been an objective of research, the increasing knowledge on combining homo- and heterogeneous catalysts (and, in general, on nanoporous materials) and on the fundamental understanding of the mechanisms of methane activation allows now to put the research in this area on new bases/prospects. DME is also a very attractive synthetic transportation fuel, because of its properties close to LPG regarding storage (with advantages in terms of explosivity), very low toxicity, and high cetane number (around 55–60). In the presence of acidic sites, methanol can be dehydrated to DME and it is thus possible to tailor the properties of a catalyst active in MTM conversion to form DME or methanol/DME mixtures.

Another challenge is the nonoxidative GTL conversion, i.e., methane to aromatics [18,19]. Many patents have been issued recently on this reaction for the valorization of stranded NG resources. For example, in the period 2006–2008, ExxonMobil issued 11 patents extended worldwide. The advantage of nonoxidative conversion is the possibility to avoid separation of NG components and the co-production of H_2. However, due to the toxicity of benzene and the concomitant ever more stringent regulation of its content in gasoline, its further alkylation is necessary. As an important side effect, this step considerably increases the octane number and lowers vapor pressure. Alkylation of benzene-rich reformate with olefins is a common practice in refinery, where the olefins are readily available. However, for applications in remote areas close to NG wells, i.e., where the olefins are not available, it would be necessary to develop a completely innovative technology using the alkanes (C_2–C_3) as feedstock (as alkylating agents), because they are recovered from the NG stream [37–39]. The aromatization of methane and the alkylation of benzene to alkyl benzenes are fully complementary steps. NG associated with crude oil always contains other light alkanes. The alkylation of benzene with ethane and propane ensures that the NG is used in total. It would neither be economic nor environmentally friendly to use just methane and to flare the rest.

14.2.2 New Technologies for Syngas Production

The most used process for the production of syngas is methane steam reforming (SR) (Figure 14.10). NG is converted by the reaction with steam over a

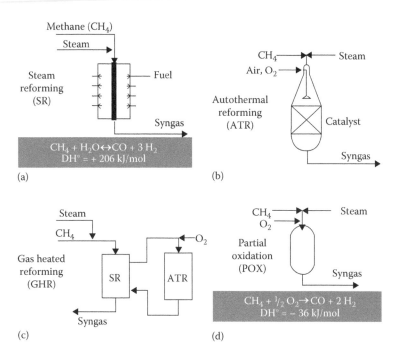

FIGURE 14.10
Syngas technologies to convert NG. (Adapted from Zennaro, R. et al., The Eni-IFP/Axens GTL technology: From R&D to a successful scale-up. *Presented at "Syngas Chemistry Symposium" (DGMK-SCI Conference)*, Dresden, Germany, October 2006.)

supported nickel catalyst at high temperatures (900°C–950°C) and moderate pressures (16–20 bar). The reactor is typically based on long, vertically hung tubes contained in a radiant furnace. Many engineering companies offer some variants of this type of design, e.g., Foster Wheeler, Kellogg, Lurgi, and Haldor Topsoe.

Before being sent to SR, the feed is preheated and desulfurized; vapor is then added and the mixture is further preheated. The reforming takes place in an oven in which tubes are located filled with catalyst, through which the reaction mixture flows. The synthesis gas at the outlet of the tubes is rapidly cooled and can be sent to water–gas shift processes and separation/purification. SR requires considerable quantities of vapor to reduce the formation of carbonaceous residues. Vapor is also necessary for increasing the H_2 content in the syngas produced, which would otherwise prevalently contain CO, in case a higher H_2/CO ratio is required.

An established alternative to SR is the partial oxidation [40], where NG is combusted in oxygen-controlling conditions. This type of technology is offered under license by, e.g., Texaco and Shell. An advance is the use of a catalyst to improve the selectivity and operate at lower temperatures, [40–43], although stability of the catalyst is still an issue, and there are not still commercial processes. Catalytic partial oxidation (CPO) with a short contact time can use either oxygen, enriched air, or air also in the absence of vapor, so the formation of carbonaceous compounds in produced syngas is strongly reduced, thus improving the efficiency of thermal recoveries and allowing the use of exchange systems (for syngas production) at a lower cost. By means of short contact time, CPO is capable of transforming hydrocarbon fractions that cannot be used by other known catalytic technologies (SR and ATR) and which can only be converted into syngas by means of the PO technology but with high energy consumptions and investment costs. It is also possible to transform liquid hydrocarbons with a high content of aromatics and sulfur into syngas, avoiding the formation of carbonaceous residues and NO_x by using air, enriched air, or oxygen as oxidizing agents. The temperature is often close to 1000°C at the catalyst inlet and the catalyst active phase and structure is subject to sintering.

A CPO/membrane reactor can then be a valuable alternative with respect to conventional SR or CPO reactors to transform methane into syngas. No attempt has been reported to integrate CPO both upstream with an O_2 membrane separator and downstream with a H_2 membrane separator (see discussion below).

A variant of this type of approach is the so-called autothermal reformer (ATR). This still uses an oxidant (oxygen or air) to carry out the reactions but the hot gases equilibrate over a fixed bed of reforming catalyst. This type of design is offered, e.g., by Lurgi, and Haldor Topsoe.

The main trends in traditional reformer designs involve increasing the reformer temperature and pressures. However, increasing the reformer temperature or pressure increases the stresses on the reformer tubes, which is the limiting factor in most reformer designs. The gas-heated reformer is a recent alternative to the conventional radiantly fired steam reformer. It is used in parallel with a conventional ATR (oxygen or air blown, depending upon the application). The hot synthesis gas from the ATR is used to provide the heat for the reforming reactions in the gas-heated reformer. Essentially, there are two variants to this type of reformer: the bayonet tube design and the open tube design.

The ATR is used for producing synthesis gas from NG for CH_3OH synthesis, FT, and carbonylation processes, due to better CO/H_2 ratio. The ATR technology requires the use of pure oxygen or strongly enriched air to limit the formation of carbonaceous residues. Membranes that selectively extract pure oxygen from air can provide oxygen at a low cost, reducing the whole syngas production charges in a remarkable way. This approach was originally developed by Air Product/Chevron for the U.S. DoE using a La–Ca–Fe perovskite-based oxygen membrane, but the membrane should operate at >900°C to have enough flux [44–46].

There is an increasing academic and industrial activity on these membranes for high-temperature air separation, because several industrial gas producers are pursuing alternatives to cryogenic and conventional non-cryogenic air separation [47]. The new processes operate at temperatures of 600°C or higher, using novel ceramic membranes or molecular sieves.

Nearest to commercialization is Air Products' ion transport membrane (ITM) system, which uses high-temperature ceramic membranes to separate oxygen from air [48]. An ITM plant is built up from a series of modules, each containing a stack of ceramic wafers. Air Products has operated a 5 t/d ITM pilot plant since 2005, and is building a 150 t/d plant on stream. Compared to a conventional cryogenic ASU, Air Products reports that an ITM plant will be smaller, about 35% cheaper, and about 35%–60% less energy intensive.

Linde's ceramic autothermal recovery (CAR) process [49] does not use ceramic membranes, but uses perovskite in the form of extruded pellets as adsorbent. Oxygen is then released by contacting with recycled flue gas or superheated vapor. The CAR process operates much like conventional process swing absorption (PSA). According to Linde, the CAR process requires little or no additional energy input. A 0.7 t/d CAR pilot plant is operating in partnership with the Western Research Institute (Laramie, Wyoming).

MTR (Membrane Techn. Res., United States) (www.mtrinc.com) offers H_2 polymeric membrane separation for the syngas process, which thus requires cooling down the feed to below 150°C and cannot be used for recycle. H_2 permselective membranes based on Pd alloy are preferable for a separation integrated with syngas reactor. A recent review discussed these aspects in detail [50]. In particular, it was evidenced that 450 trillion btu/year could be potentially saved by using H_2 membranes. The review emphasizes the flow sheet design modification with adsorption or membrane units being added downstream to the reactor for short-term impact, and an integrated membrane/reactor design for a longer-term sustainable impact.

TECHNIP-KTI [51] evaluates the production costs of a hybrid system based on a new membrane reforming reactor (MRR) concept to convert NG to hydrogen and electricity. Membrane reforming with hydrogen-selective, palladium–silver membranes pushes the chemical equilibrium and allows higher methane conversions at lower temperature such as 650°C. The new MRR concept is based on a series of modules, each module being made up of a reforming step and an external membrane separation unit. The estimates, based on utilities costs of a typical Italian refinery (end of 2006), show that the production costs for the hybrid system are 30% less than conventional tubular SR technology, and 13% less than a gas-fired cogeneration plant coupled with a conventional H_2 plant.

Although there are many academic studies on Pd-based thin film membranes for H_2 separation [35], few efforts exceed the laboratory scale. Hysep® (www.hysep.com) comprises a line of hydrogen membrane separation modules offered by ECN (www.ecn.nl) for evaluation purposes. Modules up to 0.5–1 m² geometrical membrane area are available for small-scale pilot unit tests. The technology is based on thin-film palladium membranes that are capable of separating high-purity hydrogen from a gas mixture. These membranes are based on a Pd alloy film deposited by electroless plating on a ceramic tubular support.

An effort to integrate all these advances in a new process design is made in the frame of the EU large collaborative project NEXT-GTL (started on October 2009), which is finalized at the development of the advanced nanoporous materials (catalysts and membrane) and their integration in new process layouts for low-temperature GTL production from NG, particularly in remote areas. New process architecture was proposed to produce syngas by methane partial oxidation. The reaction steps are integrated with different types of membranes for O_2, H_2, and CO_2 separation. A possible reduction of 15%–30% of costs was estimated. The use of membrane reactors allows enhancing feed conversion at lower temperature in the range of 600°C–700°C, because the selective removal of hydrogen from the reaction environment enables the multistep integrated process to overcome the thermodynamic equilibrium conversion of a single-step process carried out at higher temperature. Lowering the reaction temperature will in turn reduce the oxygen demand and together with air membrane separation not only improve the overall energy efficiency of the process but also reduce the plant capital investment. The new scheme allows operating at mild conditions, with improved life time of the catalysts, minimized overall energy consumption for syngas production and eventually, production of H_2 and CO_2 side streams, which can be used for other reactions.

Figure 14.11 reports the preliminary conceptual scheme of this advanced syngas process based on the integration of membrane and novel CPO catalysts The innovative aspects of the process can be summarized as follows:

- Use of permselective H_2 membranes to allow operations at low temperatures (600°C–700°C); at these temperature, the overall energy

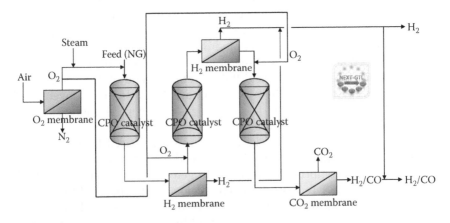

FIGURE 14.11
Advanced syngas process scheme based on integration of membrane and novel CPO catalysts.

efficiency is higher (at the usual reaction temperature, i.e., higher than 900°C, the performance is governed by heat transfer limitations), the problems of stability for CPO catalysts are largely eliminated and new catalysts formulations could be adopted.

- Introduction of permselective O_2 membrane to enrich oxygen feed and reduce the costs of air separation, one of the more costly components of the overall GTL process.

- Use of permselective CO_2 membranes to separate carbon dioxide to reduce GHG emissions and allow its reuse.

- A process layout feasible for applications in remote areas (stranded methane).

14.2.3 Advances in Fischer–Tropsch Technology

FT is an old technology, but which has received cyclic returns of interest. In 1922, Franz Fischer and Hans Tropsch used iron-based catalyst to convert an CO/H_2 mixture to mixture of hydrocarbons (HCs) and oxygenated compounds, and few years later, it was discovered that also cobalt-based catalysts are able to synthesize HCs from syngas. The catalysts are still essentially the same of those used today, although clearly, the fine tuning of their characteristics and process has greatly improved the performances. During the Second World War, FT technology was used in Germany to produce synthetic fuel, and then in South Africa (particularly, by SASOL) in conjunction with coal gasification to produce synthetic fuel due to embargo motivations. Later, SASOL stopped the use of coal and passed to NG to produce syngas for FT. SASOL was one of the major companies investing on FT development, up to when Shell (and later various other companies) started to look at this technology to produce medium distillate for clean diesel by using NG in remote areas. Shell's first demonstration plant was in Bintulu, Malaysia, while actual large plants under construction/starting operations are in Qatar, as cited before.

Current GTL via FT technology consists of three major sections (Figure 14.9), all centered on low-temperature FT synthesis. The low-temperature FT synthesis is based on R&D advancement of both catalyst and reactor technology, because both significantly influence the process characteristics such as thermal efficiency, heat removal, product selectivity, and operating costs [52–57].

The most common catalysts used in FT synthesis are still based on iron or cobalt compounds. Generally, the iron-based catalysts are cheaper and produce gasoline, hydrocarbons, and linear alpha olefins, as well as a generally unwanted mixture of oxygenates such as alcohols, aldehydes, and ketones. Iron catalysts have considerable water–gas shift activity, and therefore produce excessive amounts of carbon dioxide. Nevertheless, the water–gas shift activity can be advantageous when the syngas contains a low H_2/CO ratio. With cobalt catalyst, high yields in long-chain linear paraffins, which can be efficiently transformed into valuable products (middle distillates) by hydro-processing, are possible. To note also that the syngas step should be adapted to the type of catalysts to be used, in order to optimize the CO/H_2 ratio in reference to that optimal for the FT catalyst/reactor.

To promote FT over methanation, the FT reactor operates typically at low temperature (200°C–240°C), mid pressure (20–30 bar), and in presence of a catalyst (Co or Fe based). It is impossible to produce well-defined range of products (i.e., middle distillates). To optimize the overall liquid production, it is necessary to produce high-molecular-weight linear waxes for further hydro-processing step. The use of "cobalt catalyst" and "slurry bubble column reactor (SBCR)" is today one of the preferred technological solution.

It should be considered that FT syncrude is significantly different from crude oil [58]. When syncrude is treated as if it is a crude oil, its refining becomes inefficient. Refining technologies developed for crude oil should be thus modified to refine FT syncrude. Typical FT product composition after upgrading is 25% virgin naphtha, 25% jet fuel and 50% diesel, and minor amounts of LPG and bottom hydrocracking. The heavier fraction can be used for the production of lubricating bases or chemicals.

FT synthesis reactions are highly exothermic, and the reactor systems must be designed for removing this heat. There are two major types of reactors used for GTL applications: multi-tubular fixed bed reactor and the slurry SBCR. The most critical factors for their selection, and process economics, are the heat removal, pressure drop control, catalyst handling, and reactor capacity. In an SBCR, catalyst particles (20–200 μm) are suspended in a slurry medium formed by the product, liquid at reaction condition (20–30 bar, 200°C–240°C), and syngas fed into the bottom of the column. This solution makes possible a good control of the heat and mass transfer and allows reactor capacities around 17,000 bpd (barrels per day). Several companies have been involved in intensive R&D program to develop their own GTL technology with different technological solution.

14.2.4 Advances in Direct Methane Selective Oxidation

Even if being a holy grail for long time, various companies (in particular, in the United States) consider that knowledge in catalysis has progressed enough to address this problem of direct methane selective oxidation from new innovative prospects. There are different routes that are currently explored. It should be mentioned that by homogeneous gas-phase reactions (radical-type) under pressure (typically 50–200 bar) and high temperatures (200°C–400°C), it is possible to convert MTM, but a large number of products is formed, making highly costly the separation. The presence of ethane and propane favors the reaction, by increasing radical formation. Generally, higher pressures and temperatures, and low oxygen concentrations, favor the selectivity to methanol. Selectivities up to 75%–80% in methanol at 8%–10% conversion have been reported in a flow system under cool flame conditions

(450°C, 65 atm, <5% O_2) [59,60]. However, the scale-up and the possibility of controlling the performances in these types of reactions are extremely difficult. For this reason, in addition to separation issues, the homogeneous approach never reaches a commercial development. A different way to promote this type of high-energy reaction is by generating nonthermal plasma [61,62]. Although interesting results have been obtained, scaling up of the technology is also, in this case, extremely difficult. For these reasons, the use of catalysts for this reaction is still the preferable option. Methane may be converted also photocatalytically to methanol [63], but productivities are very low.

Another reaction of methane conversion, essentially based on the radical gas-phase chemistry, although initiated from a heterogeneous reaction on the catalyst surface (a heterogeneous–homogeneous mixed mechanism), is that of oxidative coupling of methane. In this reaction, CH_4 and O_2 react over a solid catalyst to form C_2H_6 and C_2H_4, plus C_{3+} products in minor amounts. The olefins may be then converted to liquid fuels (e.g., to alcohols by hydration) or be used as chemicals. Starting from the early work of Keller and Bhasin [64], a large research attention was dedicated to this reaction for several years [65], but never reached application. Research on this reaction is almost stopped in recent years. The main problem in oxidative coupling is that the active sites in the coupling catalysts also activate the C–H bond in C_2H_6 and C_2H_4, resulting in the formation of CO_2 by combustion. The single-pass yield of C_2 products is usually limited to <25% at a C_2 selectivity of about 80%. SrO/La_2O_3 and $Mn/Na_2WO_4/SiO_2$ are examples of some of the best catalysts reported [15]. It does not seem likely that it will be major improvements in the C_2 selectivity at reasonable conversions in the near future.

An interesting new route was developed recently by UOP [66,67]. UOP LLC, in the frame of a government co-sponsored project (NIST/ATP Award 70NANB4H3041) for selective liquid phase oxidation of MTM, first investigated the use of hydrogen peroxide to oxidize methane in acidic solvents with palladium or copper salts as catalyst to produce the Me ester corresponding to the acid solvent. Then, they observed that air is a suitable oxidant for direct methane oxidation in trifluoroacetic acid with high efficiency. Under these reaction conditions, the precatalyst was transformed into an active catalyst, which catalyzes homogeneous methane oxidation to methanol. The reduced catalyst was then reoxidized by air to complete the catalytic cycle. The chemistry is essentially related to the formation of peroxo surface species on the catalyst.

Recently, they discovered that manganese oxide is an efficient methane oxidizer [66]. When used as stoichiometric oxidant, quantitative metal oxide-based yield was observed for methane oxidation. The spent catalyst activity can be 100% regenerated with air under basic conditions. A high methane-based yield (36%) with high selectivity (>95%) was achieved when manganese oxide was used in catalytic amounts in the presence of air for methane oxidation. The issue is that it is necessary to use trifluoroacetic acid. Furthermore, there is an easy deactivation. The reaction productivity and catalyst life can be improved upon the addition of silica gel (the catalyst deactivates by manganese difluoride precipitation, which can be partially inhibited by silica addition), but still the catalyst operates on a min scale. Therefore, the results are still quite far from the possibility of applications.

An alternative approach was proposed by De Vos and Sels [68], which showed that the combination of gold as a catalyst and H_2SeO_4 as the oxidant results in a 94% selectivity for CH_3OSO_3H at 28% CH_4 conversion. Jones and Taube [69] also showed that cationic gold catalyzes the selective, low-temperature, oxidation of MTM in strong acid solvent using Se VI ions as the stoichiometric oxidant. Also, in these cases, the development on a commercial scale of the system is difficult and the process far to be sustainable. The use of gold catalysts for methane oxidation to methanol is the topic of a large project started in 2009. Dow Chemical Company awarded to Cardiff University (G. Hutchings) (together with Northwestern University—T. Marks, but which centers the activities on methane to olefins), a research grant for a total over $6.4 million as part of the 2007 Dow Methane Challenge. The challenge was initiated by Dow in March 2007 to identify collaborators and approaches in the area of methane conversion to chemicals. The Cardiff activity is focused on the design of very active oxidation catalysts (based on bimetallic gold nanoparticles) that will activate methane at low temperatures (<200°C).

The main issue in MTM conversion is the very larger reactivity of the product with respect to the reactant, and thus the easy further conversion of methanol. The use of unusual reaction medium had thus the aim to limit this consecutive reaction. This was also the central concept of the catalytica methane-to-methyl bisulphate process that uses (bpym)$PtCl_2$ in oleum for the direct MTM conversion (up to 72% yields of methanol; the process forms methyl bisulphate that can be then hydrated) [70–72]. The scheme of the reaction is illustrated in Figure 14.12. Theoretical studies [73] have shown that the Pt complex activates methane to give a Pt(II)–CH_3 complex. The latter reacts with protonated SO_3, which splits to form two new ligands, SO_2 and OH^-, thus oxidizing to a Pt(IV)–CH_3 complex. The final step in the cycle is the reductive elimination of methyl bisulphate from this complex. The catalytica team has subsequently stopped all work on this technology, due to

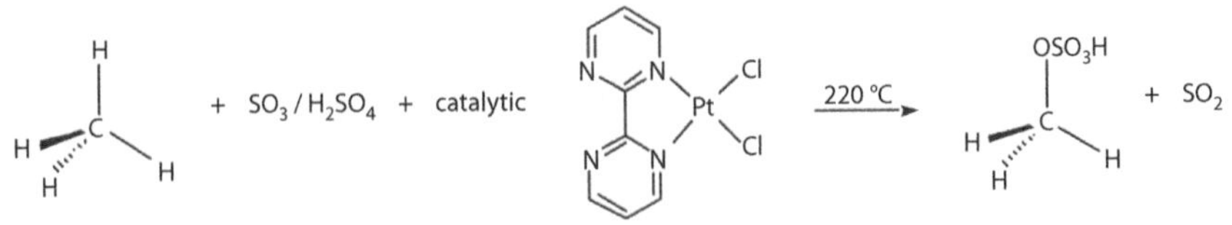

FIGURE 14.12
Catalytica methane-to-bisulfate process. (Adapted from Palkovits, R. et al., *Angew. Chem. Int. Ed.*, 48, 6909, 2009.)

the technical issues in operating with so harsh reaction medium. Recently, it was also shown by quantum mechanical approach the role of sulfuric acid solvent in facilitating the reaction between Pt(II)(bpym)Cl$_2$ (bpym = 2,2′-bipyrimidinyl) and methane [74]. Coordination of methane to the platinum catalyst was found to be catalyzed by the acidic medium.

Recently, to solve the problem of using oleum as the reaction medium, it was proposed a ternary systems of inorganic Pt salts and oxides, ionic liquids (ILs), and concentrated sulfuric acid (>96%) [75]. ILs in the homogeneous catalysis of MTM conversion not only acted as a dissolution media for those otherwise insoluble Pt salts/oxide, but also played a key role in promoting Pt reactivity, possibly through coordination and/or intermolecular interactions. However, still strongly concentrated sulfuric acid is necessary and productivities in methanol quite low (turnover number [TON] of the order of 1). Reaction temperatures around 200°C–220°C are necessary. No further studies have been subsequently reported on the use of IL for the MTM reaction. To note, however, that the use of IL as suitable reaction medium was earlier anticipated by Goddard III, based on theoretical considerations [76].

Note that supported ILs or molten salt (MS) form a rather stable thin layer on the top of an oxide support and these catalysts have been quite active/selective for the oxidative dehydrogenation of light alkanes [77]. From ethane, 80% yield of ethylene could be achieved with LiCl/MgO catalysts. The supported liquid films can also host metal complexes or metal anions such as Keggin units, and can be thus a versatile medium.

Therefore, new catalysts for direct MTM conversion may be based on supported IL/MSs, which host surface-bounded metal complexes in which strongly electron-deficient metal centers are present (to cleave the methane C–H bond via oxidative addition) or sites able to activate CH$_4$ by σ bond metathesis and hydrogen elimination under mild conditions. The IL/MS may also host other active species able to oxidize these activated methane species to a methoxy group which, by subsequent hydrolysis, forms selectively methanol. The supported IL/MS provides a well-defined volume, accessible to few reactant components, and a surface that is dynamically restructuring to give access to catalytic active components that synergistically cooperate. Therefore, it prevents further conversion of methanol. This new design in nanoporous materials for direct MTM or DME conversion is part of the cited EU NEXT-GTL project. The conceptual scheme of the new process for direct synthesis of methanol/DME from methane is reported in Figure 14.13.

Interesting new solid catalysts for the direct low-temperature oxidation of MTM have been reported recently by Max Planck (Germany) researchers [78]. They have included the Periana Pt complex (the same used in the cited catalytica methane-to-methyl bisulphate process) in a new class of high-performance polymer frameworks that are formed by the trimerization of aromatic nitriles in molten ZnCl$_2$. The materials are thermally stable up to 400°C and resist strongly oxidizing conditions. Utilizing 2,6-dicyanopyridine as monomer, a covalent triazine-based framework (CTF) with numerous

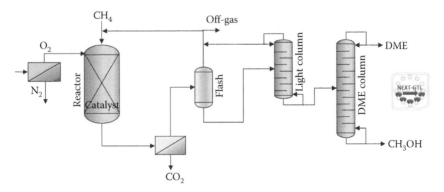

FIGURE 14.13
Scheme of the process for direct synthesis of methanol/DME from methane.

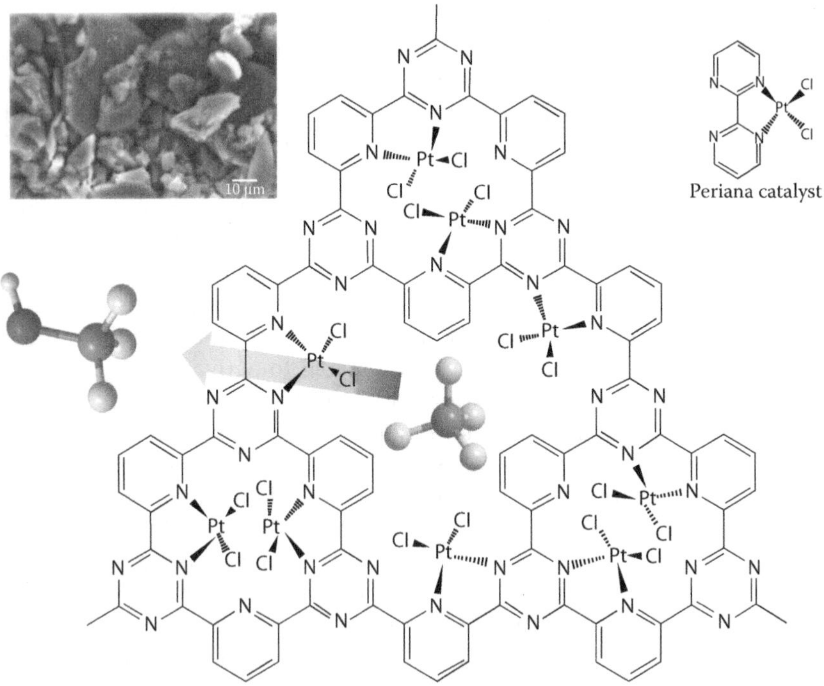

FIGURE 14.14
Pictorial view of the synthesis of methanol from methane over Pt-CFT materials.

bipyridyl structure units is accessible, which allowed coordination of platinum and resembled the coordination sites for platinum coordination in the molecular Periana catalyst (Figure 14.14). The Pt was added either by simply combining CTF and the platinum precursor in the reaction mixture for the methane oxidation reaction, or by precoordination of platinum (Pt-CTF) in a separate step. TONS of the order of 200–300 (40 bar CH_4 pressure, 215°C) are possible in liquid phase oxidation using concentrated sulfuric acid as the reaction medium. The process still remains not environmental friendly and costs high, due to the need to use special corrosion-resistant materials.

14.2.5 Advances in Direct NG Conversion to Alkylaromatics

Various review and papers have discussed recently the topic of methane nonoxidative conversion [79–83]. We may summarize the indications as follows:

- Catalysts for CH_4 aromatization are based on Mo-X or X (where X = W, Re, Ga, Ru, Pt, Cr) in ZSM-5 type zeolite (metal loading typical 3%–5% wt). MoO_x converts to a carbide species during the reaction, while less clear evidences exist on the nature/role of the other components. The role of the transition metal is indicated in dehydrogenation steps, while the zeolite acid sites catalyze cracking, oligomerization, and cyclization reactions.

- Despite several studies on the catalysts, reaction mechanism and nature of the carbonaceous deposits formed during the course of the reaction are still unknown, and relationships between the nature of the catalyst and its performance have not yet been discovered. The important aspects of the conversion can be summarized as follows:

  1. An induction period at the early stage of the conversion is present, probably related to the conversion of the MoO_x species by methane into Mo-carbide and/or Mo-oxocarbide species. However, active catalysts not containing Mo and which cannot form carbide species (Ga-ZSM5) have been also reported.

  2. The structure/activity relationships suggest that in Mo/ZSM-5 catalysts, the Mo carbide species are possibly highly dispersed on the outer surface and the partially reduced Mo species are located in the channels of the zeolite. These findings suggest that methane is dissociated on the Mo carbide cluster supported on the catalyst having optimum Brønsted acidity to form C_2-species as primary intermediates, which are then subsequently oligomerized to aromatic compounds at the interface of the Mo carbide and the catalysts having the proper Brønsted acidity. However, contradictory results on the necessary presence of a high amount of Brønsted acidity were obtained [84]. Thus, the possible role of electron acceptor sites has also to be considered.

  3. Coke deposits are formed during the conversion, leading to severe catalyst deactivation. This is a secondary reaction strongly

influenced by the pore architecture as well as the acidity of the catalyst.

4. The alkylation of aromatics by alkanes is also strongly limited by thermodynamics. Yields can be improved in packed-bed membrane reactors with hydrogen-selective membranes. However, the reaction conditions in such reactors have to be optimized so that the catalyst displays satisfactory activity, selectivity as well as stability, and the membrane is performing well. This includes the optimization of the separation conditions, e.g., the sweep gas type and flow rate.

In terms of process, literature data evidence the following aspects:

- The nonoxidative catalytic conversion of methane is usually carried out at 700°C (or higher) applying plug-flow hydrodynamics in a continuous flow system under atmospheric pressure (or slightly higher). The reaction is limited by thermodynamic equilibrium (maximum yield of 12% at 700°C); increasing the temperature increases the severity of coking and the difficulty in regeneration, being formed more polyaromatics species. Some attempts have been reported to use membrane reactors to overcome equilibrium limitations and application of measures to reduce deactivation.

- Carbonaceous deposits depend on pore structure, and are formed preferentially in the inner part of the channels. The rate of formation of these species could be limited by the addition of CO_2 in low amounts to the feed.

In order to design an advanced catalyst/process for this reaction, it is thus necessary to develop novel materials based on a hierarchical design and containing transition metal (Mo and others) in proto-zeolites or delaminated materials (ITQ-2, or analogous), and couple catalyst and membrane to improve performances both in methane aromatization and alkylation of aromatics by alkanes. The hierarchical structuring would improve the reaction rate by minimizing mass transfer limitations and reduced rate of formation of carbonaceous deposits. The use of membrane will allow overcoming thermodynamic limitations. By shifting the equilibrium through hydrogen removal with a permselective membrane, it is possible to operate at higher conversion, but maintaining the temperature as low as possible to limit the formation of carbonaceous species. This not only reduces the formation of carbonaceous species (thus higher aromatics formation) and decreases the rate of deactivation, but also reduces the severity of regeneration, hence providing an improvement of catalyst lifetime. Figure 14.15 reports the conceptual flow sheet of the new process of nonoxidative NG conversion to alkylaromatics.

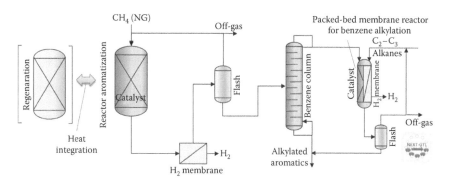

FIGURE 14.15

14.3 Conclusions

There is an increasing attention on the GTL processes to monetize stranded NG or exploit oil field associated gas. Nevertheless, being known from several years, there is still the need to develop the processes, based on the advancement in novel nanoporous materials (catalysts and membrane). We have discussed some of the opportunities, with reference in particular to the aspects investigated in the frame of the EU project NEXT-GTL, and the possible process flowsheet. The key issue is the need of an advanced design in these nanoporous materials, but which can offer a number of opportunities to improve the performances in many other large-scale energy and chemical processes.

Acknowledgment

This work was prepared in the frame of the EU large collaborative project NEXT-GTL that derives from the joint effort of two EU Networks of Excellence: IDECAT and INSIDE-PORES.

References

1. Ghaemmaghami, B. 2001. Special report—GTL: Progress and prospects: Study yields generic, coastal-based GTL plant. *Oil Gas J.*, 99(11): 64.
2. Rahmim, I.I. 2005. Stranded gas, diesel needs push GTL work. *Oil Gas J.*, 103(10): 18.

3. Chedida, R., Kobroslya, M., and Ghajarb, R. 2007. The potential of gas-to-liquid technology in the energy market: The case of Qatar. *Energy Policy* 35: 4799.
4. Stocker, M. 2008. Biofuels and biomass-to-liquid fuels in the biorefinery: catalytic conversion of lignocellulosic biomass using porous materials. *Ang. Chem. Int. Ed.*, 47(48): 9200.
5. van Steen, E., and Claeys, M. 2008. Fischer–Tropsch catalysts for the biomass-to-liquid process. *Chem. Eng. Techn.*, 31(5): 655.
6. Mikkonen, S. 2008. Second-generation renewable diesel offers advantages. *Hydrocarbon Proc.*, 87(2): 63.
7. Grobler, M. 2008. Top technologies. *Hydrocarbon Eng.*, 13(6): 93.
8. Lu, Y. and Lee, T. 2007. Influence of the feed gas composition on the Fischer–Tropsch synthesis in commercial operations. *J. Nat. Gas Chem.*, 16(4): 32.
9. Hao, X., Dong, G., Yang, Y. et al. 2007. Coal to liquid (CTL): Commercialization prospects in China. *Chem. Eng. Techn.*, 30(9): 1157.
10. Hart, W.M. 2006. Global experience with coal gasification coal-to-gas, coal-to-liquids, and IGCC coal-to-power. In: *Proceedings of the 23rd Annual International Pittsburgh Coal Conference*, Pittsburgh, PA, pp. 1.4/1–1.
11. Vogel, A., Mueller-Langer, F., and Kaltschmitt, M. 2008. Analysis and evaluation of technical and economic potentials of BtL-fuels. *Chem. Eng. Techn.*, 31(5): 755.
12. Festel, G.W. 2008. Biofuels—Economic aspects. *Chem. Eng. Techn.*, 31(5): 71.
13. Rahmin I.I. 2008. GTL, CTL finding roles in global energy supply. *Oil Gas J.*, 106(12): 22.
14. Centi, G. and van Santen, R.A. (eds). 2007. *Catalysis for Renewables*. Wiley VCH Publ., Weinheim, Germany.
15. Holmen, A. 2009. Direct conversion of methane to fuels and chemicals. *Catal. Today*, 142(1–2): 2.
16. Arutyunov, V.S., Rudakov, V.M., Savchenko, V.I. et al., 2005. Direct oxidation of methane to methanol: Its role in development and transportation of natural gas resources, DGMK Tagungsbericht, 2005-2. In: *Proceedings of the DGMK/SCI-Conference "Oxidation and Functionalization: Classical and Alternative Routes and Sources"*, Milan, 2005, p. 207.
17. Zhang, Q., He, D., and Zhu, Q. 2003. Recent progress in direct partial oxidation of methane to methanol. *J. Nat. Gas Chem.*, 12(2): 8.
18. Choudhary, T.V., Aksoylu, E., and Goodman, D.W. 2003. Nonoxidative activation of methane. *Catal. Rev. Sci. Eng.*, 45(1): 151.
19. Shu, Y. and Ichikawa, M. 2001. Catalytic dehydrocondensation of methane towards benzene and naphthalene on transition metal supported zeolite catalysts: Templating role of zeolite micropores and characterization of active metallic sites. *Catal. Today*, 71(1–2): 55.
20. Caeiro, G., Carvalho, R.H., Wang, X. et al. 2006. Activation of C_2–C_4 alkanes over acid and bifunctional zeolite catalysts. *J. Mol. Catal. A: Chem.*, 255(1–2): 131.
21. Rahmim, I.I. 2003. Gas-to-liquid technologies: Recent advances, economics, prospects. *Presented at 26th IAEE Annual International Conference*, Prague, Czech Republic, June 2003.
22. Jaramillo, P., Griffin, W.M., and Matthews, H.S., 2008. Comparative analysis of the production costs and life-cycle GHG emissions of FT liquid fuels from coal and natural gas. *Environ. Sci. Technol.*, 42 (20): 7559.
23. Carroll, J., 2009. *Natural Gas Hydrates*. Elsevier Science, the Netherlands.

24. Gudmundsson, J.S. 2008. Natural gas hydrate. *Presented at Gas Hydrates (Tekna) Conference*, Bergen (Norway), October 2008.
25. Fan, Z., Chen, W. et al. 2009. Catalytic conversion of syngas into C_2 oxygenates over Rh-based catalysts—Effect of carbon supports. *Catal. Today*, 147(2): 86.
26. Pan, X., Fan, Z. et al. 2007. Enhanced ethanol production inside carbon-nanotube reactors containing catalytic particles. *Nat. Mater.* 6: 507.
27. Pan, X. and Bao, X. 2008. Reactions over catalysts confined in carbon nanotubes. *Chem. Commun.*, 6271.
28. Chen, W., Pan, X., and Bao, X. 2007. Tuning of redox properties of iron and iron oxides via encapsulation within carbon nanotubes. *J. Am. Chem. Soc.*, 129(23): 7421.
29. Chen, W., Fan Z. et al. 2008. Effect of confinement in carbon nanotubes on the activity of Fischer–Tropsch iron catalyst. *J. Am. Chem. Soc.*, 130(29): 9414.
30. Centi, G. and Perathoner, S. 2010. Problems and perspectives in nanostructured carbon-based electrodes for clean and sustainable energy. *Catal. Today*, 150(1–2): 151.
31. Prasada, P.S.S., Baea, J.W. et al. 2008. Single-step synthesis of DME from syngas on Cu–ZnO–Al_2O_3/zeolite bifunctional catalysts: The superiority of ferrierite over the other zeolites. *Fuel Proc. Techn.*, 89(12): 1281.
32. Cheung, P., Bhan, A. et al. 2006. Selective carbonylation of dimethyl ether to methyl acetate catalyzed by acidic zeolites. *Angew. Chem. Int. Ed.*, 45: 1617.
33. Fleisch, T.H. 2006. Syngas chemistry: Key technology for the 21th century. *Presented at "Syngas Chemistry Symposium" (DGMK-SCI Conference)*, Dresden, Germany, October 2006.
34. Zennaro, R., Hugues, F., and E. Caprani, E. 2006. The Eni-IFP/Axens GTL technology: From R&D to a successful scale-up. *Presented at "Syngas Chemistry Symposium" (DGMK-SCI Conference)*, Dresden, Germany, October 2006.
35. Bernardo, P., Drioli, E., and Golemme, G. 2009. Membrane gas separation: A review/state of the art. *Ind. Eng. Chem. Res.*, 48 (10): 4638.
36. Balachandran, U., Dusek, J.T. et al. 1997. Ceramic membrane reactor for converting methane to syngas. *Catal. Today*, 36(3): 265.
37. Bottke, N., Triller, M. et al. 2006. Method for producing alkyl-aromatic compounds by direct alkylation of aromatic hydrocarbons with alkanes. WO Patent 2006-015798, Assigned to BASF Ak., Germany.
38. Kley, I., Rezai, S.A.S., and Traa, Y. 2008. Dehydroalkylation of toluene with ethane on zeolites MCM-22 and ZSM-5. *Stud Surf Sci Catal.*, 174(B): 1119.
39. Alireza, S., Rezai, S., and Traa, Y. 2008. Selectivity enhancement to the exclusive formation of ethyltoluenes and hydrogen during dehydroalkylation of toluene with ethane. *Catal. Lett.*, 122(1–2): 91.
40. Peña, M.A., Gómez, J.P., and Fierro, J.L.G. 1996. New catalytic routes for syngas and hydrogen production. *Appl. Catal. A: Gen.*, 144(1–2): 7.
41. Hu, Y.H., and Ruckenstein, E. 2004. Catalytic conversion of methane to synthesis gas by partial oxidation and CO reforming. *Adv. Catal.*, 48: 29.
42. Jing, Q.A., Liu, P., and Zheng, X. 2008. Progress on catalytic conversion of methane to syngas in the presence of oxygen. *Chem. Bull. (Huaxue Tongbao)*, 71(9): 64.
43. Veser, G., Frauhammer, J., and Friedle, U. 2000. Syngas formation by direct oxidation of methane. Reaction mechanisms and new reactor concepts. *Catal. Today*, 61 (1): 55.
44. Armor, J.N. 1998. Applications of catalytic inorganic membrane reactors to refinery products. *J. Membr. Sci.*, 147(2): 21.

45. Ishihara, T. and Takita, Y. 2000. Partial oxidation of methane into syngas with oxygen permeating ceramic membrane reactors. *Catal. Surv. Jpn.*, 4(2): 125.
46. Bouwmeeste, H.J.M. 2003. Dense ceramic membranes for methane conversion. *Catal. Today*, 82(1–4): 14.
47. Carolan, M. 2006. Syngas membrane engineering design and scale-up issues. Application of ceramic oxygen conducting membranes. In: Sammells, A.F. and Mundschau, M.V. (eds). *Nonporous Inorganic Membranes*. Wiley-VCH, Weinheim, Germany, Ch. 8, p. 215.
48. Bose, A.C., Stiegel, G.J. et al. 2009. Progress in ion transport membranes for gas separation applications. In: Bose, A.C. (ed.). *Inorganic Membranes for Energy and Environmental Applications*. Springer, New York, Ch. 1, p. 3.
49. Ekströma, C., Schwendig, F. et al. 2009. Techno-economic evaluations and benchmarking of pre-combustion CO_2 capture and oxy-fuel processes developed in the European ENCAP Project. Energy Procedia (Greenhouse Gas Control Technologies 9, *Proceedings of the 9th International Conference on Greenhouse Gas Control Technologies (GHGT-9)*, November 16–20, 2008, Washington, DC, Vol. 1(1), p. 4233.
50. Ritter, J.A. and Ebner, A.D. 2007. State-of-the-art adsorption and membrane separation processes for hydrogen production in the chemical and petrochemical industries. *Sep. Sci. Techn.*, 42(6): 1123.
51. Iaquaniello, G., Giacobbe, F. et al. 2008. Membrane reforming in converting natural gas to hydrogen: Production costs: Part II. *Int. J. Hydrogen Energy*, 33(22): 6595.
52. Botes, F.G. 2008. The effects of water and CO_2 on the reaction kinetics in the iron-based low-temperature Fischer–Tropsch synthesis: A literature review. *Catal. Rev. Sci. Eng.*, 50(4): 47.
53. Iglesia, E. 1997. Design, synthesis, and use of cobalt-based Fischer–Tropsch synthesis catalysts. *Appl. Catal. A: Gen.*, 161 (1–2): 59.
54. Dry, M.E. 2004. FT catalysts. *Stud. Surf. Sci. Catal.*, 152: 533.
55. Steynberg, A.P. 2004. Introduction to Fischer–Tropsch technology. *Stud. Surf. Sci. Catal.*, 152: 1.
56. Schulz, H. 1999. Short history and present trends of Fischer–Tropsch synthesis. *Appl. Catal. A: Gen.*, 186 (1–2): 3.
57. Geerling, J.J.C., Wilson J.H. et al. 1999. Fischer–Tropsch technology—From active site to commercial process. *Appl. Catal. A: Gen.*, 186 (1–2): 27.
58. De Klerk, A. 2008. Fischer–Tropsch refining: Technology selection to match molecules. *Green Chem.*, 10(12): 1249.
59. Olah, G.A., Goeppert, A., and Surya Prakash, G.K. 2006. *Beyond Oil and Gas: the Methanol Economy*. Wiley-VCH, Weinheim, Germany.
60. Arutyunov, V.S., Basevich, V.Ya., and Vedeneev, V.I. 1996. Direct high-pressure gas-phase oxidation of natural gas to methanol and other oxygenates. *Russ. Chem. Rev.* 65: 197.
61. Indarto, A. 2008. A review of direct methane conversion to methanol by dielectric barrier discharge. *IEEE Trans. Dielectr Electr Insulation*, 15(4): 1038.
62. Indarto, A., Choi, J.-W. et al. 2008. The kinetic studies of direct methane oxidation to methanol in the plasma process. *Chin. Sci. Bull.*, 53(18): 2783.
63. Charles, E.T. 2003. Methane conversion via photocatalytic reactions. *Catal. Today*, 84: 9.

64. Keller, G.E. and Bhasin, M.M. 1982. Synthesis of ethylene via oxidative coupling of methane. I. Determination of active catalysts. *J. Catal.*, 73(1): 9.
65. Kondrakov, E.V. and Baerns, M. Oxidative coupling of methane. In: Ertl, G., Knötzinger, H., Schüth, F., and Weitkamp, J. (eds). *Handbook of Heterogeneous Catalysis*, Vol. 6, 2008, Ch. 14.11.2, p. 3010.
66. Chen, W., Kocal, J.A. et al. 2009. Manganese oxide catalyzed methane partial oxidation in trifluoroacetic acid: Catalysis and kinetic analysis. *Catal. Today*, 140(3–4): 157.
67. Chen, W., Brandvold, T.A. et al. 2007. Catalysis and kinetic analysis of liquid phase methane oxidation. *Prepr. Am. Chem. Soc., Div. of Pet. Chem.*, 52(2): 99.
68. De Vos, D.E. and Sels, B.F. 2004. Gold Redox catalysis for selective oxidation of methane to methanol. *Angew. Chem Int. Ed.*, 44(1): 30.
69. Jones, C.J., Taube, D. et al. 2004. Selective oxidation of methane to methanol catalyzed, with C-H activation, by homogeneous, cationic gold. *Angew. Chem. Int. Ed.*, 43: 2.
70. Periana, R.A., Taube, D.J. et al. 1998. Platinum catalysts for the high-yield oxidation of methane to a methanol derivative. *Science*, 280(5363): 560.
71. Wolf, D. 1999. High yields of methanol from methane by C-H bond activation at low temperatures. *Angew. Chem. Int. Ed.*, 37(24): 3351.
72. Tijm, P.J.A., Wallera, F.J., and Browna, D.M. 2001. Methanol technology developments for the new millennium. *Appl. Catal. A: Gen.*, 221(1–2): 275.
73. Hristov, I.H. and Ziegler, T. 2003. The possible role of SO_3 as an oxidizing agent in methane functionalization by the catalytica process: A density functional theory study. *Organometallics*, 22(8): 1668.
74. Ahlquist, M., Periana, R.A., and Goddard, W.A. 2009. C-H activation in strongly acidic media: The co-catalytic effect of the reaction medium. *Chem. Commun.*, 2373.
75. Cheng, J., Li, Z. et al. 2006. Direct methane conversion to methanol by ionic liquid-dissolved platinum catalysts. *Chem. Commun.*, 4617.
76. Goddard III, W.A. 2005. Methane to methanol conversion—Compatibility study of using ionic liquids as novel reaction media. In: *2005 AIChE Spring National Meeting, Conference Proceedings*, Cincinnati, OH, p. 243.
77. Gaab, S., Machli, M. et al., 2003. Oxidative dehydrogenation of ethane over novel Li/Dy/Mg mixed oxides: Structure–activity study. *Top. Catal.* 23: 95.
78. Palkovits, R., Antonietti, M. et al. 2009. Solid catalysts for the selective low-temperature oxidation of methane to methanol. *Angew. Chem. Int. Ed.*, 48: 6909.
79. Traa, Y. 2008. Non-oxidative activation of alkanes. In: *Handbook of Heterogeneous Catalysis*, 2nd Ed. Wiley–VCH, Weinheim, Vol. 7, p. 3194.
80. Bao, X. 2008. Catalysis base for optimal utilization of natural gas. *Presented at 235th ACS National Meeting*, New Orleans, LA, USA, April 6–10, FUEL-093.
81. Zheng, H., Ma, D. et al. 2008. Direct observation of the active center for methane dehydroaromatization using an ultrahigh field ^{95}Mo NMR spectroscopy. *J. Am. Chem. Soc.*, 130: 3722.
82. Sily, P.D., Bellot Noronha, F.B. et al. 2006. Methane direct conversion on Mo/ZSM-5 catalysts modified by Pd and Ru. *J. Nat. Gas Chem.*, 15: 82.
83. Yide, X., Bao, X., and Lin, L. 2003. Direct conversion of methane under nonoxidative conditions. *J. Catal.*, 216: 386–395.
84. Sobalik, Z., Tvaruzkova, Z. et al. 2003. Acidic and catalytic properties of Mo/MCM-22 in methane aromatization: An FTIR study. *Appl. Catal. A: Gen.*, 253: 271.

15

Advanced Materials for Hydrogen Storage

Th.A. Steriotis, G.C. Charalambopoulou, and A.K. Stubos

CONTENTS

15.1 Introduction

One of the critical challenges of the ongoing intensive effort toward the acceptance and generic use of H_2 as energy carrier concerns the efficient storage of hydrogen (Felderhoff et al. 2007). Two kinds of storage functions with very different requirements are considered. Systems used for stationary applications can occupy a large area, employ multistep chemical charging/recharging cycles, operate at high temperature and pressure, and balance slow kinetics with capacity. On the other hand, H_2 storage for transportation must operate within minimum volume and weight specifications, supply enough H_2 to enable an approximately 500 km driving range, charge/recharge near ambient temperature, and provide H_2 at rates fast enough for fuel cell locomotion of vehicles. In terms of energy content, 1 kg of H_2 can replace about 3 kg of gasoline. However, hydrogen is a gas at standard pressure and temperature (SPT) and therefore it has a low volumetric density so that more than 1.3×10^4 L of H_2 gas (volume of a midsize car) are necessary to replace just 3.79 L of gasoline at SPT (Graetz 2009). The H_2 storage requirements for transportation applications are thus far more stringent and difficult to achieve than those for stationary applications. Existing technology for hydrogen storage is limited to tanks carrying compressed gas or cryogenic liquid, both of which are used at the moment in demonstration vehicles. Although gas and liquid storage are useful as temporary options in a provisional hydrogen economy, they fall far short of the aggressive EC and U.S. Department of Energy (DOE) targets (6.0 wt% and 45 g/L by the year 2010, and 9.0 wt% and 81 g/L by 2015) for on-board H_2 storage systems due to the required tank volume and energy intensity, as well as safety reasons.

It is now a common belief that the most promising H_2 storage routes are based on solid materials that chemically bind or physically adsorb H_2 at volume densities greater than that of liquid H_2. The challenge is to find a storage material that satisfies three competing requirements: high hydrogen density, reversibility of the release/charge cycle at moderate temperatures in the range of 80°C–100°C (to be compatible with the present generation of fuel cells), and fast release/charge kinetics with minimum energy barriers. The first requires strong chemical bonds and close atomic packing; the second requires weak bonds that are breakable at moderate temperature; and the third requires loose atomic packing to facilitate fast diffusion of hydrogen between the bulk and the surface, as well as adequate thermal conductivity to prevent decomposition by the heat released upon hydriding. In addition to these basic technical criteria, viable storage media must satisfy cost, weight, lifetime, and safety requirements as well (Satyapal et al. 2007).

In this respect, tremendous efforts have been devoted to the research and development of materials that can hold sufficient H_2 in terms of gravimetric and volumetric densities, and, at the same time, possess suitable thermodynamic and kinetic properties. Over the last decade, the scope of candidate materials has expanded greatly, from traditional metal hydrides to complex and chemical hydrides, and from activated carbon (AC) to zeolites and porous polymers, while the rapid progress in nanoscience in the past 5 years has opened groundbreaking directions. In solid-state storage, hydrogen is bonded by either physical (e.g., in carbon based and framework materials), or chemical forces (e.g., in hydrides, amides, and imides). In general terms, physisorption has the advantages of higher energy efficiency and faster adsorption/desorption cycles, whereas chemisorption results in the absorption of larger amounts of gas but in some cases, it is not reversible and requires a higher temperature to release the absorbed gas.

In this chapter, we briefly review the most studied material types and the recent research efforts toward developing novel materials for solid state hydrogen storage.

15.2 Metal Hydrides

Metal hydrides, depending on the type of metal(s) and the hydrogenation process followed, may be categorized into the following general groups: conventional metal hydrides, complex and chemical hydrides. Interest in them stems from the fact that in many hydride-type materials hydrogen is packed very efficiently (H–H distances as low as 2.1 Å giving rise to hydrogen densities much greater than the liquid hydrogen density).

Most of the metal hydrides decompose endothermically therefore a heat input is required to release the trapped atomic hydrogen. In other cases, hydrogen is liberated by a chemical reaction (hydrolysis) with water or alcohols. However, there are nonreversible materials having considerable storage efficiency, which do not readily hydrogenate under moderate pressure and temperature conditions. Those materials are characterized by high thermodynamic stability and slow reaction kinetics, and efforts are focused in the direction of reducing the decomposition enthalpies and improving the kinetics.

15.2.1 Conventional Metal Hydrides

One of the main reasons for pursuing conventional hydrides as storage materials is their exceptional volumetric capacity. Among the conventional metal hydrides, the simple metal hydrides (MH_n, $n = 1, 2, 3$) and the intermetallic compounds (AB_xH_n) have been studied extensively. For use in vehicles, the gravimetric capacity of these materials is generally low, or the thermodynamics of H bonding is either too strong or too weak for easy hydrogen absorption/desorption. For example, alane (AlH_3) contains a large amount of hydrogen by weight (10 wt%). However, due to a weak binding energy

(5–8 kJ/mol H_2 enthalpy [ΔH] for H_2 desorption), it is impossible to directly recharge alane from Al and H_2 using moderate pressures. Magnesium hydride (MgH_2) also possesses a reasonably high gravimetric capacity (7.6 wt%) as well as an extremely high energy density (9 MJ/kg Mg), good reversibility, heat resistance, and recyclability but suffers from the opposite problem in that the binding energy of this compound is too strong ($\Delta H = 66$–75 kJ/ mol H_2), requiring above 290°C to desorb H_2 at 0.1 MPa. Slow kinetics are also limiting its applicability. Recent studies indicate that proper mechanical treatment with certain elements may reduce the stability of the hydride, while insertion of certain catalysts may result in faster kinetics (Dornheim et al. 2007, Reule et al. 2000). An extended summary of the improvements that have been accomplished on Mg-based metal hydrides can be found in the review article of Sakintuna et al. (2007). Of particular note for the practical use of Mg-hydride in stationary storage tanks and other applications (if a suitable waste heat source is provided) are recent achievements concerning the substantial enhancement of its thermal conductivity after mixing with expanded natural graphite (Chaise et al. 2009).

On the other hand, metal hydrides with more moderate binding energies, such as VH_2 and $LaNi_5H_x$, have good thermodynamics ($\Delta H = 30$–43 kJ/mol H_2), but consist of heavy transition and rare earth metals, and therefore have limited gravimetric capacities. Efforts to improve hydrogen capacity and reaction enthalpy of conventional metal hydrides have largely focused on alloying with other elements, and several categories of conventional metal hydrides have been developed: BCC-type alloys (e.g., Fe–Ti, Ti–Mo, and V-based), AB_5 alloys (e.g., $LaNi_5$-type), AB_2 alloys (e.g., Ti(Zr)–Mn-based). Dornheim et al. (2007), Sandrock (1999), and Schlapbach (1992) provide a summary of the conventional metal hydride categories while a more detailed review of conventional metal hydrides properties can be found in Sakintuna et al. (2007) and Sandrock (1999).

It is worth noting that hybrid storage tanks combining the use of high pressure and conventional hydrides, like Ti/Cr/V alloys with 2.2 wt% capacity, have been developed offering an interesting option for vehicular applications (Mori et al. 2005).

15.2.2 Complex Hydrides

The term is used to describe a class of ionic hydrogen-containing compounds which are composed of metal cations and hydrogen-containing "complex" anions such as borohydrides (BH_4^-) (Nakamori et al. 2006a, Srinivasan et al. 2008, Vajo et al. 2005, Züttel et al. 2003a, b), alanates (AlH_4^-) (Bogdanovic and Schwickardi 1997), and amides (NH_2^-) (Chen et al. 2002, Hu and Ruckenstein 2003a). The hydrogen atoms are covalently bonded to the central atom in an anionic complex, for example, $[AlH_4]^-$, $[AlH_6]^{3-}$, $[BH_4]^-$, and $[NH_2]^-$ and stabilized by a cation (alkali, alkaline earth metal, or transition metal). Advanced complex hydrides created a new interest in the hydrogen storage research field. Their composition of light elements like Li, Na, B, Mg, and Al, their low cost, and especially their increased hydrogen density made them essential for onboard storage (Grochala and Edwards 2004, Schlapbach and Züttel 2001). A review of these compounds can be found elsewhere (Orimo et al. 2007).

15.2.2.1 Alanates

Sodium aluminohydride ($NaAlH_4$) can store reversibly up to 5.5 wt% hydrogen, while its decomposition reaction follows a two-step mechanism:

$$NaAlH_4 \leftrightarrow \frac{1}{3} NaAlH_6 + \frac{2}{3} Al + H_2 \leftrightarrow NaH + Al + \frac{3}{2} H_2$$

Hydrogen release starts above 240°C for the first step and even higher (around 300°C) for the second. A third step is possible involving the NaH decomposition but at much higher temperatures. Bogdanovic and Schwickardi (1997) showed that the Ti-doped sodium alanate decomposes at lower temperatures and pressures (first step at 33°C and second step at 110°C) (Bogdanovic et al. 2000). Also, Ti-catalyzed sodium alanate shows exceptional recyclability that exceeds 100 cycles (Srinivasan et al. 2004). Besides titanium, doping with Sc or Ce provided faster kinetics and higher storage capacities (Bogdanovic et al. 2008, Felderhoff et al. 2007). However, both materials are much more expensive than Ti.

Other alanates that can reversibly store notable amounts of hydrogen under moderate pressures and temperatures are listed in Table 15.1, which also includes the alanates that are considered non-reversible. Some of them exhibit high gravimetric theoretical capacity like the thermodynamically unstable lithium aluminohydride $LiAlH_4$ (Felderhoff et al. 2004), magnesium alanate $Mg(AlH_4)_2$ (Mamatha et al. 2006), and calcium aluminohydride $Ca(AlH_4)_2$ (Fichtner et al. 2005) (although for the latter recent studies indicate reversibility under pressures approaching practical limits).

TABLE 15.1

Hydrogen Storage Capacity of Selected Reversible and Nonreversible Alanates

	Hydrogen (wt%)	Formation Enthalpy (kJ/mol H_2)
$LiAlH_4$	8.0	−55.5
$NaAlH_4$	5.6	−54.9
$KAlH_4$	4.3	−70.0
Li_3AlH_6	5.6	−102.8
Na_3AlH_6	3.0	−69.9
K_3AlH_6	2.0	−78.5
$Mg(AlH_4)_2$	7.0	−21.1
$Ca(AlH_4)_2$	5.9	−59.4

15.2.2.2 Borohydrides

The most commonly used borohydride is $NaBH_4$ (10.8 wt% hydrogen). Despite its unfavorable thermodynamics—like every other borohydride—it operates in a PEM (Polyelectrolyte Membrane) fuel cell system when reacting with water. The decomposition product is $NaBO_2$ instead of the toxic (for the fuel cell catalyst) volatile boranes (B_xH_y) that can be formed from the usual decomposition. Other borohydrides such as $Ca(BH_4)_2$ (Kim et al. 2008a, b) and $Zn(BH_4)_2$ (Jeon and Cho 2006) can release hydrogen, but either the reaction is highly endothermic or toxic side-products are formed. It seems that the electronegativity of the metals in $M(BH_4)_n$, where ($M = $ Li, Na, Ca, Mg, Zn), associates with the stability of the hydride. Studies showed that the decomposition temperatures of borohydrides are decreased upon increasing electronegativity (Li et al. 2007, Nakamori et al. 2006c).

Nevertheless, the compound with the highest gravimetric storage density known today is $LiBH_4$, which can store up to 18.5 wt% hydrogen. The decomposition process occurs at 380°C with the formation of LiH and boron (Züttel et al. 2003a). Since the hydride is too stable, efforts have been made to achieve its destabilization (toward more stable reaction products) and dehydrogenation at lower temperatures. A number of attempts have been made involving the reaction of $LiBH_4$ with several compounds, such as $LiNH_2$ (Aoki et al. 2005, Pinkerton et al. 2005), MgH_2 (Vajo et al. 2005), and CaH_2 (Nakamori et al. 2006b, Pinkerton and Meyer 2008, Yang et al. 2007). For example, the final product of the system $LiBH_4/MgH_2$ is MgB_2, unlike the pure $LiBH_4$ where elementary boron is produced. Using this method lower hydrogen amounts are reversibly stored (up to 10 wt%). This particular system is a typical example of thermodynamic tailoring (Hanada et al. 2008, Vajo et al. 2005, 2007) where a second (or sometimes a third) component is added to a storage material thus creating new reaction pathways characterized by altered thermodynamics (in the present case the result is that $LiBH_4$ is destabilized and the decomposition enthalpy of the system is significantly reduced relative to pure $LiBH_4$).

15.2.2.3 Nanocomposites

Chemical hydrides, despite their inherent high hydrogen content, cannot offer technically viable H_2 storage solutions due to their unfavorable kinetic and thermodynamic properties as mainly described by their relatively high thermal stability (inducing significantly high hydrogen release temperatures far away from the desired application window), slow desorption/absorption kinetics, and irreversibility upon cycling (Satyapal et al. 2007, Vajo and Olson 2007). The initial attempts for tailoring the thermodynamic and/or kinetic properties of various (mainly complex) hydrides involved the formation of hydride mixtures (and/or doping with a certain catalyst) as already discussed in Sections 15.2.2.1 and 15.2.2.2. In a further step nanoscale processing targeting to a dramatic decrease of the particle size of a certain hydride phase (e.g., by mechanical milling) was explored as a potential "destabilisation" route based on the altered behaviour of matter at the nanoscale. The reduction of the size of solid-state hydride phases down to the nanoscale leads to the dramatic increase of the surface-to-volume ratio compared to the bulk materials, thus facilitating the release and uptake of hydrogen. Indeed, nanocrystalline metals and metal–hydrogen alloys with grain sizes less than 20 nm were found to possess markedly different properties from the respective bulk materials (e.g., Yamauchi et al. 2008, Züttel et al. 2000). However, conventional ball-milling approaches do not allow the accurate control of the produced particle sizes; in most cases the smallest particles produced are as large as 100 nm while the respective size distribution is broad (Brinks et al. 2005), thus limiting the possibility for evident size effects. At the same time, mechanically processed hydrides are in a metastable state and as a result it is unlikely to avoid agglomeration and sintering effects upon cycling. Quite recently, nanoconfinement, based on the incorporation or even synthesis of a hydride within the pore system of an inert (most commonly carbonaceous) matrix (scaffold), has emerged as a facile pathway to preserve the dimensions of nanosized hydrides and thus modify effectively the dehydrogenation reaction (Wu 2008). Several experimental and theoretical studies on such nanoconfined systems have indeed exhibited reduced dehydrogenation temperatures but also improved desorption kinetics (Fichtner et al. 2009, Zhang et al. 2009).

Gutowska et al. (2005) demonstrated that the incorporation of ammonia borane (H_3NBH_3) into a mesoporous silica scaffold led to improved hydrogen release rates as well as a change in thermodynamics compared to the bulk compound. Furthermore, Schüth et al. (2004a) found that $NaAlH_4$ exhibits improved kinetics and reduced dehydrogenation temperature (by 40 K) when encapsulated in a mesoporous carbon substrate. Improved hydrogen desorption and absorption kinetics for $NaAlH_4$ were also reported by Baldé et al. (2006) who used carbon nanofibers (CNFs) to support the hydride. On the other hand, cycling studies on $NaAlH_4$ nanoconfined in AC have shown equilibrium sorption properties considerably different from those of the Ti-doped and ball-milled material, as the solubility of hydrogen was greatly enhanced and the two-step behavior vanished (Fichtner et al. 2008). Faster dehydrogenation rates, lower hydrogen desorption activation energies/temperatures, and increased cycling capacities have in general been obtained from the infiltration of complex hydrides such as $LiBH_4$ and $Mg(BH_4)_2$ in various microporous and mesoporous scaffolds including AC, carbon nanotubes/nanofibers, and carbon aerogels (e.g., Fichtner et al. 2009, Gross et al. 2008, Vajo and Olson 2007, Zhang et al. 2009). These encouraging results have triggered quite an intense interest in research on the complex hydrogen desorption mechanism from the hydride nanoporous composites, that includes several diverse parameters such as the pore properties of the scaffold, the available gas diffusion paths, the interplay between the porous matrix and the confined hydride, the weight penalties, etc.

15.2.2.4 Amides

It was recently found that lithium amide $LiNH_2$ follows a stepwise decomposition procedure similar to the alanates (Chen et al. 2002). The reactions are reversible and theoretically up to 10 wt% hydrogen can be stored:

$$Li_3N + H_2 \leftrightarrow Li_2NH + LiH \quad (5.4 wt\% \ H_2)$$

$$Li_2NH + H_2 \leftrightarrow LiNH_2 + LiH \quad (6.4 \ wt\% \ H_2)$$

However, the second reaction is not practical for vehicular storage applications, since the temperatures required for complete dehydration can reach 400°C.

An important issue that inhibits the application of N-based storage materials is the generation of ammonia during the dehydrogenation process. NH_3 reduces the hydrogen capacity over each cycle and poisons the membrane of a PEM fuel cell. A significant part of the emitted amount is captured by the LiH in a very fast reaction (Hu and Ruckenstein 2003b).

Adding small amounts of $TiCl_3$ as dopant to the $LiNH_2/LiH$ mixture during ball-milling was found to have a strong accelerating effect on the reaction. 5.5 wt% hydrogen was achieved, ammonia was not produced, and the kinetics were improved after three cycles (Ichikawa et al. 2004). Improving the porous structure of the material can further optimize the function of the amide and provoke an increase in the reversible storage capacity with the number of cycles (Hu and Ruckenstein 2006a). Recent attempts have achieved to reduce the dehydrogenation temperature to almost 150°C by introducing catalytic nano-sized metals such as Pt, Pd, Ni, Co, etc. (Pinkerton et al. 2007, Tang et al. 2008).

Combined systems may achieve essential advantages over the pure phases. Their principal function is based on the presence of a stable hydride like an amide and a less stable one. The most well-known system is LiH/MgH_2, where LiH is partially substituted by MgH_2 (Luo 2004). The hydrogenation reaction is suggested to be:

$$2LiNH_2 + MgH_2 \leftrightarrow Li_2Mg(NH)_2 + 2H_2$$

The new Mg-substituted system can absorb 4.5 wt% hydrogen at about 30 bar and 200°C. Other complex systems include the reaction of $LiNH_2$ with complex hydrides. The mixture $LiNH_2/LiBH_4$ reached a hydrogen storage capacity higher than 11 wt% and the decomposition took place in the temperature range 250°C–350°C (Aoki et al. 2005, Pinkerton et al. 2005). Interesting features were also observed from the combination of the amide with alanates (Kojima et al. 2006b, Lu et al. 2006, Xiong et al. 2007).

Another system combining amide with hydrides and showing favorable thermodynamics is $Mg(NH_2)_2/LiH$ that can release 5.6 wt% hydrogen at 150°C–250°C (Leng et al. 2004, Liu et al. 2008, Xiong et al. 2004, 2006).

15.2.2.5 Chemical Hydrides

Certain complex hydrides are classified as "chemical hydrides" due to the fact that they can easily react with water solution (KOH, NaOH) or water steam releasing hydrogen (Aiello et al. 1999, Amendola et al. 2000, Kojima et al. 2002, Liu and Suda 2008, Suda et al. 2001). The most studied hydrolytic system to date is $NaBH_4$, the hydrolysis of which produces theoretically four moles of H_2 and $NaBO_2$ using only two moles of water. However, in practice the efficiency of the hydrolysis reaction is prohibited by factors like water concentration, pH, and temperature. Most of the chemical hydrides have relatively low solubility, hence a complete dissolution at 25°C requires excess water, which leads to a proportional decrease of the gravimetric efficiency (Marrero-Alfonso et al. 2007).

Despite the research efforts, the exact mechanism of hydrolysis still remains unclear (D'Ulivo 2004, Davis et al. 1962, Gardiner and Collat 1965, Kreevoy and Hutchins 1972). In any case, hydrogen production by hydrolysis is nonreversible and off-board regeneration is required, a fact that renders such systems potentially useful for niche applications, like remote power supply, but limits their large-scale applicability in transportation. Schüth et al. (2004b) provide a survey over certain interesting hydrolytic systems for hydrogen storage.

It is also worth mentioning that some organic liquid systems have been proposed based on the hydrogenation and dehydrogenation of cyclic hydrocarbons (e.g., naphthalene/decalin with 7.2 wt% capacity). Again the target application is not on-board vehicle hydrogen storage (due to the required off-board regeneration and the relatively high temperature for full conversion) but rather their use as a buffer in renewable energy systems or as a hydrogen carrier that would be dehydrogenated at hydrogen filling stations (Okada et al. 2006). Carbazoles are further organic materials investigated for their hydrogen storage properties as they exhibit capacities in the range 4–6.2 wt% but their high decomposition temperature and high melting point are considered as important disadvantages.

Likewise, materials related to ammonia borane are being actively investigated as candidate hydrogen storage systems. More detailed reviews of chemical hydrides with emphasis on ammonia borane can be found elsewhere (Hamilton et al. 2009, Marder 2007).

15.3 Carbon

Carbon atoms can bond to two (sp), three (sp^2), or four (sp^3) other carbon atoms to form a variety of structures. In respect to crystalline structure there are three bulk solid carbon polymorphs, namely diamond (sp^3 hybridized bonding), graphite (sp^2 bonding), and amorphous carbon (am-C) (mixture of bonding). On the nanoscale there are several ordered allotropes, which constitute the basis of new carbon chemistry and nanotechnology, i.e., nano-diamonds (sp³ bonding), graphene (sp² bonding), carbon nanotubes (CNTs, sp² bonding), fullerenes (perturbed sp² bonding), and carbon cones (CCs, perturbed sp² bonding). To complete the list of forms in which elemental carbon can be found, carbyne (sp hybridized bonds) should also be mentioned. The aforementioned bonding (and thus structural) variety can produce light-weight structures with high surface area and thus carbon and carbon-based materials hold significant potential for hydrogen storage, specially when combined with facile and low-cost bulk synthesis processes.

15.3.1 Graphene

Graphene is a single planar sheet of sp²-bonded carbon atoms (Figure 15.1). Graphene sheets are finite in size and are composed of carbon atoms arranged on a hexagonal 2D structure, while other elements appear at the edge in non-negligible ratios (e.g., a typical graphene formula could be $C_{62}H_{20}$). Planar single graphenes do not exist in "free-state," as they are unstable and can form graphitic stacks and curved structures such as soot, fullerenes, nanotubes, etc. However, recently, several successful attempts to isolate them have been reported (Bourlinos et al. 2009, Hernandez et al. 2008, Novoselov et al. 2004), and apparently, it becomes more and more easy to produce single graphenes either by exfoliation or mechanical peeling of layers from bulk graphite or via reduction of graphite oxide (Park and Ruoff 2009).

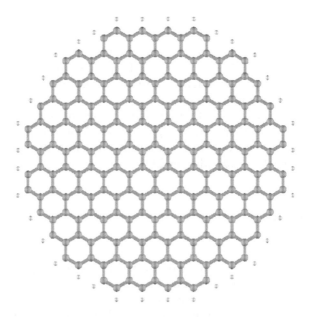

FIGURE 15.1
Hydrogen terminated single graphene sheet model.

Physisorption of hydrogen on graphenes is based on weak Lennard–Jones (LJ) interactions and the minimum energy of LJ potential curve is low (approximately 5 kJ/mol). As a consequence, weak-bonded hydrogen desorbs from graphene surfaces at temperatures well below ambient temperatures. Additionally, physisorption of supercritical hydrogen follows the Langmuir isotherm; i.e., it proceeds practically only in a monolayer and the storage capacity is limited by the available specific surface area (SSA) of graphene. An ideally completely separated graphene sheet should provide an SSA of around 2600 m²/g (Becher et al. 2003). By assuming that the structure of the adsorbed hydrogen is closed-packed face centered, the minimum surface area required for the adsorption of 1 mmol of hydrogen is around 86 m² and the storage capacity of hydrogen adsorbed on graphite is around 6 wt%. Alternatively, the theoretical concentration is 0.4 hydrogen atoms per surface carbon atom (on each side of the sheet) or 6.6 wt%, but this can be typically achieved only at cryogenic temperatures and very high pressure.

Excess sorption of hydrogen, above the limits set by physisorption, could be possible by making porous structures with graphene walls (see, e.g., pillared graphenes) or through chemisorption, i.e., formation of strong covalent bonds between C and H atoms. Chemisorption is by definition at play on the periphery of graphene plates and in fact, if chemical bonding is to be fully utilized, every carbon atom could in principle be a site for chemisorption. However, because of the strength of bonds, hydrogen can only be released at very high temperatures and thus chemisorption (as it is now understood) cannot be utilized in efficient reversible hydrogen storage.

15.3.2 Graphitic/Carbon Nanofibers

Graphitic nanofibers (GNFs), or CNFs, are layered graphitic nanostructures that initially generated significant interest and great controversy. The structures are catalytically synthesized and consist of graphene planes arranged in a variety of stacks, for example, parallel or angled arrangements, that result in a herringbone structure (Figure 15.2). Extremely high hydrogen uptake values were reported for these materials, up to H:C ratio equal to 24 (68 wt% H_2) in GNF herringbone fibers and 55 wt% H_2 in platelet stacks at 278 K and about 11 MPa (Chambers et al. 1998, Park et al. 1999). These results have never been confirmed or reproduced. On the other hand, a series of measurements in several laboratories point to values around 1 wt% (Poirier et al. 2001, Ströbel et al. 1999), or even much lower (Ahn et al. 1998, Hirscher et al. 2002, Ritschel et al. 2002, Tibbetts et al. 2001). Recently the group who made the 68 wt% claim has reported much lower capacities of less than 4 wt% H_2 at 6.5 MPa and room temperature (Lueking et al. 2004). Likewise, originally reported (yet never reproduced by other laboratories) uptakes of more than 10 wt% (Fan et al. 1999) have been recently reduced to half by the same group (Cheng et al. 2000).

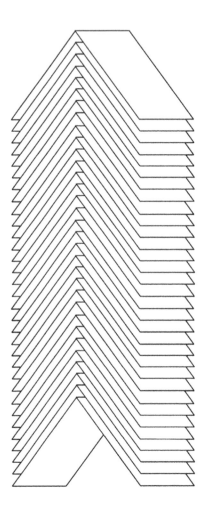

FIGURE 15.2
GNF sketch.

15.3.3 Graphite

Short 2D graphene sheets when stacked along c-axis normal to the hexagonal basal plane produce 3D nano-graphitic structures. When the hexagonal
ordering of C atoms extends in $a \times b$ basal planes, and the stacking of planes
extends along c-axis normal to basal plane, then such a massive stacking
makes graphite. The interplanar distance in regular graphite is approximately
0.34 nm making it inaccessible to dihydrogen molecules (0.41 nm). The theoretical density of graphite is 2.25 g/cm³ and its SSA 1–7 m²/g. After treatment (e.g.,
mechanical milling, chemical activation) the density can be reduced to as low
as 1.85 g/cm³ (Fukunaga et al. 1998) and the surface area increased to more
than 100 m²/g. In any case, such SSA values are considered extremely low for
significant hydrogen adsorption. For instance, the measured H_2 adsorption on

activated graphite with a surface area of around $200\,m^2/g$ was $0.125\,wt\%$ at $77\,K$ (Nijkamp et al. 2001). Even if the bulk graphite is exfoliated to double graphene sheets with SSA $1315\,m^2/g$, the amount of hydrogen stored by physisorption at cryogenic temperatures should not exceed $3\,wt\%$ (Ströbel et al. 2006). Higher values of storage capacities have been reported, but they must be attributed to concurrent chemisorption, leading to very strong chemical bonds, which break and release hydrogen gas only at high temperatures. For example, it is reported that after 80 h of ball milling under hydrogen (10 bar), graphite nanostructures contain up to 0.95 hydrogen atoms per carbon atom, or $7.4\,wt\%$ from which 80% could be desorbed at temperatures above $300°C$ (Orimo et al. 1999, 2001). However it has been shown that after desorption it is impossible to reload the graphitic sample in hydrogen atmosphere is impossible (Hirscher et al. 2003).

15.3.4 Nanodiamonds

A cluster of sp^3-bonded carbon atoms terminated with hydrogen atoms forms diamondoids ($C_{4n+6}H_{4n+12}$), which are also known as cage-hydrocarbons. Given their high hydrogen content, they have been reported to have possible interest for hydrogen storage (Varin et al. 2009), but the H–C bonds are very strong and reversibility is considered problematic if not impossible. In a different context, nanodiamond particles can be converted to nanographite, but so far there has not been any report on hydrogen storage properties of this novel nanophase (Dahl et al. 2003, Enoki et al. 2006).

15.3.5 Amorphous Carbon

Am-Cs consist of extremely small nanocrystalline graphitic carbon clusters (below 1–$2\,nm$) exhibiting a mixture of sp^2, sp^3, and even sp^1 bonding. Am-Cs require hydrogen to stabilize and thin films can contain as much as 40–60 at%. Lower content (20 at%) results in an increased sp^2 character in a phase known as *graphite-like hydrogenated am-C* (Varin et al. 2009). Am-Cs are relatively easily produced by physical- or chemical-vapor deposition or magnetron sputtering (Jariwala et al. 2009, Casiraghi et al. 2007). It has been demonstrated that along with irreversibly stored hydrogen a quasi-free hydrogen phase is adsorbed by graphite-like structural fragments (Ivanov-Omskii et al. 1998). Temperature Programmed Desorption (TPD) experiments showed that the quasi-free hydrogen began to effuse from the material at $100°C$ and peaked at $400°C$ (Kapitonov et al. 2000). In other experiments, the amount of hydrogen released at low-temperature ($<500°C$) was 10 at%, while 12 at% more was released at higher temperature ($750°C$).

15.3.6 Activated Carbons

ACs are more or less random stacks of graphene planes of various sizes and degree of disorder. Such carbons are mainly microporous (pore size $<2\,nm$); however, meso- (2–$50\,nm$) and macro-pores ($>50\,nm$) are not uncommon.

Typical (e.g., commercial) ACs usually have SSAs between 500 and 1500 m²/g and can be prepared from inexpensive raw or polymeric materials such as agricultural by-products, coal, peat, lignite, etc. through carbonisation and consequent activation, i.e., reaction with oxidizing agents (H_2O, O_2, CO_2), so that a fraction of carbon atoms is gasified. Super-ACs can be produced through reaction with potassium hydroxide and are characterized by increased SSAs (>3000 m²/g). Hydrogen adsorption, follows closely the Langmuir isotherm and in general, storage capacity correlates well with SSA. The amounts adsorbed are rather low at room temperature and moderate pressures. For instance, typical values are <1 wt% at 100 bars and 298 K, for a high (2800 m²/g) surface area carbon (Jin et al. 2007). For super-ACs (AX-21 or Maxsorb) storage capacities of around 5 wt% at 77 K and 1.3 wt% at 296 K was reported (Kojima et al. 2006a, Chahine and Bose 1994), in agreement with measurements on similar carbons (Panella et al. 2005). In all cases the measured heats of adsorption (about 5–6 kJ/mol H_2) are still lower than the desirable values (>10 kJ/mol) for reversible hydrogen storage and release at ambient temperatures. It might however be concluded that ACs are better and mainly much cheaper storage materials than most experimentally investigated carbon nanostructures (e.g., CNTs, GNFs, etc.).

15.3.7 Fullerenes and Hydrofullerenes

Fullerenes (Figure 15.3) are made of hexagonal carbon atom rings (similar to graphene), but they also contain pentagonal or sometimes heptagonal rings.

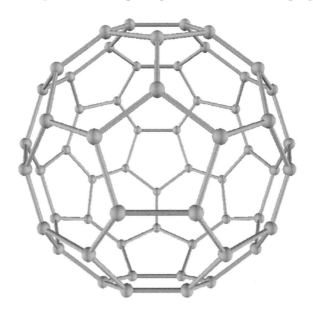

FIGURE 15.3
C_{60} fullerene model.

Fullerenes (e.g., C_{60}) are the only form of carbon that can be hydrogenated and dehydrogenated reversibly (Peera et al. 2004). As a matter of fact, hydrogenations are the simplest reactions of fullerene chemistry and *hydrofullerenes* up to $C_{60}H_{50}$ can be relatively easily produced (Brosha et al. 1999, Sui et al. 1996, Taylor 1999). Hydrogenation proceeds rapidly and oxidation back to C_{60} can be performed but requires wet chemical reactions. Hydrogen localization within the fullerene frame as a method for hydrogen capacity enhancement is of great interest. At 77 K, about 3 mol H_2/mol C_{60} can be present in the interstitial sites of $C_{60}H_x$ as H_2 molecules and can evolve from the sample at 293 K (Schur et al. 2002). Theoretically, C–H bonds are weaker than C–C bonds (Peera et al. 2004) and upon heating, hydrogen can be released while the fullerene structure is preserved. Nevertheless, in practice it has been observed that upon release, hydrogen forms strong bonds with the outer surface fullerenes, and sometimes the fullerene structure collapses (Brosha et al. 1999). Adsorption of H_2 has also been attempted in molecular crystals using C_{60} as building block but results in very low (1/20) H/C ratio (Talyzina and Klyamkinb 2004).

15.3.8 Carbon Nanotubes

CNTs are generated by rolled graphite sheets. Tubes formed by only one single layer (Figure 15.4) are called single walled nanotubes (SWNTs with diameter 0.7–3 nm), whereas those consisting of multiple concentric layers are called multi walled nanotubes (MWNTs) and consist of layers (2–50) of nested concentric SWNTs with interlayer distances close to graphite (Dillon and Heben 2001). MWNTs have inner and outer diameters that are typically 2–10 and 15–30 nm, respectively, and are generally 5–100 μm in length. Most of the shells of MWNTs are inactive for hydrogen adsorption and thus decrease the hydrogen-storage capacity.

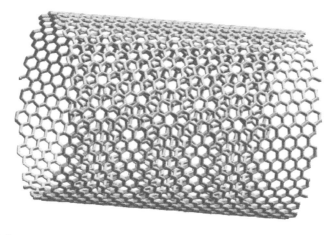

FIGURE 15.4
CNT (SWNT) model.

Interest in CNTs for hydrogen storage boomed with highly optimistic early results of 5–10 wt% (Dillon et al. 1997, Liu et al. 1999). However in the years to follow, a large series of carefully designed and performed experiments proved that the actual capacity of CNTs is rather low (<1 wt%) at ambient temperatures and in fact scales reasonably linearly with surface area as in, for example, ACs. Xu et al. (2007) investigated the hydrogen-storage capacities of various carbon materials, including AC, single-walled carbon nanohorns (SWNHs), SWNTs, and GNFs at 303 and 77 K. The hydrogen-storage capacity of carbon materials was <1 wt% at 303 K, and AC had the highest capacity (0.67 wt%). Under cryogenic conditions, hydrogen uptake ranges from 1 to 2.4 wt% and adsorption enthalpies are 4.3–4.5 kJ/mol, which is typical of carbon adsorbents (Poirier et al. 2006). In general, significant time and effort has been invested into studying CNTs as a hydrogen storage media. Since the processes to prepare, scale-up, and manufacture opened clean CNTs is difficult and very costly, a question on whether interest should focus more on cheap am-Cs arises.

15.3.9 Other Carbon Nanostructures (Horns, Cones, Scrolls, Pillared Graphenes)

The disappointing results for H storage in carbon nanostructures may be compensated in the future through the emergence of a series of different novel promising carbon nanostructures.

Carbon Cones (CCs) with the smallest apex angle (approximately 19°) were observed as cone-shaped fullerenes for the first time in 1994 (Ge and Sattler 1994), while all five possible apex angles CCs were synthesized by accident in 1995 (Krishnan et al. 1997) in the so-called Kværner Carbon Black & H_2 Process (Hugdahl et al. 1998), which decomposes hydrocarbons directly into carbon and H_2. This process under certain conditions produces a carbon material composed of microstructures, which are approximately 80% flat carbon discs, 5%–15% cones with five different apex angles and 5%–15% soot (Figure 15.5). The structures are multiwalled and there is solid experimental and theoretical evidence that CCs mainly consist of curved graphite sheets (Krishnan et al. 1997). In ordinary periodic graphite, each layer consists of hexagonally arranged carbon atoms, while in CCs the five different angles observed are consistent with the incurrence of one to five pentagons at the cone tips. It should be mentioned that the cone allotropes constitute the "intermediate" case between graphene (zero pentagons) and fullerene (six pentagons) geometry. CCs are considered a promising material for hydrogen storage as they have revealed unique H_2 uptake-release properties which are believed to be based on a peculiar H_2-cone interaction (http://www.hycones. eu, U.S. patent No. 6,290,753, Hydrogen storage in carbon material, 2001). CCs are structures with "unusual" heterogeneity originating from their topology. More specifically, their structure is non-periodic, while the local surface curvature, the confinement, and the surface electronic properties change significantly when approaching the cone tip (Heiberg-Andersen and Skjeltorp 2007,

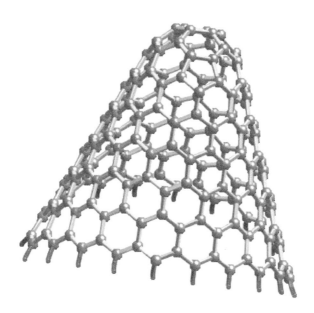

FIGURE 15.5
Carbon cone model (hydrogen terminated) with four pentagons at the tip.

Heiberg-Andersen et al. 2008). For these reasons and for the fact that CCs cannot be easily isolated (e.g., from the raw CC production which contains disks and soot) or separated (in different apex angles), they constitute a challenging material for which adsorption is currently being studied.

Single-walled carbon nanohorns (CNHs) are a (sub)category of cone structures and they consist of single-walled graphitic structures formed out of a single graphene sheet rolled up to form a conical (hornlike) shape (the five-pentagon cone) (Tanaka et al. 2005). CNHs are prepared by arc-discharge synthesis or laser ablation (Wang et al. 2004) and their average size is 2–3 nm while they aggregate to form globular "dahlia-flower" structures with sizes of about 80–100 nm (Iijima et al. 1999). CNHs exhibit very large surface areas approaching 1500 m²/g. In contrast to CNTs, their growth does not require the use of metal catalysts and low-cost and high-purity (>95%) CNH material can be produced in scaled-up process. For these reasons CNHs have become attractive candidates for hydrogen storage. As-prepared CNHs have no open intrananohorn space, and reveal low amounts of hydrogen adsorption. However, the intraparticle pores become open by oxidation pre-treatment, which significantly enhances hydrogen uptake (Xu et al. 2007). Recently, isosteric heats of H$_2$ adsorption corresponding to 100–120 meV were reported (Tanaka et al. 2004). This increased hydrogen bonding energy has been attributed to enhanced interaction of H$_2$ molecules at the *conical tip of nanohorn*, where gas-to-liquid transition and even solid-like H$_2$ was suggested to exist as a consequence of a quantum effect (Tanaka et al. 2004). Moreover, inelastic and quasielastic neutron scattering experiments showed unambiguous signature

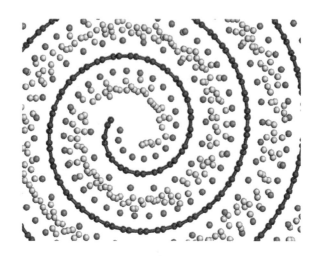

FIGURE 15.6
Snapshot from GCMC simulations of hydrogen adsorption on lithium-doped CNS structures of 7 Å interlayer distance with maximum Li ion to carbon atom ratio, at room temperature and 10 MPa pressure. (Reproduced with permission from Mpourmpakis, G., Tylianakis, E., and Froudakis, G.E., Carbon nanoscrolls: A promising material for hydrogen storage, *Nanoletters*, 7, 1893–1897, 2007. Copyright 2007 American Chemical Society.)

of strong interaction between H_2 and CNHs, and the character of this interaction is quantitatively different to that in CNTs (Fernandez-Alonso et al. 2007).

Carbon nanoscrolls (CNSs) have been synthesized in 2003 (Viculis et al. 2003). This carbon material shows a spiral form and can be schematically obtained by twisting a graphite sheet (Figure 15.6). It is very similar to MWNTs, showing similar intra-layer distance of approximately 3.6 Å. The only key difference between these materials is that in CNSs one can vary their interlayer distance while in MWNTs this is not possible. This property has been considered crucial for making CNSs suitable materials for hydrogen storage. A multiscale (ab-initio-DFT-GCMC) theoretical approach (Mpourmpakis et al. 2007) was used for the investigation of hydrogen storage in CNSs and the results showed that pure CNSs cannot accumulate hydrogen because the interlayer distance is too small. However, an opening of the spiral structure to approximately 7 Å followed by alkali doping can make them very promising materials for hydrogen storage application, reaching 3 wt% at ambient temperature.

CNT pillared graphenes (CNT-PG) is a so far imaginative robust carbon nanoporous material of large surface area and tuneable pore size. This novel 3D material proposed (Dimitrakakis et al. 2008) consists of parallel graphene layers at a variable distance, stabilized by CNTs placed vertically to the graphene planes (Figure 15.7). In this way, CNTs support the graphene layers like pillars and assemble a 3D building block reducing the empty space between the graphene layers by filling it with CNTs. Based on a multiscale theoretical investigation it was suggested that CNT-PGs are 3D nanostructures with enhanced hydrogen storage capacity. It was further deduced that this novel

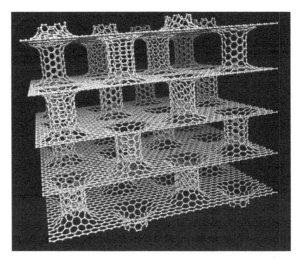

FIGURE 15.7
Pillared graphene model. (Reprinted with permission from Dimitrakakis, G.K., Tylianakis, E., and Froudakis, G.E., Pillared graphene: A new 3-D network nanostructure for enhanced hydrogen storage, *Nanoletters*, 8, 3166–3170, 2008. Copyright 2008 American Chemical Society.)

material, when doped with lithium cations, can reach the volumetric targets for mobile applications, under ambient conditions. Experimentalists are challenged to fabricate this material and validate its storage properties; indeed attempts to produce C_{60} and SiO_2 pillared graphenes (CNT-PG analogues) have already been proven successful (Gournis, personal communication).

15.3.10 Carbon-Metal Composites

Recently, a new family of novel carbon-metal (e.g., Ti, V, Pd, Pt) nanocomposites, has revealed considerable potential for hydrogen storage. The enhanced hydrogen storage capacity of metals with less than half-filled d-shells (e.g., Ti) is attributed to the sorption of hydrogen molecules via unique hybridization between metal-d, hydrogen σ^* antibonding, and carbon π-orbitals. The hydrogen adsorption of transition metals with nearly completely-filled d-shells, (Pd, Pt) on the other hand, is supposed to involve dissociative chemisorption of hydrogen followed by diffusive cascading into various adsorption sites on the carbon material. This alleged mechanism, known as spillover, seems to be a way of increasing the absolute hydrogen capacity of the carbon materials since when hydrogen moves from the metal hydride to the support, some empty sites appear in the metal structure. These sites can be occupied by other H_2 molecules from the gas phase, so the total hydrogen that is absorbed in the system (metal+carbon) is more than expected if there were no interactions between the metal and the carbon support. It has also been reported that the carbon surface might play a crucial role in this mechanism (Psofogiannakis and Froudakis 2009). Although the studies on Ti-carbon composites are mainly based on simulation results, enhanced hydrogen capacities have been measured at room temperature for Pd, Pt and several alloy-doped carbon-based materials and composites (e.g., doped carbon/metal-organic frameworks [MOFs]) (e.g., Hwang et al. 2009, Li and Yang 2006, Satyapal 2008, Yang and Wang 2009). It should nevertheless be mentioned that the phenomenon is currently under debate since results showing insignificant hydrogen storage enhancement (Stadie et al. 2010), or rather low total uptakes (Campesi et al. 2009) have been also reported.

15.4 Framework Materials

With a view to the direct correlation between physisorption and surface area, a class of materials studied extensively for hydrogen storage is that of high-surface area hydrogen sorbents based on microporous inorganic (zeolites, clathrates) or metal-organic (MOFs) frameworks. Despite the exciting progress achieved both in terms of novel structures' design and storage strategies development, the weak dispersive forces dominating physisorption processes cannot facilitate substantial hydrogen uptake in the temperature regime required for automotive applications, hampering the practical application of the respective materials as discussed below.

15.4.1 Zeolites

Zeolites comprise a large family of highly crystalline (either naturally occurring or synthetic) aluminosilicate materials, possessing rigid 3D microporous structures with high surface area (up to $1000\,m^2/g$). Their particular adsorption properties arising mainly by their tailorable pore architecture offering a multitude of different cage-channel frameworks with molecular dimensions, as well as their enhanced ion-exchange capacity, has made zeolites quite attractive candidates for numerous gas separation/storage applications. Nevertheless, and despite the intense investigations on various zeolite types, these materials have exhibited moderate performance in the area of hydrogen storage. In general it has been reported that zeolites can retain small amounts of hydrogen (<0.3 wt%) at room temperature (Kayiran and Darkrim 2002, Langmi et al. 2003, 2005), while higher hydrogen storage capacities (in the region 1–2 wt%) can be reached only at cryogenic temperatures (Dong et al. 2007, Langmi et al. 2003, 2005). A typical ZSM-5 zeolite adsorbs 0.7 wt% of hydrogen at 77 K and 1 bar, significantly less than a typical AC that adsorbs 2.1 wt% of hydrogen at similar conditions (Nijkamp et al. 2001). Likewise, zeolites Linde 5A and 13X have showed a hydrogen storage capacity of ca. 1 wt% at 77 K and 20 bar, a value five times lower than that obtained with AC AX-21 (Chahine and Bose 1994). In any case the highest

H_2 uptake reported after the systematic study of systems with different framework structures (variation of channel diameters and pore volumes) and exchangeable cations, does not practically exceed 2.2 wt%—for example, 2.07 wt% for Na-LEV (Dang et al. 2007) and 2.19 wt% for Ca-X systems (Langmi et al. 2005) measured at 77 K and 15 atm.

15.4.2 Metal-Organic Frameworks

MOFs are a rapidly growing class of crystalline hybrid microporous solids exhibiting exceptionally large surface areas (even higher than 5000 m^2/g), porosity, pore volumes, and window diameters comfortably exceeding those found in zeolites. MOFs are generally produced by simple and versatile procedures, based on the connection of metal ion clusters with rod-like organic moieties and their assembly into uniform 3D lattices (Figure 15.8).

The vast collection of building units combined with the myriad ways in which they can be connected has provided a huge number of possibilities for creating novel, highly diversified MOF structures with tuneable porosities and functionalities. Several of these materials have exhibited encouraging

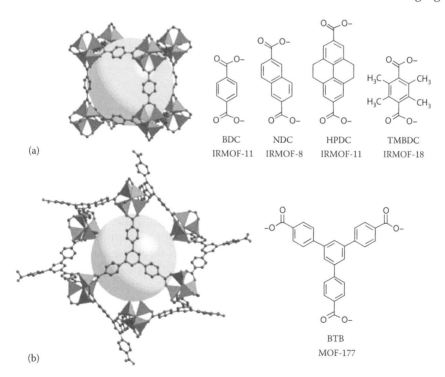

FIGURE 15.8
(a) MOF-5 (IRMOF-1) and different linear linkers and (b) MOF-177 and the respective trigonal carboxylate linker. (Reprinted with permission from Rowsell, J.L.C. et al., *J. Am. Chem. Soc.*, 126, 5666, 2004. Copyright 2004 American Chemical Society.)

TABLE 15.2

H_2 Storage Properties at 77 K of Representative MOF Materials

Compound	Specific Surface Area (m²/g)	Pressure (bar)	Gravimetric H_2 Uptake (wt%)
HKUST-1			
$Cu_3(BTC)_2$, BTC: 1,3,5-benzenetricarboxylate	1154	50	3.6 (Panella et al. 2006)
MIL-101			
$Cr_3OF(BDC)_3$, BDC: 1,4-benzenedicarboxylate	5500	80	6.1 (Latroche et al. 2006)
IRMOF-20			
$Zn_4O(TTDC)_2$, TTDC: thieno[3,2-b] thiophene-2,5-dicarboxylate	4024	78	6.25 (Wong-Foy et al. 2006)
MOF-177			
$Zn_4O(BTB)_2$, BTB: 1,3,5-benzenetribenzoate	4750	66	7.1 (Furukawa et al. 2007)
MOF-5 (IRMOF-1)			
$Zn_4O(BDC)_3$, BDC: 1,4-benzenedicarboxylate	3800	100	10.0 (Kaye et al. 2007)

hydrogen storage performance (much higher than that reported for zeolites) at 77 K and pressures up to 100 bar, and a tremendous effort is being devoted to the improvement of the attained H_2 adsorption characteristics mainly by aiming to specific structural (e.g., high porosity and surface area with appropriate pore size) and chemical properties. The higher capacities (at 77 K) have been reported so far for the representative systems listed in Table 15.2.

Unfortunately, all MOF materials investigated up to now have demonstrated poor performance at or near ambient temperature, adsorbing less than 2 wt% H_2. It is now understood that the improvement of the respective capacities cannot be merely based on strategies targeting only the increase of the surface area and porosity of the framework material. The development of MOFs with clear potential for practical applications remains a greatly challenging task (Murray et al. 2009). The utilization for instance of light main group metal ions (e.g., Li^+, Mg^{2+}, or Al^{3+}) comprises a significant prospect; nevertheless, the key to enhanced gravimetric capacities is considered to associate with the manipulation of the associated H_2 binding energy. The heat of adsorption of hydrogen at room temperature is typically between 5 and 10 kJ/mol, and it has been suggested (Gigras et al. 2007) that its slight increase (e.g., to 15 kJ/mol levels) is likely to lead to higher H_2 uptake at 298 K. In the same concept, advanced synthetic routes (e.g., for the introduction of open metal coordination sites or for inducing strong surface dipole moments) are continuously investigated in an attempt to strengthen the affinity of the porous surface for H_2. The main challenge is to design bridging ligands or surface functionalization routes that can lead to frameworks with a high concentration of open metal sites, capable of binding more than one H_2 molecule/metal centre. In such a case, the resulting charge density on the metal ions is expected to give rise to H_2 binding enthalpies as high as 20 kJ/mol.

15.4.3 Clathrates

Gas hydrates are ice-like, crystalline, non-stoichiometric materials composed of a framework of hydrogen-bonded water molecules that form cavities (cages) with specific geometry and size, inside which small molecules can be encaged (enclathrated). Depending on their crystal structure, hydrates are divided into three types: sI, sII, and sH. Each type of hydrate consists of two or three different types of cavities and the number of cavities of each type per unit cell is fixed. Gas hydrates owe their stability to the interactions between the lattice of water molecules and the enclathrated gas. This is not a chemical bond but a weak, Van der Waals-type, physical interaction. However, in the absence of the encaged gas, hydrates are not stable. On the other hand, they exhibit stability under conditions where pure ice is not thermodynamically stable (e.g., above 0°C at ambient pressure).

Hydrates have been considered as an alternative material for storing and transporting gases like hydrogen (Hu and Ruckenstein 2006b, Mao and Mao 2004). Until recently, the hydrogen molecule was considered too small to stabilize the hydrate cavities and, therefore, incapable of forming hydrate by itself. This picture changed dramatically with the synthesis of pure hydrogen hydrate (Mao et al. 2002) that was found to be of cubic sII structure, with H_2 content up to approximately 5 wt%. The unit cell of the sII hydrate consists of 136 water molecules that form small and large cavities. It was estimated that more than one H_2 molecule can enter the same cavity. However, molecular dynamics simulations (Alavi et al. 2005) and neutron diffraction studies reported (Lokshin et al. 2004) single occupancy of the small cavities and quadruple occupancy of the large cavities. The number of hydrogen molecules inside the cavities is a crucial factor that determines the total storage capacity of the hydrate (e.g., 5 wt% H_2 with double occupancy of the small cavities, while 3.9 wt% H_2 with single occupancy, assuming quadruply occupied large cavities). However, for the stability of pure H_2 hydrates, very high pressures are required (200 MPa at 280 K) (Mao et al. 2002). On the other hand, hydrogen hydrate can be stable at ambient pressure (0.1 MPa) for temperatures lower than 140 K. Both these cases can be regarded as extreme for practical applications.

Florusse et al. (2004) managed to produce hydrogen hydrates at low pressures (down to 5 MPa) by adding hydrate promoter, a substance that assists in stabilizing the hydrate structure at moderate conditions, by occupying some of the available cavities, allowing H_2 molecules to enter the remaining ones.

Obviously, the presence of the promoter reduces the storage capacity of the material. Tetrahydrofuran (THF) was found to be the most effective promoter. If all eight large cavities of the sII structure are occupied by THF and assuming single hydrogen occupancy of the small cavities, the maximum hydrogen content of the hydrate would be 1.05 wt%. Furthermore, Lee et al. (2005) and Kim et al. (2007) suggested that by adjusting the THF concentration in the formation solution, one can "tune" how many of the large cavities would be occupied by THF, with the rest of them remaining available to H_2 molecules. They found a critical value for the THF concentration which led to a binary H_2–THF hydrate having approximately 4 wt% H_2, stable at 12 MPa. Recent experimental and computational studies (Anderson et al. 2007, Hashimoto et al. 2006, 2007, Papadimitriou et al. 2008, Strobel et al. 2006) that attempted to confirm the feasibility of the tuning process of H_2–THF hydrate produced results refuting earlier suggestions (Kim et al. 2007, Lee et al. 2005). The new experimental studies, based on various experimental techniques showed a fixed THF content in the hydrate, independent of its concentration in the initial solution. However, the controversy remains as more recent tests claim again the feasibility of the tuning mechanism (Sugahara et al. 2009). The reported hydrogen content varies from 0.3 up to 1.05 wt%. These values lie far from the requirements for the practical use of the material in mobile applications.

15.5 Outlook

Recent developments in the field of solid state storage of hydrogen lend support to the potential use of this technology for stationary and automotive applications. The performance evaluation of any promising material must be made following an ensemble of criteria including gravimetric and volumetric storage capacity, kinetic and thermodynamic properties, life cycle and safety issues.

Assuming that an optimum (for use in vehicles) hydrogen storage material must release hydrogen at about 80°C–100°C temperature and 1–10 bar pressure, the decomposition enthalpy of the hypothetical reversible material can be calculated by the van't Hoff equation in the range of 35–45 kJ/mol H_2. Development of such a material is a great challenge and up to now no system has been found possessing all the required properties and satisfying all industrial targets, despite some promising achievements. In addition, if short refueling times are required, the heat exchange system must be capable of providing cooling capacities of hundreds of kW assuming enthalpies of hydrogenation around 40 kJ/mol H_2. This poses severe engineering problems limiting the practical viability of some of the available options and implies that the whole system has to be considered in the research efforts and not a single particular element of the technology chain exclusively.

TABLE 15.3

Qualitative Evaluation of the Available Hydrogen Storage Systems for
Mobile Applications

Storage Properties/ Performance	Conventional Hydrides	Complex Hydrides	Sorbent Systems	Chemical Hydrides
Volumetric capacity	+	+	~	+
Gravimetric capacity	−	+	+ (cryogenic) − (ambient)	+
Reversibility	+	~	+	−
Temperature pressure of operation	+	~	−	+
Kinetics/refueling	+	~	+	~
Costs	~	−	~	−

+, ~, and − denote positive, average, and negative performance, respectively.

Table 15.3 provides a rough qualitative overview of the most important storage properties and performance of the major categories of candidate materials.

In turn, Figure 15.9 (presented during the 2009 Annual Review Meeting of the DoE hydrogen program) shows a compilation of the large number of materials and systems investigated so far emphasizing two of the main properties considered, namely gravimetric capacity and temperature of operation.

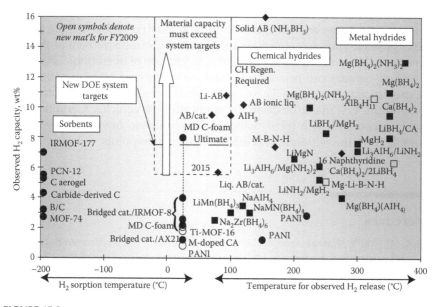

FIGURE 15.9
Material capacity vs. temperature. (Presented during the 2009 Annual Review of the DoE hydrogen program.)

Clearly, very few systems approach the targeted window in Figure 15.9 while the vast majority of them seem to possess certain interesting features but also distinct weaknesses that need to be improved through intensive R&D. The systems described can be summarized as follows:

Conventional metal hydrides exhibit high volumetric capacities, reversibility, good kinetics, and thermodynamics but suffer from low gravimetric capacities. A notable exception here is catalyzed magnesium hydride which provides 6–7 wt% but operates at a temperature range of around 300°C. Of note also is the possibility for developing hybrid high-pressure/solid storage tanks which can partially overcome the disadvantage of low gravimetric capacities in this type of hydrides.

Complex hydrides offer promises for satisfactory gravimetric capacities and reversibility. Thermodynamics and kinetics however, are in most cases insufficient and so far only sodium alanate has demonstrated at pilot scale good potential for satisfying several of the industrially set targets for automotive use. Its gravimetric capacities though are too low for mobile applications. Other types like amides and borohydrides also suffer from specific problems related to reversibility, temperatures of decomposition, and production of unwanted by-products. Destabilization of high temperature hydrides by thermodynamic tailoring and/or nanostructuring seems to be a promising route but needs much more research and upscaling efforts.

Sorbents have been found to possess sufficient gravimetric capacities and reversibility with good kinetics. To exhibit all those positive features though they require cryogenic temperatures. This key drawback needs to be overcome for the development of commercial storage tanks. Doping to enhance the interaction between hydrogen and the solid matrix and/or to take advantage of possible spillover effects offers potential for improvement of the performance of such materials in terms of temperature of operation.

Chemical hydrides can provide solutions characterized by high volumetric and gravimetric capacities. Their irreversibility (requiring costly off-board regeneration) is a major obstacle for application in vehicles. Certain liquid organic carriers though (possessing high volumetric capacities) would be favorable from the refueling and infrastructure viewpoints.

References

Ahn, C. C., Ye, Y., Ratnakumar, B. V., Witham, C. et al. 1998. Hydrogen desorption and adsorption measurements on graphite nanofibers. *Applied Physics Letters* 73:3378–3380.

Aiello, R., Sharp, J. H., and M. A. Matthews. 1999. Production of hydrogen from chemical hydrides via hydrolysis with steam. *International Journal of Hydrogen Energy* 24:1123–1130.

Alavi, S., Ripmeester, J. A., and D. D. Klug. 2005. Molecular-dynamics study of structure II hydrogen clathrates. *Journal of Chemical Physics* 123:024507–024513.

Amendola, S. C., Sharp-Goldman, S. L., Saleem Janjua, M. et al. 2000. A safe, portable, hydrogen gas generator using aqueous borohydride solution and Ru catalyst. *International Journal of Hydrogen Energy* 25:969–975.

Anderson, R., Chapoy, A., and B. Tohidi. 2007. Phase relations and binary clathrate hydrate formation in the system H_2–THF–H_2O. *Langmuir* 23:3440–3444.

Aoki, M., Miwa, K., Noritake, T. et al. 2005. Destabilization of $LiBH_4$ by mixing with $LiNH_2$. *Applied Physics A Materials Science and Processing* 80:1409–1412.

Baldé, C. P., Hereijgers, B. P. C., Bitter, J. H., and K. P. de Jong. 2006. Facilitated hydrogen storage in $NaAlH_4$ supported on carbon nanofibers. *Angewandte Chemie International Edition* 45:3501–3503.

Becher, M., Haluska, M., Hircher, M. et al. 2003. Hydrogen storage in carbon nanotubes. *Comptes Rendus Physique* 4:1055–1062.

Bogdanovic, B. and M. Schwickardi. 1997. Ti-doped alkali metal aluminium hydrides as potential novel reversible hydrogen storage materials. *Journal of Alloys and Compounds* 253–254:1–9.

Bogdanovic, B., Brand, R. A., Marjanovic, A. et al. 2000. Metal doped sodium aluminum hydrides as potential new hydrogen storage materials. *Journal of Alloys and Compounds* 302:36–58.

Bogdanovic, B., Felderhoff, M., Pommerin, A. et al. 2008. Cycling properties of Sc- and Ce-doped $NaAlH_4$ hydrogen storage materials prepared by the one-step direct synthesis method. *Journal of Alloys and Compounds* 471:383–386.

Bourlinos, A. B., Georgakilas, V., Zboril, R. et al. 2009. Liquid-phase exfoliation of graphite towards solubilized graphenes. *Small* 5:1841–1845.

Brinks, H. W., Hauback, B. C., Srinivasan, S. S., and C. M. Jensen. 2005. Synchrotron X-ray studies of Al1-yTiy formation and re-hydriding inhibition in Ti-enhanced $NaAlH_4$. *Journal of Physical Chemistry B* 109:15780–15785.

Brosha, E. L., Davey, J., Garzon, F. H., and S. Gottesfeld. 1999. Irreversible hydrogenation of solid C_{60} with and without catalytic metals. *Journal of Materials Research* 14:2138–2146.

Campesi, R., Cuevas, F., Leroy, E. et al. 2009. In situ synthesis and hydrogen storage properties of PdNi alloy nanoparticles in an ordered mesoporous carbon template. *Microporous and Mesoporous Materials* 117:511–514.

Casiraghi, C., Robertson, J., and C. Ferrari. 2007. Diamond-like carbon for data and beer storage. *Materials Today* 10:44–53.

Chahine, R. and T. Bose. 1994. Low-pressure adsorption storage of hydrogen. *International Journal of Hydrogen Energy* 19:161–164.

Chaise, A., de Rango, P., Marty, Ph. et al. 2009. Enhancement of hydrogen sorption in magnesium hydride using expanded natural graphite. *International Journal of Hydrogen Energy* 34:8589–8596.

Chambers, A., Park, C., Baker, R. T. K., and N. M. Rodriguez. 1998. Hydrogen storage in graphite nanofibers. *Journal of Physical Chemistry B* 102:4253–4256.

Chen, P., Xiong, Z. T., Luo, J. Z. et al. 2002. Interaction of hydrogen with metal nitrides and imides. *Nature* 420:302–304.

Cheng, H. M., Liu, C., Fan, Y. Y. et al. 2000. Synthesis and hydrogen storage of carbon nanofibers and single-walled carbon nanotubes. *Zeitschrift für Metallkunde* 91:306–310.

D'Ulivo, A. 2004. Chemical vapor generation by tetrahydroborate(III) and other borane complexes in aqueous medias a critical discussion of fundamental processes and mechanisms involved in reagent decomposition and hydride formation. *Spectrochimica Acta B* 59:793–825.

Dahl, J. E., Liu, S. G., and R. M. K. Carlson. 2003. Isolation and structure of higher diamondoids, nanometersized diamond molecules. *Science* 299:96–99.

Davis, R. E., Bromels, E., and C. L. Kibby. 1962. III. Hydrolysis of sodium borohydride in aqueous solution. *Journal of American Chemical Society* 84:885–892.

Dillon, A. C. and M. J. Heben. 2001. Hydrogen storage using carbon adsorbents: Past, present and future. *Applied Physics A* 72:133–142.

Dillon, A. C., Jones, K. M., Bekkeddahl, A. et al. 1997. Storage of hydrogen in single-walled carbon nanotubes. *Nature* 386:377–379.

Dimitrakakis, G. K., Tylianakis, E., and G. E. Froudakis. 2008. Pillared graphene: A new 3-D network nanostructure for enhanced hydrogen storage. *Nanoletters* 8:3166–3170.

Dong, J., Wang, X., Xu, H., Zhao, Q., and J. Li. 2007. Hydrogen storage in several microporous zeolites. *International Journal of Hydrogen Energy* 32:4998–5004.

Dornheim, M., Doppiu, S., Barkhordarian, G. et al. 2007. Hydrogen storage in magnesium-based hydrides and hydride composites. *Scripta Materialia* 56:841–846.

Enoki, T., Yu, V., Osipov, K. et al. 2006. Magnetic and high resolution TEM studies of nanographite derived from nanodiamond. *Carbon* 44:1225–1234.

Fan, Y. Y., Liao, B., Liu, M. et al. 1999. Hydrogen uyptake in vapor–grown carbon nanofibers. *Carbon* 37:1652–1654.

Felderhoff, M., Klementiev, K., Grünert, W. et al. 2004. Combined TEM-EDX and XAFS studies of Ti-doped sodium alanate. *Physical Chemistry Chemical Physics* 6:4369–4374.

Felderhoff, M., Weidenthaler, C., von Helmholt, R., and U. Eberle. 2007. Hydrogen storage: the remaining scientific and technological challenges. *Physical Chemistry Chemical Physics* 9:2643–2653.

Fernandez-Alonso, F., Bermejo, F. J., Cabrillo, C. et al. 2007. Nature of the bound states of molecular hydrogen in carbon nanohorns. *Physical Review Letters* 98:215503–215506.

Fichtner, M., Frommen, C., and O. Fuhr. 2005. Synthesis and properties of calcium alanate and two solvent adducts. *Inorganic Chemistry* 44:3479–3484.

Fichtner, M., Lohstroh, W., and O. Zabara. 2008. Hydrides on the nanoscale-physics, examples, and research directions. *Paper Presented at the International Symposium on Metal–Hydrogen Systems*, Reykjavik, Iceland.

Fichtner, M., Zhao-Karger, Z., Hu, J. et al. 2009. The kinetic properties of $Mg(BH_4)_2$ infiltrated in activated carbon. *Nanotechnology* 20:204029.

Florusse, L. J., Peters, C. J., Schoonman, J. et al. 2004. Stable low-pressure hydrogen clusters stored in a binary clathrate hydrate. *Science* 306:469–471.

Fukunaga, T., Nagano, K., Mizutani, U. et al. 1998. Structural change of graphite subjected to mechanical milling. *Journal of Non-Crystalline Solids* 232–234:416–420.

Furukawa, H., Miller, M. A., and O. M. Yaghi. 2007. Independent verification of the saturation hydrogen uptake in MOF-177 and establishment of a benchmark for hydrogen adsorption in metal–organic frameworks. *Journal of Materials Chemistry* 17:3197–3204.

Gardiner, J. A. and J. W. Collat. 1965. Kinetics of the stepwise hydrolysis of tetrahydroborate ion. *Journal of American Chemical Society* 87:1692–1700.

Ge, M. and K. Sattler. 1994. Observation of fullerene cones. *Chemical Physics Letters* 220:192–196.

Gigras, A., Bhatia, S. K., Kumar, A. V. A., and A. L. Myers. 2007. Feasibility of tailoring for high isosteric heat to improve effectiveness of hydrogen storage in carbons. *Carbon* 45:1043–1050.

Gournis, D. Personal communication.

Graetz, J. 2009. New approaches to hydrogen storage. *Chemical Society Reviews* 38:73–82.

Grochala, W. and P. P. Edwards. 2004. Thermal decomposition of the non-interstitial hydrides for the storage and production of hydrogen. *Chemical Reviews* 104:1283–1316.

Gross, A. F., Vajo, J. J., Van Atta, S. L., and G. L. Olson. 2008. Enhanced hydrogen storage kinetics of $LiBH_4$ in nanoporous carbon scaffolds. *Journal of Physical Chemistry C* 112:5651–5657.

Gutowska, A., Li, L., Shin, Y. et al. 2005. Nanoscaffold mediates hydrogen release and the reactivity of ammonia borane. *Angewandte Chemie International Edition* 44:3578–3582.

Hamilton, C. W., Baker, R. T., Staubitz, A. and I. Manners. 2009. B–N compounds for chemical hydrogen storage. *Chemical Society Reviews* 38:279–293.

Hanada, N., Lohstroh, W., and M. Fichtner. 2008. Comparison of the calculated and experimental scenarios for solid-state reactions involving $Ca(AlH_4)_2$. *Journal of Physical Chemistry C* 112:131–138.

Hashimoto, S., Murayama, T., Sugahara, T. et al. 2006. Thermodynamic and Raman spectroscopic studies on H2+tetrahydrofuran+water and H2+ tetra-n-butyl ammonium bromide+water mixtures containing gas hydrates. *Chemical Engineering Science* 61:7884–7888.

Hashimoto, S., Sugahara, T., Sato, H., and K. Ohgaki. 2007. Thermodynamic stability of H2 + tetrahydrofuran mixed gas hydrate in nonstoichiometric aqueous solutions. *Journal of Chemical & Engineering Data* 52:517–520.

Heiberg-Andersen, H. and A. T. Skjeltorp. 2007. Spectra of conic carbon radicals. *Journal of Mathematical Chemistry* 42:707–727.

Heiberg-Andersen, H., Skjeltorp, A. T., and K. Sattler. 2008. Carbon nanocones: A variety of non-crystalline graphite. *Journal of Non-Crystalline Solids* 354:5247–5249.

Hernandez, Y., Nicolosi, V., Lotya, M. et al. 2008. High-yield production of graphene by liquid-phase exfoliation of graphite. *Nature Nanotechnology* 3:563–568.

Hirscher, M., Becher, M., Haluska, M. et al. 2002. Hydrogen storage in carbon nanostructures. *Journal of Alloys and Compounds* 330–332:654–658.

Hirscher, M., Becher, M., Haluska, M. et al. 2003. Are carbon nanostructures an efficient hydrogen storage medium? *Journal of Alloys and Compounds* 356–357:433–437.

http://www.hycones.eu

Hu, Y. H. and E. Ruckenstein. 2003a. H_2 storage in Li_3N. Temperature-programmed hydrogenation and dehydrogenation. *Industrial and Engineering Chemistry Research* 42:5135–5139.

Hu, Y. H. and E. Ruckenstein. 2003b. Ultrafast reaction between LiH and NH_3 during H_2 storage in Li_3N. *Journal of Physical Chemistry A* 107:9737–9739.

Hu, Y. H. and E. Ruckenstein. 2006a. Hydrogen storage of Li_2NH prepared by reacting Li with NH_3. *Industrial and Engineering Chemistry Research* 45:182–186.

Hu, Y. H. and E. Ruckenstein. 2006b. Clathrate hydrogen hydrate—A promising material for hydrogen storage. *Angewandte Chemie International Edition* 45:2011–2013.

Hugdahl, J., Hox, K., Lynum, S. et al. 1998. Norwegian patent PCT/NO98/00093.

Hwang, S.-W. Rather, S., Naik, M. et al. 2009. Hydrogen uptake of multiwalled carbon nanotubes decorated with Pt–Pd alloy using thermal vapour deposition method. *Journal of Alloys and Compounds* 480:L20–L24.

Ichikawa, T., Isobe, S., Hanada, N., and H. Fujii. 2004. Lithium nitride for reversible hydrogen storage. *Journal of Alloys and Compounds* 365:271–276.

Iijima, S., Yudasaka, M., Yamada, R. et al. 1999. Nano-aggregates of single-walled graphitic carbon nano-horns. *Chemical Physics Letters* 309:165–170.

Ivanov-Omskii, V. I., Korobkov, M. P., Namozov, B. R. et al. 1998. Bonded and nonbonded hydrogen in diamond-like carbon. *Journal of Non-Crystalline Solids* 227–230:627–631.

Jariwala, B. N., Ciobanu, C. V., and S. Agarwal. 2009. Atomic hydrogen interactions with amorphous carbon thin films. *Journal of Applied Physics* 106:073305.

Jeon, E. and Y. W. Cho. 2006. Mechanochemical synthesis and thermal decomposition of zinc borohydride. *Journal of Alloys and Compounds* 422:273–275.

Jin, H., Lee, Y. S., and I. Hong. 2007. Hydrogen adsorption characteristics of activated carbon. *Catalysis Today* 120:399–406.

Kapitonov, I. N., Konkov, O. I., Terukov, E. I., and I. N. Trapeznikova. 2000. Amorphous carbon: how much of free hydrogen?. *Diamond and Related Materials* 9:707–710.

Kaye, S. S., Dailly, A., Yaghi, O. M., and J. R. Long. 2007. Impact of preparation and handling on the hydrogen storage properties of $Zn_4O(1,4$-benzenedicarboxylate)3 (MOF-5). *Journal of the American Chemical Society* 129:14176–14177.

Kayiran, S. B. and F. L. Darkrim. 2002. Synthesis and ionic exchanges of zeolites for gas adsorption. *Surface and Interface Analysis* 34:100–104.

Kim, D. Y., Park, Y., and H. Lee. 2007. Tuning clathrate hydrates: Application to hydrogen storage *Catalysis Today* 120:257–261.

Kim, J.-H., Jin, S.-A., Shim, J.-H., and Y. W. Cho. 2008a. Thermal decomposition behavior of calcium borohydride $Ca(BH_4)_2$. *Journal of Alloys and Compounds* 461:L20–L22.

Kim, J.-H., Jin, S.-A., Shim, J.-H., and Y. W. Cho. 2008b. Reversible hydrogen storage in calcium borohydride $Ca(BH_4)_2$. *Scripta Materialia* 58:481–483.

Kojima, Y., Suzuki, K., Fukumoto, K. et al. 2002. Hydrogen generation using sodium borohydride solution and metal catalyst coated on metal oxide. *International Journal of Hydrogen Energy* 27:1029–1034.

Kojima, Y., Kawai, Y., Koiwai, A. et al. 2006a. Hydrogen adsorption and desorption by carbon materials. *Journal of Alloys and Compounds* 421:204–208.

Kojima, Y., Matsumoto, M., Kawai, Y. et al. 2006b. Hydrogen absorption and desorption by the Li–Al–N–H system. *Journal of Physical Chemistry B* 110:9632–9636.

Kreevoy, M. and J. Hutchins. 1972. H_2BH_3 as an intermediate in tetrahydridoborate hydrolysis. *Journal of American Chemical Society* 94:6371–6376.

Krishnan, A., Dujardin, E., Treacy, M.M.J. et al. 1997. Graphitic cones and the nucleation of curved carbon surfaces. *Nature* 388:451–454.

Langmi, H. W., Walton, A., Al-Mamouri, M. M. et al. 2003. Hydrogen adsorption in zeolites A, X, Y and RHO. *Journal of Alloys and Compounds* 356–357:710–715.

Langmi, H. W., Book, D., Walton, A. et al. 2005. Hydrogen storage in ion-exchanged zeolites. *Journal of Alloys and Compounds* 404–406:637–642.

Latroche, M., Surblé, S., Serre, C. et al. 2006. Hydrogen storage in the giant-pore metal-organic frameworks MIL-100 and MIL-101. *Angewandte Chemie International Edition* 45:8227–8231.

Lee, H., Lee, J., Kim, D. Y. et al. 2005. Tuning clathrate hydrates for hydrogen storage. *Nature* 434:743–746.

Leng, H. Y., Ichikawa, T., Hino, S. et al. 2004. New metal–N–H system composed of $Mg(NH_2)_2$ and LiH for hydrogen storage. *Journal of Physical Chemistry B* 108:8763–8765.

Li, H. W., Orimo, S., Nakamori, Y. et al. 2007. Materials designing of metal borohydrides: Viewpoints from thermodynamical stabilities. *Journal of Alloys and Compounds* 446–447:315–318.

Li, Y. and R. T. Yang. 2006. Significantly enhanced hydrogen storage in metal-organic frameworks via spillover. *Journal of American Chemical Society* 128:726–727.

Liu, Y., Zhong, K., Gao, M., Wang, J. et al. 2008. Hydrogen storage in a $LiNH_2$–MgH_2 (1:1) system. *Chemistry of Materials* 20:3521–3527.

Liu, B. H. and S. Suda. 2008. Hydrogen storage alloys as the anode materials of the direct borohydride fuel cell. *Journal of Alloys and Compounds* 454:280–285.

Liu, C., Fan, Y. Y., Liu, M. et al. 1999. Hydrogen storage in single-walled carbon nanotubes at room temperature. *Science* 286:1127–1129.

Lokshin, K. A., Zhao, Y., He, D. et al. 2004. Structure and dynamics of hydrogen molecules in the novel clathrate hydrate by high pressure neutron diffraction. *Physical Review Letters* 93:125503–125506.

Lu, J., Fang, Z. Z., and H. Y. Sohn. 2006. A new Li–Al–N–H system for reversible hydrogen storage. *Journal of Physical Chemistry B* 110:14236–14239.

Lueking, A. D., Yang, R. T., Rodriguez, N. M., and R. T. K. Baker. 2004. Hydrogen storage in graphite nanofibers: Effect of synthesis catalyst and pretreatment conditions. *Langmuir* 20:714–721.

Luo, W. 2004. ($LiNH_2$–MgH_2): A viable hydrogen storage system. *Journal of Alloys and Compounds* 381:284–287.

Marder, T. B. 2007. Will we soon be fueling our automobiles with ammonia–borane? *Angewandte Chemie International Edition* 46:8116–8118.

Mao, W. L. and H. K. Mao. 2004. Hydrogen storage in molecular compounds. *Proceedings of the National Academy of Science USA* 101:708–710.

Mao, W. L., Mao, H. K., Goncharov, A. F. et al. 2002. Hydrogen clusters in clathrate hydrate. *Science* 297:2247–2249.

Mamatha, M., Bogdanović, B., Felderhoff, M. et al. 2006. Mechanochemical preparation and investigation of properties of magnesium, calcium and lithium–magnesium alanates. *Journal of Alloys and Compounds* 407:78–86.

Marrero-Alfonso, E. Y., Gray, J. R., Davis, T. A., and M. A. Matthews. 2007. Minimizing water utilization in hydrolysis of sodium borohydride: The role of sodium metaborate hydrates. *International Journal of Hydrogen Energy* 32:4723–4730.

Mori, D., Kobayashi, N., Matsunaga, T. et al. 2005. Hydrogen storage materials for fuel cell vehicles high-pressure MH system. *Journal of the Japan Institute of Metals* 69:308–311.

Mpourmpakis, G., Tylianakis, E., and G. E. Froudakis. 2007. Carbon nanoscrolls: A promising material for hydrogen storage. *Nanoletters* 7:1893–1897.

Murray, L. J., Dinca, M., and J. R. Long. 2009. Hydrogen storage in metal–organic frameworks. *Chemical Society Reviews* 38:1294–1314.

Nakamori, Y., Li, H., Miwa, K. et al. 2006a. Syntheses and hydrogen desorption properties of metal-borohydrides $M(BH_4)_n$ (M=Mg, Sc, Zr, Ti, and Zn; n=2–4) as advanced hydrogen storage materials. *Materials Transactions* 47:1898–1901.

Nakamori, Y., Orimo, S., and T. Tsutaoka. 2006b. Dehydriding reaction of metal hydrides and alkali borohydrides enhanced by microwave irradiation. *Applied Physics Letters* 88:112104–112106.

Nakamori, Y., Miwa, K., Ninomiya, A. et al. 2006c. Correlation between thermodynamical stabilities of metal borohydrides and cation electronegativites: First-principles calculations and experiments. *Physical Review B* 74:045126–045134.

Nijkamp, M. G., Raaymakers, J. E. M. J., van Dillen, A. J., and K. P. de Jong. 2001. Hydrogen storage using physisorption-materials demands. *Applied Physics A* 72:619–623.

Novoselov, K. S., Geim, A. K., Morozov, S. V. et al. 2004. Electric field effect in atomically thin carbon films. *Science* 306:666–669.

Okada, Y., Sasaki, E., Watanabe, E. et al. 2006. Development of dehydrogenation catalyst for hydrogen generation in organic chemical hydride method. *International Journal of Hydrogen Energy* 31:1348–1356.

Orimo, S., Majer, G., Fukunaga, T. et al. 1999. Hydrogen in the mechanically prepared nanostructured graphite. *Applied Physics Letters* 75:3093–3095.

Orimo, S., Matsushima, T., Fujii, H. et al. 2001. Hydrogen desorption property of mechanically prepared nanostructured graphite. *Journal of Applied Physics* 90:1545–1549.

Orimo, S., Nakamori, Y., Eliseo, J. R. et al. 2007. Complex hydrides for hydrogen storage. *Chemical Reviews* 107:4111–4132.

Panella, B., Hirscher, M., and S. Roth. 2005. Hydrogen adsorption in different carbon nanostructures. *Carbon* 43:2209–2214.

Panella, B., Hirscher, M., Pütter, H., and U. Müller. 2006. Hydrogen adsorption in metal-organic frameworks: Cu-MOFs and Zn-MOFs compared. *Advanced Functional Materials* 16:520–524.

Papadimitriou, N. I., Tsimpanogiannis, I. N., Papaioannou, A. Th., and A. K. Stubos. 2008. Evaluation of the hydrogen-storage capacity of pure H_2 and binary H_2-THF hydrates with Monte Carlo simulations. *Journal of Physical Chemistry C* 112:10294–10302.

Park, C., Anderson, P. E., Chambers, A. et al. 1999. Further studies of the interaction of hydrogen with graphite nanofibers. *Journal of Physical Chemistry B* 103:10572–10581.

Park, S. and R. S. Ruoff. 2009. Chemical methods for the production of graphemes. *Nature Nanotechnology* 4:217–224.

Peera, A. A., Alemany, L. B., and W. E. Billups. 2004. Hydrogen storage in hydrofullerides. *Applied Physics A* 78:995–1000.

Pinkerton, F. E., Meisner, G. P., Meyer, M. S., Balogh, M. P., and M. D. Kundrat. 2005. Hydrogen desorption exceeding ten weight percent from the new quaternary hydride $Li_3BN_2H_8$. *Journal of Physical Chemistry B* 109:6–8.

Pinkerton, F. E., Meyer, M. S., Meisner, G. P., and M. P. Balogh. 2007. Improved hydrogen release from $LiB_{0.33}N_{0.67}H_{2.67}$ with metal additives: Ni, Fe, and Zn. *Journal of Alloys and Compounds* 433:282–291.

Pinkerton, F. E. and M. S. Meyer. 2008. Reversible hydrogen storage in the lithium borohydride—calcium hydride coupled system. *Journal of Alloys and Compounds* 464:L1–L4.

Poirier, E., Chahine, R., and T. K. Bose. 2001. Hydrogen adsorption in carbon nano-structures. *International Journal of Hydrogen Energy* 26:831–835.

Poirier, E., Chahine, R., Benard, P. et al. 2006. Hydrogen adsorption measurements and modeling on metal-organic frameworks and single-walled carbon nano-tubes. *Langmuir* 22:8784–8789.

Psofogiannakis, G.M. and G. E. Froudakis. 2009. DFT study of enhanced hydrogen storage by spillover on graphite with oxygen surface groups. *Journal of American Chemical Society* 131:15133–15135.

Reule, H., Hirscher, M. Weißhardt, A., and H. Kronmüller. 2000. Hydrogen desorption properties of mechanically alloyed MgH_2 composite materials. *Journal of Alloys and Compounds* 305:246–252.

Ritschel, M., Uhlemann, M., Gutfleisch, O. et al. 2002. Hydrogen storage in different carbon nanostructures. *Applied Physics Letters* 80:2985–2987.

Sakintuna, B., Lamari-Darkrim, F., and M. Hirscher. 2007. Metal hydride materials for solid hydrogen storage: A review. *International Journal of Hydrogen Energy* 32:1121–1140.

Sandrock, G. 1999. A panoramic overview of hydrogen storage alloys from a gas reaction point of view. *Journal of Alloys and Compounds* 293–295:877.

Satyapal, S. 2008. U.S. Department of Energy Hydrogen Storage Sub-Program Overview. DOE Hydrogen Program Merit Review and Pear Evaluation Meeting-June 9, 2008.

Satyapal, S., Petrovic, J., Read, C. et al. 2007. The U.S. Department of Energy's National Hydrogen Storage Project: Progress towards meeting hydrogen-powered vehicle requirements. *Catalysis Today* 120:246–256.

Schlapbach, L. 1992. Hydrogen in intermetallic compounds II. In: Schlapbach, L. (ed.). *Topics in Applied Physics*. Berlin: Springer, pp. 15–258.

Schlapbach, L. and A. Züttel. 2001. Hydrogen-storage materials for mobile applications. *Nature* 414:353–358.

Schur, D. V., Tarasov, B. P., Yu. S. et al. 2002. The prospects for using of carbon nano-materials as hydrogen storage systems. *International Journal of Hydrogen Energy* 27:1063–1069.

Schüth, F., Bogdanovic, B., and M. Felderhoff. 2004b. Light metal hydrides and complex hydrides for hydrogen storage. *Chemical Communications* 2249–2258.

Schüth, F., Taguchi, A., Kounosu, S., and B. Bogdanovic. 2004a. DE Patent Specification 10332438.

Srinivasan, S. S., Brinks, H. W., Hauback, B. C. et al. 2004. Long term cycling behavior of titanium doped $NaAlH_4$ prepared through solvent mediated milling of NaH and Al with titanium dopant precursors. *Journal of Alloys and Compounds* 377:283–289.

Srinivasan, S. S., Escobar, D., Jurczyk, M. et al. 2008. Nanocatalyst doping of $Zn(BH_4)_2$ for on-board hydrogen storage. *Journal of Alloys and Compounds* 462:294–302.

Stadie, N. P., Purewal, J. J., Ahn, C. C., and B. Fultz. 2010. Measurements of hydrogen spillover in platinum doped superactivated carbon. *Langmuir* DOI: 10.1021/la9046758.

Ströbel, R., Jörissen, L., Schliermann, T. et al. 1999. Hydrogen adsorption on carbon materials. *Journal of Power Sources* 84:221–224.

Ströbel, R., Garche, J., Moseley, P. T. et al. 2006. Hydrogen storage by carbon materials. *Journal of Power Sources* 159:781–801.

Strobel, T. A., Taylor, C. J., Hester, K. C. et al. 2006. Molecular hydrogen storage in binary THF–H_2 clathrate hydrates. *Journal of Physical Chemistry B* 110:17121–17125.

Suda, S., Sun, Y.-M., Liu, B.-H. et al. 2001. Catalytic generation of hydrogen by applying fluorinated-metal hydrides as catalysts. *Applied Physics A* 72:209–212.

Sugahara, T., Haag, J. C., Prasad, P. S. R. et al. 2009. Increasing hydrogen storage capacity using tetrahydrofuran. *Journal of the American Chemical Society* 131:14616–14617.

Sui, Y., Qian, J., Zhang, J. et al. 1996. Direct and catalytic hydrogenation of buckminsterfullerene C_{60}. *Fullerenes, Nanotubes and Carbon Nanostructures* 4:813–818.

Talyzina, A. V. and S. Klyamkinb. 2004. Hydrogen adsorption in C_{60} at pressures up to 2000 atm. *Chemical Physics Letters* 397:77–81.

Tanaka, H., Kanoh, H., El-Merraoui, M. et al. 2004. Quantum effects on hydrogen adsorption in internal nanospaces of single-wall carbon nanohorns. *Journal of Chemical Physics B* 108:17457–17465.

Tanaka, H., Kanoh, H., Yudasaka, M. et al. 2005. Quantum effects on hydrogen isotope adsorption on single-wall carbon nanohorns. *Journal of the American Chemical Society* 127:7511–7516.

Tang, W. S., Wu, G., Liu, T. et al. 2008. Cobalt-catalyzed hydrogen desorption from the $LiNH_2$–$LiBH_4$ system. *Dalton Transactions* 18:2395–2399.

Taylor, R. 1999. *Lecture Notes on Fullerene Chemistry: A Handbook for Chemists*. London: Imperial College Press, p. 248.

Tibbetts, G. G., Meisner, G. P., and C. H. Olk. 2001. Hydrogen storage capacity of carbon nanotubes, filaments, and vapor-grown fibers. *Carbon* 39:2291–2301.

U.S. patent no. 6,290,753. 2001. Hydrogen storage in carbon material.

Vajo, J. J. and G. L. Olson. 2007. Hydrogen storage in destabilized chemical systems. *Scripta Materia* 56:829–834.

Vajo, J. J., Salguero, T. T., Gross, A. F. et al. 2007. Thermodynamic destabilization and reaction kinetics in light metal hydride systems. *Journal of Alloys and Compounds* 446:409–414.

Vajo, J. J., Skeith, S. L., and F. Mertens. 2005. Reversible storage of hydrogen in destabilized $LiBH_4$. *Journal of Physical Chemistry B* 109:3719–3722.

Varin, R. A., Czujko, T., and Z. S. Wronski. 2009. *Nanomaterials for Solid State Hydrogen Storage*. ISBN 978-0-387-77711-5. Springer Science+Business Media, LLC, New York.

Viculis, L. M., Mack, J. J., and R. B. Kaner. 2003. A chemical route to carbon nanoscrolls. *Science* 299:1361.

Wang, H., Chhowalla, M., Sano, N. et al. 2004. Large-scale synthesis of single-walled carbon nanohorns by submerged arc. *Nanotechnology* 15:546–550.

Wong-Foy, A. G., Matzger, A. J., and O. M. Yaghi. 2006. Exceptional H_2 saturation uptake in microporous metal-organic frameworks. *Journal of the American Chemical Society* 128:3494–3495.

Wu, H. 2008. Strategies for the improvement of the hydrogen storage properties of metal hydride materials. *ChemPhysChem* 9:2157–2162.

Xiong, Z. T., Wu, G., Hu, J., and P. Chen. 2004. Ternary imides for hydrogen storage. *Advanced Materials* 16:1522–1525.

Xiong, Z. T., Xiong, Z., Wu, G. et al. 2006. Investigations on hydrogen storage over Li–Mg–N–H complex—The effect of compositional changes. *Journal of Alloys and Compounds* 417:190–194.

Xiong, Z. T., Wu, G., Hu, J. et al. 2007. Reversible hydrogen storage by a Li–Al–N–H complex. *Advanced Functional Materials* 17:1137–1142.

Xu, W.-C., Takahashi, K., Matsuo, Y. et al. 2007. Investigation of hydrogen storage capacity of various carbon materials. *International Journal of Hydrogen Energy* 32:2504–2512.

Yamauchi, M., Ikeda, R., Kitagawa, H., and M. Takata. 2008. Nanosize effects on hydrogen storage in palladium. *Journal of Physical Chemistry C* 112:3294–3299.

Yang, J., Sudik, A., and C. Wolverton. 2007. Destabilizing $LiBH_4$ with a metal (M = Mg, Al, Ti, V, Cr, or Sc) or metal hydride ($MH_2 = MgH_2$, TiH_2, or CaH_2). *Journal of Physical Chemistry C* 111:19134–19140.

Yang, R. T. and Y. Wang. 2009. Catalyzed hydrogen spillover for hydrogen storage. *Journal of American Chemical Society* 131:4224–4226.

Zhang, S., Gross, A., Van Atta, S. L. et al. 2009. The synthesis and hydrogen storage properties of a MgH_2 incorporated carbon aerogel scaffold. *Nanotechnology* 20:204027.

Züttel, A., Nützenadel, Ch., Schmid, G. et al. 2000. Thermodynamic aspects of the interaction of hydrogen with Pd clusters. *Applied Surface Science* 162–163:571–575.

Züttel, A., Rentsch, S., Fisher, P. et al. 2003a. Hydrogen storage properties of $LiBH_4$. *Journal of Alloys and Compounds* 356:515–520.

Züttel, A., Wenger, P., Rentsch, S. et al. 2003b. $LiBH_4$ a new hydrogen storage material. *Journal of Power Sources* 118:1–7.

Index

A

Activated carbon fiber (ACF), 354
Adsorbed natural gas (ANG), 347
AlPO$_4$-5 pore system, 156–158
Aluminosilicates
    mesoporous aluminosilicates (*see*
        Mesoporous aluminosilicates)
    zeolites (*see* Zeolites)
Anomalous scattering, 14
Argon and nitrogen
    attapulgite and sepiolite, 152–153
    differential enthalpy curve, 152
    micropore filling, 153
    microporous silica gel, 149
    palygorskites, 152
    pore size distribution and surface
        chemistry, 149
    porous carbons, 150–151
    pyrogenic and Stöber silicas, 150
    two-dimensional graphite surface,
        151–152
Atomic force microscopy
    cantilever deflections, 113
    force *vs.* distance, 113
    interatomic force interactions, 112
    Lennard–Jones potential, 112
    optical spectroscopy, 113–114
    photodetector, 113–114
Autothermal reformer (ATR), 496

B

Batch method, 407–408
Brunauer, Emmett, and Teller (BET)
    method, 316–317

C

Carbide-derived carbon (CDC), 362
Carbon membranes, 410–412
Carbon nanofibers (CNF), 523–524
Carbon nanoscrolls (CNS), 530
Carbon nanotube (CNT)

bundled structure, 366–367
catalyst size distribution, 196
characteristics, 194
CH$_4$/H$_2$ mixture, 195
CVD, 194, 368
DWCNT, 365, 371–372
electron microscopy
    catalyst nanostructure, 205
    3-DOF manipulator, 209
    EBID, 209
    Fe$_3$C particle, 210
    MWCNTs, 209
    Ni-catalyzed CVD, 210–211
    SEM and TEM, 209
fullerene, 354
functional reactor elements, 195
GCMC, 368
generic "inside-tubes" technique,
    196–197
graphene structure, 365
gravimetric technique
    amorphous carbon formation,
        199–200
    carbon nanostructure,
        polymorphism, 200
    C$_2$H$_2$ decomposition, 197
    growth/monitoring approach,
        197–198
    growth rate and morphology,
        198–199
    structural and morphological
        characteristics, 197
HiPCO-SWCNT, 368
internal nanospace and monolayer
    adsorption, 368–369
internal tube, interstitial channel,
    and groove site, 367–368
laser ablation, 196
membrane growth monitoring
    AAO pores, 202
    array morphology, 202
    c-oriented AlPO$_4$-5 film, 204–208
    internal surface and hollow space
        morphology, 202–203